Signaling and Communication in Plants

More information about this series at http://www.springer.com/series/8094

P. Vidhyasekaran

Switching on Plant Innate Immunity Signaling Systems

Bioengineering and Molecular Manipulation of PAMP-PIMP-PRR Signaling Complex

Springer

P. Vidhyasekaran
Plant Pathology
Tamil Nadu Agricultural university
Coimbatore, Tamil Nadu, India

ISSN 1867-9048 ISSN 1867-9056 (electronic)
Signaling and Communication in Plants
ISBN 978-3-319-26116-4 ISBN 978-3-319-26118-8 (eBook)
DOI 10.1007/978-3-319-26118-8

Library of Congress Control Number: 2016930022

Springer Cham Heidelberg New York Dordrecht London

Printed on acid-free paper

Springer International Publishing AG Switzerland is part of Springer Science+Business Media (www.springer.com)

Contents

Chapter 1
Introduction

Abstract Plant innate immune system is a surveillance system against possible attack by pathogens. It is quiescent in normal healthy plants. It is a sleeping giant and when awakened by specific signals it triggers expression of several defense genes. Unlike, transgenic plants developed by engineering disease resistance genes against specific pathogens, plants overexpressing the plant immune system awakened by the alarm signals PAMP and PIMP trigger expression of hundreds of defense genes conferring resistance against wide range of pathogens. Both PAMPs and PIMPs are perceived by plants as alarm signals by specific receptors called pattern recognition receptors (PRRs). PAMPs activate expression of the genes encoding various PRRs. Besides PAMP molecules, pathogens secrete another type of molecules called effectors. While the pathogen-derived PAMPs are involved in switching-on the plant immune responses, the effectors are involved in switching-off the PAMP-triggered innate immunity. The effectors may also bind with PRRs and disrupt binding of PAMP with PRR in PAMP-PRR signaling complex to impede PAMP-triggered plant immunity. Effectors may bind with the PRR signal amplifier BAK1 and block the function of PAMP-PRR signaling complex. Early and robust activation of PAMP-PRR signaling complex before the pathogens invade and secrete virulence effectors seems to be necessary for triggering strong defense responses. Several PAMP formulations have been developed and foliar application of the formulations triggers the induction of plant immune responses. The time of application is very critical in enhancing the efficacy of the PAMPs in controlling diseases. The PAMPs should be applied prior to pathogen invasion. The concentration of the PAMP applied also determines the efficacy of the treatment in controlling diseases. Oligogalacturonates (OGAs), plant elicitor peptides (Peps), and PAMP-induced Peptides (PIPs) are the important PIMPs capable of switching on plant innate immune responses. Bioengineering technologies have been exploited to utilize PIMPs to develop transgenic plants expressing enhanced disease resistance. Bioengineering PRRs has been shown to be another potential technology to awaken the quiescent plant innate immunity for effective management of crop diseases. Intergeneric transfer of PRR has been achieved to develop disease-resistant crop plants. Transcription factors are the master switches, which regulate expression of defense genes in the PAMP-triggered plant immune signaling systems. Several transcription factors have been shown to trigger "priming" of defense responses and induce "Systemic Acquired resistance (SAR)" in plants. The plant defense activators benzothiadiazole, probenazole, tiadinil and ergosterol trigger the expression of

P. Vidhyasekaran, *Switching on Plant Innate Immunity Signaling Systems*,
Signaling and Communication in Plants, DOI 10.1007/978-3-319-26118-8_1

transcription factors which modulate the expression of defense genes inducing disease resistance. The rhizobacterial strain *Pseudomonas fluorescens* WCS417r induces systemic resistance (ISR) by activating the transcription factor MYB72. The MYC2 transcription factor also has been shown to be involved in *P. fluorescens* WCS417r-induced priming to trigger ISR. Several WRKY, MYB, MYC, bZIP, EREBP, and NAC transcription factors have been engineered in various crop plants to develop disease resistant plants. Some transcription factors have been found to negatively regulate the expression of defense signaling systems. Silencing of the negative regulator transcription factors may be a useful strategy in developing disease-resistant plants. Several bioengineering and molecular manipulation technologies have been developed to switch on the 'sleeping' plant innate immune system, which has potential to detect and suppress the development of a wide range of plant pathogens in economically important crop plants. Enhancing disease resistance through altered regulation of plant immunity signaling systems would be durable and publicly acceptable.

1.1 Plant Innate Immunity Is a Sleeping Giant to Fight against Pathogens

Diseases caused by oomycete, fungal, bacterial, and viral pathogens cause enormous crop losses and in some areas the crop may be completely devastated (Vidhyasekaran 2004; Byamukama et al. 2015; Cohen et al. 2015; Han et al. 2015; Handiseni et al. 2015; Holmes et al. 2015; Sharma-Poudyal et al. 2015; Strehlow et al. 2015). Chemical control is widely practiced to manage fungal and oomycete pathogens (Gent et al. 2015; Handiseni et al. 2015). Frequent development of resistance to the modern fungicides in the field population of fungal/oomycete pathogens is a challenging problem in using the fungicides to manage these diseases (Miles et al. 2012; Gudmestad et al. 2013; Tymon and Johnson 2014; Fernández-Ortuño et al. 2015; Hu et al. 2015; Keinath 2015; Saville et al. 2015; Zeng et al. 2015). Effective chemicals are still not available to control bacterial, viral, viroid, and phytoplasma diseases (Jones 2001; Vidhyasekaran 2004; Bradley 2008; Kanetis et al. 2008). Breeding varieties with built-in resistance may be the sound approach to manage diseases (Vidhyasekaran 2007; Singh et al. 2008; Tagle et al. 2015). However, new races of pathogens appear frequently and the resistance often breaks down (Sørensen et al. 2014; Kitner et al. 2015; Maccaferri et al. 2015). Breeding for quantitative resistance is useful (Yasuda et al. 2015), but it is difficult to achieve (Vidhyasekaran 2007; St Clair 2010; Zhang et al. 2015). Breeding for resistance against broad-spectrum of pathogens will be ideal, but the traditional breeding methods are inefficient (Vleeshouwers et al. 2008; Tran et al. 2015). An alternative technology based on switching on plant innate immunity using pathogen-associated molecular patterns (PAMPs) and pathogen-induced molecular patterns (PIMP)/host-associated molecular patterns (HAMPs) has been developed recently for management of viral and bacterial diseases (Li et al. 2011; Pavli et al. 2011, 2012; Choi et al. 2012; Li et al. 2012), and also for management of a wide range of biotrophic, hemibiotrophic, and necrotrophic fungal and oomycete pathogens (Miao et al. 2010; Xu et al. 2010; Choi et al. 2012; Miao and Wang 2013).

Plant innate immune system is a surveillance system against possible attack by pathogens. It is quiescent in normal healthy plants. It is a sleeping giant and when awakened by specific signals it triggers expression of several defense genes. Several hundreds of defense genes have been detected in plants and these are involved in plant immunity (Coram and Pang 2005; Vega-Sánchez et al. 2005; Wang et al. 2005; Wilkinson et al. 2005; Hermosa et al. 2006; Vidhyasekaran 2007; Sun et al. 2008). These defense genes encode different pathogenesis-related proteins (Anand et al. 2003; Zhu et al. 2006; Sun et al. 2008) most of which inhibit growth of oomycete (Lee et al. 2000), fungal (Moravčikova et al. 2004; Pervieux et al. 2004; Chen et al. 2006; Zhu et al. 2006; Hernández-Blanco et al. 2007) and bacterial pathogens (Vidhyasekaran 2002). Several defense-related genes encode enzymes involved in biosynthesis of toxic compounds such as phytoalexins (Nawar and Kuti 2003; Liu et al. 2006; Chassot et al. 2008), phenylpropanoids and isoflavonoids (Farag et al. 2008), and terpenoids (Keeling et al. 2008), or enzymes involved in cell wall-fortification (Hamiduzzaman et al. 2005; Flors et al. 2008). The triggered plant immune responses include accumulation of pathogenesis-related proteins, deposition of lignin and callose in the cell wall, and production of anti-microbial compounds (Tsuda and Katagiri 2010; Gimenez-Ibanez and Rathjen 2010). Plant innate immunity is a powerful weapon to fight against a wide range of plant pathogens. Plants have innate immunity system (Nicaise 2014; Vidhyasekaran 2014, 2015; Piasecka et al. 2015; Robinson and Bostock 2015; Schwessinger et al. 2015; Tena 2015) and this system provides basic protective functions against broadest range of pathogens (Boller and He 2009; Boutrot et al. 2010; Chen et al. 2010a, b; Dodds and Rathjen 2010; Park et al. 2010b; Shimizu et al. 2010; Segonzac and Zipfel 2011; Zamioudis and Peterse 2012; Li et al. 2014a, b; Vidhyasekaran 2014).

1.2 Potential Signals to Switch on Plant Immune System

The plant innate immune systems have high potential to fight against viral, bacterial, oomycete, and fungal pathogens and protect the crop plants against wide range of diseases (Knecht et al. 2010; Lacombe et al. 2010; D'Amelio et al. 2011; Hwang and Hwang 2011; Alkan et al. 2012). However, these plant immune systems are quiescent in healthy normal plants. Specific signals are needed to switch on the sleeping giant for exploiting the quiescent immune system for combating diseases. These signals are derived from invading pathogens and called 'pathogen-associated molecular patterns' (PAMPs). The immune system is activated on perception of the PAMP of invading pathogens (Nürnberger and Kufner 2011; Segonzac and Zipfel 2011). Potential pathogens contain several PAMPs and they serve as alarm signals to activate the plant innate immunity (Böhm et al. 2014; Vidhyasekaran 2014;

Zhang et al. 2014). PAMPs are detected not only in pathogens, but also in saprophytes, probably in all microbes. Hence the PAMPs are also called as microbe-associated molecular patterns (MAMPs) (Jeworutzki et al. 2010; Thomma et al. 2011). PAMPs/MAMPS are potential tools to activate plant immune systems and can be effectively used to manage crop diseases (Iriti et al. 2011; Choi et al. 2012; Dafermos et al. 2012; Li et al. 2012; Sanchez et al. 2012; Quang et al. 2015; Sathiyabama et al. 2014).

Besides the microbe-derived elicitors (MAMPs), some host plant-derived elicitors called pathogen-induced molecular patterns (PIMPs) or host-associated molecular patterns (HAMPs) have been shown to activate the plant innate immune system (Yamaguchi and Huffaker 2011; Vallarino and Osorio 2012; Bellincampi et al. 2014; Hou et al. 2014). The PIMPs/HAMPs (host-derived elicitors) function almost in the same fashion as the PAMPs function in switching on plant innate immunity (Denoux et al. 2008; Ferrari et al. 2013). PIMPs and PAMPs activate similar downstream responses using many of the same molecular components (Ryan et al. 2007; Krol et al. 2010; Postel et al. 2010; Qi et al. 2010; Yamaguchi et al. 2010; Huffaker et al. 2011). Both of them bind specific LRR receptors and both activate the same downstream signaling events (Yamaguchi et al. 2006; Huffaker and Ryan 2007; Krol et al. 2010).

Both PAMPs and PIMPs are perceived by plants as alarm signals by specific receptors called pattern recognition receptors (PRRs) (Nicaise et al. 2009; Brutus et al. 2010; Petutschnig et al. 2010; Shinya et al. 2010; Schulze et al. 2010; Segonzac and Zipfel 2011; Hann et al. 2014). Plants utilize the PRRs to recognize PAMPs/MAMPs (Böhm et al. 2014; Macho and Zipfel 2014; Zhang et al. 2014). Most of the PRRs identified are receptor-like kinases (RLKs) and receptor-like proteins (RLPs) (Yang et al. 2012; Zhang et al. 2013; Zipfel 2014). PAMPs activate expression of the genes encoding various PRRs (Zipfel et al. 2004, 2006; Qutob et al. 2006; Lohmann et al. 2010). PRRs bind with PAMPs for their activation (Boutrot et al. 2010; Petutschnig et al. 2010). The PRRs recognize PAMPs and PIMPs and switch on the plant innate immunity (Mentlak et al. 2012).

1.3 Pathogens Possess Weapons to Switch-Off Plant Immune Systems

Besides PAMP molecules, pathogens secrete another type of molecules called effectors (Wu et al. 2011; Vleeshouwers and Oliver 2014). While the pathogen-derived PAMPs are involved in switching-on the plant immune responses, the effectors are involved in switching-off the PAMP-triggered innate immunity (Thomma et al. 2011; Wu et al. 2011; Cheng et al. 2012). The effectors secreted by various pathogens have been shown to suppress the PAMP-triggered immunity (Shan et al. 2008; de Jonge et al. 2010). Effectors induce susceptibility, mostly by suppressing PAMP-induced immune responses. The effector proteins target basic innate immunity in plants (Boller and He 2009; Song and Yang 2010; Szczesny et al. 2010; Rajput et al. 2014; Zheng et al. 2014). Several bacterial pathogens use a specialized type III secretion system to deliver effector proteins into host cells to subvert PAMP-triggered host defense mechanisms, thereby promoting pathogenesis (Song and Yang 2010; Szczesny et al. 2010; Zhang et al. 2010; Wu et al. 2011; Akimoto-Tomiyama et al. 2012).

The effectors may also bind with PRRs and block plant defense responses in the plant cell (Xiang et al. 2008; Zeng et al. 2012; Rosli et al. 2013; Xin and He 2013). Effectors may disrupt binding of PAMP with PRR in PAMP-PRR signaling complex to impede PAMP-triggered plant immunity (Mentlak et al. 2012). Some effectors have been shown to degrade the PRRs through ubiquitin-proteasome pathway and inhibit PAMP-triggered immunity (Göhre et al. 2008; Gimenez-Ibanez et al. 2009a, b). Effectors may also target the kinase domains of PRR and inhibit the PRR receptor kinase activity to block PAMP-triggered immunity (Shan et al. 2008; Xiang et al. 2008; Zipfel and Rathjen 2008; Xiang et al. 2011). The effectors may also inhibit the autophosphorylation of PRRs to suppress the PAMP-triggered immune system (Xiang et al. 2008). The effectors may prevent the activation of PRR signaling complex by inhibiting the autophosphorylation of PRRs (Hann and Rathjen 2007). Effectors may bind with the PRR signal amplifier BAK1 and block the function of PAMP-PRR signaling complex (Shan et al. 2008; Xiang et al. 2008; Hann et al. 2010).

1.4 Bioengineering and Molecular Manipulation Technologies to Switch on the Sleeping Quiescent Plant Immune System to Win the War against Pathogens

Several bioengineering and molecular manipulation technologies have been developed for management of a broad-spectrum of diseases caused by a wide-range of viral, bacterial, fungal and oomycete pathogens, exploiting the potential of plant innate immunity (Ferrari et al. 2008; Lacombe et al. 2010; Hwang and Hwang 2011; Volpi et al. 2011; Alkan et al. 2012; Ferrari et al. 2012; Li et al. 2012; Wang et al. 2013; Fu et al. 2014; Lloyd et al. 2014; Macho and Zipfel 2014; Trouvelot et al. 2014). The crop diseases can be controlled by switching on plant innate immunity by manipulating PAMP-PIMP-PRR signaling complex. Early, rapid and strong activation of plant innate immune system is necessary to induce strong defense responses against pathogens. Early and robust activation of PAMP-PRR signaling complex before the pathogens invade and secrete virulence effectors seems to be necessary for triggering strong defense responses and for effective management of crop diseases (Orlowska et al. 2011; Aghnoum and Niks 2012; Lanubile et al. 2012).

Several PAMP formulations have been developed and foliar application of the formulations triggered the induction of plant immune responses (Dong et al. 2004;

Elmer and Reglinski 2006; de Capdeville et al. 2008; Shao et al. 2008; Iriti et al. 2011; Dafermos et al. 2012; Chuang et al. 2014). Several factors such as environment, genotype, and crop nutrition determine the efficacy of the PAMPs in controlling diseases under field conditions (Walters et al. 2005). The time of application is very critical in enhancing the efficacy of the PAMPs in controlling diseases (de Capdeville et al. 2002; Agostini et al. 2003). The PAMPs should be applied 2–5 days prior to pathogen invasion (Qiu et al. 2001; de Capdeville et al. 2002, 2003). The concentration of the PAMP applied also determines the efficacy of the treatment in controlling diseases (de Capdeville et al. 2002; Chen et al. 2008). Variability in structure and function has been reported among various PAMPs (Che et al. 2000; Tanaka et al. 2003; Fujiwara et al. 2004). The time of induction (Luna et al. 2011), intensity of induction (Lecourieux et al. 2002, 2005), and duration of induction (Aziz et al. 2007) of the defense signals may vary depending on the type of PAMPs. Efficacy of PAMPs in controlling diseases may also vary depending on the challenging pathogens (Agostini et al. 2003). Hence, suitable PAMPs have to be selected for management of various crop diseases.

Bioengineering PAMP genes has been shown to be powerful tool to trigger plant immune responses (Keller et al. 1999; Li and Fan 1999; Belbahri et al. 2001; Choi et al. 2004; Donghua et al. 2004; Peng et al. 2004; Takakura et al. 2004, 2008; Malnoy et al. 2005; Jang et al. 2006; Ren et al. 2006a, b; Cai et al. 2007; Sohn et al. 2007; Shao et al. 2008; Qiu et al. 2009; Huo et al. 2010; Miao et al. 2010; Xu et al. 2010; Pavli et al. 2011, 2012; Choi et al. 2012; Li et al. 2012; Miao and Wang 2013; Quang et al. 2015). Levels of PAMP gene expression may vary among different transgenic plant lines developed by bioengineering technologies (Peng et al. 2004). The line, which shows high level of PAMP gene expression, shows very high level of resistance against pathogens, while the line, which shows low level of expression of the PAMP gene shows only low level of resistance (Peng et al. 2004). Hence, the transgenic lines should be carefully selected to generate highly useful disease-resistant cultivars. Expression of PAMP genes can be enhanced by properly selecting the promoter for gene transcription (Takakura et al. 2004). The transgenic plants expressing introduced gene may have side effects, showing retardation of plant growth and reduced crop yield potential. However, transgenic plants expressing the PAMP harpin gene show good agronomic characters (Xu et al. 2010; Li et al. 2011; Pavli et al. 2011). Selection of suitable pathogen-inducible promoter for expressing the PAMP gene appears to be a perquisite for developing disease-resistant plants without any reduction in yield potential (Choi et al. 2004; Donghua et al. 2004)

Bioengineering technologies have been exploited to utilize PIMPs/HAMPs to develop transgenic plants expressing enhanced disease resistance. Oligogalacturonides are the best-characterized plant cell wall-derived PIMPs/HAMPs (Vallarino and Osorio 2012). However, not all OGAs are capable of eliciting a defense response. Their ability to elicit defense responses depends on length (degree of polymerization), degree of methyl esterification and the level of acetylation (Côté and Hahn 1994; Vidhyasekaran 1997, 2007; Wiethölter et al. 2003; Aziz et al. 2004; Ferrari et al. 2007; Osorio et al. 2008; Vallarino and Osorio 2012).

Both the degrees of substitution (methylesterification and/or acetylation) and polymerization can be controlled by specific enzymes such as pectin methylesterases (PMEs), pectin acetylesterases (PAEs), polygalacturonases (PGs), or pectate lyases-like (PLLs) (Sénéchal et al. 2014). PME can modify the structure of OGAs and the modified OGAs will be highly active in triggering plant innate immune signaling systems. Transgenic plants expressing pectin methyl esterase gene (*PME*) generate oligogalacturonides, which act as host-derived elicitor/PIMP/HAMP. These transgenic plants show enhanced expression of plant immune responses and enhanced disease resistance (Lionetti et al. 2007, 2014; Osorio et al. 2008, 2011). PME activity is tightly regulated by an inhibitor protein called pectin methylesterase inhibitor protein (PMEI) (Giovane et al. 2004; Di Matteo et al. 2005). Transgenic plants overexpressing genes encoding PME inhibitor proteins show enhanced disease resistance (Lionetti et al. 2007, 2014; An et al 2008).

Transgenic plants expressing PG gene show enhanced disease resistance (Ferrari et al. 2008). Polygalacturonase-inhibiting proteins (PGIPs) play important role in switching on plant immune signaling systems (Manfredini et al. 2005; Federici et al. 2006; Alexandersson et al. 2011). Transgenic plants expressing *PGIP* genes also show enhanced disease resistance (Joubert et al. 2006, 2007; Alexandersson et al. 2011; Borras-Hidalgo et al. 2012; Nguema-Ona et al. 2013; Wang et al. 2013). The expression of *PGIP* genes does not affect the agronomic characters of the transformed plants (Powell et al. 2000; Capodicasa et al. 2004; Agüero et al. 2005; Borras-Hidalgo et al. 2012; Nguema-Ona et al. 2013).

Plant elicitor peptides (Peps) are the other group of PIMPs/HAMPs (Huffaker et al. 2011; Logemann et al. 2013; Hann et al. 2014). The Pep proteins are processed from the precursor PROPEP proteins. Transgenic plants overexpressing *PROPEP* genes show enhanced disease resistance (Huffaker et al. 2006). Transgenic plants overexpressing PIP (PAMP-induced Peptides) and systemin switch on plant innate immunity and show enhanced disease resistance (Coppola et al. 2014; Hou et al. 2014). PIMPs/HAMPs appear to be powerful tools to engineer disease resistance in field crops.

Bioengineering PRRs has been shown to be another potential technology to awaken the quiescent plant innate immunity for effective management of crop diseases. Intergeneric transfer of PRR from the weed plant Arabidopsis to various crop species has been achieved to develop disease-resistant crop plants. EFR is a Brassicaceae-specific PRR (Zipfel et al. 2006). Transfer of EFR from *Arabidopsis* to various crop plants is highly useful for crop disease management. Pathogens that are adapted to a particular host plant may be adept at suppressing the PRRs of that host by their effectors. The effectors of the pathogens might not recognize PRRs from other host plants and development of transgenic plants expressing PRRs from other plant species may provide good resistance against various bacterial pathogens possessing the PAMP EF-Tu (Lacombe et al. 2010). Transgenic tomato plants expressing *EFR* gene from *Arabidopsis* show enhanced resistance against the tomato wilt pathogen *Ralstonia solanacearum* (Lacombe et al. 2010). Transgenic tobacco expressing EFR also show resistance against *Agrobacterium. tumefaciens* (Brutus et al. 2010). Transgenic banana plants expressing the rice PRR *XA21* gene showed complete resistance to *Xanthomonas campestris* pv. *musacearum* (Tripathi et al. 2014). Transgenic *Citrus sinensis* plants expressing the rice PRR *XA21* gene enhance resistance against the citrus canker pathogen *Xanthomonas axonopodis* pv. *citri* (Mendes et al. 2010). Transgenic rice plants overexpressing *Xa21* gene showed enhanced resistance against the rice bacterial blight pathogen *Xanthomonas oryzae* pv. *oryzae* (Park et al. 2008, 2010a, b; Chen et al. 2014). Transgenic *Arabidopsis* plants expressing *FLS2* gene showed increased resistance against *Pseudomonas syringae* pv. *tomato* DC3000 (De Lorenzo et al. 2011). The transgenic rice plants overexpressing the HAMP receptor *WAK1* show enhanced disease resistance (Li et al. 2009).

Transcription factors are the master switches, which regulate expression of defense genes in the PAMP-triggered plant immune signaling systems (Century et al. 2008; Moreau et al. 2012). PAMPs and PIMPs/HAMPs switch on the expression of various transcription factor genes involved in plant defense responses (Denoux et al. 2008; Higashi et al. 2008; Chujo et al. 2013; McLellan et al. 2013). Several transcription factors have been shown to trigger "priming" of defense responses and induce "Systemic Acquired resistance (SAR)" in plants (Chavan and Kamble 2013; Nakayama et al. 2013). The plant defense activators benzothiadiazole, probenazole, and tiadinil trigger the expression of transcription factors which modulate the expression of defense genes inducing disease resistance (Shimono et al. 2007, 2012). DL-3-aminobutyric acid (β-aminobutyric acid, BABA) has been found to induce priming of WRKY transcription factors and trigger systemic resistance (Jakab et al. 2001). Foliar spray with BABA led to a significant reduction of lesion development in *Brassica carinata* caused by *Alternaria brassicae* (Chavan and Kamble 2013). Ergosterol treatment triggered a 23-fold increase of *VvWRKY* gene expression in grape plantlets and induced resistance against the necrotrophic fungal pathogen *Botrytis cinerea* (Laquitaine et al. 2006). The rhizobacterial strain *Pseudomonas fluorescens* WCS417r induces systemic resistance (ISR) in *A. thaliana* by activating the transcription factor MYB72 (Van der Ent et al. 2008). The MYC2 transcription factor also has been shown to be involved in *P. fluorescens* WCS417r-induced priming to trigger ISR (Pozo et al. 2008).

Several WRKY, MYB, MYC, bZIP, EREBP, and NAC transcription factors, have been engineered in various crop plants to develop disease resistant plants (He et al. 2001; Shin et al. 2002; Fischer and Dröge-Laser 2004; Guo et al. 2004; Cao et al. 2006; Waller et al. 2006; Chujo et al. 2007; Kim et al. 2007; Marchive et al. 2007; Mzid et al. 2007; Qiu et al. 2007; Wang et al. 2007; Zuo et al. 2007; Dai et al. 2008; Zhang et al. 2008; Bahrini et al. 2011a, b; Fan et al. 2011; Abbruscato et al. 2012; Liu et al. 2012; Peng et al. 2012; Shimono et al. 2012; Yu et al. 2012; Zhu et al. 2012; Han et al. 2013; Lee et al. 2013; Marchive et al. 2013; Wei et al. 2013; Chujo et al. 2014; Dang et al. 2014; Yokotani et al. 2014; Cao et al. 2016; Cheng et al. 2015; Jisha et al. 2015; Li et al. 2015; Shan et al. 2015).

Most of the successful stories in management of crop diseases using transcription factors are in rice plants. Transgenic rice plants overexpressing *OsWRKY13* showed enhanced blast (*Magnaporthe oryzae*) and bacterial blight (*Xanthomonas oryzae* pv. *oryzae*) disease resistance (Qiu et al. 2007; Cheng et al. 2015). WRKY30-

overexpressing rice plants show resistance against the rice blast pathogen *M. oryzae*, the sheath blight pathogen *Rhizoctonia solani* and the bacterial blight pathogen *X. oryzae* pv. *oryzae* (Peng et al. 2012; Han et al. 2013; Lee et al. 2013) Transgenic rice plants overexpressing *OsWRKY31* (Zhang et al. 2008), *WRKY45* (Shimono et al. 2007, 2012; Goto et al. 2015), *OsWRKY47* (Wei et al. 2013), *WRKY53* (Chujo et al. 2007, 2014), OsWRKY89 (Wei et al. 2013) show enhanced resistance to *M. oryzae*. Transgenic rice plants overexpressing *OsWRKY71* showed enhanced resistance to the bacterial blight pathogen *Xanthomonas oryzae* pv. *oryzae* (Liu et al. 2007).

Some transcription factors have been found to negatively regulate the expression of defense signaling systems. Silencing of the negative regulator transcription factors may be a useful strategy in developing disease-resistant plants. The *TaNAC1* gene-silenced wheat plants showed enhanced resistance against the stripe rust pathogen (Wang et al. 2015). *WRKY42*-suppressing (*WRKY42*-RNA interference [RNAi]) rice plants were developed and these plants showed increased resistance to *M. oryzae* (Cheng et al. 2015).

1.5 Switching on Plant Innate Immunity Using PAMP-PIMP-PRR-Transcription Factor Is the Most Potential Biotechnological Approach for Management of Crop Diseases

Susceptibility and resistance are two sides of the same coin (Vidhyasekaran 2007). The plant immune system is induced faster and to a higher level in resistant interactions (Makandar et al. 2006). The major differences between susceptible and resistant interactions are the magnitude and timing of induction of plant immune signaling system (Makandar et al. 2006; Rinaldi et al. 2007; Asselbergh et al. 2008). Higher and faster expression of genes involved in signal transduction systems has been found to be associated with improved tolerance to pathogens (Coppinger et al. 2004; Waller et al. 2006; Yamamizo et al. 2006; Zhang et al. 2006; Brader et al. 2007; Qiu et al. 2007). PAMP-induced defense in susceptible host plants is insufficient to stop infection; nonetheless, it is referred to as basal resistance (Nürnberger and Lipka 2005; Fung et al. 2008). Early, rapid and strong activation of plant innate immune system is necessary to induce strong defense responses against pathogens (Orlowska et al. 2011; Aghnoum and Niks 2012; Lanubile et al. 2010, 2012, 2014; Groβkinsky et al. 2012). Strong signals are needed to switch on early and strong activation of plant immunity. Engineering and/or proper application of PAMP/PRR products much before pathogen invasion results in early switching on the plant immune system.

Recently several bioengineering and molecular manipulation technologies have been developed to switch on the 'sleeping' plant innate immune system, which has potential to detect and suppress the development of a wide range of plant pathogens in economically important crop plants (Lacombe et al. 2010). Enhancing disease resistance through altered regulation of plant immunity signaling systems would be durable and publicly acceptable (Yamamizo et al. 2006; Shao et al. 2008; Gust et al. 2010; Lacombe et al. 2010). The plant innate immune systems have high potential to fight against viral, bacterial, oomycete, and fungal pathogens and protect the crop plants against wide range of diseases (Knecht et al. 2010; Lacombe et al. 2010; D'Amelio et al. 2011; Hwang and Hwang 2011; Alkan et al. 2012). This book describes various bioengineering and molecular manipulation technologies employed to trigger defense responses and manage crop diseases.

References

Abbruscato P, Nepusz T, Mizzi L, Del Corvo M, Morandini P, Fumasoni I, Michel C, Paccanaro A, Guiderdoni E, Schaffrath U, Morel J-B, Piffanelli P, Faivre-Rampant O (2012) *OsWRKY22*, a monocot *WRKY* gene, plays a role in the resistance response to blast. Mol Plant Pathol 13:828–841

Aghnoum R, Niks RE (2012) Compatible *Puccinia hordei* infection in barley induces basal defense to subsequent infection by *Blumeria graminis*. Physiol Mol Plant Pathol 77:17–22

Agostini JP, Bushong PM, Timmer LW (2003) Greenhouse evaluation of products that induce host resistance for control of scab, melanose, and Alternaria brownspot of citrus. Plant Dis 87:69–74

Agüero CB, Uratsu SL, Greve C, Powell ALT, Labavitch JM, Meredith CP, Dandekar AM (2005) Evaluation of tolerance to Pierce's disease and *Botrytis* in transgenic plants of *Vitis vinifera* L. expressing the pear PGIP gene. Mol Plant Pathol 6:43–51

Akimoto-Tomiyama C, Furutani A, Tsuge S, Washington EJ, Nishizawa Y, Minami E, Ochiai H (2012) XopR, a type III effector secreted by *Xanthomonas oryzae* pv. *oryzae*, suppresses microbe-associated molecular pattern-triggered immunity in *Arabidopsis thaliana*. Mol Plant Microbe Interact 25:505–514

Alexandersson E, Becker JVW, Jacobson D, Nguema-Ona E, Steyn C, Denby KJ, Vivier MA (2011) Constitutive expression of a grapevine polygalacturonase-inhibiting protein affects gene expression and cell wall properties in uninfected tobacco. BMC Res Notes 4:493

Alkan N, Fluhr R, Prusky D (2012) Ammonium secretion during *Colletotrichum coccodes* infection modulates salicylic and jasmonic acid pathways of ripe and unripe tomato fruit. Mol Plant Microbe Interact 25:85–96

An SH, Sohn KH, Choi HW, Hwang IS, Lee SC, Hwang BK (2008) Pepper pectin methylesterase inhibitor protein CaPMEI1 is required for antifungal activity, basal resistance and abiotic stress tolerance. Planta 228:61–78

Anand A, Schmelz EA, Muthukrishnan S (2003) Development of a lesion-mimic phenotype in a transgenic wheat line over expressing genes for pathogenesis-related (PR) proteins is dependent on salicylic acid concentration. Mol Plant Microbe Interact 16:916–925

Asselbergh B, De Vleesschauwer D, Höfte M (2008) Global switches and fine-tuning ABA modulates plant pathogen defense. Mol Plant Microbe Interact 21:709–719

Aziz A, Heyraud A, Lambert B (2004) Oligogalacturonide signal transduction, induction of defence-related responses and protection of grapevine against *Botrytis cinerea*. Planta 218:767–774

Aziz A, Gauthier A, Bezier A, Poinssot B, Joubert JM, Pugin A, Heyraud A, Baillieu F (2007) Elicitor and resistance-inducing activities of β-1,4 cellodextrins in grapevine, comparison with β-1,3 glucans and α-1,4 oligogalacturonides. J Exp Bot 58:1463–1472

Bahrini I, Ogawa T, Kobayashi F, Kawahigashi H, Handa H (2011a) Overexpression of the pathogen-inducible wheat *TaWRKY45* gene confers disease resistance to multiple fungi in transgenic wheat plants. Breed Sci 61:319–326

Bahrini I, Sugisawa M, Kikuchi R, Ogawa T, Kawahigashi H, Ban T, Handa H (2011b) Characterization of a wheat transcription factor, *TaWRKY45*, and its effect on *Fusarium* head blight resistance in transgenic wheat plants. Breed Sci 61:121–129

Belbahri L, Boucher C, Candresse T, Nicole M, Ricci P, Keller HA (2001) A local accumulation of the *Ralstonia solanacearum* PopA protein in transgenic tobacco renders a compatible interaction incompatible. Plant J 28:419–430

Bellincampi D, Cervone F, Lionetti V (2014) Plant cell wall dynamics and wall-related susceptibility in plant-pathogen interactions. Front Plant Sci 5:228. doi:10.3389/fpls.2014.00228

Böhm H, Albert I, Fan L, NürnbergerT RA (2014) Immune receptor complexes at the plant surface. Curr Opin Plant Biol 20:47–54

Boller T, He SH (2009) Innate immunity in plants: an arms race between pattern recognition receptors in plants and effectors in microbial pathogens. Science 324:742–744

Borras-Hidalgo O, Caprari C, Hernandez-Estevez I, De Lorenzo G, Cervone F (2012) A gene for plant protection: expression of a bean polygalacturonase inhibitor in tobacco confers a strong resistance against *Rhizoctonia solani* and two oomycetes. Front Plant Sci 3:1–6

Boutrot F, Segonzac C, Chang KN, Qiao H, Ecker JR, Zipfel C (2010) Direct transcriptional control of the *Arabidopsis* immune receptor FLS2 by the ethylene-dependent transcription factors EIN3 and EIL1. Proc Natl Acad Sci U S A 107:14502–14507

Brader G, Djamei A, Teige M, Palva ET, Hirt H (2007) The MAP kinase kinase MKK2 affects disease resistance in *Arabidopsis*. Mol Plant Microbe Interact 20:589–596

Bradley CA (2008) Effect of fungicide seed treatments on stand establishment, seedling disease, and yield of soybean in North Dakota. Plant Dis 92:120–125

Brutus A, Sicilia F, Macone A, Cervone F, De Lorenzo G (2010) A domain swap approach reveals a role of the plant wall-associated kinase 1 (WAK1) as a receptor of oligogalacturonides. Proc Natl Acad Sci U S A 107:9452–9457

Byamukama E, Robertson AE, Nutter FW Jr (2015) *Bean pod mottle virus* time of infection influences soybean yield, yield components, and quality. Plant Dis 99:1026–1032

Cai X-Z, Zhou X, Xu Y-P, Joosten MHAJ, deWit PJGM (2007) *Cladosporium fulvum* CfHNNI1 induces hypersensitive necrosis, defence gene expression and disease resistance in both host and nonhost plants. Plant Mol Biol 64:89–101

Cao Y, Wu Y, Zheng Z, Song F (2006) Overexpression of the rice EREBP-like gene *OsBIERF3* enhances disease resistance and salt tolerance in transgenic tobacco. Physiol Mol Plant Pathol 67:202–211

Cao W-L, Chu R-Z, Zhang Y, Luo J, Su Y-Y, Xie L-J, Zhang H-S, Wang J-F, Bao Y-M (2016) *OsJAMyb*, a R2R3-type MYB transcription factor, enhanced blast resistance in transgenic rice. Physiol Mol Plant Pathol (in press)

Capodicasa C, Vairo D, Zabotina O, McCartney L, Caprari C, Mattei B, Manfredini C, Aracri B, Benen J, Knox JF, De Lorenzo G, Cervone F (2004) Targeted modification of homogalacturonan by transgenic expression of a fungal polygalacturonase alters plant growth. Plant Physiol 135:1294–1304

Century K, Reuber TL, Ratcliffe OJ (2008) Regulating the regulators: the future prospects for transcription-factor-based agricultural biotechnology products. Plant Physiol 147:20–29

Chassot C, Buchala A, Schoonbeek H-J, Metraux J-P, Lamotte O (2008) Wounding of Arabidopsis leaves causes a powerful but transient protection against *Botrytis* infection. Plant J 55:555–567

Chavan V, Kamble A (2013) β-Aminobutyric acid primed expression of WRKY and defence genes in *Brassica carianta* against Alternaria blight. J Phytopathol 161:859–865

Che FS, Nakajima Y, Tanaka N, Iwano M, Yoshida T, Takayama S, Kadota I, Isogai A (2000) Flagellin from an incompatible strain of *Pseudomonas avenae* induces a resistance response in cultured rice cells. J Biol Chem 275:32347–32356

Chen ZY, Brown RL, Rajasekaran K, Damann KE, Cleveland TE (2006) Identification of a maize kernel pathogenesis-related protein and evidence for its involvement in resistance to *Aspergillus flavus* infection and aflatoxin production. Phytopathology 96:87–95

Chen L, Zhang S-J, Zhang S-S, Qu S, Ren X, Long J, Yin Q, Qian J, Sun F, Zhang C, Wang L, Wu X, Wu T, Zhang Z, Cheng Z, Hayes M, Beer SV, Dong H (2008) A fragment of the *Xanthomonas oryzae* pv. *oryzicola* harpin $HpaG_{Xooc}$ reduces disease and increases yield of rice in extensive grower plantings. Phytopathology 98:792–802

Chen F, Gao J, Miao S, Yuan Y-X, Wang M-Y, Li Q, Mao B-Z, Jiang L-W, He ZH (2010a) Plasma membrane localization and potential endocytosis of constitutively expressed XA21 proteins in transgenic rice. Mol Plant 3:917–926

Chen X, Chern M, Canlas PE, Jiang C, Ruan D, Cao P, Ronald PC (2010b) A conserved threonine residue in the juxtamembrane domain of the XA21 pattern recognition receptor is critical for kinase autophosphorylation and XA21-mediated immunity. J Biol Chem 285:10454–10463

Chen X, Zuo S, Schwessinger B, Chern M, Canlas PE, Ruan D, Zhou X, Wang J, Daudi A, Petzold CJ (2014) An XA21-associated kinase (OsSERK2) regulates immunity mediated by the XA21 and XA3 immune receptors. Mol Plant 7:874–892

Cheng B, Yu X, Ma Z, Dong S, Dou D, Wang Y, Zheng X (2012) *Phytophthora sojae* effector Avh331 suppresses the plant defence response by disturbing the MAPK signalling pathway. Physiol Mol Plant Pathol 77:1–9

Cheng H, Liu H, Deng Y, Xiao J, Li X, Wang S (2015) The WRKY45-2 WRKY13 WRKY42 transcriptional regulatory cascade is required for rice resistance to fungal pathogen. Plant Physiol 167:1087–1099

Choi JJ, Klosterman SJ, Hadwiger LA (2004) A promoter from pea gene DRR206 is suitable to regulate an elicitor-coding gene and develop disease resistance. Phytopathology 94:651–660

Choi M-S, Heu S, Paek N-C, Koh H-J, Lee J-S, Oh C-S (2012) Expression of *hpa1* gene encoding a bacterial harpin protein on *Xanthomonas oryzae* pv. *oryzae* enhances disease resistance to both fungal and bacterial pathogens in rice and *Arabidopsis*. Plant Pathol J 28:364–372

Chuang H-W, Chang P-Y, Syu Y (2014) Harpin protein, an elicitor of disease resistance, acts as a growth promoter in *Phalaenopsis* orchids. J Plant Growth Regul 33:788–797. doi:10.1007/s00344-014-9425-1

Chujo T, Takai R, Akimoto-Tomiyama C, Ando S, Minami E, Nagamura Y, Kaku H, Shibuya N, Yasuda M, Nakashita H, Umemura K, Okada K, Nojiri H, Yamane H (2007) Involvement of the elicitor-induced gene OsWRKY53 in the expression of defense-related genes in rice. Biochim Biophys Acta 1769:497–505

Chujo T, Miyamoto K, Shimogawa T, Shimizu T, Otake Y, Yokotani N, Nishizawa Y, Shibuya N, Nojiri H, Yamane H, Minami E, Okada K (2013) OsWRKY28, a PAMP-responsive transrepressor, negatively regulates innate immune responses in rice against blast fungus. Plant Mol Biol 82:23–37

Chujo T, Miyamoto K, Ogawa S, Masuda Y, Shimizu T, Kishi-Kaboshi M, Takahashi A, Nishizawa Y, Minami E, Norjiri H, Yamane H, Okada K (2014) Overexpression of phosphomimic mutated OsWRKY53 leads to enhanced blast resistance in rice. PLoS One 9(6):e98737. doi:10.1371/journal.pone.0098737

Cohen Y, Van den Langenberg KM, Wehner TC, Ojiambo PS, Hausbeck M, Quesada-Ocampo LM, Lebeda A, Sierotzki H, Gisi U (2015) Resurgence of *Pseudoperonospora cubensis*: the causal agent of cucurbit downy mildew. Phytopathology 105:998–1012

Coppinger P, Repetti PP, Day B, Dahlbeck D, Mehlert A, Staskawicz BJ (2004) Overexpression of the plasma membrane-localized NDR1 protein results in enhanced bacterial disease resistance in *Arabidopsis thaliana*. Plant J 40:225–237

Coppola M, Corrado G, Coppola V, Cascone P, Martinelli R, Digilio MC, Pennacchio F, Rao R (2014) Prosystemin overexpression in tomato enhances resistance to different biotic stresses by activating genes of multiple signaling pathways. Plant Mol Biol Rep 33:1270–1285. doi:10.1007/s11105-014-0834-x

Coram TE, Pang ECK (2005) Isolation and analysis of candidate ascochyta blight defence genes in chickpea, Part 1. Generation and analysis of an expressed sequence tag (EST) library. Physiol Mol Plant Pathol 66:192–200

Côté F, Hahn MG (1994) Oligosaccharins: structures and signal transduction. Plant Mol Biol 26:1379–1411

D'Amelio R, Berta G, Gamalero E, Massa N, Avidano L, Cantamessa S, D'Agostino G, Bosco D, Marzachi C (2011) Increased plant tolerance against chrysanthemum yellows phytoplasma (*Candidatus Phytoplasma asteris*) following double inoculation with *Glomus mosseae* BEG12 and *Pseudomonas putida* SIPflRif. Plant Pathol 60:1014–1022

Dafermos NG, Kasselaki AM, Goumas DE, Spantidakis K, Eyre MD, Leifert C (2012) Integration of elicitors and less-susceptible hybrids for the control of powdery mildew in organic tomato crops. Plant Dis 96:1506–1512

Dai S, Wei X, Alfonso AA, Pei L, Duque UG, Zhang Z, Babb GM, Beachy RN (2008) Transgenic rice plants that overexpress transcription factors RF2a and RF2b are tolerant to rice tungro virus replication and disease. Proc Natl Acad Sci U S A 105:21012–21016

Dang F, Wang Y, She J, Lei Y, Liu Z, Eulgem T, Lai Y, Lin J, Yu L, Lei D, Guan D, Li X, Yuan Q, He S (2014) Overexpression of *CaWRKY27*, a subgroup IIe transcription factor of *Capsicum annuum*, positively regulates tobacco resistance to *Ralstonia solanacearum* infection. Physiol Plant 150:397–411

de Capdeville G, Beer SV, Wilson CL, Aist JR (2002) Alternative disease control agents induced resistance to blue mold in harvested 'Red Delicious' apple fruit. Phytopathology 92:900–908

de Capdeville G, Beer SV, Watkins CB, Wilson CL, Tedeschi LO, Aist JR (2003) Pre- and post-harvest harpin treatments of apples induce resistance to blue mold. Plant Dis 87:39–44

de Capdeville GD, Beer SV, Wilson CL, Aist JR (2008) Some cellular correlates of harpin-induced resistance to blue mold of apples. Trop Plant Pathol 33:103–113

de Jonge R, van Esse HP, Kombrink A, Shinya T, Desaki Y, Bours R, van der Krol S, Shibuya N, Joosten MHAJ, Thomma BPHJ (2010) Conserved fungal LysM effector Ecp6 prevents chitin-triggered immunity in plants. Science 329:953–955

De Lorenzo G, Brutus A, Savatin DY, Sicilia F, Cervone F (2011) Engineering plant resistance by constructing chimeric receptors that recognize damage-associated molecular patterns (DAMPs). FEBS Lett 585:1521–1526

Denoux C, Galletti R, Mammarella N, Gopalan S, Werck D, De Lorenzo G, Ferrari S, Ausubel FM, Dewdney J (2008) Activation of defense response pathways by OGs and Flg22 elicitors in Arabidopsis seedlings. Mol Plant 1:423–445

Di Matteo A, Giovane A, Raiola A, Camardella L, Bonivento D, De Lorenzo G, Cervone F, Bellincampi D, Tsernoglou D (2005) Structural basis for the interaction between pectin methylesterase and a specific inhibitor protein. Plant Cell 17:849–858

Dodds PN, Rathjen JP (2010) Plant immunity: towards an integrated view of plant-pathogen interactions. Nat Rev Genet 11:539–548

Dong H-P, Peng J, Bao Z, Meng X, Bonasera JM, Chen G, Beer SV, Dong H (2004) Downstream divergence of the ethylene signaling pathway for harpin-stimulated Arabidopsis growth and insect resistance. Plant Physiol 136:3628–3638

Donghua J, Zejian G, Xujun C, Zhiqiang C, Zhong Z (2004) Study on tobacco disease resistances mediated by the elicitor gene cryptogein from *Phytophthora cryptogea*. Chin J Agric Biotechnol 1:125–131

Elmer PAG, Reglinski T (2006) Biosuppression of *Botrytis cinerea* in grapes. Plant Pathol 55:155–177

Fan H, Wang F, Gao H, Wang L, Xu J, Zhao Z (2011) Pathogen-induced *MdWRKY1* in 'Qinguan' apple enhances disease resistance. J Plant Biol 54:150–158

Farag MA, Huhman DV, Dixon RA, Sumner LW (2008) Metabolomics reveals novel pathways and differential mechanistic and elicitor-specific responses in phenylpropanoid and isoflavonoid biosynthesis in *Medicago truncatula* cell cultures. Plant Physiol 1486:387–402

Federici L, Di Matteo A, Fernandez-Recio J, Tsernoglou D, Cervone F (2006) Polygalacturonase-inhibiting proteins: players in plant immunity? Trends Plant Sci 11(2):65–70
Fernández-Ortuño D, Grabke A, Li X, Schnabel G (2015) Independent emergence of resistance to seven chemical classes of fungicides in *Botrytis cinerea*. Phytopathology 105:424–432
Ferrari S, Galletti R, Denoux C, De Lorenzo G, Ausubel FM, Dewdney J (2007) Resistance to *Botrytis cinerea* induced in Arabidopsis by elicitors is independent of salicylic acid, ethylene, or jasmonate signaling but requires PHYTOALEXIN DEFICIENT3. Plant Physiol 144:367–379
Ferrari S, Galletti R, Pontiggia D, Manfredni D, Lionetti V, Bellincampi D, Cervone F, De Lorenzo G (2008) Transgenic expression of a fungal *endo*-polygalacturonase increases plant resistance to pathogens and reduces auxin sensitivity. Plant Physiol 146:669–681
Ferrari S, Sella L, Janni M, De Lorenzo G, Favaron F, D'Ovidio R (2012) Transgenic expression of polygalacturonase-inhibitimg proteins in *Arabidopsis* and wheat increases resistance to the flower pathogen *Fusarium graminearum*. Plant Biol 14:31–38
Ferrari S, Savatin DV, Sicilia F, Gramegna G, Cervone F, Lorenzo GD (2013) Oligogalacturonides: plant damage-associated molecular patterns and regulators of geowth and development. Front Plant Sci 4:49. doi:10.3389/fpls.2013.00049
Fischer U, Dröge-Laser W (2004) Overexpression of *Nt*ERF5, a new member of the tobacco ethylene response transcription factor family enhances resistance to *Tobacco mosaic virus*. Mol Plant Microbe Interact 17:1162–1171
Flors V, Ton J, van Doorn R, Garcia-Agustin P, Mauch-Mani B (2008) Interplay between JA, SA, and ABA signaling during basal and induced resistance against *Pseudomonas syringae* and *Alternaria brassicicola*. Plant J 54:81–92
Fu M, Xu M, Zhou T, Wang D, Tian S, Han L, Dong H, Zhang C (2014) Transgenic expression of a functional fragment of harpin protein Hpa1 in wheat induces the phloem-based defence against English grain aphid. J Exp Bot 65:1439–1453
Fujiwara S, Tanaka N, Kaneda T, Takayama S, Isogai A, Che F-S (2004) Rice cDNA microarray-based gene expression profiling of the response to flagellin perception in cultured rice cells. Mol Plant Microbe Interact 17:986–998
Fung RWM, Gonzalo M, Fekete C, Kovacs LG, He Y, Marsh E, McIntyre LM, Schchtman DP, Qiu W (2008) Powdery mildew induces defense-oriented reprogramming of the transcriptome in a susceptible but not in a resistant grapevine. Plant Physiol 146:236–249
Gent DH, Twomey MC, Wolfenbarger SN, Woods JL (2015) Pre- and postinfection activity of fungicides in control of hop downy mildew. Plant Dis 99:858–865
Gimenez-Ibanez S, Rathjen JP (2010) The case for the defense: plants versus *Pseudomonas syringae*. Microbes Infect 12:428–437
Gimenez-Ibanez S, Hann DR, Ntoukakis V, Petutschnig E, Lipka V, Rathjen JP (2009a) AvrPtoB targets the LysM receptor kinase CERK1 to promote bacterial virulence on plants. Curr Biol 19:423–429
Gimenez-Ibanez S, Ntoukakis V, Rathjen JP (2009b) The LysM receptor kinase CERK1 mediates bacterial perception in Arabidopsis. Plant Signal Behav 4:539–541
Giovane A, Servillo L, Balestrieri C, Raiola A, D'Arvino R, Tamburrini M, Ciardiello MA, Camardella L (2004) Pectin methylesterase inhibitor. Biochim Biophys Acta 1696:245–252
Göhre V, Spallek T, Häweker H, Mersmann S, Mentzel T, Boller T, de Torres M, Mansfield JW, Robatzek S (2008) Plant pattern-recognition receptor FLS2 is directed for degradation by the bacterial ubiquitin ligase AvrPtoB. Curr Biol 18:1824–1832
Goto S, Sasakura-Shimoda F, Suetsugu M, Selvaraj MG, Hayashi N, Yamazaki M, Ishitani M, Shimono M, Sugano S, Matsushita A, Tanabata T, Takatsuji H (2015) Development of disease-resistant rice by optimized expression of WRKY45. Plant Biotechnol J 13:753–765
Groβkinsky DK, Koffler BE, Roitsch T, Maier R, Zechmann B (2012) Compartment-specific antioxidative defense in *Arabidopsis* against virulent and avirulent *Pseudomonas syringae*. Phytopathology 102:662–673

Gudmestad NC, Arabiat S, Miller JS, Pasche JS (2013) Prevalence and impact of SDHI fungicide resistance in *Alternaria solani*. Plant Dis 97:952–960

Guo Z, Chen XJ, Wu XL, Ling JQ, Xu P (2004) Overexpression of the AP2/EREBP transcription factor OPBP1 enhances disease resistance and salt tolerance in tobacco. Plant Mol Biol 55:607–618

Gust AA, Brunner F, Nürnberger T (2010) Biotechnological concepts for improving plant innate immunity. Curr Opin Biotechnol 21:204–210

Hamiduzzaman MM, Jakeb G, Barnavon L, Neuhaus JM, Mauch-Mani B (2005) β-Aminobutyric acid-induced resistance against downy mildew in grapevine acts through the potentiation of callose formation and jasmonic acid signaling. Mol Plant Microbe Interact 18:819–829

Han M, Ryu H-S, Kim C-Y, Park D-S, Ahn Y-K, Jeon J-S (2013) *OsWRKY30* is a transcription activator that enhances rice resistance to the *Xanthomonas oryzae* pathovar oryzae. J Plant Biol 56:258–265

Han DJ, Wang QL, Chen XM, Zeng QD, Wu JH, Xue WB, Zhan GM, Huang LL, Kang ZS (2015) Emerging Yr26-virulent races of *Puccinia striiformis* f. *tritici* are threatening wheat production in the Sichuan Basin, China. Plant Dis 99:754–760

Handiseni M, Jo Y-K, Zhou X-G (2015) Integration of *Brassica* cover crop with host resistance and azoxystrobin for management of rice sheath blight. Plant Dis 99:883–885

Hann DR, Rathjen JP (2007) Early events in the pathogenecity of *Pseudomonas syringae* on *Nicotiana benthamiana*. Plant J 49:607–618

Hann DR, Gimenez-Ibanez S, Rathjen JP (2010) Bacterial virulence effectors and their activities. Curr Opin Plant Biol 13:388–393

Hann CT, Bequette CJ, Dombrowski JE, Stratmann JVV (2014) Methanol and ethanol modulate responses to danger- and microbe-associated molecular patterns. Front Plant Sci 5:550. doi:10.3389/fpls.2014.00550

He P, Warren RF, Zhao TH, Shan LB, Zhu LH, Tang XY, Zhou JM (2001) Overexpression of Pti5 in tomato potentiates pathogen-induced defense gene expression and enhances disease resistance to *Pseudomonas syringae* pv. *tomato*. Mol Plant Microbe Interact 14:1453–1457

Hermosa MR, Turra D, Fogliano V, Monte E, Lorito M (2006) Identification and characterization of potato protease inhibitors able to inhibit pathogenicity and growth of *Botrytis cinerea*. Physiol Mol Plant Pathol 68:138–148

Hernández-Blanco C, Feng DX, Hu J, Sanchez-Vallet A, Deslandes L, Llorente F, Berrocal-Lobo M, Keller H, Barlet X, Sánchez-Rodriguez C, Anderson LK, Somerville S, Marco Y, Molina A (2007) Impairment of cellulose synthase required for *Arabidopsis* secondary cell wall formation enhances disease resistance. Plant Cell 19:890–903

Higashi K, Ishiga Y, Inagaki Y, Toyoda K, Shiraishi T, Ichinose Y (2008) Modulation of defense signal transduction by flagellin-induced WRKY41 transcription factor in *Arabidopsis thaliana*. Mol Genet Genomics 279:303–312

Holmes GJ, Ojiambo PS, Hausbeck MK, Quesada-Ocampo L, Keinath AP (2015) Resurgence of cucurbit downy mildew in the United States: a watershed event for research and extension. Plant Dis 99:428–441

Hou S, Wang X, Chen D, Yang X, Wang M, Turră D, Di Pietro A (2014) The secreted peptide PIP1 amplifies immunity through receptor-like kinase 7. PLoS Pathog 10(9):e1004331. doi:10.1371/journal.ppat.1004331

Hu M-J, Grabke A, Dowling ME, Holstein HJ, Schnabel G (2015) Resistance in *Colletotrichum siamense* from peach and blueberry to thiophanate-methyl and azoxystrobin. Plant Dis 99:806–814

Huffaker A, Ryan CA (2007) Endogenous peptide defense signals in *Arabidopsis* differentially amplify signaling for the innate immune response. Proc Natl Acad Sci U S A 104:10732–10736

Huffaker A, Pearce G, Ryan CA (2006) An endogenous peptide signal in *Arabidopsis* activates components of the innate immune response. Proc Natl Acad Sci U S A 103:10098–10103

Huffaker A, Dafoe NJ, Schmelz EA (2011) ZmPep1, an ortholog of Arabidopsis elicitor peptide 1, regulates maize innate immunity and enhances disease resistance. Plant Physiol 155:1325–1338

Huo R, Wang Y, Ma L, Qiao J, Shao M, Gao X (2010) Assessment of inheritance pattern and agronomic performance of transgenic rapeseed having harpinXooC-encoding *hrf2* gene. Transgenic Res 19:841–847

Hwang IS, Hwang BK (2011) The pepper mannose-binding lectin gene *CaMBL1* is required to regulate cell death and defense responses to microbial pathogens. Plant Physiol 155:447–463

Iriti M, Vitalini S, Di Tommaso G, D'Amico S, Borgo M, Faoro F (2011) New chitosan formulation prevents grapevine powdery mildew infection and improves polyphenol content and free radical scavenging activity of grape and wine. Aust J Grape Wine Res 17:263–269

Jakab G, Cottier V, Toquin V, Rigoli G, Zimmerli L, Metraux JP, Mauch-Mani B (2001) β-aminobutyric acid-induced resistance in plants. Eur J Plant Pathol 107:29–37

Jang YS, Sohn SI, Wang MH (2006) The *hrpN* gene of *Erwinia amylovora* stimulates tobacco growth and enhances resistance to *Botrytis cinerea*. Planta 223:449–456

Jeworutzki E, Roelfsema MR, Anschütz U, Krol E, Elzenga JT, Felix G, Boller T, Hedrich R, Becker D (2010) Early signaling through the *Arabidopsis* pattern recognition receptors FLS2 and EFR involves Ca^{2+}-associated opening of plasma membrane anion channels. Plant J 62:367–378

Jisha V, Dampanaboina L, Vadassery J, Mithofer A, Kappara S, Ramanan R (2015) Overexpression of an AP2/ERF type transcription factor *OsEREBP1* confers biotic and abiotic stress tolerance in rice. PLoS One 10(6):e0127831. doi:10.1371/journal.pone.0127831

Jones AL (2001) Challenges of controlling fire blight. Compact Fruit Tree 34(3):86–91

Joubert DA, Slaughter AR, Kemp G, Becker JV, Krooshof GH, Bergmann C, Benen J, Pretorius IS, Vivier MA (2006) The grapevine polygalacturonase-inhibiting protein (VvPGIP1) reduces *Botrytis cinerea* susceptibility in transgenic tobacco and differentially inhibits fungal polygalacturonases. Transgenic Res 15:702–887

Joubert DA, Kars I, Wagemakers L, Bergmann C, Kemp G, Vivier MA, van Kan JA (2007) A polygalacturonase-inhibiting protein from grapevine reduces the symptoms of the endopolygalacturonase BcPG2 from *Botrytis cinerea* in *Nicotiana benthamiana* leaves without any evidence for in vitro interaction. Mol Plant Microbe Interact 20:392–402

Kanetis L, Förster H, Adaskaveg JE (2008) Optimizing efficacy of new postharvest fungicides and evaluation of sanitizing agents for managing citrus green mold. Plant Dis 92:261–269

Keeling CI, Weisshaar S, Lin RPC, Bohlmann J (2008) Functional plasticity of paralogous diterpene synthases involved in conifer defense. Proc Natl Acad Sci U S A 105:1090–1985

Keinath AP (2015) Baseline sensitivity of *Didymella bryoniae* to cyprodinil and fludioxonil and field efficacy of these fungicides against isolates resistant to pyraclostrobin and boscalid. Plant Dis 99:836–841

Keller H, Pamboukdjian N, Ponchet M, Poupet A, Delon R, Verrier J-L, Roby D, Ricci P (1999) Pathogen-induced elicitin production in transgenic tobacco generates a hypersensitive response and nonspecific disease resistance. Plant Cell 11:223–236

Kim S-Y, Kim Y-C, Seong ES, Lee Y-H, Park M, Choi D (2007) The chili pepper CaATL1: an AT-hook motif-containing transcription factor implicated in defence responses against pathogens. Mol Plant Pathol 8:761–771

Kitner M, Lebeda A, Sharma R, Runge F, Dvořák P, Tahir A, Choi Y-J, Sedláková B, Thines M (2015) Coincidence of virulence shifts and population genetic changes of *Pseudoperonospora cubensis* in the Czech Republic. Plant Pathol 64:1461–1470

Knecht K, Seyffarth M, Desel C, Thurau T, Sherameti I, Lou B, Oelmüller R, Cai D (2010) Expression of *BvGLP-1* encoding a germin-like protein from sugar beet in *Arabidopsis thaliana* leads to resistance against phytopathogenic fungi. Mol Plant Microbe Interact 23:446–457

Krol E, Mentzel T, Chinchilla D, Boller T, Felix G, Kemmerling B, Postel S, Arents M, Jeworutzki E, Al-Rasheid KAS, Becker D, Hedrich R (2010) Perception of the Arabidopsis danger signal peptide 1 involves the pattern recognition receptor AtPEPR1 and its close homologue AtPEPR2. J Biol Chem 285:13471–13479

Lacombe S, Rougon-Cardoso A, Sherwood E, Peeters N, Dahlbeck D, van Esse HP, Smoker M, Rallapalli G, Thomma BPHJ, Stakawicz B, Jones JDG, Zipfel C (2010) Interfamily transfer of a plant pattern-recognition receptor confers broad-spectrum bacterial resistance. Nat Biotechnol 28:365–369

Lanubile A, Pasini L, Marocco A (2010) Differential gene expression in kernels and silks of maize lines with contrasting levels of ear rot resistance after *Fusarium verticillioides* infection. J Plant Physiol 167:1398–1406

Lanubile A, Bernardi J, Marocco A, Logrieco A, Paciolla C (2012) Differential activation of defense genes and enzymes in maize genotypes with contrasting levels of resistance to *Fusarium verticillioides*. Environ Exp Bot 78:39–46

Lanubile A, Ferrarini A, Maschietto V, Delledonne M, Marocco A, Bellin D (2014) Functional genomic analysis of constitutive and inducible defense responses to *Fusarium verticillioides* infection in maize genotypes with contrasting ear rot resistance. BMC Genomics 15:710

Laquitaine L, Gomès E, Francois J, Marchive C, Pascal S, Hamdi S, Atanassova R, Delrot S, Coutos-Thévenot P (2006) Molecular basis of ergosterol-induced protection of grape against *Botrytis cinerea*: induction of type 1 LTP promoter activity, WRKY, and stilbene synthase gene expression. Mol Plant Microbe Interact 19:1103–1112

Lecourieux D, Mazars C, Pauly N, Ranjeva R, Pugin A (2002) Analysis and effects of cytosolic free calcium elevations in response to elicitors in *Nicotiana plumbaginifolia* cells. Plant Cell 14:2627–2641

Lecourieux D, Lamotte O, Bourque S, Wendehenne D, Mazars C, Ranjeva R, Pugin A (2005) Proteinaceous and oligosaccharidic elicitors induce different calcium signatures in the nucleus of tobacco cells. Cell Calcium 38:527–538

Lee YK, Hong JK, Hippe-Sanwald S, Hwang BK (2000) Histological and ultrastructural comparisons of compatible, incompatible and DL-β-amino-*n* –butyric acid-induced resistance responses of pepper stems to *Phytophthora capsici*. Physiol Mol Plant Pathol 57:269–280

Lee H, Ko YJ, Cha J-Y, Park R, Ahn I, Hwang D-J (2013) The C-terminal region of *Os*WRKY30 is sufficient to confer enhanced resistance to pathogen and activate the expression of defense-related genes. Plant Biotechnol Rep 7:221–230

Li R, Fan Y (1999) Reduction of lesion growth rate of late blight plant disease in transgenic potato expressing harpin protein. Sci China (SerC) 40:96–101

Li Y-C, Sun X-J, Bi Y, Ge Y-H, Wang Y (2009) Antifungal activity of chitosan on *Fusarium sulphureum* in relation to dry rot of potato tuber. Agric Sci China 8:597–604

Li JG, Cao J, Sun FF, Niu DD, Yan F, Liu HX, Guo JH (2011) Control of *Tobacco mosaic virus* by PopW as a result of induced resistance in tobacco under greenhouse and field conditions. Phytopathology 101:1202–1208

Li G, Meng X, Wang R, Mao G, Han L, Liu Y, Zhang S (2012) Dual-level regulation of ACC synthase activity by MPK3/MPK6 cascade and its downstream WRKY transcription factor during ethylene induction in Arabidopsis. PLoS Genet 8(6):e1002767. doi:10.1371/journal.pgen.1002767

Li C, He X, Luo X, Xu L, Liu L, Min L, Jin L, Zhu L, Zhang X (2014a) Cotton WRKY1 mediates the plant defense-to-development transition during infection of cotton by *Verticillium dahliae* by activating JASMONATE ZIM-DOMAIN1 expression. Plant Physiol 166:2179–2194

Li F, Cheng C, Cui F, de Oliveira MVV, Yu X, Meng X, Intorne AC, Babilonia K, Li M, Li B, Chen S, Ma X, Xiao S, Zheng Y, Fei Z, Metz RP, Johnson CD, Koiwa H, Sun W, Li Z, de Souza Filho GA, Shan L, He P (2014b) Modulation of RNA polymerase II phosphorylation downstream of pathogen perception orchestrates plant immunity. Cell Host Microbe 16:748–758

Li J, Wang J, Wang N, Guo X, Gao Z (2015) *GhWRKY44*, a WRKY transcription factor of cotton, mediates defense responses to pathogen infection in transgenic *Nicotiana benthamiana*. Plant Cell Tissue Organ Cult 121:127–140

Lionetti V, Raiola A, Camardella L, Giovane A, Obel N, Pauly M, Favaron F, Cervone F, Bellincampi D (2007) Overexpression of pectin methylesterase inhibitors in Arabidopsis restricts fungal infection by *Botrytis cinerea*. Plant Physiol 143:1871–1880

Lionetti V, Raiola A, Cervone F, Bellincampi D (2014) Transgenic expression of pectin methylesterase inhibitors limits tobamovirus spread in tobacco and Arabidopsis. Mol Plant Pathol 15:265–274

Liu B, Quang Z, Zhang Y, Li X, Hong Y, Huang L, Liu S, Zhang H, Li D, Song F (2014) Tomato NAC transcription factor SISRN1 positively regulates defense response against biotic stress but negatively regulates abiotic stress response. PLoS One 9(7):e102067. doi:10.1371/journal.pone.0102067

Liu C-J, Deavours BE, Richard SB, Ferrer J-L, Blount JW, Huhman D, Dixon RA, Noel JP (2006) Structural basis for dual functionality of isoflavonoid *O*-methyltransferases in the evolution of plant defense responses. Plant Cell 18:3656–3669

Liu D, Chen X, Liu J, Ye J, Guo Z (2012) The rice transcription factor *OsERF922* negatively regulates resistance to *Magnaporthe oryzae* and salt tolerance. J Exp Bot 53:3899–3911

Liu X, Bai X, Wang X, Chu C (2007) OsWRKY71, a rice transcription factor, is involved in rice defense response. J Plant Physiol 164:969–979

Lloyd SR, Schoonbeek H, Trick M, Zipfel C, Ridout CJ (2014) Methods to study PAMP-triggered immunity in *Brassica* species. Mol Plant Microbe Interact 27:286–295

Logemann E, Birkenbihl RP, Rawat V, Schneeberger K, Schmelzer E, Somssich IE (2013) Functional dissection of the *PROPEP2* and *PROPEP3* promoters reveals the importance of WRKY factors in mediating microbe-associated molecular pattern-induced expression. New Phytol 198:1165–1177

Lohmann GV, Shimoda Y, Nielsen W, Jørgensen FG, Grossmann C, Sandal N, Sørensen K, Thirup S, Madsen LH, Tabata S, Sato S, Stougaard J, Radutoiu S (2010) Evolution and regulation of the *Lotus japonica* LysM receptor gene family. Mol Plant Microbe Interact 23:510–521

Luna E, Pastor V, Robert J, Flors V, Mauch-Mani B, Ton J (2011) Callose deposition: a multifaceted plant defense response. Mol Plant Microbe Interact 24:183–193

Maccaferri M, Zhang J, Bulli P, Abate Z, Chao S, Canter D, Bossolini E, Chen X, Pumphrey M, Dubcovsky J (2015) A genome-wide association study of resistance to stripe rust (*Puccinia striiformis* f. sp *tritici*) in a worldwide collection of hexaploid spring wheat (*Triticum aestivum* L.). G3 (Bethesda) 5:449–465

Macho AP, Zipfel C (2014) Plant PRRs and the activation of innate immune signalling. Mol Cell 54:263–272

Makandar R, Essig JS, Schapaugh MA, Trick HJ, Shah J (2006) Genetically engineered resistance to Fusarium head blight in wheat by expression of *Arabidopsis NPR1*. Mol Plant Microbe Interact 19:123–129

Malnoy M, Venisse JS, Chevreau E (2005) Expression of a bacterial effector, harpin N, causes increased resistance to fire blight in *Pyrus communis*. Tree Genet Genomes 1:41–49

Manfredini C, Sicilla F, Ferrari S, Pontiggia D, Salvi G, Caprari C, Lorito M, De Lorenzo G (2005) Polygalacturonase-inhibiting protein 2 of *Phaseolus vulgaris* inhibits BcPG1, a polygalacturonase of *Botrytis cinerea* important for pathogenicity, and protects transgenic plants from infection. Physiol Mol Plant Pathol 67:108–115

Marchive C, Mzid R, Deluc L, Barrieu F, Pirrello J, Gauthier A, Corio-Costet M-F, Regard F, Cailleteau B, Hamdi S, Lauvergeat V (2007) Isolation and characterization of a *Vitis vinifera* transcription factor, VvWRKY1, and its effect on responses to fungal pathogens in transgenic tobacco plants. J Exp Bot 58:1999–2010

Marchive C, Léon C, Kappel C, Coutos-Thévenot P, Corio-Costet M-F, Delrot S, Lauvergeat V (2013) Over-expression of VvWRKY1 in grapevines induces expression of jasmonic acid pathway-related genes and confers higher tolerance to the downy mildew. PLoS One 8(1):e54185. doi:10.1371/journal.pone.0054185

McLellan H, Boevink PC, Armstrong MR, Pritchard L, Gomez S, Morales J, Whisson SC, Beynon JL, Birch PR (2013) An RxLR effector from *Phytophthora infestans* prevents relocalisation of two plant NAC transcription factors from the endoplasmic reticulum to the nucleus. PLoS Pathog 9(10):e1003670. doi:10.1371/journal ppat 1003670

Mendes BMJ, Cardoso SC, Boscariol-Camargo RL, Cruz RB, Filho M, Bergamin Filho A (2010) Reduction in susceptibility to *Xanthomonas axonopodis* pv. *citri* in transgenic *Citrus sinensis* expressing the rice *Xa21* gene. Plant Pathol 59:68–75

Mentlak TA, Kombrink A, Shinya T, Ryder LS, Otomo I, Saitoh H, Terauchi R, Nishizawa Y, Shibuya N, Thomma BPHJ, Talbot NJ (2012) Effector-mediated suppression of chitin-triggered immunity by *Magnaporthe oryzae* is necessary for rice blast disease. Plant Cell 24:322–335

Miao W, Wang J (2013) Genetic transformation of cotton with a harpin-encoding gene hpa_{Xoo} confers defense response against *Verticillium dahliae* Kieb. Methods Mol Biol 958:223–246

Miao W, Wang X, Li M, Song C, Wang Y, Hu D, Wang J (2010) Genetic transformation of cotton with a harpin-encoding gene hpa_{Xoo} confers an enhanced defense response against different pathogens through a priming mechanism. BMC Plant Biol 10:67

Miles LA, Miles TD, Kirk WW, Schilder AMC (2012) Strobilurin (QoI) resistance in populations of *Erysiphe necator* on grapes in Michigan. Plant Dis 96:1621–1628

Moravčikova J, Matušikova I, Libantová J, Bauer M, Mlynárová L (2004) Expression of a cucumber class III chitinase and *Nicotiana plumbaginifolia* class I glucanase genes in transgenic potato plants. Plant Cell Tissue Organ Cult 79:161–168

Moreau M, Tian M, Klessig DF (2012) Salicylic acid binds NPR3 and NPR4 to regulate NPR1-dependent defense responses. Cell Res 22:1631–1633. doi:10.1038/cr 2012.100

Mzid R, Marchive C, Blancard D, Deluc L, Barrieu F, Corio-Costet MF, Drira N, Hamdi S, Lauvergeat V (2007) Overexpression of *VvWRKY2* in tobacco enhances broad resistance to necrotrophic fungal pathogens. Physiol Plant 131:434–447

Nakayama A, Fukushima S, Goto S, Matsushita A, Shimono M, Sugano S, Jiang C-J, Akagi A, Yamazaki M, Inoue H, Takatsuji H (2013) Genome-wide identification of WRKY45-regulated genes that mediate benzothiadiazole-induced defense responses in rice. BMC Plant Biol 13:150

Nawar HF, Kuti JO (2003) Wyerone acid phytoalexin synthesis and peroxidase activity as marker for resistance of broad beans to chocolate spot disease. J Phytopathol 151:564–570

Nguema-Ona E, Moore JP, Fagerström AD, Fangel JU, Willats WG, Hugo A, Vivier MA (2013) Overexpression of the grapevine PGIP1 in tobacco results in compositional changes in the leaf arabinoxyloglucan network in the absence of fungal infection. BMC Plant Biol 13:46. doi:10.1186/1471-2229-13-46

Nicaise V (2014) Crop immunity against viruses: outcomes and future challenges. Front Plant Sci 5:660. doi:10.3389/fpls2014.00660

Nicaise V, Roux M, Zipfel C (2009) Recent advances in PAMP-triggered immunity against bacteria: pattern recognition receptors watch over and raise the alarm. Plant Physiol 150:1638–1647

Nürnberger T, Kufner I (2011) The role of the plant plasma membrane in microbial sensing and innate immunity. Plant Cell Monogr 19:471–483

Nürnberger T, Lipka V (2005) Non-host resistance in plants – new insights into an old phenomenon. Mol Plant Pathol 6:335–345

Orlowska E, Fiil A, Kirk HG, Llorente B, Cvitanich C (2011) Differential gene induction in resistant and susceptible potato cultivars at early stages of infection by *Phytophthora infestans*. Plant Cell Rep 31:187–203

Osorio S, Castillejo C, Quesada MA, Medina-Escobar N, Brownsey GJ, Suau R, Heredia A, Botella MA, Valpuesta V (2008) Partial demethylation of oligogalacturonides by pectin methylesterase 1 is required for eliciting defence responses in wild strawberry (*Fragaria vesca*). Plant J 54:43–55

Osorio S, Bombarely A, Giavalisco P, Usadel B, Stephens C, Aragüez I, Medina-Escobar N, Botella MA, Fernie AR, Valpuesta V (2011) Demethylation of oligogalacturonides by *FaPE1* in the fruits of the wild strawberry *Fragaria vesca* triggers metabolic and transcriptional changes associated with defence and development of the fruit. J Exp Bot 62:2855–2873

Park C-J, Peng Y, Chen X, Dardick C, Ruan D, Bart R, Canlas PE, Ronald PC (2008) Rice XB15 a protein phosphatase 2C negatively regulates cell death and XA21-mediated innate immunity. PLoS Biol 6:e231

Park C-J, Bart R, Chern M, Canlas PE, Bai W, Ronald PC (2010a) Overexpression of the endoplasmic reticulum chaperone BiP3 regulates XA21-mediated innate immunity in rice. PLoS One 5(2):e9262

Park C-J, Han S-W, Chen X, Ronald PC (2010b) Elucidation of XA21-mediated innate immunity. Cell Microbiol 12:1017–1025

Pavli OI, Kelaidi GI, Tampakaki AP, Skaracis GN (2011) The *hrpZ* gene of *Pseudomonas syringae* pv *phaseolicola* enhances resistance to rhizomania disease in transgenic *Nicotiana benthamiana* and sugar beet. PLoS One 6(3):e17306, 9 pp

Pavli OI, Tampakaki AP, Skaracis GN (2012) High level resistance against rhizomania disease by simultaneously integrating two distinct defense mechanisms. PLoS One 7(12):e51414

Peng J-L, Bao Z-L, Ren H-Y, Wang J-S, Dong HS (2004) Expression of harpin$_{Xoo}$ in transgenic tobacco induces pathogen defense in the absence of hypersensitive cell death. Phytopathology 94:1048–1055

Peng X, Hu Y, Tang X, Zhou P, Deng X, Wang H, Guo Z (2012) Constitutive expression of rice *WRKY30* gene increases the endogenous jasmonic acid accumulation, PR gene expression and resistance to fungal pathogens in rice. Planta 236:1485–1498

Pervieux I, Bourassa M, Laurans F, Hamelin R, Séguin A (2004) A spruce defensin showing strong antifungal activity and increased transcript accumulation after wounding and jasmonate treatments. Physiol Mol Plant Pathol 64:331–341

Petutschnig EK, Jones AME, Serazetdinova L, Lipka U, Lipka V (2010) The Lysin motif receptor-like kinase (LysM-RLK) is a major chitin binding protein in *Arabidopsis thaliana* and subject to chitin-induced phosphorylation. J Biol Chem 285:28902–28911

Piasecka A, Jedrzejczak-Rey N, Bednarek P (2015) Secondary metabolites in plant innate immunity: conserved function of divergent chemicals. New Phytol 206:948–964

Postel S, Kufner I, Beueter C, Mazzotta S, Schwedt A, Borlotti A, Halter T, Kemmerling B, Nürnberger T (2010) The multifunctional leucine-rich repeat receptor kinase BAK1 is implicated in *Arabidopsis* development and immunity. Eur J Cell Biol 89:169–174

Powell AL, van Kan J, ten Have A, Visser J, Greve LC, Bennett AB, Labavitch JM (2000) Transfenic expression of pear PGIP in tomato limits fungal colonization. Mol Plant Microbe Interact 13:942–950

Pozo MJ, Van Der Ent S, Van Loon LC, Pieterse CMJ (2008) Transcription factor MYC2 is involved in priming for enhanced defense during rhizobacteria-induced systemic resistance in *Arabidopsis thaliana*. New Phytol 180:511–523

Qi Z, Verma R, Gehring C, Yamaguchi Y, Zhao Y, Ryan CA, Berkowitz GA (2010) Ca^{2+} signaling by plant *Arabidopsis thaliana* Pep peptides depends on AtPepR1, a receptor with guanyl cyclase activity, and CGMP-activated Ca^{2+} channels. Proc Natl Acad Sci U S A 107:21193–21198

Qiu D, Clayton K, Wei Z-M (2001) Bioassay development for Messenger® activity. Phytopathology 91:S74

Qiu D, Xiao J, Ding X, Xiong M, Cai M, Cao Y, Li X, Xu C, Wang S (2007) OsWRKY13 mediates rice disease resistance by regulating defense-related genes in salicylate- and jasmonate-dependent signaling. Mol Plant Microbe Interact 20:492–499

Qiu D, Mao J, Yang X, Zeng H (2009) Expression of elicitor-encoding gene from *Magnaporthe grisea* enhances resistance against blast disease in transgenic rice. Plant Cell Rep 28:925–933

Quang Z, Li X, Huang L, Hong Y, Zhang Y, Zhang H, Li D, Song F (2015) Elicitin-like proteins Oli-D1 and Oli-D2 from *Pythium oligandrum* trigger hypersensitive response in *Nicotiana benthamiana* and induce resistance against *Botrytis cinerea* in tomato. Mol Plant Pathol 16:238–250

Qutob D, Kemmerling B, Brunner F, Küfner I, Engelhardt S, Gust AA, Luberacki B, Seitz HU, Stahl D, Rauhut T, Glawischnig E, Schween G, Lacombe B, Watanabe N, Lam E, Schlichting

R, Scheel D, Nau K, Dodt G, Hubert D, Gijzen M, Nürnberger T (2006) Phytotoxicity and innate immune responses induced by NEP1-like proteins. Plant Cell 18:3721–3744

Rajput NA, Zhang M, Ru Y, Liu T, Mafurah JJ, Dou D (2014) *Phytophthora sojae* effector PsCRN70 suppresses plant defenses in *Nicotiana benthamiana*. PLoS One 9(5):e98114. doi:10.1371/jornal pone 0098114

Ren H, Gu G, Long J, Yin Q, Wu T, Song T, Zhang S, Chen Z, Dong H (2006a) Combinative effects of a bacterial type-III effector and a biocontrol bacterium on rice growth and disease resistance. J BioSci 31:617–627

Ren H, Song T, Wu T, Sun L, Liu Y, Yang F, Chen Z, Dong H (2006b) Effects of a biocontrol bacterium on growth and defence of transgenic rice plants expressing a bacterial type-III effector. Ann Microbiol 56:281–287

Rinaldi C, Kohler A, Frey P, Duchaussoy F, Ningre N, Couloux A, Wincker P, Le Thiec D, Fluch S, Martin F, Duplessis S (2007) Transcript profiling of poplar leaves upon infection with compatible and incompatible strains of the foliar rust *Melampsora larici-populina*. Plant Physiol 144:347–366

Robinson SM, Bostock RM (2015) β-Glucans and eicosapolyenoic acids as MAMPs in plant-oomycete interactions: past and present. Front. Plant Sci 5:797. doi:10.3389/fpls.2014.00797

Rosli HG, Zheng Y, Pombo MA, Zhong S, Bombarely A, Fei Z, Collmer A, Martin GB (2013) Transcriptomics-based screen for genes induced by flagellin and repressed by pathogen effectors identifies a cell wall-associated kinase involved in plant immunity. Genome Biol 14:R139

Ryan CA, Huffaker A, Yamaguchi Y (2007) New insights into innate immunity in Arabidopsis. Cell Microbiol 9:1902–1908

Sanchez L, Courteaux B, Hubert J, Kauffmann S, Renault J-H, Clément C, Baillieul F, Dorey S (2012) Rhamnolipids elicit defense responses and induce disease resistance against biotrophic, hemibiotrophic, and necrotrophic pathogens that require different signaling pathways in Arabidopsis and highlight a central role for salicylic acid. Plant Physiol 160:1530–1641

Sathiyabama M, Akila G, Einstein Charles R (2014) Chitosan-induced defense responses in tomato plants against early blight disease caused by *Alternaria solani* (Ellis and Martin) Sorauer. Arch Phytopathol Plant Protect 47:1963–1973

Saville A, Graham K, Grünwald NJ, Myers K, Fry WE, Ristaino JB (2015) Fungicide sensitivity of U.S. genotypes of *Phytophthora infestans* to six oomycete-targeted compounds. Plant Dis 99:659–666

Schulze B, Mentzel T, Jehle A, Mueller K, Beeler S, Boller T, Felix G, Chinchilla D (2010) Rapid heteromerization and phosphorylation of ligand-activated plant transmembrane receptors and their associated kinase BAK1. J Biol Chem 285:9444–9451

Schwessinger B, Bart R, Krasileva KV, Coaker G (2015) Focus issue on plant immunity: from model systems to crop species. Front Plant Sci 6:195. doi:10.3389/fpls.2015.00195

Segonzac C, Zipfel C (2011) Activation of plant pattern-recognition receptors by bacteria. Curr Opin Microbiol 14:54–61

Sénéchal F, Wattier C, Rustérucci C, Pelloux J (2014) Homogalacturonan-modifying enzymes: structure, expression, and roles in plants. J Exp Bot 65:5125–5160

Shan L, He P, Li J, Heese A, Peck SC, Nümberger T, Martin GB, Sheen J (2008) Bacterial effectors target the common signaling partner BAK1 to disrupt multiple MAMP receptor-signaling complexes and impede plant immunity. Cell Host Microbe 4:17–27

Shan W, Chen J-Y, Kuang J-F, Lu W-J (2015) Banana fruit NAC transcription factor MaNAC5 cooperates with MaWRKYs to enhance the expression of pathogenesis-related genes against *Colletotrichum musae*. Mol Plant Pathol. doi:10.1111/mpp.12281

Shao M, Wang J, Dean RA, Lin Y, Gao X, Hu S (2008) Expression of a harpin-encoding gene in rice confers durable nonspecific resistance to *Magnaporthe grisea*. Plant Biotechnol J 6:73–81

Sharma-Poudyal D, Paulitz TC, du Toit LJ (2015) Stunted patches in onion bulb crops in Oregon and Washington: etiology and yield loss. Plant Dis 99:648–658

Shimono M, Sugano S, Nakayama A, Jiang C-J, Ono K, Toki S, Takatsuji H (2007) Rice WRKY45 plays a crucial role in benzothiadiazole-inducible blast resistance. Plant Cell 19:2064–2076

Shimono M, Koga H, Akagi A, Hayashi N, Goto S, Sawada M, Kurihara T, Matsushita A, Sugano S, Jiang CJ, Kaku H, Inoue H, Takatsuji H (2012) Rice WRKY45 plays important roles in fungal and bacterial disease resistance. Mol Plant Pathol 13:83–94

Shimizu T, Nakano T, Takamizawa D, Desaki Y, Ishi-Minami N, Nishizawa Y, Minami E, Okada K, Yamane H, Kaku H, Shibuya N (2010) Two LysM receptor molecules, CEBiP and OsCERK1, cooperatively regulate chitin elicitor signaling in rice. Plant J 64:204–214

Shin R, Park JM, An JM, Paek KH (2002) Ectopic expression of *Tsi1* in transgenic hot pepper plants enhances host resistance to viral, bacterial, oomycete pathogens. Mol Plant Microbe Interact 15:983–989

Shinya T, Osada T, Desaki Y, Hatamoto M, Yamanaka Y, Hirano H, Takai R, Che F-S, Kaku H, Shibuya N (2010) Characterization of receptor proteins using affinity cross-linking with biotinylated ligands. Plant Cell Physiol 51:262–270

Singh S, Bockus WW, Sharma I, Bowden RL (2008) A novel source of resistance in wheat to *Pyrenophora tritici-repentis* race 1. Plant Dis 92:91–95

Sohn SI, Kim YH, Kim BR, Lee SY, Lim CK, Hur JH, Lee JY (2007) Transgenic tobacco expressing the $hrpN_{EP}$ gene from *Erwinia pyrifoliae* triggers defense responses against *Botrytis cinerea*. Mol Cells 24:232–239

Song C, Yang B (2010) Mutagenesis of 18 type II effectors reveals virulence function of $XopZ_{Pxo99}$ in *Xanthomonas oryzae* pv *oryzae*. Mol Plant Microbe Interact 23:893–902

Sørensen CK, Hovmoller MS, Leconte M, Dedryver F, de Vallavieille-Pope C (2014) New races of *Puccinia striiformis* found in Europe reveal race specificity of long-term effective adult plant resistance in wheat. Phytopathology 104:1042–1051

St Clair DA (2010) Quantitative disease resistance and quantitative resistance loci in breeding. Annu Rev Phytopathol 48:247–268

Strehlow B, de Mol F, Struck C (2015) Risk potential of clubroot disease on winter oilseed rape. Plant Dis 99:667–675

Sun J-Y, Gaudet DA, Lu Z-X, Frick M, Puchalski B, Laroche A (2008) Characterization and antifungal properties of wheat nonspecific lipid transfer proteins. Mol Plant Microbe Interact 21:346–360

Szczesny R, Buttner B, Escolar L, Schulze S, Seiferth A, Bonas U (2010) Suppression of the AvrBs1-specific hypersensitive response by the YopJ effector homolog AvrBsT from *Xanthomonas* depends on a SNF1-related kinase. New Phytol 187:1058–1074

Tagle AG, Chuma I, Tosa Y (2015) *Rmg7*, a new gene for resistance to *Triticum* isolates of *Pyricularia oryzae* identified in tetraploid wheat. Phytopathology 105:495–499

Takakura Y, Ishida Y, Inoue Y, Tsutsumi F, Kuwata S (2004) Induction of hypersensitine response-like reaction by powdery mildew in transgenic tobacco expressing harpin $_{pss}$. Physiol Mol Plant Pathol 64:83–89

Takakura Y, Che F-S, Ishida Y, Tsutsumi F, Kuotani K-I, Usami S, Isogai A, Imaseki H (2008) Expression of a bacterial flagellin gene triggers plant immune responses and confers disease resistance in transgenic rice plants. Mol Plant Pathol 9:525–529

Tanaka N, Che F-S, Watanabe N, Fujiwara S, Takayama S, Isogai A (2003) Flagellin from an incompatible strain of *Acidovorax avenae* mediates H_2O_2 generation accompanying hypersensitive cell death and expression of *PAL Cht-1* and *PBZ1* but not of *LOX* in rice. Mol Plant Microbe Interact 16:422–428

Tena G (2015) Innate immunity: defence by transcription. Nat Plants 1:14018. doi:10.1038/nplants.2014./18

Thomma BPHJ, Nürnberger T, Joosten M (2011) Of PAMPs and effectors: the blurred PTI-ETI dichotomy. Plant Cell 23:4–15

Tran HS, You MP, Khan TN, Barbetti MJ (2015) Relative host resistance to black spot disease in field pea (*Pisum sativum*) is determined by individual pathogens. Plant Dis 99:580–587

Tripathi JN, Lorenzen J, Bahar O, Ronald P, Tripathi L (2014) Transgenic expression of the rice *Xa21* pattern-recognition receptor in banana (*Musa* sp.) confers resistance to *Xanthomonas campestris* pv. *musacearum*. Plant Biotechnol J 12:663–673

Trouvelot S, Hѐloir M-C, Poinssot B, Gauthier A, Paris F, Guillier C, Combier M, Trdȧ L, Daire X, Adrian M (2014) Carbohydrates in plant immunity and plant protection: roles and potential application as foliar sprays. Front Plant Sci 5:14. Article 592. doi:10.3389/fpls.2014.00592

Tsuda K, Katagiri P (2010) Comparing signaling mechanisms engaged in pattern-triggered and effector-triggered immunity. Curr Opin Plant Biol 13:459–465

Tymon L, Johnson DA (2014) Fungicide resistance of two species of *Alternaria* from potato in the Columbia basin of Washington. Plant Dis 98:1648–1653

Vallarino JG, Osorio S (2012) Signaling role of oligogalacturonides derived during cell wall degradation. Plant Signal Behav 7:1447–1449

Van der Ent S, Verhagen BWM, Van Doorn R, Bakker D, Verlaan MG, Pel MJC, Joosten RG, Proveniers MCG, Van Loon LC, Ton J, Pieterse CMJ (2008) *MYB72* is required in early signaling steps of rhizobacteria-induced systemic resistance in *Arabidopsis*. Plant Physiol 146:1293–1304

Vega-Sánchez ME, Redinbaugh MG, Costanzo S, Dorrance AF (2005) Spatial and temporal expression analysis of defense-related genes in soybean cultivars with different levels of partial resistance to Phytophthora sojae. Physiol Mol Plant Pathol 66:175–182

Vidhyasekaran P (1997) Fungal pathogenesis in plants and crops: molecular biology and host defense mechanisms. Marcel Dekker, New York, p 553

Vidhyasekaran P (2002) Bacterial disease resistance in plants: molecular biology and biotechnological applications. Haworth Press, Binghamton, p 452

Vidhyasekaran P (2004) Concise encyclopedia of plant pathology. Haworth Press, Binghamton, p 619

Vidhyasekaran P (2007) Fungal pathogenesis in plants and crops: molecular biology and host defense mechanisms, 2nd edn. CRC Press, Taylor Francis Group, Boca Raton, p 510

Vidhyasekaran P (2014) PAMP signals in plant innate immunity: signal perception and transduction. Springer, Dordrecht, p 442

Vidhyasekaran P (2015) Plant hormone signaling systems in plant innate immunity. Springer, Dordrecht, p 458

Vleeshouwers VGAA, Oliver RP (2014) Effectors as tools in disease resistance breeding against biotrophic, hemibiotrophic, and necrotrophic plant pathogens. Mol Plant Microbe Interact 27:196–206

Vleeshouwers VGAA, Rietman H, Krenek P, Champouret N, Young C, Oh S-K, Wang M, Bouwmeester K, Vosman B, Visser RGF, Jacobsen E, Govers F, Kamoun S, Vander Vossen EAG (2008) Effector genomics accelerates discovery and functional profiling of potato disease resistance and *Phytophthora infestans* avirulence genes. PLoS One 3(8):e2875. doi:10.1371/journal.pone 0002875

Volpi C, Janni M, Lionetti V, Bellincampi D, Favaron F, D'Ovidio R (2011) The ectopic expression of a pectin methyl esterase inhibitor increases pectin methyl esterification and limits fungal diseases in wheat. Mol Plant Microbe Interact 24:1012–1019

Waller F, Müller A, Chung K-M, Yap Y-K, Nakamura K, Weiler E, Sano H (2006) Expression of a WIPK-activated transcription factor results in increase of endogenous salicylic acid and pathogen resistance in tobacco plants. Plant Cell Physiol 47:1169–1174

Walters D, Walsh D, Newton A, Lyon G (2005) Induced resistance for plant disease control: maximizing the efficacy of resistance elicitors. Phytopathology 95:1368–1373

Wang X, El Hadrami A, Adam LR, Daayf F (2005) Genes encoding pathogenesis-related proteins PR-2, PR-3 and PR-9, are differentially regulated in potato leaves inoculated with isolates from US-1 and US-8 genotypes of *Phytophthora infestans* (Mont.) de Bary. Physiol Mol Plant Pathol 67:49–56

Wang H, Hao J, Chen X, Hao Z, Wang X, Lou Y, Peng Y, Guo Z (2007) Overexpression of rice WRKY89 enhances ultraviolet B tolerance and disease resistance in rice plants. Plant Mol Biol 65:799–815

Wang X, Zhu X, Tooley P, Zhang X (2013) Cloning and functional analysis of three genes encoding polygalacturonase-inhibiting proteins from *Capsicum annuum* and transgenic CaPGIP1 in tobacco in relation to increased resistance to fungal pathogens. Plant Mol Biol 81:379–400

Wang F, Lin R, Feng J, Chen W, Qiu D, Xu S (2015) TaNAC1 acts as a negative regulator of stripe rust resistance in wheat, enhances susceptibility to *Pseudomonas syringae*, and promotes lateral root development in transgenic *Arabidopsis thaliana*. Front Plant Sci 6:108. doi:10.3389/fpls.2015 00108

Wei T, Ou B, Li J, Zhao Y, Guo D, Zhu Y, Chen Z, Gu H, Li C, Qin G, Qu LJ (2013) Transcriptional profiling of rice early response to *Magnaporthe oryzae* identified OsWRKYs as important regulators in rice blast resistance. PLoS One 8:e59720

Wiethölter N, Graessner B, Mierau M, Mort AJ, Moerschbacher BM (2003) Differences in the methyl ester distribution of homogalacturonans from near-isogenic wheat lines resistant and susceptible to the wheat stem rust fungus. Mol Plant Microbe Interact 16:945–952

Wilkinson JR, Spradling KD, Yoder DW, Pirtle IL, Pirtle RM (2005) Molecular cloning and analysis of a cotton gene cluster of two genes and two pseudogenes for the PR5 protein osmotin. Physiol Mol Plant Pathol 67:68–82

Wu S, Lu D, Kabbage M, Wei HL, Swingle B, Records AR, Dickman M, He P, Shan L (2011) Bacterial effector HopF2 suppresses arabidopsis innate immunity of the plasma membrane. Mol Plant Microbe Interact 24:585–593

Xiang T, Zong N, Zou Y, Wu Y, Zhang J, Xing W, Li Y, Tang X, Zhu L, Chai J, Zhou J-M (2008) *Pseudomonas syringae* effector AvrPto blocks innate immunity by targeting receptor kinases. Curr Biol 18:74–80

Xiang T, Zong N, Zhang J, Chen J, Chen M, Zhou J-M (2011) BAK1 is not a target of the *Pseudomonas syringae* effector AvrPto. Mol Plant Microbe Interact 24:100–107

Xin XF, He SY (2013) *Pseudomonas syringae* pv. *tomato* DC3000: a model pathogen for probing disease susceptibility and hormone signalling in plants. Annu Rev Phytopathol 51:1003–1011

Xu G, Chen S, Chen F (2010) Transgenic chrysanthemum plants expressing a $harpin_{Xoo}$ gene demonstrate induced resistance to alternaria leafspot and accelerated development. Russ J Plant Physiol 57:548–553

Yamaguchi Y, Huffaker A (2011) Endogenous peptide elicitors in higher plants. Curr Opin Plant Biol 14:351–357

Yamaguchi Y, Pearce G, Ryan CA (2006) The cell surface leucine-rich repeat receptor for *At*Pep1, an endogenous peptide elicitor in *Arabidopsis*, is functional in transgenic tobacco cells. Proc Natl Acad Sci U S A 103:10104–10109

Yamaguchi Y, Huffaker A, Bryan AC, Tax FE, Ryan CA (2010) PEPR2 is a second receptor for the Pep1 and Pep2 peptides and contributes to defense responses in *Arabidopsis*. Plant Cell 22:508–522

Yamamizo C, Kuchimura K, Kobayashi A, Katou S, Kawakita K, Jones JDG, Doke N, Yoshioka H (2006) Rewiring mitogen-activated protein kinase cascade by positive feedback confers potato blight resistance. Plant Physiol 140:681–692

Yang X, Deng F, Ramonell KM (2012) Receptor-like kinases and receptor – like proteins; keys to pathogen recognition and defense signaling in plant innate immunity. Front Biol 7(2):155–166. doi:10.1007/s11515-011-1185-8

Yasuda N, Mitsunaga T, Hayashi K, Koizumi S, Fujita Y (2015) Effects of pyramiding quantitative resistance genes *pi21*, *Pi34*, and *Pi35* on rice leaf blast disease. Plant Dis 99:904–909

Yokotani N, Tsuchida-Mayama T, Ichikawa H, Mitsuda N, Ohme-Takagi M, Kaku H, Minami E, Nishizawa Y (2014) OsNAC111, a blast disease-responsive transcription factor in rice, positively regulates the expression of defense-related genes. Mol Plant Microbe Interact 27:1027–1034

Yu F, Huaxia Y, Lu W, Wu C, Cao X, Guo X (2012) *GhWRKY15*, a member of the WRKY transcription factor family identified from cotton (*Gossypium hirsutum* L.), is involved in disease resistance and plant development. BMC Plant Biol 12:144

Zamioudis C, Peterse CMJ (2012) Modulation of host immunity by beneficial microbes. Mol Plant Microbe Interact 25:139–150

Zeng L, Velasquez AC, Munkvold KR, Zhang J, Martin GB (2012) A tomato LysM receptor-like kinase promotes immunity and its kinase activity is inhibited by AvrPtoB. Plant J 69:92–103

Zeng F, Arnao E, Zhang G, Olaya G, Wullschleger J, Sierotzki H, Ming R, Bluhm BH, Bond JP, Fakhoury M, Bradley CA (2015) Characterization of quinone outside inhibitor fungicide resistance in *Cercospora sojina* and development of diagnostic tools for its identification. Plant Dis 99:544–550

Zhang J, Peng Y, Guo Z (2008) Constitutive expression of pathogen-inducible OsWRKY31 enhances disease resistance and root growth and auxin response in transgenic rice plants. Cell Res 18:508–521

Zhang J, Li W, Xiang T, Liu Z, Laluk K, Ding X, Zou Y, Gao M, Zhang X, Chen S, Mengiste T, Zhang Y, Zhou JM (2010) Receptor-like cytoplasmic kinases integrate signaling from multiple plant immune receptors and are targeted by a *Pseudomonas syringae* effector. Cell Host Microbe 7:290–301

Zhang W, Fraiture M, Kolb D, Löffelhardt B, Desaki Y, Boutrot FFG, Tör M, Zipfel C, Gust AA, Brunner F (2013) Arabidopsis RECEPTOR-LIKE PROTEIN30 and receptor-like kinase SUPPRESSOR OF BIR1-1/EVERSHED mediate innate immunity to necrotrophic fungi. Plant Cell 25:4227–4241

Zhang L, Kars I, Essenstam B, Liebrand TWH, Wagemakers L, Elberse J, Tagkalaki P, Tjoitang D, Ackerveken G, van Kan JAL (2014) Fungal endoploygalacturonases are recognized as microbe-associated molecular patterns by the Arabidopsis receptor-like protein RESPONSIVENESS TO BOTRYTIS POLYGALACTURONASES1. Plant Physiol 164:353–364

Zhang H, Feng J, Shu Z, Shi Z, Li F, Strelkov SE, Sun R, Hwang S-F (2015) Resistance to *Plasmodiophora brassicae* in *Brassica juncea* genotypes from China. Plant Dis 99:776–779

Zhang Y, Guenzi AC, Anderson MP, Taliaferro CM, Gonzales RA (2006) Enrichment of bermudagrass genes associated with tolerance to the spring dead spot fungus *Ophiosphaerella herpotricha*. Physiol Mol Plant Pathol 68:105–118

Zheng X, McLellan H, Fraiture M, Liu X, Boevink PC, Gilroy EM, Chen Y, Kandel K, Sessa G, Birch PRJ, Brunner F (2014) Finctionally redundant RXLR effectors from *Phytophthora infestans* act at different steps to suppress early flg22-triggered immunity. PLoS Pathog 10(4):e1004057

Zhu T, Song F, Zheng Z (2006) Molecular characterization of the rice pathogenesis-related protein, OsPR-4b, and its antifungal activity against *Rhizoctonia solani*. J Phytopathol 154:378–384

Zhu Z, Shi J, Cao J, He M, Wang Y (2012) VpWRKY3, a biotic and abiotic stress-related transcription factor from the Chinese wild *Vitis pseudoreticulata*. Plant Cell Rep 31:2109–2120

Zipfel C (2014) Plant pattern-recognition receptors. Trends Immunol 35:345–351

Zipfel C, Rathjen P (2008) Plant immunity: AvrPto targets the frontline. Curr Biol 18:R218–R220

Zipfel C, Robatzek S, Navarro L, Oakeley EJ, Jones JDG, Felix G, Boller T (2004) Bacterial disease resistance in *Arabidopsis* through flagellin perception. Nature 428:764–767

Zipfel C, Kunze G, Chinchilla D, Caniard A, Jones JD, Boller T, Felix G (2006) Perception of the bacterial PAMP EF-Tu by the receptor EFR restricts *Agrobacterium*-mediated transformation. Cell 125:749–760

Zuo K-J, Qin J, Zhao J-Y, Ling H, Zhang L-D, Cao Y-F, Tang K-X (2007) Over-expression of *GbERF2* transcription factor in tobacco enhances brown spots disease resistance by activating expression of downstream genes. Gene 391:80–90

Yu F, Huaxia Y, Lu W, Wu C, Cao X, Guo X (2012) GhWRKY15, a member of the WRKY transcription factor family identified from cotton (Gossypium hirsutum L.), is involved in disease resistance and plant development. BMC Plant Biol 12:144

Zamioudis C, Pieterse CMJ (2012) Modulation of host immunity by beneficial microbes. Mol Plant Microbe Interact 25:139–150

[illegible]

[illegible]

[illegible]

[illegible]

[illegible]

[illegible] flagellin perception. Nature 428:764–767

Zipfel C, Kunze G, Chinchilla D, Caniard A, Jones JD, Boller T, Felix G (2006) Perception of the bacterial PAMP EF-Tu by the receptor EFR restricts Agrobacterium-mediated transformation. Cell 125:749–760

[illegible]

Chapter 2
Role of Plant Immune Signals and Signaling Systems in Plant Pathogenesis

Abstract Plants possess innate immune system to resist pathogen attack. Innate immunity is the first line of defense against invading microorganisms. Pathogens possess pathogen-associated molecular patterns (PAMPs). The PAMPs are primary danger/alarm signal molecules to switch on the plant immune systems. PAMPs are evolutionarily conserved building blocks of microbial surfaces that directly bind to plant pattern recognition receptors (PRRs). Plants use the PRRs to defend themselves from microbial pathogens. The PRRs are localized at the plasma membrane and the PAMPs activate expression of the genes encoding various PRRs. When activated by the PAMP, the PRR is translocated to endocytic compartments and endocytosis of the PRR is important for activation of several downstream signaling events. The plant immune system uses several second messengers to encode information generated by the PAMPs and deliver the information downstream of PRRs to proteins which decode and interpret the signals and initiate defense gene expression. G-proteins act as molecular switches in signal transduction system. Calcium ion is an important intracellular second messenger and carries the PAMP signal downstream to initiate immune responses. Reactive oxygen species (ROS) serve as second messengers transmitting the message. ROS appears to interact with various defense signaling systems. It plays a central role in launching the defense response. Nitric oxide (NO) is a diffusible molecular messenger that plays an important role in plant immune response signal transduction. Mitogen-activated protein kinase (MAPK) cascades are major pathways downstream of PAMP/PRR signaling complex that transduce extracellular stimuli into intracellular responses in plants. The plant hormones salicylic acid (SA), jasmonates (JA), ethylene (ET), abscisic acid (ABA), auxin (AUX), cytokinin (CK), gibberellin (GA), and brassinosteroid (BR) play important role in intercellular and systemic signaling systems triggering expression of various defense-responsive genes. SA signaling is involved in triggering systemic acquired resistance (SAR). SAR is associated with priming of defense responses and the priming results in a faster and stronger induction of defense responses after pathogen attack. The priming can be inherited epigenetically and descendants of primed plants exhibit next-generation systemic acquired resistance. Thus when pathogens land on the plant surface, the PAMPs trigger highly complex defense responses against the pathogens and suppress disease development. However virulent pathogen may modify its PAMP structure during its pathogenesis

P. Vidhyasekaran, *Switching on Plant Innate Immunity Signaling Systems*,
Signaling and Communication in Plants, DOI 10.1007/978-3-319-26118-8_2

to reduce its elicitor activity. Virulent pathogens may also contain inefficient PAMPs and trigger subdued defense responses favoring disease development. The reduced activity of PAMPs might facilitate the virulent pathogens to cause disease. Besides PAMP molecules, pathogens produce effectors, which play an important role in pathogenesis. Effectors specifically contribute to virulence of pathogens by targeting host plant innate immunity. The effectors secreted by various pathogens have been shown to suppress the PAMP-triggered immunity. Effectors disrupt binding of PAMP with PRR in the PAMP-PRR signaling complex. Effectors may promote ubiquitin-proteasome-mediated degradation of PRRs to impede PAMP-triggered plant immunity. Effectors have been shown to target the receptor kinase activity of the PRRs and inhibit the kinase activity to block PAMP-triggered immunity. Autophosphorylation of PRRs results in activation of PRRs and the effectors may inhibit the autophosphorylation of PRRs to suppress the PAMP-triggered immune system. Some effectors have been shown to block the action of the PRR signal amplifier BAK1. Several receptor-like cytoplasmic kinases (RLCK) including BIK1, PBS1, and PBS1-like (PBL) proteins play important role in regulation of the signaling pathways downstream of PAMP-PRR-BAK1 signaling complex and the effectors have been shown to block the action of these RLCKs. Effectors may also suppress the MAPK signaling cascade triggered by PAMPs. Effectors have been shown to suppress SA signaling system, which is involved in triggering defense responses against a broad range of plant pathogens. Pathogens may induce specific signaling systems, which may favor disease development. Pathogen hijacks ABA signaling system to suppress SA-mediated defense responses promoting disease development. JA signaling system has been reported to confer susceptibility against some pathogens and pathogens may hijack JA signaling system to cause disease. Necrotrophic pathogens use SA signaling pathway to promote disease development by suppressing JA signaling pathway. Auxin signaling system has been shown to be involved in promoting susceptibility to pathogens and inducing disease development. Pathogens hijack the host auxin metabolism leading to the accumulation of a conjugated form of the hormone, indole-3-acetic acid (IAA) – Asp, to promote disease development. Pathogens may hijack BR signaling machinery to interfere with effectual SA- and GA-controlled defenses. These studies suggest that various signals and signaling systems in plants modulate the pathogenesis inducing susceptibility and disease resistance and precise manipulation of these signaling systems will be an ideal tool to manage crop diseases.

2.1 Susceptibility and Disease Resistance Are Two Sides of the Same Coin Modulated by Plant Immune System Signals and Signaling Systems

Pathogens contain or secrete 'pathogen-associated molecular patterns (PAMP)' (Vidhyasekaran 2014; Zhang et al. 2014) and 'effector' molecules (McLellan et al. 2013; King et al. 2014; Vleeshouwers and Oliver 2014; Rajput et al. 2014; Zheng et al. 2014). While PAMPs are involved in triggering disease resistance, effectors are involved in inducing susceptibility and in disease development (Thomma et al. 2011; Denance et al. 2013). Plants are endowed with innate immune system and signals emanated from virulent or avirulent pathogens activate the dormant immune systems. Defense signalling systems are activated in both susceptible and resistant interactions (Fung et al. 2008; Mitsuhara et al. 2008; Spoel and Dong 2008; Van Damme et al. 2008; Beckers et al. 2009; Alazem et al. 2014; Rahman et al. 2014). Plant disease resistance and susceptibility are two sides of the same coin (Vidhyasekaran 2007, 2014).

Several signals and signaling systems are involved in induction of susceptibility or resistance during plant pathogenesis. PAMP-plant pattern recognition receptor (PRR) signaling complex, G-proteins, calcium signature, reactive oxygen species (ROS), nitric oxide (NO), mitogen-activated protein kinase (MAPK) signaling cascade, protein kinases, ubiquitin-proteasome system, salicylic acid (SA), jasmonic acid (JA), ethylene (ET), abscisic acid (ABA), auxins, cytokinins, gibberellins, and brassinosteroids are involved in triggering susceptibility or disease resistance (Besson-Bard et al. 2008; Katsir et al. 2008; De Vleesschauwer et al. 2010; Zhu et al. 2011b; Bethke et al. 2012; Makandar et al. 2012; Zeng et al. 2012; Yang et al. 2013; Rahman et al. 2014; Vidhyasekaran 2014; Vleeshouwers and Oliver 2014). Relative speed and intensity of expression of specific plant hormone signaling systems determines resistance or susceptibility against specific pathogens (Denance et al. 2013; Yang et al. 2013; Vleeshouwers and Oliver 2014). SA signaling system induces resistance against biotrophic and hemibiotrophic pathogens but this signaling system has been shown to promote development of disease caused by the necrotrophic fungal pathogen *Botrytis cinerea* (El Rahman et al. 2012). Several plant hormone signaling systems trigger defense responses against various pathogens, but these hormone signaling pathways are targeted by virulent pathogens to evade plant defense responses (Denance et al. 2013). Some signaling systems are involved in promoting pathogenesis. The wilt pathogen *Fusarium oxysporum* f. sp. *raphani* has been shown to hijack ET signaling to promote disease development in Arabidopsis plants (Pantelides et al. 2013).

Potential pathogens may suppress defense signaling systems to cause disease. The bacterial pathogen *Pseudomonas syringae* promotes systemic susceptibility by suppressing SA signaling (Zheng et al. 2012). DNA viruses selectively interfere with ubiquitin-proteasome system to cause infection in plants (Lozano-Durán and Bejarano 2011; Lozano-Durán et al. 2011). Suppression of JA signaling pathway induces enhanced susceptibility to the maize seed-infecting *Aspergillus* spp. while it increases resistance to other fungal pathogens (*Fusarium*, *Colletotrichum*, *Cochliobolus*, and *Exserohilum* spp.) (Gao et al. 2009). The resistance against the bacterial blight pathogen *Xanthomonas oryzae* pv. *oryzae* was associated with increased accumulation of SA and JA and suppressed accumulation of ethylene and expression of ethylene-related genes in rice plants (Shen et al. 2011). These studies suggest the importance of various signals and signaling systems in pathogenesis and disease resistance.

2.2 Signals and Signaling Systems Involved in Triggering Immune Responses

2.2.1 PAMP-PRR Signaling Complex in Triggering Immune Responses

Plant innate immune system is a surveillance system against possible attack by pathogens and is activated on perception of the pathogen-associated molecular patterns (PAMP; the pathogen's signature) of invading pathogens (Vidhyasekaran 2004, 2007; Boller and He 2009; Boutrot et al. 2010; Chen et al. 2010a, c; Dodds and Rathjen 2010; Park et al. 2010; Shimizu et al. 2010; Ronald and Beutler 2010; Nürnberger and Küfner 2011; Segonzac and Zipfel 2011; Zamioudis and Pieterse 2012; Vidhyasekaran 2014). Potential pathogens contain several PAMPs and they serve as alarm signals to activate the plant innate immunity (Vidhyasekaran 2014). PAMPs are evolutionarily conserved building blocks of microbial surfaces that directly bind to the plant pattern recognition receptors (PRRs) and induce defense responses (Tsuda and Katagiri 2010; Zhang and Zhou 2010; Thomma et al. 2011). The plant innate immune systems have high potential to fight against viral, bacterial, oomycete, and fungal pathogens and protect the crop plants against wide range of diseases (Knecht et al. 2010; Lacombe et al. 2010; Molloy 2010; D'Amelio et al. 2011; Hwang and Hwang 2011; Alkan et al. 2012).

Plants use the PRRs to defend themselves against microbial pathogens. These receptors recognize PAMPs and activate signaling pathways that lead to immunity (Mentlak et al. 2012). Most of the PRRs identified are receptor kinases and these protein kinase PRRs are known to be activated by PAMPs (Segonzac and Zipfel 2011). PAMPs activate expression of the genes encoding various PRRs (Zipfel et al. 2004, 2006; Qutob et al. 2006; Lohmann et al. 2010). PRRs bind with PAMPs for their activation (Boutrot et al. 2010; Petutschnig et al. 2010). The binding of PAMP induces a conformational alteration in PRRs leading to their activation (Ali et al. 2007). The PRR FLS2 is found localized at the plasma membrane. When activated by the PAMP flg22, FLS2 is translocated to endocytic compartments (Robatzek et al. 2006). Endocytosis of the PRR FLS2 has been shown to be important for the PAMP flg22-induced defense signaling system (Chinchilla et al. 2007b).

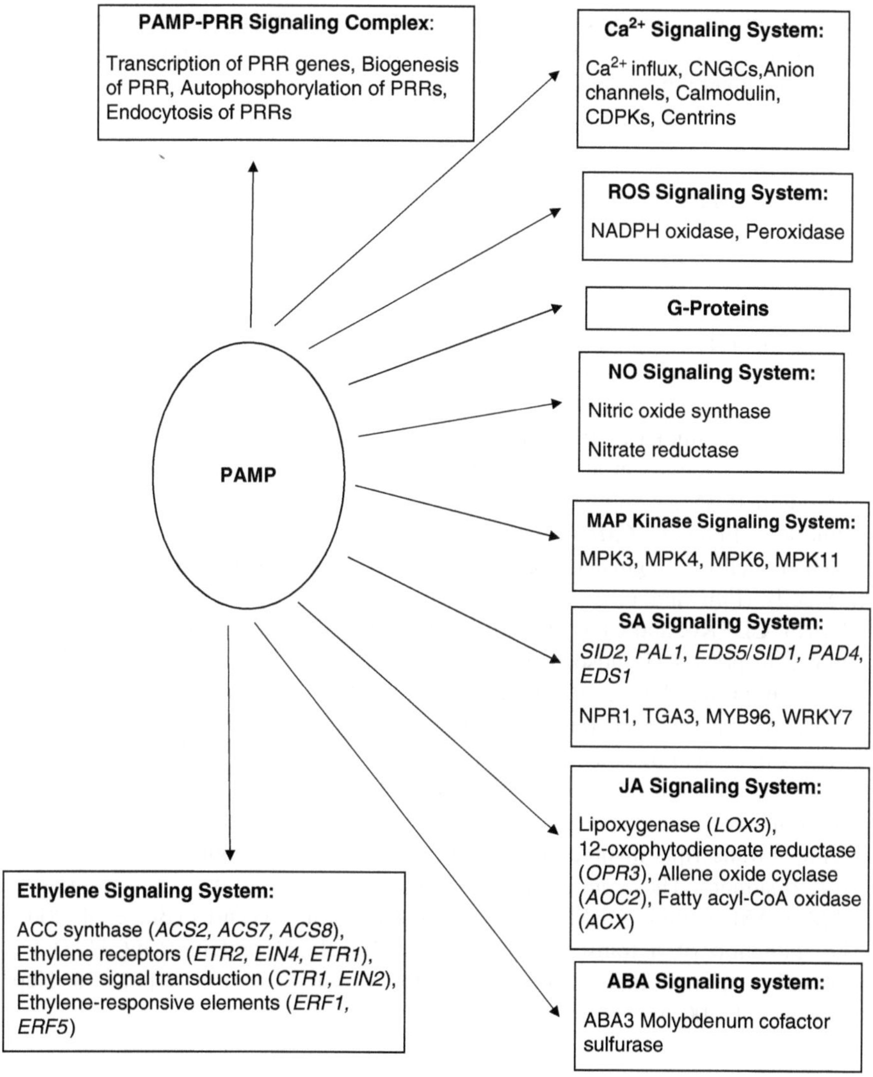

Fig. 2.1 PAMPs-triggered plant defense signaling systems

The plant immune system uses several second messengers to encode information generated by the PAMPs and deliver the information downstream of PRRs to proteins which decode/interpret signals and initiate defense gene expression (van Verk et al. 2008; Mersmann et al. 2010; Boudsocq et al. 2010; Hwang and Hwang 2011). Analysis of the *Arabidopsis* transcriptome revealed that more than 1000 genes were significantly upregulated or downregulated within 30 min after the PAMP flg22 treatment (Zipfel et al. 2004). Denoux et al. (2008) observed 4413 genes with altered expression in response to the PAMP flg22 in *Arabidopsis* seedlings. These genes are involved in activation of several distinct signaling systems in *Arabidopsis thaliana* (Denoux et al. 2008). PAMPs activate several distinct signaling systems to trigger the plant innate immune responses (Fig. 2.1).

2.2.2 PAMPs Activate Ca^{2+} Signaling Systems

PAMPs activate Ca^{2+}-signaling system (Lecourieux et al. 2006; Aslam et al. 2009). PAMPs elicit calcium ion influx within 15–30 min after PAMP treatment in plant cells (Le-Courieux-Ouaked et al. 2000; Aslam et al. 2008). Several Ca^{2+}-permeable channels have been found in plant plasma membranes. Cyclic nucleotide-gated ion channels (CNGCs) are involved in the Ca^{2+}-dependent signaling pathways (Talke et al. 2003; Yoshioka et al. 2006). The PAMP flg22 activates various CNGCs in *Arabidopsis* (Denoux et al. 2008). These CNGC channels have been found to be calmodulin (CaM)-binding proteins (Borsics et al. 2007). The PAMP flg22 activates calmodulin-like gene *CML41* within 1 h after treatment (Denoux et al. 2008). The PAMP flagellin upregulated the gene encoding Ca^{2+}-dependent protein kinase in rice cells (Fujiwara et al. 2004). Calcium-dependent protein kinases CDPK4, CDPK5, CDPK6, and CDPK11 were shown to mediate the PAMP flg22-triggered defense responses (Boudsocq et al. 2010). The pathway involving the calcium-dependent protein kinases (CDPK) 4/5/6/11 has been proposed to act in parallel to the MAPK pathways to control flg22-dependent gene expression (Boudsocg et al. 2010). Flg22 also activated *CCD1* in cultured rice cells (Fujiwara et al. 2004). *CCD-1* encodes a Ca^{2+}-binding protein that shares homology with the C-terminal half domain of centrin and centrins are involved in Ca^{2+} signaling (Takezawa 2000). Eighteen transcripts involved in calcium sensing were up-regulated by the PAMP Nep1 treatment. These genes encoded calcium-binding EF hand family proteins, CAM, CAM-binding, CAM-related proteins, and Ca^{2+}-ATPases (Bae et al. 2006). The induction of a transcript encoding CAM-related protein (*CmCAL-1*) was reported in spotted knapweed and dandelion within 15 min in response to Nep1 (Keates et al. 2003).

2.2.3 PAMPs Activate G-Protein Signaling

G-proteins (guanosine triphosphate-binding proteins) act as molecular switches in signal transduction system (Yalovsky et al. 2010; Zhang et al. 2011, 2012a). Plant cells contain many G-proteins (Zeng et al. 2007; Yong et al. 2010; Zhang et al. 2012a). Two classes of signaling G-proteins have been reported. These include heterotrimeric G-proteins and small monomeric G-proteins (Gu et al. 2004; Perfus-Barboch et al. 2004). G-proteins are involved in defense signaling initiated by PAMPs (Park et al. 2000; Schiene et al. 2000; Zhang et al. 2011, 2012a). Both heteromeric G-proteins and small G-proteins play an important role in activating various signal transduction systems initiated by PAMPs (Gao et al. 2010a, b; Kiirika et al. 2012). G-proteins trigger changes in cytosolic Ca^{2+} concentrations (Schultheiss et al. 2003). The G-proteins induce Ca^{2+} channel opening in plants through the action of PAMPs (Gelli et al. 1997). Protein phosphorylation precedes Ca^{2+} influx in tobacco cells treated with a PAMP isolated from the oomycete pathogen *Phytophthora cryptogea* (Tavernier et al. 1995). The G-proteins modulate the phosphorylation/dephosphorylation system in the plasma membrane of tomato cells and transduce the signal (Vera-Estrella et al. 1994). The *Arabidopsis* G-protein GPA1 has been demonstrated to be involved in the regulation of inward K^+ channels and slow anion channels (Wu and Assmann 1994; Wang et al. 2001; Zhang et al. 2008). G-proteins are involved in PAMP-activated ROS-mediated signaling system (Park et al. 2000; Suharsono et al. 2002). The PAMP flg22 induces G-protein-activated ROS signaling systems. The gene *AGB1*, encoding the β-subunit of G-protein in *Arabidopsis*, is highly induced after flg22 treatment (Zipfel et al. 2004). The *agb1* mutants are impaired in the oxidative burst triggered by flg22, suggesting the importance of G-proteins in PAMP-induced ROS signaling system (Ishikawa 2009).

2.2.4 PAMPs Activate ROS Signaling System

PAMPs trigger rapid and transient production of reactive oxygen species (ROS) in plants (Asada 2006; Sagi and Fluhr 2006; Lehtonen et al. 2012). The PAMP flg22 activates NADPH oxidase (RBOH), which is responsible for ROS production (Asai et al. 2008; Zhu et al. 2009; Zhang et al. 2011; Kiirika et al. 2012). Bolwell et al. (1998) showed that in bean cells treated with a fungal elicitor, H_2O_2 was derived directly from cell wall peroxidases following extracellular alkalinization and the appearance of a reductant. Production of apoplastic ROS by chitin treatment in *Physcomitrella patens* has been shown to require peroxidase. The fungal elicitor chitin caused an immediate oxidative burst in wild-type *P. patens* but not in the Δ*Prx34* mutants lacking the chitin-responsive secreted class III peroxidase (Prx34), suggesting the requirement of peroxidase for the production of ROS (Lehtonen et al. 2012). The PAMP flg22 has been shown to trigger the activity of cell wall peroxidase involved in ROS production (Denoux et al. 2008).

2.2.5 PAMPs Activate NO Signaling System

PAMPs are known to trigger nitric oxide (NO) burst within minutes in plant cells (Foissner et al. 2000; Lamotte et al. 2004). The bacterial PAMP lipopolysaccharide (Lipid A) generates a rapid burst of NO production in *Arabidopsis* cells (Zeidler et al. 2004). Treatment of tomato cell cultures with the fungal PAMP xylanase resulted in rapid NO accumulation (Laxalt et al. 2007). The bacterial PAMP harpin induces NO generation in *Arabidopsis* cells (Krause and Durner 2004). NO is synthesized predominantly by nitric oxide synthase (NOS) (Zhang et al. 2007a; Zhao et al. 2007; Zottini et al. 2007). The PAMP flg22 triggered the expression of *AtNOS1* encoding NOS (Denoux et al. 2008). NOS catalyses NO production (Crawford et al. 2006), suggesting that the PAMP activates NO signaling system. NO may be formed also from nitrite by the action of nitrate reductase (Rockel et al. 2002; Bethke et al. 2004; Wilson et al. 2008). Nitrate reductase transcript and protein levels increase in response to a PAMP in potato tubers, suggesting a role for nitrate reductase in the synthesis of NO during the plant immune response (Delledonne 2005).

2.2.6 PAMPs Activate Mitogen-Activated Protein Kinase Signaling System

Mitogen-activated protein kinase (MAPK) cascades are major pathways downstream of PAMP/PRR signaling complex that transduce extracellular stimuli into intracellular responses in plants (Liu et al. 2003; Pedley and Martin 2005). Different PAMPs, including bacterial flagellin, elongation factor Tu, peptidoglycan, lipopolysaccharide, HrpZ1 harpin, and fungal chitin activate MAP kinase signaling system (Pitzschke et al. 2009b; Wu et al. 2011; Bethke et al. 2012). Flg22 activates the expression of *MPK3* in *Arabidopsis* (Denoux et al. 2008). The PAMP flg22 triggers a rapid and strong activation of MPK3, MPK4 and MPK6 (Droillard et al. 2004).

2.2.7 PAMPs Activate Salicylic Acid Signaling System

Salicylic acid (SA) is an important signal induced by PAMPs in plant innate immune system (Garcion et al. 2008; Mukherjee et al. 2010; Makandar et al. 2012). The PAMP Flg22 induces production of salicylic acid in plant cells (Mishina and Zeier 2007; Tsuda et al. 2008). It induces accumulation of SA in *Arabidopsis* (Mishina and Zeier 2007). SA is synthesized via both the isochorismate pathway and phenylalanine pathway (Wildermuth et al. 2001; Ferrari et al. 2003; Ogawa et al. 2006; Dempsey et al. 2011). The induction of *SID2* gene, encoding the key SA biosynthesis enzyme isochrorismate synthase (ICS), was initiated 2 h after the treatment and reached the peak after flg22 application. The *SID2* gene was significantly upregulated in the activation-tagging *myb96-1d* line but slightly suppressed in the *myb96-1* mutant (Seo and Park 2010). The MYB gene *MYB96* has been shown to be involved in SA biosynthetic pathway (Seo and Park 2010). The *MYB96* gene was rapidly induced by the bacterial PAMP flg22 treatment and the transcript abundance reached the peak within 1 h (Seo and Park 2010).

SA is also synthesized from phenylalanine by phenylalanine ammonia lyase (PAL) activity (Lee et al. 1995). The *PAL1* gene was also induced moderately in the activation tagging *myb96-1d* line (Seo and Park 2010). The PAMP flg22 has been shown to activate the expression of PAL and SID2 involved in SA biosynthesis (Denoux et al. 2008). The PAMP also activates EDS1, EDS5, and PAD4 involved in SA signaling system (Denoux et al. 2008). The regulatory proteins SID2, EDS1, EDS4, EDS5, and PAD4 are involved in triggering SA production in *Arabidopsis*. SID2 is an isochorismate synthase (Wildermuth et al. 2001), whereas EDS5/SID1 exhibits homology to multidrug and toxin extrusion (MATE) transporter proteins from animals (Nawrath et al. 2002). EDS5 is involved in the transport of precursors for SA biosynthesis (Nawrath et al. 2002). EDS1 is another regulatory protein (Moreau et al. 2012a) and it controls SA production to amplify defense signals (Eulgem et al. 2004; Song et al. 2004). PAD4 displays similarity to triacyl glycerol lipases and other esterases (Jirage et al. 1999). EDS1 forms distinct complexes with PAD4 (Wiermer et al. 2005; Xing and Chen 2006) and EDS1 and PAD4 complex transduces ROS-derived signals leading to SA production (Mateo et al. 2004; Wiermer et al. 2005). SA signaling consists of a linear pathway in which EDS1, PAD4 and EDS4 activate EDS5 and SID2, which produce SA (Glazebrook et al. 2003). NPR1 is an important regulator of responses downstream of SA (Mou et al. 2003; Zhang et al. 2003). The PAMP flg22 activates NPR1 (Denoux et al. 2008). NPR1 is a co-factor of TGA transcription factor and it enhances binding of TGA transcription factors to the promoter of *PR1* gene to activate transcription of *PR1* gene (Fan and Dong 2002; Johnson et al. 2008). The TGA transcription factor TGA3 is activated by flg22 treatment in *Arabidopsis* (Denoux et al. 2008).

2.2.8 PAMPs Activate Jasmonate Signaling System

PAMPs have been shown to trigger accumulation of jasmonic acid (Wang et al. 2000; Fabro et al. 2008). The fungal PAMP chitosan treatment induced accumulation of jasmonic acid in tomato leaves (Doares et al. 1995b). Several enzymes including lipoxygenase, allene oxide synthase, allene oxide cyclase, OPDA reductase 3 (OPR3) and acyl-CoA oxidase (ACX) are involved in biosynthesis of jasmonic acid (JA) (Mei et al. 2006; Schilmiller et al. 2006; Balbi and Devoto 2008; Delker et al. 2007; Vidhyasekaran 2007). Flg22 enhanced the expression of *LOX3* and *LOX4* genes encoding lipoxygenases (LOX), *OPR3* gene encoding 12-oxophytodienoate reductase (OPR) and *ACX1* gene encoding acyl-CoA oxidase (ACX) (Denoux et al. 2008). The fungal PAMP chitosan also activates lipoxygenase, the key enzyme in JA-mediated signaling system (Rakwal et al. 2002). Lipoxygenase activity significantly increased in chitosan-treated carrot plants (Jayaraj et al. 2009). The PAMP β-1,3-glucan induces expression of *LOX* gene encoding lipoxygenase in grapevine (Aziz et al. 2003; Balbi and Devoto 2008) and tobacco cells (Klarzynski et al. 2000). The PAMP Nep1 rapidly induces genes involved in JA biosynthesis (Bae et al. 2006). It triggered genes encoding lipoxygenases, 12-oxophytodienoate reductase, and allene oxide cyclase (*AOC2*), which are involved in JA biosynthesis (Bae et al. 2006).

2.2.9 PAMPs Activate Ethylene Signaling System

Ethylene (ET) signaling system is another important component in plant's innate immune system (Al-Daoud and Cameron 2011; Zhu et al. 2011b; Sun et al. 2012). Several PAMPs are known to induce production of ethylene in plant cells. The oomycete PAMP cryptogein induced production of ethylene in tobacco cells (Milat et al. 1991). The bacterial PAMP flg22 up-regulated *ACS* genes encoding ACC synthase, which is the key enzyme involved in biosynthesis of ethylene (Denoux et al. 2008). Flg22 induces ET production through activation of ACS6, an ET biosynthetic enzyme (Liu and Zhang 2004). The PAMP Nep1 induced the expression of transcripts encoding ACC synthase and ethylene-responsive element binding factors (ERF), which are involved in ethylene signaling (Bae et al. 2006). Expression of the gene encoding ACC oxidase also responded to the PAMP-induced host-associated molecular pattern (HAMP) treatment (Huffaker et al. 2011).

In *Arabidopsis*, ethylene is perceived by membrane-bound receptors such as ETR1 and EIN4, which transmit the signal to downstream effectors (Qu et al. 2007; Grefen et al. 2008). Flg22 induced expression of the ethylene receptors ETR1 and EIN4 (Denoux et al. 2008). Flg22 treatment up-regulated the expression of the ET-responsive transcription factor ERF1 (Ethylene-Responsive element-binding Factor 1) in *Arabidopsis* (Clay et al. 2009). ERF1 is a downstream component of ethylene signaling system (Berrocal-Lobo and Molina 2004). ET signaling is required for the full induction of ERF1 in response to flg22 (Clay et al. 2009). ERF5, another ethylene-responsive element-binding factor, is induced by the fungal PAMP chitin. It induces defense against *Pseudomonas syringae* pv. *tomato* in *Arabidopsis* (Son et al. 2012). The transcription factor MYB51 is also induced by ET and MYB51 acts downstream of ET signaling for the callose response (Clay et al. 2009). MYB51 induced expression of all known indole glucosinolate (IGS) biosynthetic enzymes. MYB51 is involved in the transcriptional activation of IGS biosynthetic gene *ASA1*. ASA1 expression is ET-inducible (Clay et al. 2009). *ASA1* gene expression was also induced by Flg22 treatment (Clay et al. 2009).

2.2.10 PAMPs Trigger ABA Signaling System

Abscisic acid (ABA) signaling system is another important component in plant immune system activated by PAMPs. PAMPs trigger increases in ABA concentrations (Koga et al. 2004; Schmidt et al. 2008). The PAMP flg22 triggers ABA biosynthesis in plants (Melotto et al. 2006). The HAMP oligogalacturonates (OG) induced the enzyme molybdenum cofactor sulfurase (ABA3), which is involved in the biosynthesis of ABA (Denoux et al. 2008).

2.2.11 PAMPs Trigger Expression of Transcription Factors

Transcription factors are the master regulators of expression of genes and they play important role in activating plant immune responses either positively (Liu et al. 2007; Qiu et al. 2007; Moreau et al. 2012b) or negatively (Sun et al. 2010; Moreau et al. 2012b). After Flg22 treatment WRKY70 was consistently induced around 20-fold in *A. thaliana*. WRKY70 induces expression of SA-responsive PR genes (Li et al. 2004). WRKY70 has been shown to function downstream of ROS and SA (Knoth et al. 2007). The WRKY70 transcription factor acts as a node of convergence for jasmonate-mediated and salicylate-mediated signals in plant defense (Li et al. 2004). It modulates the selection of signaling pathways in plant defense (Li et al. 2006).

Flg22 induces activation of several other transcription factors in *Arabidopsis*. It activates WRKY33 and WRKY40 transcription factors (Denoux et al. 2008). These transcription factors function as activators of JA signaling system and repressors of SA signaling system (Zheng et al. 2006). WRKY33 positively regulates JA-mediated responses, while it retards the SA-mediated PR genes expression (Zheng et al. 2006). The transcription factors WRKY22 and WRKY29 are known to function downstream of the FLS2-mediated immune response in *Arabidopsis*. Overexpression of the *AtWRKY29* constitutively activates the plant defense response against bacterial invasion (Asai et al. 2002). Nine *WRKY* genes were induced by the PAMP Nep1 treatment, while *WRKY65* was repressed in *A. thaliana* (Bae et al. 2006). Keates et al. (2003) detected the induction of WRKY18 within 15 min after treatment with the PAMP Nep1. These results suggest that PAMPs may trigger the expression of several transcription factors and modulate the plant innate immune responses.

2.3 Reduced Activity of PAMPs May Facilitate the Virulent Pathogens to Cause Disease

2.3.1 Pathogen May Modify Its PAMP Structure during Its Pathogenesis to Reduce Its Elicitor Activity

Pathogens contain/secrete multiple PAMPs and structure of individual PAMPs also differs (Silipo et al. 2008; Siipo and Molinaro 2010). Variation in the structure of the PAMP alters function of the PAMP to trigger the defense responses (Robatzek et al. 2007; Takai et al. 2008; Wan et al. 2008b). The bacterial lipopolysaccharide (LPS) is the important PAMP triggering plant innate immune system. The bacterial cell envelope components lipopolysaccharides have been reported to be involved in triggering induced systemic resistance (ISR) in many plant-pathogen systems. Alterations in lipid A or other structures within LPS occur during plant pathogenesis (Molinaro et al. 2009). The LPS of a mutant of *Xanthomonas campestris* pv. *campestris* defective in core completion have shown that loss of the glycosyl residues of the outer core is accompanied by alterations in the lipid A, which becomes penta-acylated (rather than hexa-acylated) and is substituted with phosphoethanolamine moieties (Silipo et al. 2008). This modified lipid A was not able to induce defense-related responses such as *PR* gene induction in *Arabidopsis* (Molinaro et al. 2009). This suggests that the bacterial pathogen has the capacity to modify the structure of the lipid A in order to reduce its activity as a PAMP.

2.3.2 Virulent Pathogen May Contain Inefficient PAMP and Trigger Subdued Defense Responses Favoring Disease Development

Both avirulent and virulent pathogens contain or secrete several PAMPs (Denoux et al. 2008). However, the PAMPs detected in the different strains of pathogens may differ in their efficacy to trigger defense responses. Flagellin purified from the incompatible *Acidovorax avenae* N1141 strain induced immune responses, whereas flagellin from the compatible K1 strain induced no responses in rice (Takai et al. 2008). Flagellin purified from the K1 strain was identical to that of the N141 flagellin, suggesting that N1141 flagellin has an epitope in addition to the flg22 region capable of eliciting immune responses (Takai et al. 2008). Flagellins purified from *P. syringae* pv. *glycinea*, an incompatible pathogen for tobacco, induced immune responses in tobacco, whereas flagellin from *P. syringae* pv. *tabaci*, a compatible pathogen, does not, despite complete amino acid identity (Taguchi et al. 2003a). Flagellins derived from nonadapted bacteria but having identical protein sequences differentially induce strong defense responses in nonhost plants, suggesting that other domains and/or posttranslational modifications of flagellin may be involved in triggering immune responses (Taguchi et al. 2003a, b, 2006, Takeuchi et al. 2003, 2007).

The amount of PAMP available in the infection court may determine the intensity of induced defense gene expression (Lecourieux et al. 2002; Denoux et al. 2008). Transgenic tobacco plants expressing the $hpaG_{Xoo}$ gene encoding the PAMP $\text{harpin}_{\text{Xoo}}$ were obtained by Peng et al. (2004). The harpin gene expression varied in the transgenic lines. The line 7 showed highest expression of the harpin gene while the line 15 showed lowest expression of the gene. The transgenic tobacco line, which showed highest PAMP expression, showed greatest expression of the regulatory gene *NPR1* and the defense gene *Chia5* encoding chitinase. By contrast, the tobacco line which showed lowest PAMP expression, showed lowest expression of the defense gene *Chia5* and the regulatory gene *NPR1* (Peng et al. 2004). High and low expression of other defense genes *PR-1a* and *PR-1b* was also correlated with the high and low expression of the PAMP gene in the transgenic plants, The tobacco line which showed highest PAMP expression showed highest disease resistance, while the line which showed lowest PAMP expression showed lowest disease resistance (Peng et al. 2004). These results suggest a link between lower level of PAMP expression in plants and lower level of induction of immune responses. In the compatible interactions, the PAMP expression may be low, probably by the suppressive action of effectors and this low level of PAMP expression may facilitate the pathogens to cause disease.

2.4 Pathogen-Secreted Effectors Suppress PAMP-Triggered Plant Immune Responses

2.4.1 Pathogen-Secreted Effector Molecules

Besides PAMP molecules, pathogens produce another type of molecules called effectors (Hogenhout et al. 2009; Kwon 2010; Wu et al. 2011; Vleeshouwers and Oliver 2014). Effectors are pathogen molecules that manipulate host cell structure and function thereby facilitating infection (Hogenhout et al. 2009). Effectors play an important role in pathogenesis of bacterial, fungal, and oomycete pathogens (Melotto et al. 2006; Sohn et al. 2007; de Jonge et al. 2010; Khang et al. 2010; Valent and Khang 2010; Thomma et al. 2011; Wu et al. 2011). Different groups of effectors have been recognized and these include PAMPs (harpins, Nep1-like proteins) (Kufner et al. 2009; Cabral et al. 2012), *Avr* gene products (Dodds et al. 2006; Qutob et al. 2009; Bos et al. 2009, 2010; Engelhardt et al. 2012; Na et al. 2013; Yin et al. 2013), toxins (Wolpert et al. 2002; Tada et al. 2005; McDonald et al. 2013), and cell wall-degrading enzymes (Rotblat et al. 2002; Boudart et al. 2003; Poinssot et al. 2003). Plant innate immune systems have high potential to fight against a wide range of viral, bacterial, oomycete, and fungal pathogens (Lacombe et al. 2010; Hwang and Hwang 2011; Alkan et al. 2012). However, potential pathogens produce several effectors to nullify the defense responses induced by the innate immune system (Wu et al. 2011; Cheng et al. 2012). To avoid or suppress or delay the expression of the defense gene-activating signaling systems, the pathogens secrete several effectors into the host cell (Göhre et al. 2008; Kim et al. 2010; Wu et al. 2011; Cheng et al. 2012). Circumvention of innate immunity is crucial for pathogenesis in plants and effectors play important role in suppression of plant immunity (Block et al. 2008; Schornack et al. 2009). While PAMPs are essential for microbial fitness and survival, effectors specifically contribute to virulence by targeting host plant innate immunity (Thomma et al. 2011).

2.4.2 Effectors Suppress PAMP-Triggered Plant Immunity

The effectors secreted by various pathogens have been shown to suppress the PAMP-triggered immunity. Several bacterial pathogens use a specialized type III secretion system to deliver effector proteins into host cells to subvert

PAMP-triggered host defense mechanisms, thereby promoting pathogenesis (Hauck et al. 2003; DebRoy et al. 2004; He et al. 2006; Jelenska et al. 2007; Göhre et al. 2008; Song and Yang 2010; Szczesny et al. 2010; Zhang et al. 2010; Wu et al. 2011; Akimoto-Tomiyama et al. 2012). The bacterial pathogen *Pseudomonas syringae* secretes the effector HopF2, which has been found to be a potent suppressor of early immune gene transcription signaling activated by multiple PAMPs, including bacterial flagellin, ef-Tu, peptidoglycan, lipopolysaccharide and HrpZ1 harpin, and fungal chitin (Wu et al. 2011). Expression of HopF2 in transgenic plants compromised plant nonhost immunity to bacterial pathogen *P. syringae* pv. *phaseolicola* and plant immunity to the fungal pathogen *Botrytis cinerea* (Wu et al. 2011). It suggests that HopF2 plays important role in suppression of function of multiple PAMP signaling and the suppression of function of PAMP signaling suppresses PAMP-triggered immunity and makes the plants susceptible to various pathogens.

P. syringae effector protein *AvrPto* suppresses host defenses by directly targeting the transmembrane receptor kinases involved in bacterial perception (Zipfel and Rathjen 2008). *P. syringae* injects two sequence-distinct effectors, AvrPto and AvrPtoB, to intercept convergent defense responses stimulated by PAMPs (Shan et al. 2008). The AvrPto binds the PRRs *Arabidopsis* FLS2 and EFR and tomato LeFLS2, to block plant defense responses in the plant cell (Xiang et al. 2008). AvrPto and AvrPtoB act as suppressors of early-defense gene transcription. These effectors intercept multiple PAMP-mediated signaling upstream of MAPKKK at the plasma membrane linked to the PRR (He et al. 2006). The effectors AvrPto and AvrPtoB of the bacterial pathogen *Pseudomonas syringae* pv. *tomato* have been reported to undermine the FLS2-BAK1 pathway (Xin and He 2013). About 30 % of the genes induced by flgII-28 are suppressed by the effectors AvrPto and AvrPtoB, suggesting that the effectors also disrupt the flgII-28 receptor complex (Rosli et al. 2013). A tomato LysM receptor-like kinase promotes immunity and its kinase activity is inhibited by the effector AvrPtoB of *P. syringae* pv. *tomato* (Zeng et al. 2012).

Effectors induce susceptibility, mostly by suppressing PAMP-induced immune responses. The effector proteins target basic innate immunity in plants (Block et al. 2008; Bartetzko et al. 2009; Boller and He 2009; Song and Yang 2010; Szczesny et al. 2010). The effector AvrBsT from *Xanthomonas campestris* pv. *vesicatoria* has been identified as a suppressor of specific plant defense in pepper plants (Szczesny et al. 2010). The effector $XopZ_{Pxo99}$ from the rice bacterial blight pathogen interferes with host innate immunity during the pathogen infection process in rice (Song and Yang 2010). The *P. syringae* pv. *tomato* DC3000 effector HopF2 intercepts PAMP signaling at the plasma membrane immediately after signal perception. It acts as a potent suppressor of early immune-response gene transcription and mitogen-activated protein kinase signaling activated by multiple PAMPs (Wu et al. 2011). RXLR effectors from *Phytophthora infestans* suppress the PAMP flg22-triggered immunity (Zheng et al. 2014). *Phytophthora sojae* effector PsCRN70 suppresses plant defenses in *Nicotiana benthamiana* (Rajput et al. 2014).

2.4.3 *Effectors May Disrupt Binding of PAMP with PRR in PAMP-PRR Signaling Complex to Impede PAMP-Triggered Plant Immunity*

Plants use pattern recognition receptors (PRRs) to defend themselves against pathogens. These receptors recognize PAMPs and activate signaling pathways that lead to immunity (Iriti and Faoro 2009; Mentlak et al. 2012). Chitin, (β-1→4-linked polymer of *N*-acetylglucosamine; GlcNAc), is a major component of fungal cell walls. It is not found in plants; however the plants secrete chitin-degrading enzymes (Vidhyasekaran 2007). Fungal infection induces the expression of chitinases in plant cells, and these chitin-degrading enzymes accumulate at the sites of invasion. The chitinases release chitin fragments (chitin oligomers or chitooligosaccharides) from fungal cell walls (Eckardt 2008). The chitooligosaccharides are the classical PAMPs detected in fungi (Miya et al. 2007; Hamel and Beaudoin 2010; Lizasa et al. 2010) and they are known to trigger the plant innate immune responses in a wide range of plants including both monocots and dicots (Shimizu et al. 2010; Son et al. 2012). Mitogen-activated protein kinases play important role in chitin signaling (Zhang et al. 2002; Wan et al. 2004). Perception of PAMPs by receptors leads to the rapid activation of MAP kinases including MPK3, MPK4, and MPK6. In particular, MAP kinase 3 and 6 (MPK3/MPK6) were shown to be rapidly activated by chitin in *Arabidopsis* and their activation depended on upstream MAPK kinases (MKK4 and MKK5) (Wan et al. 2004).

LysM domain-containing effector protein Ecp6 of the fungal plant pathogen *Cladosporium fulvum* mediates virulence through perturbation of the PAMP chitin-triggered host immunity. During infection, Ecp6 sequesters chitin oligosaccharides that are released from the cell walls of invading hyphae to prevent elicitation of host immunity. Since LysM effectors are widely conserved in the fungal kingdom, this type of effector action may be a common strategy of fungal pathogens in suppressing host immune systems (de Jonge et al. 2010).

The rice blast pathogen *Magnaporthe oryzae* overcomes this first line of plant defense by secreting an effector protein, Slp1 (Secreted LysM Protein1) during invasion of rice cells. Slp1 accumulates in the interface between the fungal cell wall and the rice plasma membrane. Slp1 competes with the rice PRR CEBiP for binding of the PAMP chitin oligosaccharide and this may facilitate the pathogen to suppress the PAMP-induced plant immune responses. Slp1 has been shown to be required for full virulence of *M. oryzae*. Gene silencing of CEBiP in rice allows *M. oryzae* to cause disease. The results suggest that the fungal effector Slp1 sequesters the fungal PAMP to prevent PAMP-triggered immunity in rice and it facilitates the development of the disease (Mentlak et al. 2012).

2.4.4 Effectors May Promote Ubiquitin-Proteasome-Mediated Degradation of PRRs to Impede PAMP-Triggered Plant Immunity

Some effectors have been shown to degrade the PRRs through ubiquitin-proteasome pathway and inhibit PAMP-triggered immunity. The *Pseudomonas syringae* pv. *tomato* effector AvrPtoB interacts with the PRR FLS2 through its N terminus and the interaction is enhanced by the PAMP flg22 activation. The AvrPtoB has been shown to be an E3 ubiquitin ligase and it structurally and functionally mimics E3 ubiquitin ligase (Abramovitch et al. 2006; Janjusevic et al. 2006). The C-terminal region (residues 400–550) of AvrPtoB encodes an ubiquitin E3-ligase domain and it ubiquitinates the PRR FLS2 to promote its degradation. E3 ligases play a key role in the ubiquitin-proteasome-mediated degradation of protein. The effector catalyzes polyubiquitination of the kinase domain of FLS2, resulting in degradation of the PRR. Degradation of the PRR has been shown to occur during the bacterial infection in *Arabidopsis thaliana* (Göhre et al. 2008). AvrPtoB targets another PRR CERK1 for degradation (Gimenez-Ibanez et al. 2009a, b). The LysM receptor kinase CERK1 has shown to be required for chitin signaling (Miya et al. 2007; Wan et al. 2008a). AvrPtoB ubiquitinates the CERK1 kinase domain in vitro and targets CERK1 for degradation in vivo (Gimenez-Ibanez et al. 2009a). The results show that the effector AvrPtoB suppresses PAMP-triggered immunity by degrading the PRR and inactivating the PAMP-PRR signaling complex.

2.4.5 Effectors May Target the Kinase Domains of PRR and Inhibit the PRR Receptor Kinase Activity to Block PAMP-Triggered Immunity

PAMPs are perceived by cognate PRRs and the PAMP-PRR complex activates the plant immune system (Altenbach and Robatez 2007; He et al. 2007; Wan et al. 2008a; Iriti and Faoro 2009). Most of the PRRs identified are LRR-RLKs (Goff and Ramonelli 2007). The PRRs identified to date are modular proteins harbouring an extracellular domain consisting of leucine-rich repeat (LRR) or lysine motifs (LysM) (Saijo 2010; Segonzac and Zipfel 2011), a transmembrane domain and a Ser/Thr protein kinase domain related to *Drosophila* Pelle (Shiu and Bleecker 2001). The sensors for extracellular molecules consisting of an extracellular ligand-binding domain, a single transmembrane domain, and a cytosolic protein kinase domain have been termed RLKs (Seifert and Blaukopf 2010). The extracellular domains of RLKs bind directly to legands to perceive extracellular signals (PAMPs), whereas the cytoplasmic kinase domains transduce these signals into the cell (Bi et al. 2010).

Effectors have been shown to target the receptor kinase activity of the PRRs. *Pseudomonas syringae* injects two sequence-distinct effectors, AvrPto and AvrPtoB, to suppress activity of PRR signaling complex. AvrPto is a kinase inhibitor that inhibits the PRR receptor kinase activity (Xiang et al. 2008). AvrPto directly targets the transmembrane receptor kinases involved in bacterial perception and suppresses host defense responses (Zipfel and Rathjen 2008). The AvrPto binds receptor kinases, including *Arabidopsis* FLS2 and EFR and tomato LeFLS2, to block plant defense responses in the plant cell (Xiang et al. 2008). Xiang et al. (2011) showed that FLS2 is targeted by the *P. syringae* effector AvrPto in plants. The bacterial effector AvrPtoB directly targets the LysM receptor kinase CERK1 (Gimenez-Ibanez et al. 2009a). It targets the kinase domains of CERK1and FLS2 (Shan et al. 2008). Collectively these studies suggest that virulent pathogens secrete effectors, which inhibit the PRR receptor kinase activity resulting in suppression of host defense responses to facilitate disease development.

2.4.6 Effectors May Inhibit Autophosphorylation of PRRs

PAMPs are known to activate PRRs in the PAMP-PRR signaling complex (Boutrot et al. 2010; Segonzac and Zipfel 2011). Most of the PRRs identified are receptor kinases (Zipfel et al. 2006; Kanzaki et al. 2008; Chen et al. 2010a; Wang et al. 2010a). Before activation by PAMPs, the protein kinases are frequently autophosphorylated (Schlessinger 2000; Gomez-Gomez et al. 2001; Wang et al. 2005; Kanzaki et al. 2008; Xiang et al. 2008; Lee et al. 2009; Chen et al. 2010a, b). Autophosphorylation of PRRs results in activation of PRRs (Chen et al. 2010a, b; Park et al. 2010; Petutschnig et al. 2010). The effectors may inhibit the autophosphorylation of PRRs to suppress the PAMP-triggered immune system. The effector AvrPto produced by *P. syringae* pv. *tomato* interacts *in vivo* with the PRRs FLS2 and EFR and inhibits their autophosphorylation in the dose-dependent manner (Xiang et al. 2008). The results suggest that the effectors may prevent the activation of PRR signaling complex by inhibiting the autophosphorylation of PRRs (Hann and Rathjen 2007).

2.4.7 Effectors May Bind With the PRR Signal Amplifier BAK1 and Block the Function of PAMP-PRR Signaling Complex

PRRs have been shown to interact with some additional transmembrane proteins that act as signaling adapters or amplifiers to achieve full functionality of PRRs (Zipfel 2009; Postel et al. 2010; Zhang and Zhou 2010; Zhang et al. 2010). BAK1 (for BRI1 (BRASSINOSTEROID INSENSITIVE1) is a transmembrane protein acting as adaptor of the PRRs FLS2, EFR, PEPR1 and PEPR2 and it is required for proper functionality of these PRRs (Chinchilla et al. 2007b; Ryan et al. 2007; Postel et al. 2010; Schulze et al. 2010; Zhang and Zhou 2010). BAK1 is a positive regulator of PAMP-triggered plant immunity (Saijo 2010). BAK1 appears to be a key component in plant innate immune system and suppression of *BAK1* gene expression results in suppression of the plant innate immunity (Chinchilla et al. 2007a, b; Heese et al. 2007; Kemmerling et al. 2007). BAK1 interacts with PRRs and acts downstream of PRRs (Chinchilla et al. 2007a, b; Heese et al. 2007).

BAK1 acts as a co-activator of the PAMP-PRR receptor complex (Wang et al. 2008). Upon PAMP perception, PRR rapidly associates with BAK1 and initiates downstream signaling (Nicaise et al. 2009). Unlike the direct binding of the PAMP to PRR (Chinchilla et al. 2007b; Kinoshita et al. 2005), BAK1 functions as an adaptor or signaling partner for the regulation of PRRs. (Heese et al. 2007; Shan et al. 2008). PRRs form tight complexes with the receptor kinase BAK1 instantaneously after ligand binding (Schulze et al. 2010). PRRs appear to interact with some transmembrane proteins that act as signaling adapters or amplifiers to achieve their full functionality (Zipfel 2009). BAK1 is an important transmembrane protein, which interacts with PRRs and acts downstream of PRRs (Chinchilla et al. 2007b; Heese et al. 2007). BAK1 is a signal amplifier (Nicaise et al. 2009). BAK1 has been shown to be required for full function of PAMP-PRR signaling complex to activate plant immune responses (Postel et al. 2010; Schulze et al. 2010; Zhang and Zhou 2010).

Some effectors have been shown to block the action of BAK1. The *P. syringae* effectors AvrPto and AvrPtoB bind to BAK1 and block the interaction of BAK1 with the PRR FLS2 resulting in suppression of immunity (Shan et al. 2008). AvrPtoB binds BAK1 during infection and directly targets the kinase domains of BAK1. The inhibition of kinase activity of BAK1 impedes BAK1-dependent plant defense responses (Shan et al. 2008). AvrPtoB targets BAK1 to disrupt the PRR signaling complex (Hann et al. 2010). AvrPto blocks PAMP-triggered immunity by targeting BAK1 (Xiang et al. 2008).

2.4.8 *Effectors May Target the Receptor-Like Cytoplasmic Kinases BIK1 and PBL1*

Several receptor-like cytoplasmic kinases (RLCK) including BIK1 (Botrytis-induced kinase 1), PBS1 (AvrPphB susceptible 1), and PBS1-like (PBL) proteins play important role in regulation of the signaling pathways downstream of PAMP-PRR-BAK1 signaling complex (Lu et al. 2010; Zhang et al. 2010a). BIK1 links the PAMP-PRR signaling complex to downstream intracellular signaling (Lu et al. 2010). BIK1 forms a complex with the unstimulated PRR FLS2 in plants, and the PAMP flg22 induces a rapid phosphorylation of BIK1 in both an FLS2- and BAK1-dependent manner (Lu et al. 2010; Zhang et al. 2010a; Wu et al. 2011). PBS receptor-like kinases are required for signaling from multiple PAMPs and act downstream of FLS2, EFR, and CERK1 to trigger immune responses. BIK1 and PBS proteins integrate immune signaling from multiple PRRs (Zhang et al. 2010). PBS1-like (PBL) cytoplasmic receptor-like kinases act additively with BIK1 in plant innate immune system (Zhang et al. 2010).

Effectors have been shown to target the cytoplasmic receptor kinases to suppress the PAMP-triggered plant innate immunity (PTI) (Zhang et al. 2010; Lu et al. 2010). The effector AvrPphB is a cysteine protease that cleaves PBS1 kinase (Ade et al. 2007). AvrPphB cleaves BIK1, PBS1, PBL1, and PBL2 (Zhang et al. 2010). The *P. syringae* effector HopF2 has been found to be a potent suppressor of early immune gene transcription signaling activated by multiple PAMPs (Wu et al. 2011). The plasma membrane-associated cytoplasmic kinase BIK1 is phosphorylated within 1 min upon flagellin perception. Expression of HopF2 in plants potently diminished the flagellin-induced phosphorylation of BIK1 (Wu et al. 2011). These results suggest that HopF2 likely intercepts PAMP signaling at the plasma membrane immediately of signal perception by PRR by inhibiting the phosphorylation of the protein kinase BIK1 (Wu et al. 2011). These studies suggest that the effectors interfere with the function of various cytoplasmic receptor kinases downstream of PRR-BAK1 signaling complex.

2.4.9 Effector Suppresses MAPK Signaling to Promote Disease Development

Mitogen-activated protein kinase (MAPK) cascades are major pathways downstream of PAMP-PRR signaling complex triggering plant defense responses (Hettenhausen et al. 2012; Zhang et al. 2012b). The MAPKs transduce extracellular stimuli into intracellular transcription factors through activation of Ca^{2+}, ROS, SA, JA, and ethylene-dependent signaling systems, and enhance expression of defense-related genes in plant innate immune system (Vidhyasekaran 2014). *Arabidopsis* MPK3 and MPK6 have been shown to be positive regulators of plant immune responses and these MAPKs are activated by pathogens and also by PAMP elicitors (Asai et al. 2002; Bethke et al. 2012). The bacterial PAMP flg22 and the oligogalacturonides elicitor of host plant origin activated MPK3 and MPK6 in *Arabidopsis* (Galletti et al. 2011). Root treatment with *N*-3-oxo-tetradecanoyl-*L*-homoserine lactone (HSL) induced resistance against *Pseudomonas syringae* pv. *tomato* DC3000 in *Arabidopsis*. HSL treatment promoted a stronger activation of MPK3 and MPK6 when challenged with flg22, followed by a higher expression of the defense-related transcription factors *WRKY22* and *WRKY29*, and the *PR-1* gene (Schikora et al. 2011). These studies show that MPK3 and MPK6 are involved in triggering induced resistance.

MPK3 and MPK6 have been found to be important components for full priming in *Arabidopsis* and the prestress deposition of these inactive kinases may be a possible mechanism of priming during development of systemic acquired resistance

(Beckers et al. 2009). The resistance-inducing avirulent strains of *P. syringae* pv. *tomato* DC3000 and *P. syringae* pv. *phaseolicola* induced SA accumulation and MPK3 expression. Both SA and the SA-related compounds benzothidiazole (BTH) and 4-chloro-SA activate *MPK3* gene expression and induce priming and systemic acquired resistance (SAR). This strong correlation between the ability of avirulent bacteria and various SA-related compounds to activate *MPK3* gene expression and their capacity to prime plants for augmented defense gene activation and induced resistance suggests that MPK3 plays a role in priming (Beckers et al. 2009). The *Arabidopsis* MAP kinases, MPK3 and MPK6, have been implicated in positive plant immune responses (Gao et al. 2008; Ren et al. 2008; Pitzschke et al. 2009b; Liu et al. 2011a). MPK3 has been shown to be required for camalexin accumulation upon *Botrytis cinerea* infection (Ren et al. 2008).

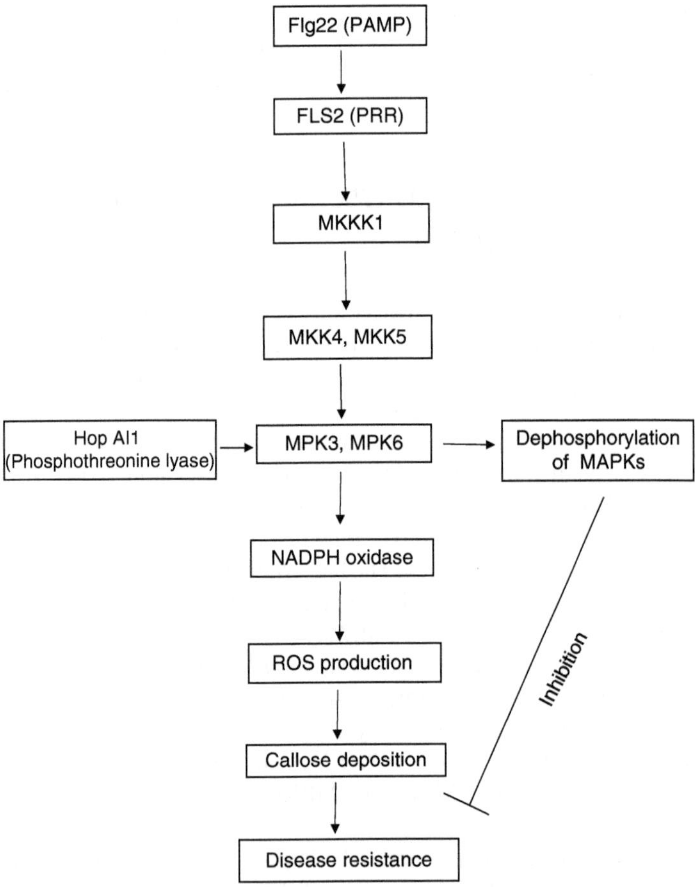

Fig. 2.2 Effector HopAI1-triggered dephosphorylation of MAPKs suppresses host defense responses to promote disease development (Adapted from Shan et al. 2007; Zhang et al. 2007b)

Various bacterial effectors have been shown to suppress the MAPK signaling cascade triggered by PAMPs. The effector HopAl1 of *P. syringae* inhibits the defense-related MAP kinase signaling system activated by PAMPs (Shan et al. 2007; Zhang et al. 2007b). Inactivation of MPK3 and MPK6 by the *P. syringae* effector HopAI1 and inactivation of MKKs by the *P. syringae* effector HopF2 severely impair PAMP-induced defenses and render plants highly susceptible to nonpathogenic *P. syringae* bacteria (Zhang et al. 2007b; Wang et al. 2010c). The HopAl1 plays an important role in virulence of *P. syringae* by directly targeting the MAPKs MPK3 and MPK6. HopAl1 inactivates MAPKs through a phosphothreonine lyase activity unique to bacterial effectors. HopAl1 directly dephosphorylates MAPKs through its phosphothreonine lyase activity, thereby inactivating MPK3 and MPK6 and suppressing the PAMP-induced defense responses (Fig. 2.2; Shan et al. 2007; Zhang et al. 2007b). These results suggest that the bacterial pathogen suppresses specific plant defense signaling system (MAPK cascade) and invades the host tissues.

2.4.10 Effectors May Suppress SA Signaling System to Facilitate Pathogenesis

The effector Hopl1 isolated from *P. syringae* pv. *maculicola* strain PmaES4326 suppresses SA accumulation in plant cells (Jelenska et al. 2007). Hopl1 expression in Arabidopsis plant cells resulted in a 60 % decrease in the level of SA-inducible *PR1* gene transcript and approximately 50 % lower free and total SA levels. It suggests that Hopl1 suppresses SA accumulation and SA-dependent defenses (Jelenska et al. 2007). Expression of *HopAl1* in plants enhances disease susceptibility to *Pseudomonas syringae* and suppresses the PAMP flg22-induced transcription of *Arabidopsis NHO1*, a gene required for resistance to *P. syringae* (Li et al. 2005). The effectors HopPtoM and avrE of *P. syringae* pv. *tomato* suppresses the SA-dependent basal immunity in *Arabidopsis* (DebRoy et al. 2004). *P. syringae* pv. *tomato* effector AvrPtoB suppresses basal defense in *Arabidopsis* (de Torres et al. 2006). Effectors of HopPtoM/AvrE family appear to be widespread in plant-pathogenic bacteria and they appear to target the SA-dependent defense signaling pathways (DebRoy et al. 2004).

2.4.11 Effector May Subvert Ubiquitin-Proteasome System to Suppress PAMP-Triggered Immunity

Some pathogens interfere with the ubiquitin-proteasome system and subvert the defense responses induced by the proteolysis to cause disease. Park et al. (2012) showed that the rice blast pathogen *Magnaporthe oryzae* produces an effector called

AvrPiz-t. The effector targets the RING E3 ubiquitin ligase APIP6 to suppress PAMP-triggered immunity in rice. AvrPiz-1 accumulates in the specialized structure called the biotrophic interfacial complex and is then translocated into rice cells. AvrPiz-t suppressed the ubiquitin ligase activity of the rice RING E3 ubiquitin ligase. Silencing of the ubiquitin ligase in transgenic rice enhanced susceptibility of rice plants to *M. oryzae*, suggesting the role of the ubiquitin ligase in disease resistance (Park et al. 2012).

2.5 Host Plants May Manipulate the Defense Signaling Systems to Suppress the Disease Development

Plants are endowed with innate immune system with high potential to trigger defense responses against wide range of pathogens (Vidhyasekaran 2014). Specific signals and signaling systems are required to activate the defense systems against biotrophic, hemibiotrophic, and necrotrophic fungal, oomycete, and bacterial pathogens and against various virus diseases. Plant pathogens are known to regulate different signaling systems by employing their 'PAMP' and 'Effector' molecules to activate or suppress defense responses against them. Infection with both the hemibiotrophic bacterial pathogen *Pseudomonas syringae* pv. *tomato* and necrotrophic fungal pathogen *Alternaria brassicicola* activates SA, JA, and ET signaling systems in *Arabidopsis thaliana* (De Vos et al. 2005). *P. syringae* pv. *tomato* signaled induction of biosynthesis of SA and the induced SA levels were more than that induced by the necrotrophic pathogen *A. brassicicola*. In contrast, the necrotrophic pathogen induced more ET than that induced by the hemibiotrophic bacterial pathogen (Fig. 2.3; De Vos et al. 2005). SA is known to induce resistance against *P. syringae* pv. *tomato* (Son et al. 2012), while ET is known to confer resistance against the necrotrophic pathogens (El Rahman et al. 2012). The results suggest that the host plant may manipulate the defense signaling systems to effectively control the invading pathogens.

2.6 Specificity of Plant Hormone Signaling Systems in Conferring Resistance against Various Pathogens

Different hormone signaling systems have been shown to be involved in triggering defense responses against specific pathogens. In the model plant *Arabidopsis thaliana* SA signaling plays a key role in triggering defense responses against biotrophic and hemibiotrophic pathogens (Table 2.1). SA induces resistance against biotrophic pathogens such as the powdery mildew fungal pathogen *Erysiphe orontii* (Thomma et al. 2001a) and the downy mildew oomycete pathogen *Hyaloperonospora arabidopsidis* (formerly known as *Hyaloperonospora parasitica*) (Thomma et al. 2001a,

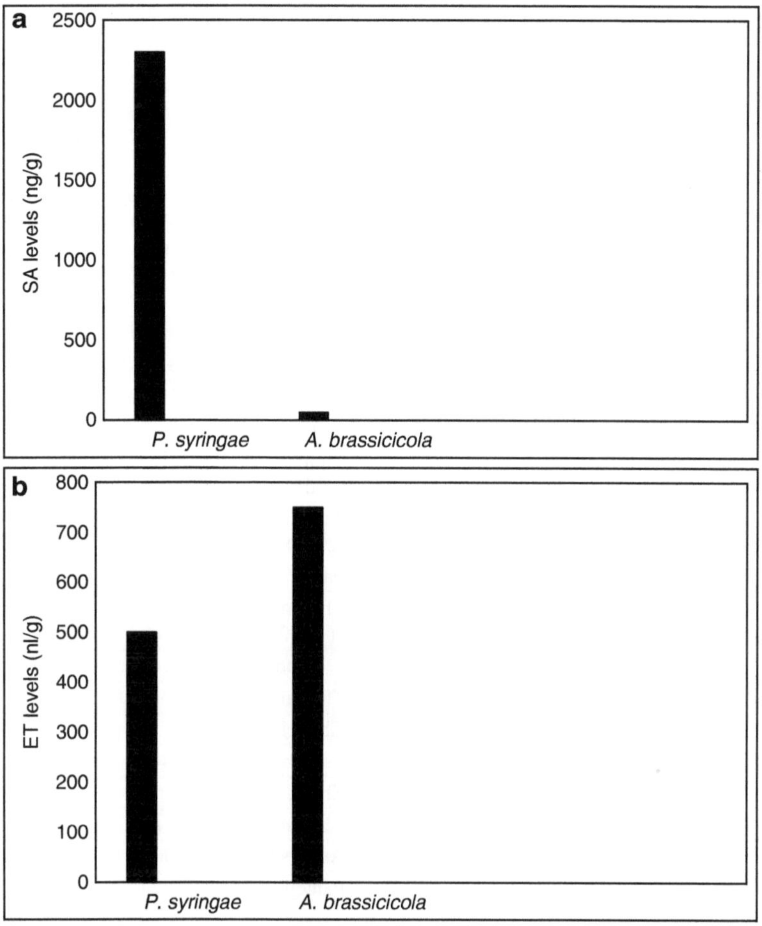

Fig. 2.3 Changes in endogenous SA, JA, and ET levels in *Arabidopsis* plants after inoculation with *Pseudomonas syringae* pv. *tomato* or *Alternaria brassicicola* (Adapted from De Vos et al. 2005)

b; Genger et al. 2008; Fan et al. 2009; Slaughter et al. 2012). SA induces resistance against the hemibiotrophic bacterial pathogen *Pseudomonas syringae* pv. *tomato* (Thomma et al. 2001a, b; Choi et al. 2010; Naseem et al. 2012; Son et al. 2012). SA triggers resistance against several viral pathogens including *Cauliflower mosaic virus* (Love et al. 2005) and *Turnip crinkle virus* (Kachroo et al. 2000). JA signaling induces resistance against several necrotrophic pathogens in Arabidopsis. JA induces resistance against the necrotrophic fungal pathogens *Botrytis cinerea* (Mendez-Bravo et al. 2011; Wild et al. 2012), *Alternaria brassicicola* (Thomma et al. 2001a, b), *Plectosphaerella cucumerina* (Zheng et al. 2006), *Fusarium oxysporum* (Anderson et al. 2004), and *Verticillium longisporum* (Johansson et al. 2006). JA signaling system is also involved in triggering resistance against the necrotrophic bacterial pathogen *Erwinia carotovora* in Arabidopsis (Thomma et al. 2001a).

Table 2.1 Signaling systems involved in triggering defense responses against various pathogens in *Arabidopsis thaliana*

Pathogen	Signaling pathway required for inducing resistance	References
Pseudomonas syringae	SA	Thomma et al. (2001a, b), Kim et al. (2006), Melotto et al. (2006), Xu et al. (2006), Xing et al. (2008), Canet et al. (2012), Slaughter et al. (2012), Son et al. (2012), and Zhang et al. (2012c)
	Cytokinin	Choi et al. (2010) and Naseem et al. (2012)
	GA	Navarro et al. (2008) and Xia et al. (2010)
Erwinia carotovora	JA/ET	Thomma et al. (2001a, b)
Ralstonia solanacearun	ABA	Feng et al. (2012)
Hyaloperonospora parasitica	SA	Thomma et al. (2001a, b), Genger et al. (2008, and Slaughter et al. (2012)
Hyaloperonospora arabidopsidis	Cytokinin	Argueso et al. (2012) and Slaughter et al. (2012)
	ABA	Fan et al. (2009)
Botrytis cinerea	JA	Thomma et al. (2001a, b), Mendez-Bravo et al. (2011), and Wild et al. (2012)
	Auxin	Llorente et al. (2008)
Alternaria brassicicola	JA	Thomma et al. (2001a, b)
	Cytokinin	Choi et al. (2010)
Plectosphaerella cucumerina	SA	Thomma et al. (2001)
	JA	Zheng et al. (2006)
	Auxin	Llorente et al. (2008)
Fusarium oxysporum	JA	Anderson et al. (2004)
Fusarium oxysporum f. sp. raphani	ET	Pantelides et al. (2013)
Verticillium longisporum	JA	Johansson et al. (2006)
Turnip crinkle virus	SA	Kachroo et al. (2000)
Cauliflower mosaic virus	SA	Love et al. (2005)

ET signaling confers resistance against *E. carotovora* in Arabidopsis (Thomma et al. 2001a). ABA signaling is a key signaling component in triggering resistance against the bacterial pathogen *Ralstonia solanacearum* in Arabidopsis (Feng et al. 2012). Auxin signaling also is involved in augmenting disease resistance against necrotrophic pathogens *B. cinerea* and *P. cucumerina* (Llorente et al. 2008). Cytokinin signaling confers resistance against the necrotrophic fungal pathogen *A. brassicicola* (Choi et al. 2010), against the biotrophic oomycete pathogen *Hyaloperonospora arabidopsodis* (Argueso et al. 2012) and against the hemibiotrophic bacterial pathogen *P. syringae* pv. *tomato* in tomato (Choi et al. 2010; Naseem et al. 2012). GA signaling is involved in activation of defense responses against *P. syringae* pv. *tomato* in *Arabidopsis* (Navarro et al. 2008; Xia et al. 2010).

Table 2.2 Signaling systems involved in triggering defense responses against pathogens in crop plants

Crop plants	Pathogen	Signaling pathway required for triggering resistance	References
Rice	*Magnaporthe oryzae*	ET	De Vleesschauwer et al. (2010)
		SA	Oostendorp et al. (2001)
		Auxin	Fu et al. (2011)
		BR	Nakashita et al. (2003)
		JA	Schweizer et al. (1998)
	Xanthomonas oryzae pv.*oryzae*	SA	Qiu et al. (2007), Wang et al. (2008), and Xu et al. (2013)
		BR	Nakashita et al. (2003)
		Auxin	Fu et al. (2011)
	Xanthomonas oryzae pv. oryzicola	Auxin	Fu et al. (2011)
	Cochliobolus miyabeanus	ABA	De Vleesschauwer et al. (2010)
	Pythium graminicola	GA	De Vleesschauwer et al. (2012)
Maize	*Pythium* spp.	JA	Yan et al. (2012)
Tomato	*Botrytis cinerea*	JA	El Rahman et al. 2012
	Alternaria solani	ET	Nambeesan et al. (2012)
		JA	El Rahman et al. (2012)
	Cucumber mosaic virus	SA	Anfoka (2000)
Potato	*Potato virus X*	SA	Sånchez et al. (2010)
Tobacco	*Tobacco mosaic virus*	SA	Waller et al. (2006)
		ET	Gerrats et al. (2003)
	Tomato spotted wilt virus	SA	Mandal et al. (2008)
	Cucumber mosaic virus	SA	Naylor et al. (1998)
	Plum pox virus	SA	Alamillo et al. (2006)
	Tobacco necrosis virus	Cytokinin	Pogåny et al. (2004)
	Cercospora nicotianae	ET	Gerrats et al. (2003)
	Oidium sp.	SA, BR	Nakashita et al. (2002, 2003)
	Erwinia carotovora	ET	Geraats et al. (2003)
	P. syringae pv *tabaci*	Cytokinin	Groβkinsky et al. (2011)
		BR	Nakashita et al. (2003)
Bean	*Tobacco necrosis virus*	ABA	Iriti and Faoro (2008)
		Cytokinin	Fayza and Sabrey (2006)
	White clover mosaic virus	Cytokinin	Clarke et al. (1998)
Apple	*Botrytis cinerea*	ET	Akagi et al. (2011)
	Erwinia amylovora	SA	Sklodowsk et al. (2010)

Various plant hormone signaling systems differentially regulate plant defense responses against biotrophic, hemibiotrophic, necrotrophic, fungal, oomycete, bacterial, and viral pathogens in crop plants (Table 2.2). Blast disease caused by *Magnaporthe oryzae* is the most devastating disease in rice. Ethylene signaling system has been shown to confer resistance against the disease (Iwai et al. 2006; De Vleesschauwer et al. 2008). SA (Oostendorp et al. 2001), JA (Schweizer et al. 1998), auxin (Fu et al. 2011), and BR (Nakashita et al. 2003) signaling systems have also been reported to be involved in triggering defense responses against *M. oryzae* in rice. ABA signaling is involved in triggering resistance against the rice brown spot pathogen *Cochliobolus miyabeanus* (De Vleesschauwer et al. 2010). SA signaling system triggers resistance against the rice bacterial blight pathogen *Xanthomonas oryzae* pv. *oryzae* (Chern et al. 2005; Wang et al. 2008, Xu et al. 2013). BR signaling also is involved in conferring resistance against the bacterial pathogen (Nakashita et al. 2003). Auxin modulates defense responses against both *X. oryzae* pv. *oryzae* and *X. oryzae* pv. *oryzicola* in rice (Fu et al. 2011). GA signaling has been shown to be involved in triggering defense responses against the rice root rot oomycete pathogen *Pythium graminicola* (De Vleesschauwer et al. 2012).

SA signaling system confers resistance against *Cucumber mosaic virus* (CMV) in tomato plants (Anfoka 2000). JA signaling system confers resistance against the necrotrophic fungal pathogens *Botryti cinerea* and *Alternaria solani* in tomato (El Rahman et al. 2012). ET signaling induces resistance against *B. cinerea* in tomato (Nambeesan et al. 2012). In potato, SA signaling system triggers defense responses against *Potato virus X* (PVX) (Sänchez et al. 2010) (Table 2.2).

In tobacco SA signaling induces resistance against *Tobacco mosaic virus* (TMV) (Waller et al. 2006) *Tomato spotted wilt virus* (TSWV) (Momol et al. 2004; Mandal et al. 2008), *Cucumber mosaic virus* (Naylor et al. 1998), and *Plum pox virus* (Alamillo et al. 2006). SA signaling confers resistance against *Agrobacterium tumefaciens* in *Nicotiana benthamiana* (Anand et al. 2008). ET signaling is involved in triggering defense responses against *Cercospora nicotianae* and *Erwinia carotovora* in tobacco (Gerrats et al. 2003). Cytokinin signaling modulates defense responses against the bacterial pathogen *P. syringae* pv. *tabaci* (Groβkinsky et al. 2011) and against *Tobacco necrosis virus* (TNV) (Pogány et al. 2004) in tobacco. BR signaling confers resistance against the biotrophic powdery mildew pathogen *Oidium* sp., the hemibiotrophic bacterial pathogen *P. syringae* pv. *tabaci*, and *Tobacco mosaic virus* (TMV) in tobacco (Nakashita et al. 2003). BR signaling also triggers resistance against the oomycete pathogen *Phytophthora infestans* in *N. benthamiana* (Chaparro-Garcia et al. 2011).

SA signaling system is involved in triggering resistance against the hemibiotrophic fungal pathogen *Leptosphaeria maculans* and the hemibiotrophic bacterial pathogen *Pseudomonas syringae* pv. *maculicola* in *Brassica napus* plants (Potlakayala et al. 2007). JA signaling induces resistance against the necrotrophic fungal pathogen *Sclerotinia sclerotiorum* in *B. napus* (Wang et al. 2009). ET induces systemic resistance against *Turnip mosaic virus* (TuMV) infection in leafy mustard plants (Zhu et al. 2011a). SA signaling induces resistance against the biotrophic rust fungal pathogen *Puccinia helianthi* in sunflower plants (Amzalek and Cohen 2007).

JA induces resistance against the root-rotting oomycetes *Pythium* spp. in maize (Yan et al. 2012).

SA signaling is involved in triggering resistance against the hemibiotrophic fungal pathogens *Colletotrichum destructivum* in cowpea (Latunde-Dada and Lucas 2001) and *Colletotrichum orbiculare* in cucumber (Cools and Ishii 2002; Deepak et al. 2006). It also confers resistance against the hemibiotrophic oomycete pathogen *Phytophthora palmivora* in papaya (Zhu et al. 2003). SA signaling system induces resistance against the fire blight pathogen *Erwinia amylovora* in apple leaf tissues (Sklodowska et al. 2010). Ethylene enhances resistance to *B. cinerea* infection in apple (Akagi et al. 2011).

2.7 Plant Hormone Signaling Systems May Also Induce Susceptibility against Pathogens

Plant hormone signaling systems which are involved in triggering plant defense responses, may also be involved in enhancing susceptibility against some pathogens (Table 2.3). SA signaling is not effective in the control of necrotrophic pathogens. It confers enhanced susceptibility to the necrotrophic fungal pathogen *Botrytis cinerea* (Kachroo et al. 2001, 2003a, b; Nandi et al. 2003, 2005; Genger et al. 2008). SA has been shown to promote disease development caused by *B. cinerea* in tomato by suppressing the JA- dependent defense genes (El Rahman et al. 2012). SA signaling pathway contributes to disease development caused by another necrotrophic pathogen *Alternaria solani* in tomato (El Rahman et al. 2012). SA signaling has been shown to be associated with susceptibility in pea-*Clover yellow vein virus* interaction. SA enhanced disease symptoms induced by the wild-type virulent strain of *Clover yellow vein virus* (ClYVV) (Atsumi et al. 2009).

Ethylene signaling system increases susceptibility to the hemibiotroph *Pseudomonas syringae* in Arabidopsis (Berrocal-Lobo et al. 2002). ET signaling system induces susceptibility against the oomycete *Peronospora tabacina* and *Tobacco mosaic virus* in tobacco (Gerrats et al. 2003). Ethylene induces susceptibility against brown spot disease caused by *Cochliobolus miyabeanus* in rice plants (De Vleesschauwer et al. 2010). Ethylene signaling system induces susceptibility to *Clavibacter michiganensis* subsp. *michiganensis* and increased synthesis of ethylene aids in disease development in tomato plants (Balaji et al. 2008).

ABA has been shown to promote susceptibility to the rice bacterial blight pathogen *Xanthomonas oryzae* pv. *oryzae* by suppressing SA-mediated defenses (Xu et al. 2013). ABA enhances host susceptibility against the rice blast pathogen *M. oryzae* in rice plants (De Vleesschauwer et al. 2010). The GA-overproducing plants were more susceptible to *M. oryzae* and *X. oryzae* pv. *oryzae* compared with the wild-type plants. The susceptibility of wild-type plants to *X. oryzae* pv. *oryzae* was increased by exogenous application of GA3 (Qin et al. 2013). Cytokinin has been shown to induce susceptibility against *Botrytis cinerea* in tobacco leaves (Großkinsky

Table 2.3 Signaling systems involved in promoting susceptibility against various pathogens

Host plant	Pathogen	Signaling pathway promoting susceptibility	References
Rice	*Magnaporthe oryzae*	ABA	De Vleesschauwer et al. (2010)
		GA	Qin et al. (2013)
	Cochliobolus miyabeanus	ET	De Vleesschauwer et al. (2010)
	Xanthomonas oryzae pv. oryzae	Auxin	Ding et al. (2008)
		ABA	Xu et al. (2013)
		GA	Qin et al. (2013)
	Pythium graminicola	BR	De Vleesschauwer et al. (2012)
Barley	*Pyrenophora teres*	Cytokinin	Angra-Sharma and Sharma (1999)
Maize	*Drechslera maydis*	Cytokinin	Angra-Sharma and Sharma (1999)
Tomato	*Botrytis cinerea*	ABA	Curvers et al. (2010)
	Clavibacter michiganensis subsp. *michiganensis*	ET	Balaji et al. (2008)
	Erwinia chrysanthemi	ABA	Asselbergh et al. (2008a)
Tobacco	*Botrytis cinerea*	Cytokinin	Groβkinsky et al. (2011)
	Peronospora tabacina	ET	Gerrats et al. (2003)
	Tobacco mosaic virus	ET	Gerrats et al. (2003)
Pepper	*Pseudomonas syringae*	ABA	Choi and Hwang (2011)
Pea	*Clover yellow vein virus*	SA	Atsumi et al. (2009)
Arabidopsis	*Pseudomonas syringae* pv. *tomato*	ABA	Mohr and Cahill (2003) and Fan et al. (2009)
		Auxin	Navarro et al. (2006) and Kazan and Manners (2009)
		JA	Laurie-Berry et al. (2006) and Wild et al. (2012)
	Hyaloperonospora parasitica	Auxin	Wang et al. (2007)
	Hyaloperonospora arabidopsidis	ABA	Fan et al. (2009)
	Fusarium oxysporum	Auxin	Kidd et al. (2011)
	Fusarium oxysporum f. sp. *raphani*	ET	Pantelides et al. (2013)

et al. 2011). Cytokinins have been shown to play a key role in the development of clubroot disease caused by the obligate protist *Plasmodiophora brassicae* in *Arabidopsis* (Siemens et al. 2006). Auxin was found to promote development of citrus canker caused by *Xanthomonas axonopodis* pv. *citri* (Cernadas and Benedetti 2009). Brassinosteroids (BRs) induce susceptibility against *Pythium graminicola* in rice (De Vleesschauwer et al. 2012).

2.8 Pathogens May Hijack Specific Signaling Pathways to Cause Disease

2.8.1 Pathogens May Hijack ABA Signaling Pathway to Cause Disease

Pathogens may induce specific signaling systems, which may favor disease development. *Pseudomonas syringae* pv. *tomato* has been shown to hijack the *Arabidopsis* ABA signaling pathway to cause disease (de Torres-Zabala et al. 2007). The bacterial pathogen infection resulted in activation of ABA signaling system. The pathogen triggered ABA accumulation (Sanchez-Vallet et al. 2012; Alazem et al. 2014; Vidhyasekaran 2014). In the compatible interaction leading to disease, *NCED*, the gene encoding 9-*cis*-epoxycarotenoid dioxygenase, the key enzyme involved of ABA biosynthesis was found to be strongly induced (de Torres-Zabala et al. 2007). In the presence of accumulated ABA, the PYR/PYL/RCAR receptor proteins disrupt the interaction between the SnRK2s (protein kinases) and PP2Cs (phosphatases), thus preventing the PP2Cs-mediated dephosphorylation of the SnRK2s and resulting in the activation of the SnRK2 kinases (Fujii et al. 2009). ABA binds with the receptor and the binding of ABA to the receptor induces a conformational change in the receptor that allows it to sequester the PP2Cs. This excludes the PP2Cs from the negative regulation of the downstream ABA-activated SnRK protein kinases (Miyazono et al. 2009; Umezawa et al. 2009; Vlad et al. 2009; Raghavendra et al. 2010). The recognition of ABA by the PYR/PYL/RCAR receptor results in the dissociation of the SnRK2/PP2C complex and the subsequent release of SnRK2s. The released SnRK2s phosphorylate and activate downstream transcription factors, which initiate transcription at ABA-responsive promoter elements (Sheard and Zheng 2009). These results suggest that the PP2Cs play important role in ABA signaling pathway.

The *abscisic acid insensitive 1* (*ABI1*) clade of genes encoding protein phosphatases type 2C (PP2C) involved in the ABA signaling were strongly induced by *P. syringae* pv. *tomato* infection in *Arabidopsis*. These events led to the accumulation of ABA and activation of ABA signaling. ABA accumulation suppressed the expression of defense genes and enhanced the susceptibility of *A. thaliana* to *P. syringae* pv. *tomato* (de Torres-Zabala et al. 2007). ABA suppresses the deposition of lignin, which reinforces the cell wall to prevent the bacterial invasion (Mohr and Cahill

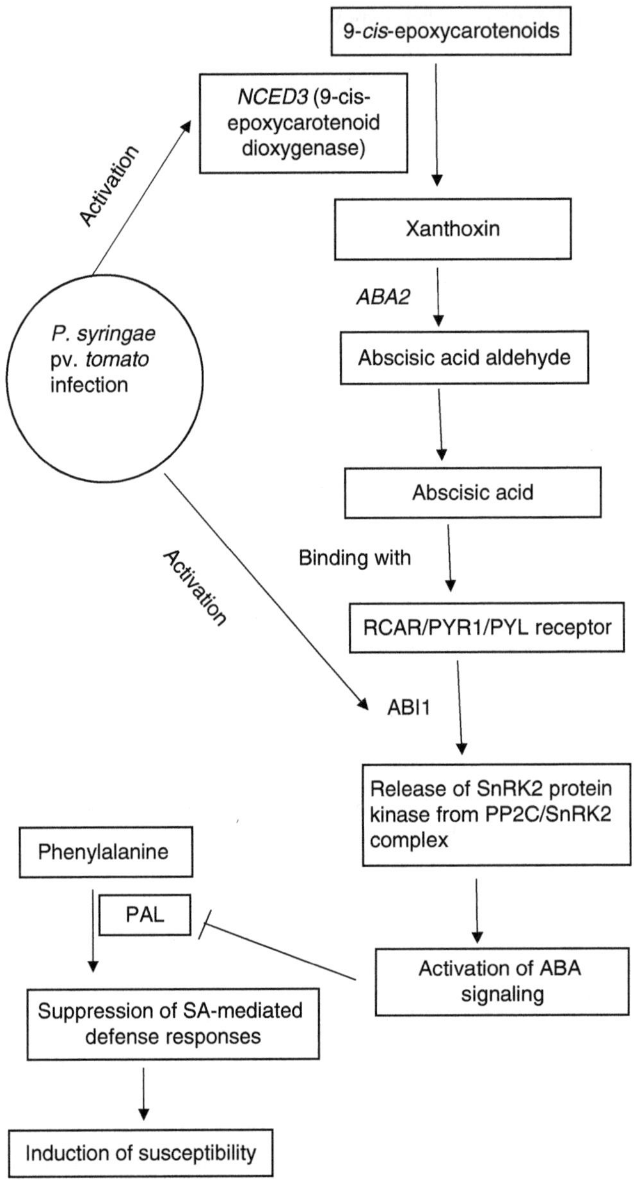

Fig. 2.4 Pathogen hijacks ABA signaling system to suppress SA-mediated defense responses promoting disease development (Adapted from de Torres-Zabala et al. 2007; Schmidt et al. 2008)

2007). ABA inhibits the accumulation of SA and the expression of SA-mediated defense genes (Mohr and Cahill 2007; Spoel and Dong 2008). SA signaling has been reported to induce resistance against *P. syringae* pv. *tomato* (Son et al. 2012). These results suggest that the virulent bacterial pathogen hijacks ABA signaling system to promote disease development by suppressing SA-signaling system (Fig. 2.4).

The concentration of ABA increased in sugar beet (*Beta vulgaris*) leaves during *Cercospora beticola* infection (Schmidt et al. 2008). After an initial burst of ABA induced by inoculation, elevated ABA concentrations were detected during the fungal penetration and colonization phases. Expression of the basic leucine zipper transcription factor AREB1 was induced by the fungal infection. Activation of the ABA-dependent signaling cascade resulted in reduced expression of the phenylalanine ammonia lyase (*PAL*) gene, the key enzyme involved in SA signaling system (Schmidt et al. 2008). These results suggest that activation of ABA-dependent signaling may be involved in promoting the fungal disease development by suppressing SA pathway (Fig. 2.4).

The rice bacterial blight pathogen *Xanthomonas oryzae* pv. *oryzae* has been shown to hijack the rice ABA machinery to cause disease (Xu et al. 2013). Successful *X. oryzae* pv. *oryzae* infection was found to be associated with extensive reprogramming of ABA biosynthesis and response genes (Xu et al. 2013). ABA negatively regulated resistance to the bacterial pathogen. Exogenously applied ABA rendered rice hypersusceptible to infection, suggesting that ABA suppresses the host immune responses. It has been suggested that this immune-suppressive effect of ABA may be due to suppression of SA-mediated defenses that normally serve to limit growth of the bacterial pathogen. These results suggest that the pathogen hijacks ABA signaling system to cause the rice bacterial blight disease.

2.8.2 Pathogens May Hijack ET Signaling System to Cause Disease

Ethylene signaling system has been shown to be involved in conferring susceptibility to some diseases. Ethylene induces susceptibility against brown spot disease caused by *Cochliobolus miyabeanus* in rice plants (De Vleesschauwer et al. 2010). The ethylene system induced susceptibility against the oomycete *Peronospora tabacina* and *Tobacco mosaic virus* in tobacco (Gerrats et al. 2003). ERF5 is an ethylene-responsive element-binding factor. It negatively regulates chitin signaling and plant defense against the fungal pathogen *Alternaria brassicicola* (Son et al. 2012). Ethylene signaling system induces susceptibility to *Clavibacter michiganensis* subsp. *michiganensis* and increased synthesis of ethylene aids in disease development (Balaji et al. 2008). ET pathway induces susceptibility to *P. syringae* pv. *tomato* (Berrocal-Lobo et al. 2002).

In Arabidopsis, ETR1 is a membrane-bound receptor and ethylene is perceived by the ETR receptor, which transmits the signal to downstream effectors (Wang et al. 2006). The ET signal receptor ETR1 negatively regulates plant immune responses against *Fusarium oxysporum* f. sp. *raphani* and induces susceptibility to the pathogen in *Arabidopsis* (Pantelides et al. 2013). The pathogen infection triggered ET signaling pathway and the plants inoculated with *F. oxysporum* f. sp. *raphani* showed increased expression of ET-responsive PR genes *PR3, PR4* and *PDF1.2* while expression of SA-responsive PR genes *PR1*, *PR2* and *PR5* genes was suppressed in the pathogen-inoculated plants (Pantelides et al. 2013). The results suggest that *F. oxysporum* hijacks ethylene signaling system to suppress SA signaling system, which is involved in activation of defense responses against *F. oxysporum.* The results also suggest that *F. oxysporum* f. sp. *raphani* hijacks ETR1-mediated ET signaling to promote disease development in plants (Pantelides et al. 2013).

2.8.3 Pathogens May Hijack JA Signaling System to Cause Disease

JA signaling system has been reported to confer susceptibility against some pathogens. In the infection process, pathogen may suppress the particular signaling system which may be able to induce the expression of specific defense genes conferring resistance against the particular pathogen. Activation of SA-mediated pathway appears to confer resistance to *Pseudomonas syringae* pv. *tomato* DC3000 (Slaughter et al. 2012; Zhang et al. 2012c). The bacterial pathogen *Pseudomonas syringae* produces coronatine that mimics jasmonic acid isoleucine and promotes opening of bacterial entry, bacterial growth in the apoplast, and systemic susceptibility (Zheng et al. 2012). Coronotine was found to activate three homologous NAC transcription factor (TF) genes through direct activity of the transcription factor MYC2. The activated NAC TFs repress the *isochorismate synthase 1* (ICS1) gene, which is involved in SA biosynthesis. These TFs also activate expression of *BSMT1* gene encoding SA methyltransferase, which converts SA to the inactive volatile methyl SA (MeSA) (Zheng et al. 2012). These results suggest that the toxin produced by the bacterial pathogen promotes systemic susceptibility by suppressing SA signaling (Zheng et al. 2012). The pathogen-produced coronatine functions as a JA mimic and induces JA- signaling pathway (Katsir et al. 2008). The induced JA signaling system suppresses effectual SA-dependent defenses (Brooks et al. 2005; Cui et al. 2005; Glazebrook 2005; Laurie-Berry et al. 2006; Uppalapati et al. 2007). These results suggest that the pathogen is able to suppress the signaling pathway which may suppress its pathogenesis and diverts the signals for activation of an inefficient pathway resulting in defense response that cannot inhibit its pathogenesis (Spoel et al. 2003, 2007; Uppalapati et al. 2007). The PR-1 protein is required for the inhibition of *P. syringae* pv. *tomato* (Laurie-Berry et al. 2006) and induction of this pathogenesis-related protein is dependent on SA accumulation (van Verk et al. 2008). The toxin produced by the pathogen triggers JA signaling, which suppresses SA-mediated defenses. Thus, it appears that the bacterial pathogen utilizes the toxin to manipulate JA signaling both to suppress SA-mediated responses and to promote symptom development (Brooks et al. 2005; Laurie-Berry et al. 2006; Koornneef and Pieterse 2008).

Fusarium oxysporum, a root-infecting hemibiotrophic fungal pathogen has been shown known to hijack jasmonate responses for disease development in *Arabidopsis thaliana* (Thatcher et al. 2009). The JA signal is perceived by the F-box protein COI1, which plays a key role in JA signaling (Katsir et al. 2008). Jasmonate signaling mediated through COI1 in *A. thaliana* may be responsible for susceptibility to wilt disease caused by *F. oxysporum*. In contrast to leaves of wild-type plants, leaves from *coi1* mutant plants showed no necrosis following the application of *F. oxysporum* culture filtrate, and showed reduced expression of senescence-associated genes during disease development, suggesting that *coi1* resistance is most likely achieved through the inhibition of *F. oxysporum*-incited lesion development and plant senescence (Thatcher et al. 2009). These results indicate that *F. oxysporum* hijacks non-defensive aspects of the JA-signaling pathway to cause wilt-disease symptoms that lead to plant death in *Arabidopsis*.

2.8.4 Pathogen May Hijack Auxin Metabolism to Cause Disease

The plant hormone auxin, indole-3-acetic acid (IAA), is a simple signaling molecule involved in modulation of different signaling pathways (Hayashi 2012; Peer et al. 2013; Sauer et al 2013; Tatsuki et al. 2013). Auxin signaling system has been shown to be involved in promoting susceptibility to pathogens and inducing disease development (O'Donnell et al. 2003; Kidd et al. 2011). Auxin signaling system has been shown to promote susceptibility to the wilt pathogen *Fusarium oxysporum* in *Arabidopsis*. Several auxin-signaling mutants, *axr1*, *axr2*, *axr3*, and *sgt1b*, showed increased resistance against *F. oxysporum* in *Arabidopsis. AXR2* and *AXR3* encode the IAA/AUX proteins IAA7 and IAA17, respectively (Kidd et al. 2011). AXR1 protein is required for post-translational modification of CUL1, a component of the SCF complex involved in the degradation of AUX/IAA proteins (Chapman and Estelle 2009). The degradation of AUX and IAA proteins is compromised in the *axr1* mutant (Kidd et al. 2011). SGT1B encodes a regulatory protein involved in SCF^{TIR1}-mediated degradation of Aux/IAA proteins (Gray et al. 2003). *SGT1B* is required for SCF^{TIR1}-mediated protein degradation and the *sgt1b* mutant is compromised in protein ubiquitination (Gray et al. 2003). These results suggest that Aux/IAA proteins, downstream of the auxin receptor in the auxin signaling system, contribute to the susceptibility of *Arabidopsis* to *Fusarium oxysporum*.

One of the responses induced by the activation of auxin signaling is the formation of the conjugated forms of auxin through the action of GH3 (Gtretchen Hagen3) proteins (Fu et al. 2011; González-Lamothe et al. 2012). Auxin conjugation is mostly controlled by GH3 family proteins (Woodward and Bartel 2005). *GH3s* encode auxin conjugating enzymes which show acyl acid amido synthetase activity (Ding et al. 2008; Fu et al. 2011; Westfall et al. 2012). IAA-Asp is the important conjugated form of auxin, which is involved in triggering susceptibility and disease development.

Manipulation of auxin signaling may be one of the strategies used by pathogens to overcome plant immune pathways to cause the disease (González-Lamothe et al. 2012). Fungal and bacterial pathogens hijack the host auxin metabolism in *Arabidopsis thaliana* leading to the accumulation of a conjugated form of the hormone, indole-3-acetic acid (IAA) –Asp, to promote disease development. IAA-Asp increases pathogen progression in the plant by regulating the transcription of virulence genes (González-Lamothe et al. 2012). The fungal and bacterial pathogens may manipulate auxin metabolism to produce a conjugated auxin and perceive it as a signal to activate its virulence. IAA-Asp accumulates in *Arabidopsis* plants in response to the necrotrophic fungal pathogen *Botrytis cinerea* and the hemibiotroph bacterial pathogen *Pseudomonas syringae* pv. *tomato* strain DC3000 (González-Lamothe et al. 2012). IAA-Asp treatment enhanced both the fungal and bacterial disease development in *Arabidopsis*. IAA-Asp treatment promoted *P. syringae* pv. *tomato* bacterial growth in both tomato and *Nicotiana benthamiana* and these plants had around 8–10 times more bacteria than control plants (González-Lamothe et al. 2012). It is suggested that the effect of IAA-Asp on disease susceptibility might be exerted on the pathogen, by contributing to the regulation of the transcription of virulence genes. The expression of the effector genes *AvrPto* and *HopAO1* of *P. syringae* pv. *tomato* and *BcatrB* gene of *Botrytis cinerea* was significantly reduced in *gh3.2* mutant plants (González-Lamothe et al. 2012). These results suggest that the GH3.2 is involved in activating expression of the virulence genes in both the pathogens and this action is mediated by IAA-Asp.

2.8.5 Pathogen Hijacks Brassinosteroid Signaling Machinery to Cause Disease

The rice root-infecting oomycete pathogen *Pythium graminicola* exploits brassinosteroids (BRs) as virulence factors and hijacks the rice BR machinery to cause disease (De Vlesschauwer et al. 2012). BRs suppress root immunity to *P. graminicola*, resulting in increased pathogen proliferation and substantially enhanced disease susceptibility. *P. graminicola* co-opts the plant BR machinery as a decoy strategy to tap into the immune signaling circuitry and interfere with effectual SA- and GA-controlled defenses (De Vleesschauwer et al. 2012).

The most active brassinosteroid, brassinolide (BL) markedly promoted susceptibility to *P. graminicola*, whereas treatment with brassinazole (BRZ), a highly specific inhibitor of BR biosynthesis, induced substantial levels of resistance in rice plants. The extensive transcriptional reprogramming of various BR biosynthetic, signaling, and response genes in pathogen-inoculated roots and the overrepresentation of BR-responsive *cis*-elements in the promoters of *P. graminicola*-dependent rice genes strongly suggest that *P. graminicola* hijacks the rice BR biosynthesis and signaling machinery to cause diseases (De Vleesschauwer et al. 2012).

2.9 Pathogens May Suppress Specific Signaling System to Promote Disease Development

2.9.1 Pathogens May Suppress SA Signaling System to Promote Disease Development

Salicylic acid (SA) is an important endogenous immune signal in the induction of disease resistance response against biotrophic and hemibiotrophic pathogens in plants (Anand et al. 2008; Fung et al. 2008; Mukherjee et al. 2010; Dempsey et al. 2011; Liu et al. 2011b, Argueso et al. 2012; Denancé et al. 2013; Gimenez-Ibanez and Solano 2013; Yang et al. 2013). SA signaling system activates not only local resistance, but also systemic acquired resistance (SAR) observed in distal (systemic) tissues. SAR is a SA-dependent heightened defense to a broad spectrum of pathogens that is activated throughout a plant following local infection (Liu et al. 2011b). SAR is associated with priming of defense (Kohler et al. 2002; Jung et al. 2009; Luna et al. 2012) and the priming results in a faster and stronger induction of defense mechanisms after pathogen attack (Conrath 2011; Po-Wen et al. 2013). The priming can be inherited epigenetically from disease-exposed plants (Pastor et al. 2013) and descendants of primed plants exhibit next-generation systemic acquired resistance (Slaughter et al. 2012; Luna et al. 2012). The transgenerational SA-induced SAR has also been reported (Luna et al. 2012; Pieterse 2012).

The activation of the JA-signaling pathway has been shown to be required for resistance against necrotrophic pathogens; however to combat biotrophic pathogens, the plants activate mainly the SA-signaling pathway (El Rahman et al. 2012). Necrotrophic pathogens seem to use the SA signaling pathway to promote disease development (El Rahman et al. 2012). SA- and JA-dependent pathways are reciprocally antagonistic (Mur et al. 2006; Spoel et al. 2007; Leon-Reyes et al. 2010; Van der Does et al. 2013). SA pathway inhibits JA signaling by targeting GCC-box motifs in JA-responsive promoters (Gimenenz-Ibanez and Solano 2013; Van der Does et al. 2013). Salicylic acid inhibits synthesis of proteinase inhibitors in tomato leaves induced by JA (Doares et al. 1995a). *NPR1 (NONEXPRESSER OF PR GENES1)*, the transcription cofactor, has been shown to be involved in the suppressive action of SA against JA signaling (Spoel et al. 2003). SA-induced suppression of the JA response is controlled by the function of NPR1 in the cytosol (Spoel et al. 2003; Koornneef and Pieterse 2008). Suppression of JA-inducible gene depended on the presence of TGA factors (Ndamukong et al. 2007). SA promotes disease development caused by the necrotrophic pathogens *Botrytis cinerea* and *Alternaria solani* through NPR1, by suppressing expression of two JA-dependent defense genes, *proteinase inhibitors I and II* in tomato (El Oirdi and Bouarab 2007;

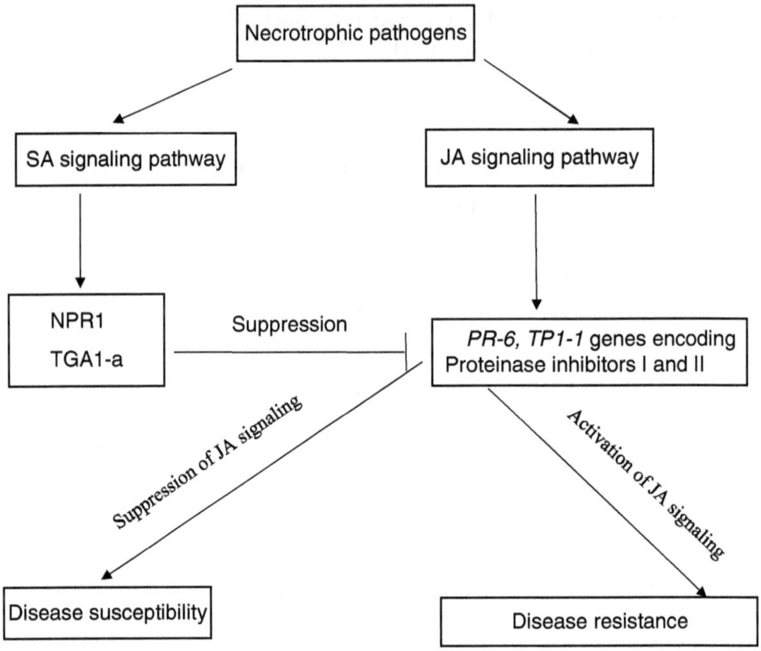

Fig. 2.5 Necrotrophic pathogens use SA signaling pathway to promote disease development by suppressing JA signaling pathway (Adapted from Hondo et al. 2007; El Rahman et al. 2012)

El Rahman et al. 2012). NPR1 interacts with several TGA transcription factors which bind to the promoter of the SA-dependent defense gene *PR1* and activate its expression (Durrant and Dong 2004; Ndamukong et al. 2007). It has been shown that the two necrotrophic pathogens, *B. cinerea* and *A. solani*, use the SA-signaling pathway component TGA1-a to promote their disease development by inhibiting the JA signaling-inducible proteinase inhibitors in tomato (Fig. 2.5; El Rahman et al. 2012). Proteinase inhibitors are pathogenesis-related proteins which confer resistance against necrotrophic pathogens (Vidhyasekaran 2007). Methyl jasmonate treatment induced *pin1* gene encoding a proteinase inhibitor in *Brachypodium distachyon* (Mur et al. 2004). Expression of the basic *PR-6* and *TPI-1* genes that encode proteinase inhibitor II and I, respectively was found to be dependent on JA-mediated signaling pathway in tomato (Hondo et al. 2007).

When *Arabidopsis* plants were inoculated with *P. syringae* pv. *tomato*, there was substantial increase in *WRKY33* transcript level (Dong et al. 2003; Zheng et al. 2006). WRKY33 induces JA-mediated signaling system and the induced JA signaling system reduces the SA-mediated signaling system by its antagonistic function (Zheng et al. 2006; van Verk et al. 2008). It suggests that the pathogen suppresses the SA signaling pathway, which is needed to induce resistance against the pathogen, to cause disease.

Activation of SA-mediated pathway appears to confer resistance to *Pseudomonas syringae* pv. *tomato* DC3000. The pathogen produces the toxin coronatine, which induces JA- signaling pathway while suppressing the SA-dependent pathway (Katsir et al. 2008). It suggests that the pathogen is able to suppress the signaling pathway which may suppress its pathogenesis and diverts the signals for activation of an inefficient pathway resulting in defense response that cannot inhibit its pathogenesis (Spoel et al. 2003, 2007; Uppalapati et al. 2007). The PR-1 protein is required for the inhibition of *P. syringae* pv. *tomato* (Laurie-Berry et al. 2006) and induction of this pathogenesis-related protein is dependent on SA accumulation (van Verk et al. 2008). The toxin produced by the pathogen triggers JA signaling, which suppresses SA-mediated defenses. Thus, it appears that the bacterial pathogen utilizes the toxin to manipulate JA signaling both to suppress SA-mediated responses and to promote symptom development (Brooks et al. 2005; Laurie-Berry et al. 2006; Koornneef and Pieterse 2008).

Bacterial entry into host tissue is a critical first step in causing infection by bacterial pathogens in plants (Melotto et al. 2006). Stomata serve as passive ports of bacterial entry during infection. The innate immune response in *Arabidopsis* activated by PAMPs involves stomatal closure, which occurs within the first hour of the bacterial contact with the plant tissue (Melotto et al. 2006). The stomatal defense against bacterial pathogens observed in *Arabidopsis* was shown to be under the control of SA signaling. The regulation of stomatal defense by SA suggests that stomatal defense is an integrated part of the SA-regulated immune system (Melotto et al. 2006). The bacterial pathogen *P. syringae* pv. *tomato* reopens the closed stomata and counteracts the SA-mediated stomatal closure, by producing the phytotoxin coronatine. The toxin overcomes or suppresses SA-dependent defenses (Brooks et al. 2005) and probably suppresses the stomatal defense to facilitate bacterial disease development (Melotto et al. 2006).

Some transcription factors involved in specific signaling pathways may act as negative regulators of SA-mediated disease resistance. Pathogens may induce such transcription factors, which suppress the SA-signaling pathway and cause disease development. *Arabidopsis thaliana WRKY38* and *WRKY62*, encoding two structurally similar type III WRKY transcription factors, are induced by virulent *Pseudomonas syringae* or SA treatment. WRKY38 and WRKY62 function additively as negative regulators of SA signaling system. Overexpression of *WRKY38* and *WRKY62* reduces disease resistance and SA-mediated *PR-1* expression (Kim et al. 2007). Collectively these results suggest that the pathogen activates the negative regulator transcription factors in the SA signaling pathway and suppresses the defense mechanisms resulting in susceptibility.

The transcription factor WRKY48 is a negative regulator of basal resistance to the bacterial pathogen *P. syringae* pv. *tomato* in *Arabidopsis*. The transcription factor is induced in the plant infected by the bacterial pathogen. The transcription factor suppressed the expression of SA-regulated *PR-1*, *PR-2*, and *PR-5* genes (Xing et al. 2008). The transgenic *WRKY48*-overexpressing plants showed enhanced susceptibility to the bacterial pathogen (Xing et al. 2008). These results suggest that the bacterial pathogen induces the transcription factor WRKY48 that represses SA-mediated defense response and promotes disease development in *Arabidopsis*.

Exogenous application of BL decreased the pathogen-induced strong upregulation of the transcription of the SA regulatory genes *OsNPR1* and *OsWRKY45*. The temporal expression of both the SA regulatory genes was inversely correlated with that of the BR marker genes *OsBLE2*, *OsCPD1*, and *OsRAVL1* (De Vleesschauwer et al. 2012). These studies suggest that BR-mediated suppression of SA defenses occurs downstream of SA biosynthesis, but upstream of NPR1 and OsWRKY45 in the SA signaling pathway (De Vleesschauwer et al. 2012).

2.9.2 Pathogens May Suppress JA Signaling System to Promote Disease Development

DNA viruses selectively interfere with CUL1-based SCF ubiquitin E3 ligases to cause infection in plants (Lozano-Durán and Bejarano 2011; Lozano-Durán et al. 2011). The DNA viruses redirect ubiquitination by interfering with the activity of the COP9 signalosome (CSN) complex. The geminiviral C2 protein interacts with CSN5 (COP9 signalosome5), and its expression in transgenic plants compromises CSN activity on CUL1. Several responses regulated by the CUL1-based SCF ubiquitin E3 ligases (including responses to jasmonates, auxins, gibberellins, ethylene, and abscisic acid) are altered in these plants (Lozano-Durán et al. 2011). Transcriptomic analysis of the transgenic plants showed that the response to jasmonates is the main SCF-dependent process affected by geminiviral C2 protein. Exogenous JA treatment of *Arabidopsis* plants disrupts geminivirus infection suggesting that the suppression of the JA response might be crucial for infection. SCFs are key regulators of JA signaling. The capability of viruses to selectively interfere with or hijack the activity of these complexes may be a powerful strategy in viral infection (Lozano-Durán et al. 2011).

2.9.3 Pathogen May Suppress ABA Signaling System to Promote Pathogenesis

ABA signaling system plays important role in plant immune responses (Adie et al. 2007; Asselbergh et al. 2008a, b, Fan et al. 2009; Garcia-Andrade et al. 2011) ABA is known to be involved in stomatal pore closure which is a key component in plant immune responses against bacterial pathogens (Takahashi et al. 2007; Hettenhausen et al. 2012; Uraji et al. 2012). Oxalic acid is a virulence factor of *Sclerotinia sclerotiorum*, which causes foliar wilting of *Vicia faba* (Guimarães and Stotz 2004). Oxalate causes foliar dehydration by disturbing guard cell function. It triggers stomatal opening, which causes increase in transpiration and wilting. ABA signaling plays important role in stomatal function and confers resistance to the fungal pathogen. It induces stomatal closure, thus preventing excess transpiration, dehydration, and foliar wilting. For successful pathogenesis, the pathogen produces oxalate, which inhibits the ABA signaling system -induced stomatal closure and keeps the stomata open. The induction of stomatal opening favors disease development (Guimarães and Stotz 2004). The results suggest that the virulent pathogen suppresses ABA-signaling system and causes disease.

2.9.4 Pathogens May Suppress GA Signaling Pathway to Cause Disease

GA has been shown to play important role in inducing resistance against the root-infecting oomycte pathogen *Pythium graminicola* in rice (De Vleesschauwer et al. 2012). GA modulates plant disease resistance by inducing the degradation of DELLA proteins, a class of nuclear growth-repressing proteins that act as central suppressors of GA signaling (Navarro et al. 2008). The rice DELLA gene, *SLENDER RICE1* (*SLR1*) plays an important role in GA signaling. Several GA-deficient and/or insensitive rice mutants have been identified and all of these mutants were shown to overaccumulate *SLR1* (Ueguchi-Tanaka et al. 2008). *SLR1* is transiently up-regulated in response to pathogen inoculation, suggesting that pathogen triggers overexpression of *SLR1* to suppress GA-regulated defense responses (De Vleesschauwer et al. 2012).

P. graminicola has been shown to suppress GA signaling system to cause infection. *P. graminicola* strongly induced the GA-degrading enzyme *GIBBERELLIN 2-β-DIOXYGENASE3* (*OsGA2ox3*) in roots of rice plants grown in the presence or absence of the most active brassinosteroid, brassinolide (BL). In contrast, the GA-biosynthesis enzyme *GIBBERELLIN 20-OXIDASE3* (*OsGA20ox3*) was severely downregulated following *P. graminicola* inoculation (De Vleesschauwer et al. 2012). *P. graminicola* recruits the BR pathway to actively suppress GA biosynthesis in rice. Exogenously administered BR inhibited *GA20ox3* and induced *GA2ox3* within 8 h of inoculation with *P. graminicola*. When rice roots were treated with GA, a fast and strong downregulation of the BR response gene *OsBLE2* and the BR biosynthesis gene *OsDWARF2* was observed (De Vleesschauwer et al. 2012).

2.9.5 Pathogens May Suppress ROS Signaling System to Promote Disease Development

ROS is involved in host plant defense. It plays a central role in launching the defense response against invading pathogens. ROS induces Ca^{2+} signaling system, reversible phosphorylation, ubiquitin-proteasome signaling pathway, NO signaling system, salicylic acid signaling system, ethylene-mediated signaling system, jasmonic acid signaling system, and abscisic acid-mediated signaling system. ROS may act as a second messenger activating Ca^{2+} signaling (Kwak et al. 2003). H_2O_2 activates MAP kinases that modulate gene expression and transduce cellular responses to extracellular stimuli (Desikan et al. 2001; Kovtun et al. 2000; Pitzschke and Hirt 2006, 2009; Zhang et al. 2006; Pitzschke et al. 2009a, b). The ROS-activated MAPK modulates nitric oxide biosynthesis in *Arabidopsis* (Wang et al. 2010b). Ubiquitin- and proteasome-mediated degradation of proteins plays an important role in plant defense signaling system (Dreher and Callis 2007; Goritsching et al. 2007). H_2O_2 triggers the up-regulation of ubiquitin precursor proteins, ubiquitin-conjugating enzymes (E2), and ubiquitin-protein ligases (E3) in tobacco (Vandenbeele et al. 2003).

ROS signaling system may act both in upstream and downstream of SA signaling (Slaymaker et al. 2002). ROS metabolism can affect the function of NPR1 by controlling NPR1 redox state (Mou et al. 2003). ROS may function in ethylene signaling system. An increase in H_2O_2 was able to trigger transcriptional changes of genes involved in the biosynthesis of ethylene. S-adenosyl-L-methionine synthetase is the first enzyme in the ethylene biosynthesis pathway and 1-aminocyclopropane-1-carboxylate (ACC) oxidase is the final stage enzyme involved in conversion of ACC to ethylene. Transcript tags coding for both the enzymes were up-regulated within 2 h of increase in H_2O_2, followed by an increase in ethylene-responsive proteins, such as EREBP/AP2 domain proteins (Vandenabeele et al. 2003). H_2O_2 mediates, at least in part, ABA responses including stomatal closure and gene expression (Guan et al. 2000; Pei et al. 2000).

In the genome of the maize smut pathogen *Ustilago maydis*, an ortholog of *YAP1* (for *Yeast AP-1-like*) from *Saccharomyces cerevisae* has been identified (Molina and Kahmann 2007). The gene is involved in degradation of H_2O_2. The *Yap1*-regulated genes include peroxidase genes and peroxidases are known to degrade H_2O_2. *Yap1* gene was found to be essential for virulence of *U. maydis* and deletion mutants of this gene were attenuated in virulence. These results suggest that the biotrophic pathogen *U. maydis* causes the disease by interfering with the H_2O_2 signaling pathway (Molina and Kahmann 2007). The biotrophic fungal pathogen *Blumeria graminis* f. sp. *hordei* elicits a burst of H_2O_2 in its host barley at sites of germ tube invasion. The fungus produces catalase during the infection process. The fungal *catB* gene encoding catalase has been characterized. Enhanced numbers of *catB* transcripts were detected at mature primary germ tube and appressorium germ tube stages in a susceptible host. Areas of H_2O_2 clearing were observed at sites of fungal invasion (Zhang et al. 2004). Collectively, these results suggest that the catalase activity of *B. graminis* f. sp. *hordei* may contribute to the fungal pathogenicity, probably by interfering with ROS signaling system (Zhang et al. 2004).

2.9.6 *Viral Pathogens May Inhibit Ubiquitin-Proteasome System to Induce Disease Development*

Ubiquitin- proteasome system plays an important role in plant defense signaling system (Dreher and Callis 2007; van den Burg et al. 2008; Sahana et al. 2012; Yao and Ndoja 2012; Yao et al. 2012; Zhang et al. 2012d). Proteasomes regulate the concentration of particular proteins and degrade misfolded proteins (Pickart and Eddins 2004). Proteins are targeted for degradation by the proteasome (Lee et al. 2011). Sahana et al. (2012) showed that a cell permeable proteasomal inhibitor facilitated an increase in *Papaya ringspot virus* (PRSV) accumulation in the host papaya plants. The PRSV viral protein HcPro was found to interact with the α1 subunit of the 20S proteasome, inhibiting the action of the 20S proteasome. The results suggest that inhibition of the host proteasome facilitates the virus accumulation in the host plant and the proteosomal catalytic activity is modulated by the viral protein. Collectively, these studies suggest that proteasome is involved in virus disease resistance and the potential viral pathogens inhibit the proteasome activity and induce the disease incidence.

References

Abramovitch RB, Janjusevic R, Stebbins CE, Martin GB (2006) Type III effector AvrPtoB requires intrinsic E3 ubiquitin ligase activity to suppress plant cell death and immunity. Proc Natl Acad Sci U S A 103:2851–2856

Ade J, DeYoung BJ, Golstein C, Innes RW (2007) Indirect activation of a plant nucleotide binding site-leucine-rich repeat protein by a bacterial protease. Proc Natl Acad Sci U S A 104:2531–2536

Adie BA, Perez-Perez J, Godoy M, Sanchez-Serrano JJ, Schmelz EA, Solano R (2007) ABA is an essential signal for plant resistance to pathogens affecting JA biosynthesis and activation of defenses in *Arabidopsis*. Plant Cell 19:1665–1681

Akagi A, Dandekar AM, Stotz HU (2011) Resistance of *Malus domestica* fruit to *Botrytis cinerea* depends on endogenous ethylene biosynthesis. Phytopathology 101:1311–1321

Akimoto-Tomiyama C, Furutani A, Tsuge S, Washington EJ, Nishizawa Y, Minami E, Ochiai H (2012) XopR, a type III effector secreted by *Xanthomonas oryzae* pv. *oryzae*, suppresses microbe-associated molecular pattern-triggered immunity in *Arabidopsis thaliana*. Mol Plant-Microbe Interact 25:505–514

Alamillo JM, Saénz P, Garcia JA (2006) Salicylic acid-mediated and RNA-silencing defense mechanisms cooperate in the restriction of systemic spread of plum pox virus in tobacco. Plant J 48:217–227

Alazem M, Lin K-Y, Lin N-S (2014) The abscisic acid pathway has multifaceted effects on the accumulation of *Bamboo mosaic virus*. Mol Plant Microbe Interact 27:177–189

Al-Daoud F, Cameron RK (2011) ANAC055 and ANAC092 comtribute non-redundantly in an EIN2-dependent manner to age-related resistance in Arabidopsis. Physiol Mol Plant Pathol 76:212–222

Ali GS, Prasad KVSK, Day I, Reddy ASN (2007) Ligand-dependent reduction in the membrane mobility of FLAGELLIN SENSITIVE2, an Arabidopsis receptor-like kinase. Plant Cell Physiol 48:1601–1611

Alkan N, Fluhr R, Prusky D (2012) Ammonium secretion during *Colletotrichum coccodes* infection modulates salicylic and jasmonic acid pathways of ripe and unripe tomato fruit. Mol Plant-Microbe Interact 25:85–96

Altenbach DD, Robatzek S (2007) Pattern recognition receptors: from the cell surface to intracellular dynamics. Mol Plant-Microbe Interact 20:1031–1039

Amzalek E, Cohen Y (2007) Comparative efficacy of systemic acquired resistance-inducing compounds against rust infection in sunflower plants. Phytopathology 97:179–186

Anand A, Uppalapati SR, Ryu C-M, Allen SN, Kang L, Tang Y, Mysore KS (2008) Salicylic acid and systemic acquired resistance play a role in attenuating crown gall disease caused by *Agrobacterium tumefaciens*. Plant Physiol 146:703–715

Anderson JP, Badruzsaufari E, Schenk PM, Manners JM, Desmond OJ, Ehlert C, Maclean DJ, Ebert PR, Kazan K (2004) Antagonistic interaction between abscisic acid and jasmonate-ethylene signaling pathways modulates defense gene expression and disease resistance in *Arabidopsis*. Plant Cell 16:3460–3479

Anfoka GH (2000) Benzo-(1,2,3)-thiadiazole-7-carbothioic acid-*S*-methyl ester induces systemic resistance in tomato (*Lycopersicon esculentum* Mill. cv. volledung) to cucumber mosaic virus. Crop Prot 19:401–405

Angra-Sharma R, Sharma DK (1999) Cytokinins in pathogenesis and disease resistance of *Pyrenophora teres*-barley and *Drechslera maydis*-maize interactions during early stages of infection. Mycopathogia 148:87–95

Argueso CT, Ferreira FJ, Epple P, To JPC, Hutchison CE, Schaller GE, Dangl JL, Kieber JJ (2012) Two-component elements mediate interactions between cytokinin and salicylic acid in plant immunity. PLoS Genet 8(1):e1002448

Asada K (2006) Production and scavenging of reactive oxygen species in chloroplasts and their functions. Plant Physiol 141:391–396

Asai T, Tena G, Plotnikova J, Willmann MR, Chiu W-L, Gómez-Gómez L, Boller T, Ausubel FM, Sheen J (2002) MAP kinase signaling cascade in *Arabidopsis* innate immunity. Nature 415:977–983

Asai S, Ohta K, Yoshioka H (2008) MAPK signaling regulates nitric oxide and NADPH oxidase-dependent oxidative bursts in *Nicotiana benthamiana*. Plant Cell 20:1390–1406

Aslam SN, Newman MA, Erbs G, Morissey KL, Chinchilla D, Boller T, Jensen TT, De Castro C, Iearno T, Molinaro A, Jackson RW, Knight MR, Cooper RM (2008) Bacterial polysaccharides suppress induced innate immunity by calcium chelation. Curr Biol 18:1078–1083

Aslam SN, Erbs G, Morrissey K, Newman M-A, Chinchilla D, Boller T, Molinaro A, Jackson RW, Cooper RM (2009) Microbe-associated molecular pattern (MAMP) signatures, synergy, size, and charge: influences on perception or mobility and host defense responses. Mol Plant Pathol 10:375–387

Asselbergh B, Achuo AE, Höfte M, Van Gijsegem F (2008a) Absicisic acid deficiency leads to rapid activation of tomato defence responses upon infection with *Erwinia chrysanthemi*. Mol Plant Pathol 9:11–24

Asselbergh B, De Vleesschauwer D, Höfte M (2008b) Global switches and fine-tuning ABA modulates plant pathogen defense. Mol Plant Microbe Interact 21:709–719

Atsumi G, Kagaya U, Kitazawa H, Nakahara KS, Uyeda I (2009) Activation of the salicylic acid signaling pathway enhances *Clover yellow vein virus* virulence in susceptible pea cultivars. Mol Plant-Microbe Interact 22:166–175

Aziz A, Poinssot B, Daire X, Adrian M, Bezier A, Lambert B, Joubert JM, Pugin A (2003) Laminarin elicits defense responses in grapevine and induces protection against *Botrytis cinerea* and *Plasmopara viticola*. Mol Plant-Microbe Interact 16:1118–1128

Bae H, Kim MS, Sicher RC, Bae H-J, Bailey BA (2006) Necrosis- and ethylene-inducing peptide from *Fusarium oxysporum* induces a complex cascade of transcripts associated with signal transduction and cell death in *Arabidopsis*. Plant Physiol 141:1056–1067

Balaji V, Mayrose M, Scherf O, Jacob-Hirsch J, Eichenlaub R, Iraki N, Manulis-Sasson S, Rechavi G, Barash I, Sessa G (2008) Tomato transcriptional changes in response to *Clavibacter michiganensis* reveal a role for ethylene in disease development. Plant Physiol 146:1797–1809

Balbi V, Devoto A (2008) Jasmonate signaling network in *Arabidopsis thaliana*: crucial regulatory nodes and new physiological scenarios. New Phytol 177:301–318

Bartetzko V, Sonnewald S, Vogel F, Hartner K, Stadler R, Hammes UZ, Bornke F (2009) The *Xanthomonas campestris* pv. *vesicatoria* type III effector protein XopJ inhibits protein secretion: evidence for interference with cell wall-associated defense responses. Mol Plant-Microbe Interact 22:655–664

Beckers GJM, Jaskiewicz M, Liu Y, Underwood WR, He Y, Zhang S, Conrath U (2009) Mitogen-activated protein kinases 3 and 6 are required for full priming of stress responses in *Arabidopsis thaliana*. Plant Cell 21:944–953

Berrocal-Lobo M, Molina A (2004) Ethylene response factor 1 mediates Arabidopsis resistance to the soilborne fungus *Fusarium oxysporum*. Mol Plant-Microbe Interact 17:763–770

Berrocal-Lobo M, Molina A, Solano R (2002) Constitutive expression of ETHYLENE-RESPONSE-FACTOR1 in *Arabidopsis* confers resistance to several necrotrophic fungi. Plant J 29:23–32

Besson-Bard A, Courtois C, Gauthier A, Dahan J, Dobrowolska G, Jeandroz S, Pugin A, Wendehenne D (2008) Nitric oxide in plants: production and cross-talk with Ca^{2+} signaling. Mol Plant 1:218–228

Bethke PC, Badger MR, Jones RL (2004) Apoplastic synthesis of nitric oxide by plant tissues. Plant Cell 16:332–341

Bethke G, Pecher P, Eschen-Lippold L, Tsuda K, Katagiri F, Glazebrook J, Scheel D, Lee J (2012) Activation of the *Arabidopsis thaliana* mitogen-activated protein kinase MPK11 by the flagellin-derived elicitor peptide, flg22. Mol Plant-Microbe Interact 25:471–480

Bi D, Cheng YT, Li X, Zhang Y (2010) Activation of plant immune responses by a gain-of-function mutation in an atypical receptor-like kinase. Plant Physiol 153:1771–1779

Block A, Li G, Fu ZQ, Alfano JR (2008) Phytopathogen type III effector weaponry and their plant targets. Curr Opin Plant Biol 11:396–403

Boller T, He SY (2009) Innate immunity in plants: an arms race between pattern recognition receptors in plants and effectors in microbial pathogens. Science 324:742–744

Bolwell GP, Davies DR, Gerrish C, Auh C-K, Murphy TM (1998) Comparative biochemistry of the oxidative burst produced by rose and French bean cells reveals two distinct mechanisms. Plant Physiol 116:1379–1385

Borsics T, Webb D, Andeme-Ondzihi C, Staehelin A, Christopher DA (2007) The cyclic nucleotide-gated calmodulin-binding channel AtCNGC10 localizes to the plasma membrane and influences numerous growth responses and starch accumulation in *Arabidopsis thaliana*. Planta 225:563–573

Bos JI, Chaparro-Garcia A, Quesada-Ocampo LM, McSpadden Gardener BB, Kamoun S (2009) Distinct amino acids of the *Phytophthora infestans* effector AVR3a condition activation of R3a hypersensitivity and suppression of cell death. Mol Plant-Microbe Interact 22:269–281

Bos JIB, Armstrong MR, Gilroy EM, Boevink PC, Hein I, Taylor RM, Zhendong T, Engelhardt S, Vetukuri RR, Harrower B, Dixelius C, Bryan G, Sadanandom A, Whisson SC, Kamoun S, Birch PRJ (2010) *Phytophthora infestans* effector AVR3a is essential for virulence and manipulates plant immunity by stabilizing host E3 ligase CMPG1. Proc Natl Acad Sci U S A 107:9909–9914

Boudart G, Charpentier M, Lafitte C, Martinez Y, Jauneau A, Gaulin E, Esquerré-Tugayé M-T, Dumas B (2003) Elicitor activity of a fungal endopolygalacturonase in tobacco requires a functional catalytic site and cell wall localization. Plant Physiol 131:93–101

Boudsocq M, Willmann MR, McCormack M, Lee H, Shan L, He P, Bush J, Cheng SH, Sheen J (2010) Differential innate immune signaling via Ca^{2+} sensor protein kinases. Nature 464:418–422

Boutrot F, Segonzac C, Chang KN, Qiao H, Ecker JR, Zipfel C (2010) Direct transcriptional control of the *Arabidopsis* immune receptor FLS2 by the ethylene-dependent transcription factors EIN3 and EIL1. Proc Natl Acad Sci U S A 107:14502–14507

Brooks DM, Bender CL, Kunkel BN (2005) The *Pseudomonas syringae* phytotoxin coronatine promotes virulence by overcoming salicylic acid-dependent defenses in *Arabidopsis thaliana*. Mol Plant Pathol 6:629–639

Cabral A, Oome S, Sander N, Küfner I, Nürnberger T, van den Ackerveken G (2012) Nontoxic Nep-1 like proteins of the downy mildew pathogen *Hyaloperonospora arabidopsis*: repression of necrosis-inducing activity by a surface-exposed region. Mol Plant-Microbe Interact 25:697–708

Canet JV, Dobón A, Tornero P (2012) Non-recognition-of BTH4, an *Arabidopsis* mediator subunit homolog, is necessary for development and response to salicylic acid. Plant Cell 24:4220–4235

Cernadas RA, Benedetti CE (2009) Roles of auxin and gibberellin in citrus canker development and in the transcriptional control of cell-wall remodeling genes modulated by *Xanthomonas axonopodis* pv. *citri*. Plant Sci 177:190–195

Chaparro-Garcia A, Wilkinson RC, Gimenez-Ibanez S, Findlay K, Coffey MD, Zipfel C, Rathjen JP, Kamoun S, Schornack S (2011) The receptor-like kinase SERK3/BAK1 is required for basal resistance against the late blight pathogen *Phytophthora infestans* in *Nicotiana benthamiana*. PLoS One 6(1):e16608

Chapman EJ, Estelle M (2009) Mechanism of auxin-regulated gene expression in plants. Annu Rev Genet 43:265–285

Chen X, Chern M, Canlas PE, Jiang C, Ruan D, Cao P, Ronald PC (2010a) A conserved threonine residue in the juxtamembrane domain of the XA21 pattern recognition receptor is critical for kinase autophosphorylation and XA21-mediated immunity. J Biol Chem 285:10454–10463

Chen X, Chern M, Canlas PE, Jiang C, Ruan D, Cao P, Ronald PC (2010b) A conserved threonine residue in the juxtamembrane domain of the XA21 pattern recognition receptor is critical for kinase autophosphorylation and XA21-mediated immunity. J Biol Chem 285:10454–10463

Chen X, Chern M, Canlas PE, Ruan D, Jiang C, Ronald PC (2010c) An ATPase promotes autophosphorylation of the pattern recognition receptor XA21 and inhibits XA21-mediated immunity. Proc Natl Acad Sci U S A 107:8029–8034

Cheng B, Yu X, Ma Z, Dong S, Dou D, Wang Y, Zheng X (2012) *Phytophthora sojae* effector Avh331 suppresses the plant defence response by disturbing the MAPK signaling pathway. Physiol Mol Plant Pathol 77:1–9

Chern M, Fitzgerald HA, Canlas PE, Navarre DA, Ronald PC (2005) Overexpression of a rice NPR1 homolog leads to constitutive activation of defense response and hypersensitivity to light. Mol Plant-Microbe Interact 18:511–520

Chinchilla D, Boller T, Robatzek S (2007a) Flagellin signaling in plant immunity. Adv Exp Med Biol 598:358–371

Chinchilla D, Zipfel C, Robatzek S, Kemmerling B, Nürnberger T, Jones JDG, Felix G, Boller T (2007b) A flagellin-induced complex of the receptor FLS2 and BAK1 initiates plant defence. Nature 448:497–500

Choi DS, Hwang BK (2011) Proteomics and functional analyses of pepper *Abscisic acid-responsive 1 (ABR1)*, which is involved in cell death and defense signaling. Plant Cell 23:823–842

Choi J, Huh SU, Kojima M, Sakakibara H, Paek K-H, Hwang I (2010) The cytokinin-activated transcription factor ARR2 promotes plant immunity via TGA3/NPR1-dependent salicylic acid signaling in *Arabidopsis*. Dev Cell 19:284–295

Clarke SF, Burritt DJ, Guy PL (1998) Influence of plant hormones on virus replication and pathogenesis-related proteins in *Phaseolus vulgaris* L infected with white clover mosaic potexvirus. Physiol Mol Plant Pathol 53:195–207

Clay NK, Adio AM, Denoux C, Jander G, Ausubel FM (2009) Glucosinolate metabolites required for an *Arabidopsis* innate immune response. Science 323:95–101

Conrath U (2011) Molecular aspects of defence priming. Trends Plant Sci 16:524–531

Cools HJ, Ishii H (2002) Pre-treatment of cucumber plants with acibenzolar-S-methyl systemically primes a phenylalanine ammonia lyase gene (*PAL1*) for enhanced expression upon attack with a pathogenic fungus. Physiol Mol Plant Pathol 61:273–280

Crawford NM, Galli M, Tischner R, Heimer YM, Okamoto M, Mack A (2006) Response to Zemojtel *et al*: plant nitric oxide synthase: back to square one. Trends Plant Sci 11:526–527

Cui J, Bahrami AK, Pringle EG, Hernandez-Guzman G, Bender CL, Pierce NE, Ausubel FM (2005) *Pseudomonas syringae* manipulates systemic plant defenses against pathogens and herbivores. Proc Natl Acad Sci U S A 102:1791–1796

Curvers K, Seifi H, Mouille G, de Rycke R, Asselbergh B, Van Hecke A, Vanderschaeghe D, Höfte H, Callewaert N, Van Brueusegem F, Höfte M (2010) Abscisic acid deficiency causes changes in cuticle permeability and pectin composition that influence tomato resistance to *Botrytis cinerea*. Plant Physiol 154:847–860

D'Amelio R, Berta G, Gamalero E, Massa N, Avidano L, Cantamessa S, D'Agostino G, Bosco D, Marzachi C (2011) Increased plant tolerance against chrysanthemum yellows phytoplasma (*Candidatus Phytoplasma asteris*) following double inoculation with *Glomus mosseae* BEG12 and *Pseudomonas putida* SIPflRif. Plant Pathol 60:1014–1022

DebRoy S, Thilmony R, Kwack Y-B, Nomura K, He S-Y (2004) A family of conserved bacterial effectors inhibits salicylic acid-mediated basal immunity and promotes disease necrosis in plants. Proc Natl Acad Sci U S A 101:9927–9932

Deepak SA, Ishii H, Park P (2006) Acibenzolar-S-methyl primes cell wall strengthening genes and reactive oxygen species forming/scavenging enzymes in cucumber after fungal pathogen attack. Physiol Mol Plant Pathol 69:52–61

de Jonge R, van Esse HP, Kombrink A, Shinya T, Desaki Y, Bours R, van der Krol S, Shibuya N, Joosten MHAJ, Thomma BPHJ (2010) Conserved fungal LysM effector Ecp6 prevents chitin-triggered immunity in plants. Science 329:953–955

de Torres M, Mansfield JW, Grabov N, Brown IR, Ammouneh H, Tsiamis G, Forsyth A, Robatzek S, Grant M, Boch J (2006) *Pseudomonas syringae* effector AvrPtroB suppresses basal defence in Arabidopsis. Plant J 47:368–382

de Torres-Zabala M, Truman W, Bennett MH, Lafforgue G, Mansfield JW, Rodriguez Egea P, Bogre L, Grant M (2007) *Pseudomonas syringae* pv. *tomato* hijacks the *Arabidopsis* abscisic acid signaling pathway to cause disease. EMBO J 26:1434–1443

De Vleesschauwer D, Djavaheri M, Bakker PAHM, Höfte M (2008) *Pseudomonas fluorescens* WCS374r-induced systemic resistance in rice against *Magnaporthe oryzae* is based on pseudobactin-mediated priming for a salicylic acid-repressable multifaceted defense response. Plant Physiol 148:1996–2012

De Vleesschauwer D, Yang Y, Vera Cruz C, Höfte M (2010) Abscisic acid –induced resistance against the brown spot pathogen *Cochliobolus miyabeanus* in rice involves MAP kinase-mediated repression of ethylene signaling. Plant Physiol 152:2036–2052

De Vleesschauwer D, Van Buyten E, Satoh K, Balidion J, Mauleon R, Choi I-R, Vera-Cruz C, Kikuchi S, Höfte M (2012) Brassinosteroids antagonize gibberellin- and salicylate-mediated root immunity in rice. Plant Physiol 158:1833–1846

De Vos M, Van Oosten VR, Van Poecke RMP, Van Pelt JA, Pozo MJ, Mueller MJ, Buchala AJ, Métraux J-P, Van Loon LC, Dicke M, Pieterse CMJ (2005) Signal signature and transcriptome changes of *Arabidopsis* during pathogen and insect attack. Mol Plant-Microbe Interact 18:923–937

Delker C, Zolman BK, Miersch O, Wasternack C (2007) Jasmonate biosynthesis in *Arabidopsis thaliana* requires peroxisomal β-oxidation enzymes – Additional proof by properties of *pex6* and *aim1*. Phytochemistry 68:1542–1650

Delledonne M (2005) NO news is good news for plants. Curr Opin Plant Biol 8:390–396

Dempsey DA, Vlot AC, Wildermuth MC, Klessig DF (2011) Salicylic acid biosynthesis and metabolism. Tha Arabidopsis Book e0156. doi:10.1199/tab.0156

Denancé N, Sánchez-Vallet A, Goffner D, Molina A (2013) Disease resistance or growth: the role of plant hormones in balancing immune responses and fitness costs. Front Plant Sci 4:Article 155. p1–12

Denoux C, Galletti R, Mammarella N, Gopalan S, Werck D, De Lorenzo G, Ferrari S, Ausubel FM, Dewdney J (2008) Activation of defense response pathways by OGs and Flg22 elicitors in Arabidopsis seedlings. Mol Plant 1:423–445

Desikan R, Hancock JT, Ichimura K, Shinozaki K, Neill SJ (2001) Harpin induces activation of the *Arabidopsis* mitogen-activated protein kinases AtMPK4 and AtMPK6. Plant Physiol 126:1579–1587

Ding X, Cao Y, Huang L, Zhao C, Li X, Wang S (2008) Activation of the indole-3-acetic acid-amido synthetase GH3-8 suppresses expansin expression and promotes salicylate- and jasmonate-independent basal immunity in rice. Plant Cell 20:228–240

Doares SH, Narvaez-Vasquez J, Conconi A, Ryan CA (1995a) Salicylic acid inhibits synthesis of proteinase inhibitors in tomato leaves induced by systemin and jasmonic acid. Plant Physiol 108:1741–1746

Doares SH, Syrovets T, Weiler EW, Ryan CA (1995b) Oligogalacturonides and chitosan activate plant defensive genes through the octadecanoid pathway. Proc Natl Acad Sci U S A 92:4095–4098

Dodds PN, Rathjen JP (2010) Plant immunity: towards an integrated view of plant-pathogen interactions. Nat Rev Genet 11:539–548

Dodds PN, Lawrence GJ, Catanzariti AM, Teh T, Wang CIA, Ayliffe MA, Kobe B, Ellis JG (2006) Direct protein interaction underlies gene-for-gene specificity and coevolution of the flax resistance genes and flax rust avirulence genes. Proc Natl Acad Sci U S A 103:8888–8893

Dong J, Chen C, Chen Z (2003) Expression profiles of the Arabidopsis WRKY gene superfamily during plant defense response. Plant Mol Biol 51:21–37

Dreher K, Callis J (2007) Ubiquitin, hormones and biotic stress in plants. Ann Bot 99:787–822

Droillard MJ, Boudsocq M, Barbier-Brygoo H, Laurière C (2004) Involvement of MPK4 in osmotic stress response pathways in cell suspensions and plantlets of *Arabidopsis thaliana*: activation by hypoosmolarity and negative role in hyperosmolarity tolerance. FEBS Lett 574:42–48

Durrant WE, Dong X (2004) Systemic acquired resistance. Annu Rev Phytopathol 42:185–209

Eckardt NA (2008) Chitin signaling in plants: insights into the perception of fungal pathogens and rhizobacterial strains. Plant Cell 20:241–243

El Oirdi M, Bouarab K (2007) Plant signaling components EDS1 and SGT1 enhance disease caused by the necrotrophic pathogen *Botrytis cinerea*. New Phytol 175:131–139

El Rahman TA, El Oirdi M, Gonzalez-Lamothe R, Bouarab K (2012) Necrotrophic pathogens use salicylic acid signaling pathway to promote disease development in tomato. Mol Plant-Microbe Interact 25:1584–1593

Engelhardt S, Boevink PC, Armstrong MR, Ramos MB, Hein I, Birch PRJ (2012) Relocalization of late blight resistance protein R3a to endosomal compartments and required for the immune response. Plant Cell 24:5142–5158

Eulgem T, Weighman VJ, Chang H-S, McDowell JM, Holub EB, Glazebrook J, Zhu T, Dangl JL (2004) Gene expression signatures from three genetically separable resistance gene signaling pathways for downy mildew resistance. Plant Physiol 135:1129–1144

Fabro G, Di Rienzo JA, Voigt CA, Savchenko T, Dehesh K, Somerville S, Alvarez ME (2008) Genome-wide expression profiling Arabidopsis at the stage of *Golovinomyces cichoracearum* haustorium formation. Plant Physiol 146:1421–1439

Fan W, Dong X (2002) In vivo interaction between NPR1 and transcription factor TGA2 leads to salicylic acid-mediated gene activation in *Arabidopsis*. Plant Cell 14:1377–1389

Fan J, Hill L, Crooks C, Doerner P, Lamb C (2009) Abscisic acid has a key role in modulating diverse plant-pathogen interactions. Plant Physiol 150:1750–1761

Fayza AF, Sabrey YMM (2006) Induction of resistance in *Phaseolus vulgaris* against TNV by salicylic acid and kinetin. Int J Agric Biol 8:47–51

Feng DX, Tasset C, Hanemian M, Barlet X, Hu J, Trèmousaygue D, Deslandes L, Marco Y (2012) Biological control of bacterial wilt in *Arabidopsis thaliana* involves abscissic acid signalling. New Phytol 194:1035–1045

Ferrari S, Plotnokova JM, De Lorenzo G, Ausubel FM (2003) *Arabidopsis* local resistance to *Botrytis cinerea* involves salicylic acid and camalexin and requires *EDS4* and *PAD2*, but not *SID2*, *EDS5* or *PAD4*. Plant J 35:193–205

Foissner L, Wendehenne D, Langebartels C, Durner J (2000) In vivo imaging of an elicitor-induced nitric oxide burst in tobacco. Plant J 23:817–824

Fu J, Liu H, Li Y, Yu H, Li X, Xiao J, Wang S (2011) Manipulating broad-spectrum disease resistance by suppressing pathogen-induced auxin accumulation in rice. Plant Physiol 155:589–602

Fujii H, Chinnusamy V, Rodrigues A, Rubio S, Antoni R, Park SY, Cutler SR, Sheen J, Rodriguez PL, Zhu JK (2009) *In vitro* reconstitution of an abscisic acid signaling pathway. Nature 462:660–664

Fujiwara S, Tanaka N, Kaneda T, Takayama S, Isogai A, Che F-S (2004) Rice cDNA microarray-based gene expression profiling of the response to flagellin perception in cultured rice cells. Mol Plant-Microbe Interact 17:986–998

Fung RWM, Gonzalo M, Fekete C, Kovacs LG, He Y, Marsh E, McIntyre LM, Schchtman DP, Qiu W (2008) Powdery mildew induces defense-oriented reprogramming of the transcriptome in a susceptible but not in a resistant grapevine. Plant Physiol 146:236–249

Galletti R, Ferrari S, De Lorenzo G (2011) *Arabidopsis* MPK3 and MPK6 play different roles in basal and oligogalacturonide- or flagellin-induced resistance against *Botrytis cinerea*. Plant Physiol 157:804–814

Gao M, Liu J, Zhang Z, Cheng F, Chen S, Zhang Y (2008) MEKK1, MKK1/MKK2 and MPK4 function together in a mitogen-activated protein kinase cascade to regulate innate immunity in plants. Cell Res 18:1190–1198

Gao X, Brodhagen M, Isakeit T, Brown SH, Göbel C, Betran J, Feussner I, Keller NP, Kolomiets MV (2009) Inactivation of the lipoxygenae ZmLOX3 increases susceptibility of maize to *Aspergillus* spp. Mol Plant Microbe Interact 22:222–231

Gao Y, Li T, Liu Y, Ren C, Zhao Y, Wang M (2010a) Isolation and characterization of gene encoding G protein α subunit protein responsive to plant hormones and abiotic stresses in *Brassica napus*. Mol Biol Rep 37:3957–3965

Gao Y, Li T, Zhao Y, Ren C, Liu Y, Wang M (2010b) Cloning and characterization of a G protein β subunit gene responsive to plant hormones and abiotic stresses in *Brassica napus*. Plant Mol Biol Rep 28:450–459

Garcia-Andrade J, Ramirez V, Flors V, Vera P (2011) Arabidopsis *ocp3* mutant reveals a mechanism linking ABA and JA to pathogen-induced callose deposition. Plant J 67:783–794

Garcion C, Lohmann A, Lamodiere E, Catinot J, Buchala A, Doermann P, Métraux J-P (2008) Characterization and biological function of the *ISOCHORISMATE SYNTHASE2* gene of the Arabidopsis. Plant Physiol 147:1279–1287

Gelli A, Higgins VJ, Blumwald E (1997) Activation of plant plasma membrane Ca^{2+}-permeable channels by race-specific fungal elicitors. Plant Physiol 113:269–279

Genger RK, Jurkowski GI, McDowell JM, Lu H, Jung HW, Greenberg JT, Bent AF (2008) Signaling pathways that regulate the enhanced disease resistance of *Arabidopsis* "Defense, No death" mutants. Mol Plant-Microbe Interact 21:1285–1296

Geraats BPJ, Bakker PAHM, Lawrence CB, Achuo EA, Höfte M, van Loon LC (2003) Ethylene-insensitive tobacco shows differentially altered susceptibility to different pathogens. Phytopathology 93:813–821

Gimenez-Ibanez S, Solano R (2013) Nuclear jasmonate and salicylate signaling and crosstalk in defense against pathogens. Front Plant Sci 4: Article 72:1–11

Gimenez-Ibanez S, Hann DR, Ntoukakis V, Petutschnig E, Lipka V, Rathjen JP (2009a) AvrPtoB targets the LysM receptor kinase CERK1 to promote bacterial virulence on plants. Curr Biol 19:423–429

Gimenez-Ibanez S, Ntoukakis V, Rathjen JP (2009b) The LysM receptor kinase CERK1 mediates bacterial perception in Arabidopsis. Plant Signal Behav 4:539–541

Glazebrook J (2005) Contrasting mechanisms of defense against necrotrophic pathogens. Annu Rev Phytopathol 43:205–227

Glazebrook J, Chen W, Estes B, Chang HS, Nawrath C, Metraux JP, Zhu T, Katagiri F (2003) Topology of the network integrating salicylate and jasmonate signal transduction derived from global expression phenotyping. Plant J 34:217–228

Goff KE, Ramonell KM (2007) The role and regulation of receptor-like kinases in plant defense. Gene Regul Syst Biol 1:167–175

Göhre V, Spallek T, Häweker H, Mersmann S, Mentzel T, Boller T, de Torres M, Mansfield JW, Robatzek S (2008) Plant pattern-recognition receptor FLS2 is directed for degradation by the bacterial ubiquitin ligase AvrPtoB. Curr Biol 18:1824–1832

Gómez-Gómez L, Bauer Z, Boller T (2001) Both the extracellular leucine-rich repeat domain and the kinase activity of FLS2 are required for flagellin binding and signaling in *Arabidopsis*. Plant Cell 13:1155–1163

González-Lamothe R, El Oirdi M, Brisson N, Bouarab K (2012) The conjugated auxin indole-3-acetic acid-aspartic acid promotes plant disease development. Plant Cell 24:762–777

Goritsching S, Zhang Y, Li X (2007) The ubiquitin pathway is required for innate immunity in *Arabidopsis*. Plant J 49:540–551

Gray WM, Muskett PR, Chuang H-W, Parker JE (2003) *Arabidopsis* SGT1b is required for SCF^{TIR1}-mediated auxin response. Plant Cell 15:1310–1319

Grefen C, Städele K, Ruzicka K, Obrdlik P, Harter K, Horák J (2008) Subcellular localization and in vivo interactions of the *Arabidopsis thaliana* ethylene receptor family members. Mol Plant 1:308–320

Großkinsky DK, Naseem M, Abdelmohsen UR, Plickert N, Engelke T, Griebel T, Zeier J, Novák O, Strnad M, Pfeifhofer H, Graaff EVD, Simon U, Roitsch T (2011) Cytokinins mediate resistance against *Pseudomonas syringae* in tobacco through increased antimicrobial phytoalexin synthesis independent of salicylic acid signaling. Plant Physiol 157:815–830

Gu Y, Wang Z, Yang Z (2004) ROP/RAC GTPase: an old new master regulator for plant signaling. Curr Opin Plant Biol 7:527–536

Guan LM, Zhao J, Scandalios JG (2000) *Cis*-elements and *trans*-factors that regulate expression of the maize *Cat1* antioxidant gene response to ABA and osmotic stress: H_2O_2 is the likely intermediary signaling molecule for the response. Plant J 22:87–95

Guimarães RL, Stotz HU (2004) Oxalate production by *Sclerotinia sclerotiorum* deregulates guard cells during infection. Plant Physiol 136:3703–3711

Hamel L-P, Beaudoin N (2010) Chitooligosaccharide sensing and downstream signaling contrasted outcomes in pathogenic and beneficial plant-microbe interactions. Planta 232:787–806

Hann DR, Rathjen JP (2007) Early events in the pathogenecity of *Pseudomonas syringae* on *Nicotiana benthamiana*. Plant J 49:607–618

Hann DR, Gimenez-Ibanez S, Rathjen JP (2010) Bacterial virulence effectors and their activities. Curr Opin Plant Biol 13:388–393

Hauck P, Thilmony R, He SY (2003) A *Pseudomonas syringae* type III effector suppresses cell wall-based extracellular defense in susceptible Arabidopsis plants. Proc Natl Acad Sci U S A 100:8577–8582

Hayashi K-I (2012) The interaction and integration of auxin signaling components. Plant Cell Physiol 53:965–975

He P, Shan L, Lin NC, Martin GB, Kemmerling B, Nurnberger T, Sheen J (2006) Specific bacterial suppressors of MAMP signaling upstream of MAPKKK in *Arabidopsis* innate immunity. Cell 125:563–575

He P, Shan L, Sheen J (2007) Elicitation and suppression of microbe-associated molecular pattern-triggered immunity in plant-microbe interactions. Cell Microbiol 9:1385–1396

Heese A, Hann DR, Gimenez-Ibanez S, Jones AME, He K, Li J, Schroeder JI, Peck SC, Rathjen JP (2007) The receptor-like kinase SERK3/BAK1 is a central regulator of innate immunity in plants. Proc Natl Acad Sci U S A 104:12217–12222

Hettenhausen C, Baldwin IT, Wu J (2012) Silencing *MPK4* in *Nicotiana attenuata* enhances photosynthesis and seed production but compromises abscisic acid-induced stomatal closure and guard cell-mediated resistance to *Pseudomonas syringae* pv. *tomato* DC3000. Plant Physiol 158:759–776

Hogenhout SA, van der Hoorn RAI, Terauchi R, Kamou S (2009) Emerging concepts in effector biology of plant-associated organisms. Mol Plant-Microbe Interact 22:115–122

Hondo D, Hase S, Kanayama Y, Yoshikawa N, Takenaka S, Takahashi H (2007) The *LeATL6*-associated ubiquitin/proteasome system may contribute to fungal elicitor-activated defense response via the jasmonic acid-dependent signaling pathway in tomato. Mol Plant-Microbe Interact 20:72–81

Huffaker A, Dafoe NJ, Schmelz EA (2011) ZmPep1, an ortholog of Arabidopsis elicitor peptide 1, regulates maize innate immunity and enhances disease resistance. Plant Physiol 155:1325–1338

Hwang IS, Hwang BK (2011) The pepper mannose-binding lectin gene *CaMBL1* is required to regulate cell death and defense responses to microbial pathogens. Plant Physiol 155:447–463

Iriti M, Faoro F (2008) Abscisic acid is involved in chitosan-induced resistance to *Tobacco necrosis virus* (TNV). Plant Physiol Biochem 46:1106–1111

Iriti M, Faoro F (2009) Chitosan as a MAMP, searching for a PRR. Plant Signal Behav 4:66–68

Ishikawa A (2009) The Arabidopsis G-protein beta-subunit is required for defense response against *Agrobacterium tumefaciens*. Biosci Biotechnol Biochem 73:47–52

Iwai T, Miyasaka A, Seo S, Ohashi Y (2006) Contribution of ethylene biosynthesis for resistance to blast fungus infection in young rice plants. Plant Physiol 142:1202–1215

Janjusevic R, Abramovitch RB, Martin GB, Stebbino CE (2006) A bacterial inhibitor of host programmed cell death defenses is an E3 ubiquitin ligase. Science 311:222–226

Jayaraj J, Rahman M, Wan A, Punja ZK (2009) Enhanced resistance to foliar fungal pathogens in carrot by application of elicitors. Ann Appl Biol 155:71–80

Jelenska J, Yao N, Vinatzer BA, Wright CM, Brodsky JL, Greenberg JT (2007) A J domain virulence effector of *Pseudomonas syringae* remodels host chloroplasts and suppresses defenses. Curr Biol 17:499–508

Jirage D, Tootle TL, Reuber TL, Frost LN, Feys BJ, Parker JE (1999) *Arabidopsis thaliana* PAD4 encodes a lipase-like gene that is important for salicylic acid signaling. Proc Natl Acad Sci U S A 96:13583–13588

Johansson A, Staal J, Dixelius C (2006) Early responses in the *Arabidopsis- Verticillium longisporum* pathosystem are dependent on *NDR1*, JA- and ET-associated signals via cytosolic NPR1 and *RFO1*. Mol Plant Microbe Interact 19:958–969

Johnson C, Mhatre A, Arias J (2008) NPR1 preferentially binds to the DNA-inactive form of *Arabidopsis* TGA2. Biochim Biophys Acta 1779:583–589

Jung HW, Tschaplinski TJ, Wang L, Glazebrook J, Greenberg JT (2009) Priming in systemic plant immunity. Science 324:89–91

Kachroo P, Yoshioka K, Shah J, Dooner HK, Klessig DF (2000) Resistance to turnip crinkle virus in *Arabidopsis* is regulated by two host genes and is salicylic acid dependent but NPR1, ethylene and jasmonate independent. Plant Cell 12:677–690

Kachroo P, Shanklin J, Shah J, Whittle EJ, Klessig DF (2001) A fatty acid desaturase modulates the activation of defense signaling pathways in plants. Proc Natl Acad Sci U S A 98:9448–9453

Kachroo P, Kachroo A, Lapchyk L, Hildebrand D, Klessig DF (2003a) Restoration of defective cross talk in *ssi2* mutants: role of salicylic acid, jasmonic acid, and fatty acids in *SSI2*-mediated signaling. Mol Plant-Microbe Interact 16:1022–1029

Kachroo A, Lapchyk L, Fukushige H, Hildebrand D, Klessig D, Kachroo P (2003b) Plastidial fatty acid signaling modulates salicylic acid- and jasmonic acid-mediated defense pathways in the Arabidopsis *ssi2* mutant. Plant Cell 15:2952–2965

Kanzaki H, Saitoh H, Takahashi Y, Berberich T, Ito A, Kamoun S, Terauchi R (2008) NbLRK1, a lectin-like receptor kinase protein of *Nicotiana benthamiana*, interacts with *Phytophthora infestans* INF1 elicitin and mediates INF1-induced cell death. Planta 228:977–987

Katsir L, Schilmiller A, Staswick P, He SY, Howe GA (2008) COI1 is a critical component of a receptor for jasmonate and the bacterial virulence factor coronatine. Proc Natl Acad Sci U S A 105:7100–7105

Kazan K, Manners JM (2009) Linking development to defense: auxin in plant-pathogen interactions. Trends Plant Sci 14:373–382

Keates SE, Kostman TA, Anderson JD, Bailey BA (2003) Altered gene expression in three plant species in response to treatment with Nep1, a fungal protein that causes necrosis. Plant Physiol 132:1610–1622

Kemmerling B, Schwedt A, Rodriguez P, Mazzotta S, Frank M, Qamar SA, Mengiste T, Betsuyaku S, Parker JE, Müssig C, Thomma BPHJ, Albrecht C, de Vries SC, Hirt H, Nürnberger T (2007) The BRI1-associated kinase 1, BAK1, has a brassinolide-independent role in plant cell-death control. Curr Biol 17:1116–1122

Khang CH, Berruyer R, Giraldo MC, Kankanala P, Park SY, Czymmek K, Kang S, Valent B (2010) Translocation of *Magnaporthe oryzae* effectors into rice cells and their subsequent cell-to-cell movement. Plant Cell 22:1388–1403

Kidd BN, Cahill DM, Manners JM, Schenk PM, Kazan K (2011) Diverse roles of the Mediator complex in plants. Semin Cell Dev Biol 22:741–748

Kiirika LM, Bergmann HF, Schikowsky C, Wimmer D, Korte J, Schmitz U, Niehaus K, Colditz F (2012) Silencing of the Rac1 GTPase *MtROP9* in *Medicago truncatula* stimulates early mycorrhizal and oomycete root colonizations but negatively affects rhizobial infection. Plant Physiol 159:501–516

Kim KC, Fan B, Chen Z (2006) Pathogen-induced *Arabidopsis* WRKY7 is transcriptional repressor and enhances plant susceptibility to *Pseudomonas syringae*. Plant Physiol 142:1180–1192

Kim S-Y, Kim Y-C, Seong ES, Lee Y-H, Park M, Choi D (2007) The chili pepper CaATL1: an AT-hook motif-containing transcription factor implicated in defence responses against pathogens. Mol Plant Pathol 8:761–771

Kim TH, Böhmer M, Hu H, Nishimura N, Schroeder JI (2010) Guard cell signal transduction network: advances in understanding abscisic acid, CO_2, and Ca^{2+} signaling. Annu Rev Plant Biol 61:561–591

King SR, McLellan H, Boevink PC, Armstrong MR, Bukharova T, Sukarta O, Win J, Kamoun S, Birch PR, Banfield MJ (2014) *Phytophthora infestans* RXLR effector PexRD2 interacts with host MAPKKK$_\varepsilon$ to suppress plant immune signaling. Plant Cell 26:1345–1359

Kinoshita T, Caño-Delgado A, Seto H, Hiranuma S, Fujioka S, Yoshida S, Chory J (2005) Binding of brassinosteroids to the extracellular domain of plant receptor kinase BRI1. Nature 433:167–171

Klarzynski O, Plesse B, Joubert J-M, Yvin J-C, Kopp M, Kloareg B, Fritig B (2000) Linear β-1,3 glucans are elicitors of defense responses in tobacco. Plant Physiol 124:1027–1037

Knecht K, Seyffarth M, Desel C, Thurau T, Sherameti I, Lou B, Oelmüller R, Cai D (2010) Expression of *BvGLP-1* encoding a germin-like protein from sugar beet in *Arabidopsis thaliana* leads to resistance against phytopathogenic fungi. Mol Plant-Microbe Interact 23:446–457

Knoth C, Ringler J, Dangl JL, Eulgem T (2007) *Arabidopsis* WRKY70 is required for full *RPP4*-mediated disease resistance and basal defense against *Hyaloperonospora parasitica*. Mol Plant-Microbe Interact 20:120–128

Koga H, Dohi K, Mori M (2004) Abscisic acid and low temperatures suppress the whole plant-specific resistance reaction of rice plants to the infection of *Magnaporthe grisea*. Physiol Mol Plant Pathol 65:3–9

Kohler A, Schwindling S, Conrath U (2002) Benzothiadiazole-induced priming for potentiated responses to pathogen infection, wounding, and infiltration of water into leaves requires the *NPR1/NIM1* gene in *Arabidopsis*. Plant Physiol 128:1046–1056

Koornneef A, Pieterse CMJ (2008) Cross talk in defense signaling. Plant Physiol 146:839–844

Kovtun Y, Chiu WL, Tena G, Sheen J (2000) Functional analysis of oxidative stress-activated-mitogen-activated protein kinase cascade in plants. Proc Natl Acad Sci U S A 2940–2945

Krause M, Durner J (2004) Harpin inactivates mitochondria in Arabidopsis suspension cells. Mol Plant-Microbe Interact 17:131–139

Kufner I, Ottmann C, Oecking C, Nürnberger T (2009) Cytolytic toxins as triggers of plant immune response. Plant Signal Behav 10:977–979

Kwak JM, Mori IC, Pei ZM, Leonhardt N, Torres MA, Dangl JL, Bloom RE, Bodde S, Jones JD, Schroeder JI (2003) NADPH oxidase *AtrbohD* and *AtrbohF* genes function in ROS-dependent ABA signaling in *Arabidopsis*. EMBO J 22:2623–2633

Kwon C (2010) Plant defense responses coming to shape. Plant Pathol J 26:115–120

Lacombe S, Rougon-Cardoso A, Sherwood E, Peeters N, Dahlbeck D, van Esse HP, Smoker M, Rallapalli G, Thomma BPHJ, Stakawicz B, Jones JDG, Zipfel C (2010) Interfamily transfer of a plant pattern-recognition receptor confers broad-spectrum bacterial resistance. Nat Biotechnol 28:365–369

Lamotte O, Gould K, Lecourieux D, Sequeira-Legrand A, Lebrun-Garcia A, Durner J, Pugin A, Wendehenne D (2004) Analysis of nitric oxide signaling functions in tobacco cells challenged by the elicitor cryptogein. Plant Physiol 135:516–529

Latunde-Dada AO, Lucas JA (2001) The plant defense activator acibenzalor-S-methyl primes cowpea [*Vigna unguiculata* (L.) Walp.] seedlings for rapid induction of resistance. Physiol Mol Plant Pathol 58:199–208

Laurie-Berry N, Joardar V, Street JH, Kunkel BN (2006) The *Arabidopsis thaliana JASMONATE INSENSITIVE 1* gene is required for suppression of salicylic acid-dependent defenses during infection by *Pseudomonas syringae*. Mol Plant-Microbe Interact 19:789–800

Laxalt AM, Raho N, ten Have A, Lamattina L (2007) Nitric oxide is critical for inducing phosphatidic acid accumulation in xylanase-elicited cells. J Biol Chem 282:21160–21168

Lecourieux D, Mazars C, Pauly N, Ranjeva R, Pugin A (2002) Analysis and effects of cytosolic free calcium elevations in response to elicitors in *Nicotiana plumbaginifolia* cells. Plant Cell 14:2627–2641

Lecourieux D, Ranjeva R, Pugin A (2006) Calcium in plant defence-signaling pathways. New Phytol 171:249–269

Lecourieux–Ouaked D, Pugin A, Lebrun-Garcia A (2000) Phosphoproteins involved in the signal transduction of cryptogein, an elicitor of defense reactions in tobacco. Mol Plant Microbe Interact 13:821–829

Lee H-I, León J, Raskin I (1995) Biosynthesis and mechanism of salicylic acid. Proc Natl Acad Sci U S A 92:4076–4079

Lee C-W, Efetova M, Engelmann JC, Kramell R, Wasternack C, Ludwig-Müller J, Hedrich R, Deeken R (2009) *Agrobacterium tumefaciens* promotes tumor induction by modulating pathogen defense in *Arabidopsis thaliana*. Plant Cell 21:2948–2962

Lee W-S, Fu S-F, Verchot-Lubicz J, Carr JP (2011) Genetic modification of alternative respiration in *Nicotiana benthamiana* affects basal and salicylic acid-induced resistance to potato virus X. BMC Plant Biol 11:41

Lehtonen NT, Akita M, Frank W, Reski R, Valkonen JPT (2012) Involvement of a class III peroxidase and the mitochondrial protein TSPO in oxidative burst upon treatment of moss plants with a fungal elicitor. Mol Plant Microbe Interact 25:363–371

Leon-Reyes A, Du Y, Koornneef A, Proietti S, Körbes AP, Memelink J, Pieterse CMJ, Ritsema T (2010) Ethylene signaling renders the jasmonate response of *Arabidopsis* insensitive to future suppression by salicylic acid. Mol Plant-Microbe Interact 23:187–197

Li J, Brader G, Palva ET (2004) The WRKY70 transcription factor: a node of convergence for jasmonate-mediated and salicylate-mediated signals in plant defense. Plant Cell 16:319–331

Li XY, Lin H, Zhang W, Zou Y, Zhang J, Tang X, Zhou JM (2005) Flagellin induces innate immunity in nonhost interactions that is suppressed by *Pseudomonas syringae* effectors. Proc Natl Acad Sci U S A 102:12990–12995

Li J, Brader G, Kariola T, Tapio Palva E (2006) WRKY70 modulates the selection of signaling pathways in plant defense. Plant J 46:477–491

Liu Y, Zhang S (2004) Phosphorylation of 1-aminocyclopropane-1-carboxylic acid synthase by MPK6, a stress-responsive mitogen-activated protein kinase, induces ethylene biosynthesis in *Arabidopsis*. Plant Cell 16:3386–3399

Liu J-J, Ekramoddoullah AKM, Yu X (2003) Differential expression of multiple PR10 proteins in western white pine following wounding, fungal infection and cold-hardening. Physiol Plant 119:544–553

Liu X, Bai X, Wang X, Chu C (2007) OsWRKY71, a rice transcription factor, is involved in rice defense response. J Plant Physiol 164:969–979

Liu J-Z, Horstman HD, Braun E, Graham MA, Zhang C, Navarre D, Qiu W-L, Lee Y, Nettleton D, Hill JH, Whitham SA (2011a) Soybean homologs of MPK4 negatively regulate defense responses and positively regulate growth and development. Plant Physiol 157:1363–1378

Liu P-P, von Dahl CC, Klessig DF (2011b) The extent to which methyl salicylate is required for signaling systemic acquired resistance is dependent on exposure to light after infection. Plant Physiol 157:2216–2226

Lizasa E, Mitsutomi M, Nagano Y (2010) Direct binding of a plant LysM receptor-like kinase LysM RLK1/CERK1 to chitin in vitro. J Biol Chem 285:2996–3004

Llorente E, Muskett P, Sáuchez_Vallet A, Lopez G, Ramos B, Sánchez Rodriguez C, Jordá L, Parker J, Molina A (2008) Repression of the auxin response pathway increases *Arabidopsis* susceptibility to necrotrophic fungi. Mol Plant 1:496–509

Lohmann GV, Shimoda Y, Nielsen W, Jørgensen FG, Grossmann C, Sandal N, Sørensen K, Thirup S, Madsen LH, Tabata S, Sato S, Stougaard J, Radutoiu S (2010) Evolution and regulation of the *Lotus japonica* LysM receptor gene family. Mol Plant-Microbe Interact 23:510–521

Love AJ, Yun B-W, Laval V, Loake GJ, Milner JL (2005) *Cauliflower mosaic virus*, a compatible pathogen of Arabidopsis, engages three distinct defense-signaling pathways and activates rapid systemic generation of reactive oxygen species. Plant Physiol 139:935–948

Lozano-Durán R, Bejarano ER (2011) Geminivirus C2 protein might be the key player for geminiviral co-option of SCF-mediated ubiquitination. Plant Signal Behav 6:999–1001

Lozano-Durán R, Rosas-Diaz T, Gusmaroli G, Luna AP, Taconnat L, Deng XW, Bejarano ER (2011) Geminiviruses subvert ubiquitination by altering CSN-mediated derubylation of SCF E3 ligase complexes and inhibit jasmonate signaling in *Arabidopsis thaliana*. Plant Cell 23:1014–1032

Lu D, Wu S, Gao X, Zhang Y, Shan L, He P (2010) A receptor-like cytoplasmic kinase, BIK1, associates with a flagellin receptor complex to initiate plant innate immunity. Proc Natl Acad Sci U S A 107:495–501

Luna E, Bruce TJA, Roberts MR, Flors V, Ton J (2012) Next generation systemic acquired resistance. Plant Physiol 158:844–853

Makandar R, Nalam VJ, Lee H, Trick HN, Dong Y, Shah J (2012) Salicylic acid regulates basal resistance to Fusarium head blight in wheat. Mol Plant-Microbe Interact 25:431–439

Mandal B, Mandal S, Csinos AS, Martinez N, Culbreath AK, Pappu HR (2008) Biological and molecular analyses of the acibenzolar-S-methyl-induced systemic acquired resistance in flue-cured tobacco against *Tomato spotted wilt virus*. Phytopathology 98:196–204

Mateo A, Muhlenbock P, Rustérucci C, Chang CC, Miszalski Z, Karpinska B, Parker JE, Mullineaux PM, Karpinski S (2004) *LESION SIMULATING DISEASE 1* is required for acclimation to conditions that promote excess excitation energy. Plant Physiol 136:2818–2830

McDonald MC, Oliver RP, Friesen TL, Brunner PC, McDonald BA (2013) Global diversity and distribution of three necrotrophic effectors in *Phaeosphaeria nodorum* and related species. New Phytol 199:241–251

McLellan H, Boevink PC, Armstrong MR, Pritchard L, Gomez S, Morales J, Whisson SC, Beynon JL, Birch PR (2013) An RXLR effector from *Phytophthora infestans* prevents relocalisation of two plant NAC transcription factors from the endoplasmic reticulum to the nucleus. PLoS Pathog 9(10) e 1003670. doi:10.1371/journal ppat 1003670

Mei C, Qi M, Sheng G, Yang Y (2006) Inducible overexpression of a rice allene oxide synthase gene increases the endogenous jasmonic acid level, PR gene expression, and host resistance to fungal infection. Mol Plant-Microbe Interact 19:1127–1137

Melotto M, Underwood W, Koczan J, Nomura K, He SY (2006) Plant stomata function in innate immunity against bacterial invasion. Cell 126:969–980

Mentlak TA, Kombrink A, Shinya T, Ryder LS, Otomo I, Saitoh H, Terauchi R, Nishizawa Y, Shibuya N, Thomma BPHJ, Talbot NJ (2012) Effector-mediated suppression of chitin-triggered immunity by *Magnaporthe oryzae* is necessary for rice blast disease. Plant Cell 24:322–335

Mersmann S, Bourdais G, Rietz S, Robatzek S (2010) Ethylene signaling regulates accumulation of the FLS2 receptor and is required for the oxidative burst contributing to plant immunity. Plant Physiol 154:391–400

Milat ML, Ricci P, Blein JP (1991) Capsidiol and ethylene production by tobacco cells in response to cryptogein, an elicitor from *Phytophthora cryptogea*. Phytochemistry 30:2171–2173

Mishina TE, Zeier J (2007) Pathogen-associated molecular pattern recognition rather than development of tissue necrosis contributes to bacterial induction of systemic acquired resistance in Arabidopsis. Plant J 50:500–513

Mitsuhara I, Iwai T, Seo S, Yanagawa Y, Kawahigasi H, Hirose S, Ohkawa Y, Ohashi Y (2008) Characteristic expression of twelve *PR1* family genes in response to pathogen infection, wounding, and defense-related signal compounds. Mol Genet Genomics 279:415–427

Miya A, Albert P, Shinya T, Desaki Y, Ichimura K, Shirasu K, Narusaka Y, Kawakami N, Kaku H, Shibuya N (2007) CERK1, a LysM receptor kinase, is essential for chitin elicitor signaling in *Arabidopsis*. Proc Natl Acad Sci U S A 104:19613–19618

Miyazono K, Miyakawa T, Sawano Y, Kubota K, Kang HJ, Asano A, Miyauchi Y, Takahashi M, Zhi Y, Fujita Y, Yoshida T, Kodaira KS, Yamaguchi-Shinozaki K, Tanokura M (2009) Structural basis of abscisic acid signaling. Nature 462:609–614

Mohr PG, Cahill DM (2003) Abscisic acid influences the susceptibility of *Arabidopsis thaliana* to *Pseudomonas syringae* pv. *tomato* and *Peronospora parasitica*. Funct Plant Biol 30:461–469

Mohr PG, Cahill DM (2007) Suppression by ABA of salicylic acid and lignin accumulation and the expression of multiple genes, in *Arabidopsis* infected with *Pseudomonas syringae* pv. *tomato*. Funct Integr Genomics 7:181–191

Molina L, Kahmann R (2007) A *Ustilago maydis* gene involved in H_2O_2 detoxification is required for virulence. Plant Cell 19:2293–2309

Molinaro A, Newman M-A, Lanzetta R, Parrilli M (2009) The structures of lipopolysaccharides from plant-associated gram-negative bacteria. Eur J Org Chem 34:5887–5896

Molloy S (2010) Plant immunity: helping plants to fight back. Nat Rev Microbiol 8:312–313

Momol MT, Olson SM, Funderburk JE, Stavisky J, Marois JJ (2004) Integrated management of tomato spotted wilt on field-grown tomatoes. Plant Dis 88:882–890

Moreau M, Degrave A, Vedel R, Bitton F, Patrit O, Renou J-P, Barny M-A, Fagard M (2012a) EDS1 contributes to nonhost resistance of *Arabidopsis thaliana* against *Erwinia amylovora*. Mol Plant-Microbe Interact 25:421–430

Moreau M, Tian M, Klessig DF (2012b) Salicylic acid binds NPR3 and NPR4 to regulate NPR1-dependent defense responses. Cell Res. doi:10.1038/cr 2012.100

Mou Z, Fan W, Dong X (2003) Inducers of plant systemic acquired resistance regulate NPR1 function through redox changes. Cell 113:935–944

Mukherjee M, Larrimore KE, Ahmed NJ, Bedick TS, Barghouthi NT, Traw MB, Barth C (2010) Ascorbic acid deficiency in *Arabidopsis* induces constitutive priming that is dependent on hydrogen peroxide, salicylic acid, and the *NPR1* gene. Mol Plant-Microbe Interact 23:340–351

Mur LAJ, Renlin X, Casson SA, Stoddart WM, Routledge APM, Draper J (2004) Characterization of a proteinase inhibitor from *Brachypodium distachyon* suggests the conservation of defense signaling pathways between dicotyledonous plants and grasses. Mol Plant Pathol 5:267–280

Mur LAJ, Kenton P, Atzorn R, Miersch O, Wasternack C (2006) The outcomes of concentration-specific interactions between salicylate and jasmonate signaling include synergy, antagonism, and oxidative stress leading to cell death. Plant Physiol 140:249–262

Na R, Yu D, Qutob D, Zhao J, Gijzen M (2013) Deletion of the *Phytophthora sojae* avirulence gene *Avr1d* causes gain of virulence on *Rps1d*. Mol Plant Microbe Interact 26:969–976

Nakashita H, Yoshioka K, Yasuda M, Nitta T, Arai Y, Yoshida S, Yamaguchi I (2002) Probenazole induces systemic acquired resistance in tobacco through salicylic acid accumulation. Physiol Plant Pathol 61:197–203

Nakashita H, Yasuda M, Okage R, Nishioka M, Arie T, Yoshida S (2003) A pyrazole derivative induces systemic acquired resistance with a new type of action. Plant Cell Physiol 44:S179–S179

Nambeesan S, AbuQamar S, Laluk K, Mattoo AK, Mickelbart MV, Ferruzzi MG, Mengiste T, Handa AK (2012) Polyamines attenuate ethylene –mediated defense responses to abrogate resistance to *Botrytis cinerea* in tomato. Plant Physiol 158:1034–1045

Nandi A, Kachroo P, Fukushige H, Hildebrand DF, Klessig DF, Shah J (2003) Ethylene and jasmonic acid signaling pathways affect the NPR1-independent expression of defense genes without impacting resistance to *Pseudomonas syringae* and *Peronospora parasitica* in the *Arabidopsis ssi1* mutant. Mol Plant Microbe Interact 18:588–599

Nandi A, Moeder W, Kachroo P, Klessig DF, Shah J (2005) Arabidopsis *ssi-2* conferred susceptibility to *Botrytis cinerea* is dependent on *EDS5* and *PAD4*. Mol Plant-Microbe Interact 18:363–370

Naseem M, Philippi N, Hussain A, Wangorsch G, Ahmed N, Dandekar T (2012) Integrated systems view on networking by hormones in *Arabidopsis* immunity reveals multiple crosstalk for cytokinin. Plant Cell 24:1793–1814

Navarro L, Dunoyer P, Jay F, Arnold B, Dhamasiri N, Estelle M, Voinnet O, Jones JDG (2006) A plant miRNA contributes to antibacterial resistance by repressing auxin signaling. Science 312:436–439

Navarro L, Bari R, Achard P, Lison P, Nemri A, Harberd NP, Jones JDG (2008) DELLAs control plant immune responses by modulating the balance of jasmonic acid and salicylic acid signaling. Curr Biol 18:650–655

Nawrath C, Heck S, Parinthawong N, Metraux JP (2002) EDS5, an essential component of salicylic acid-dependent signaling for disease resistance in *Arabidopsis*, is a member of MATE transporter family. Plant Cell 14:275–286

Naylor M, Murphy AM, Berry JO, Carr JP (1998) Salicylic acid can induce resistance in plant virus movement. Mol Plant-Microbe Interact 11:860–868

Ndamukong I, Abdallat AA, Thurow C, Fode B, Zander M, Weigel R, Gatz C (2007) SA-inducible *Arabidopsis* glutaredoxin interacts with TGA factors and suppresses JA-responsive *PDF1.2* transcription. Plant J 50:128–139

Nicaise V, Roux M, Zipfel C (2009) Recent advances in PAMP-triggered immunity against bacteria: pattern recognition receptors watch over and raise the alarm. Plant Physiol 150:1638–1647

Nürnberger T, Küfner I (2011) The role of the plant plasma membrane in microbial sensing and innate immunity: the plant plasma membrane. Plant Cell Monogr 19:471–483

O'Donnell PJ, Schmelz EA, Moussatche P, Lund ST, Jones JB, Klee HJ (2003) Susceptible to intolerance – a range of hormonal actions in a susceptible *Arabidopsis* pathogen response. Plant J 33:245–257

Ogawa D, Nakajima N, Seo S, Mitsuhara I, Kamada H, Ohashi Y (2006) The phenylalanine pathway is the main route of salicylic acid biosynthesis in *Tobacco mosaic virus*-infected tobacco leaves. Plant Biotechnol 23:395–398

Oostendorp M, Kunz W, Dietrich B, Staub T (2001) Induced disease resistance in plants by chemicals. Eur J Plant Pathol 107:19–28

Pantelides S, Tjamos SE, Pappa S, Kargakis M, Paplomatas EJ (2013) The ethylene receptor ETR1 is required for *Fusarium oxysporum* pathogenicity. Plant Pathol 62:1302–1309

Park J, Choi HJ, Lee S, Lee T, Yang Z, Lee Y (2000) Rac-related GTP-binding protein in elicitor-induced reactive oxygen generation by suspension-cultured soybean cells. Plant Physiol 124:725–732

Park C-J, Han S-W, Chen X, Ronald PC (2010) Elucidation of XA21-mediated innate immunity. Cell Microbiol 12:1017–1025

Park CH, Chen S, Shirsekar G, Zhou B, Khang CH, Songkumam P, Afzal AJ, Ning Y, Wang R, Bellizzi M, Valent B, Wang GL (2012) The *Magnaporthe oryzae* effector AvrPiz-t targets the RING E3 ubiquitin ligase APIP6 to suppress pathogen-associated molecular pattern-triggered immunity in rice. Plant Cell 24:4748–4762

Pastor V, Luna E, Mauch-Mani B, Ton J, Flors V (2013) Primed plants do not forget. Environ Exp Bot 94:46–50

Pedley KF, Martin GB (2005) Role of mitogen-activated protein kinases in plant immunity. Curr Opin Plant Biol 8:541–547

Peer WA, Cheng Y, Murphy AS (2013) Evidence of oxidative attenuation of auxin signaling. J Expt Bot 64:2629–2639

Pei Z-M, Murata Y, Benning G, Thomine S, Klüsener B, Allen GJ, Grill E, Schroeder JI (2000) Calcium channels activated by hydrogen peroxide mediate abscisic acid signaling in guard cells. Nature 406:731–734

Peng J-L, Bao Z-L, Ren H-Y, Wang J-S, Dong HS (2004) Expression of $harpin_{Xoo}$ in transgenic tobacco induces pathogen defense in the absence of hypersensitive cell death. Phytopathology 94:1048–1055

Perfus-Barbeoch I, Jones AM, Assmann SM (2004) Plant heterotrimeric G protein function: insights from *Arabidopsis* and rice mutants. Curr Opin Plant Biol 7:719–731

Petutschnig EK, Jones AME, Serazetdinova L, Lipka U, Lipka V (2010) The Lysin motif receptor-like kinase (LysM-RLK) is a major chitin binding protein in *Arabidopsis thaliana* and subject to chitin-induced phosphorylation. J Biol Chem 285:28902–28911

Pickart CM, Eddins MJ (2004) Ubiquitin: structures, functions, mechanisms. Biochim Biophys Acta 1695:55–72

Pieterse CMJ (2012) Prime time for transgenerational defense. Plant Physiol 158:545

Pitzschke A, Hirt H (2006) Mitogen-activated protein kinases and reactive oxygen species signaling in plants. Plant Physiol 141:351–356

Pitzschke A, Hirt H (2009) Disentangling the complexity of mitogen-activated protein kinases and reactive oxygen species signaling. Plant Physiol 149:606–615

Pitzschke A, Djamei A, Bitton F, Hirt H (2009a) A major role of the MEKK1-MKK1/2-MPK4 pathway in ROS signaling. Mol Plant 2:120–137

Pitzschke A, Schikora A, Hirt H (2009b) MAPK cascade signaling networks in plant defence. Curr Opin Plant Biol 12:1–6

Pogány M, Koehl J, Heiser I, Elstner EF, Barna B (2004) Juvenility of tobacco induced by cytokinin gene introduction decreases susceptibility to *Tobacco necrosis virus* and confers tolerance to oxidative stress. Physiol Mol Plant Pathol 65:39–47

Poinssot B, Vandelle E, Bentejac M, Adrian M, Levis C, Brygoo Y, Garin J, Sicilia F, Coutos-Thevenot P, Pigin A (2003) The endopolygalacturonase I from *Botrytis cinerea* activates grapevine defense reactions unrelated to its enzymatic activity. Mol Plant-Microbe Interact 16:553–564

Postel S, Kufner I, Beueter C, Mazzotta S, Schwedt A, Borlotti A, Halter T, Kemmerling B, Nürnberger T (2010) The multifunctional leucine-rich repeat receptor kinase BAK1 is implicated in *Arabidopsis* development and immunity. Eur J Cell Biol 89:169–174

Potlakayala SD, Reed DW, Covello PS, Fobert PR (2007) Systemic acquired resistance in canola is linked with pathogenesis-related gene expression and requires salicylic acid. Phytopathology 97:794–802

Po-Wen C, Singh P, Zimmerli L (2013) Priming of the Arabidopsis pattern-triggered immunity response upon infection by necrotrophic *Pectobacterium carotovorum* bacteria. Mol Plant Pathol 14:58–70

Qin X, Liu JH, Zhao WS, Chen XJ, Guo ZJ, Peng YL (2013) Gibberellin 20-oxidase gene *OsGA20ox3* regulates plant stature and disease development in rice. Mol Plant-Microbe Interact 26:227–239

Qiu D, Xiao J, Ding X, Xiong M, Cai M, Cao Y, Li X, Xu C, Wang S (2007) OsWRKY13 mediates rice disease resistance by regulating defense-related genes in salicylate- and jasmonate-dependent signaling. Mol Plant-Microbe Interact 20:492–499

Qutob D, Kemmerling B, Brunner F, Küfner I, Engelhardt S, Gust AA, Luberacki B, Seitz HU, Stahl D, Rauhut T, Glawischnig E, Schween G, Lacombe B, Watanabe N, Lam E, Schlichting R, Scheel D, Nau K, Dodt G, Hubert D, Gijzen M, Nürnberger T (2006) Phytotoxicity and innate immune responses induced by NEP1-like proteins. Plant Cell 18:3721–3744

Qutob D, Tedman-Jones J, Dong S, Kufluk K, Pham H, Wang Y, Dou D, Kale SD, Arredondo FD, Tyler BM, Gijzen M (2009) Copy number variation and transcriptional polymorphisms of *Phytophthora sojae* RXLR effector genes *Avr1a* and *Avr3a*. PLoS One 4, e5066. doi:10.1371/journalpone0005066

Raghavendra AS, Gonugunta VK, Christmann A, Grill L (2010) ABA perception and signaling. Trends Plant Sci 15:395–401

Rahman A, Kuldau A, Uddin W (2014) Induction of salicylic acid-mediated defense response in perennial ryegrass against infection by *Magnaporthe oryzae*. Phytopathology 104:614–625

Rajput NA, Zhang M, Ru Y, Liu T, Mafurah JJ, Dou D (2014) *Phytophthora sojae* effector PsCRN70 suppresses plant defenses in *Nicotiana benthamiana*. PLoS One 9(5):e98114. doi:10.1371/jornal pone 0098114

Rakwal R, Tamogani S, Agrawal GK, Iwahashi H (2002) Octadecanoid signaling component 'burst' in rice (*Oryza sativa* L) seedlings leaves upon wounding by cut and treatment with fungal elicitor chitosan. Biochem Biophys Res Commun 295:1041–1045

Ren D, Liu Y, Yang K-Y, Han L, Mao G, Glazebrook J, Zhang S (2008) A fungal-responsive MAPK cascade regulates phytoalexin biosynthesis in Arabidopsis. Proc Natl Acad Sci U S A 105:5638–5643

Robatzek S, Chinchilla D, Boller T (2006) Ligand-induced endocytosis of the pattern recognition receptor FLS2 in *Arabidopsis* Genes. Dev 20:537–542

Robatzek S, Bittel P, Chinchilla D, Köchner P, Felix G, Shiu S, Boller T (2007) Molecular identification and characterization of the tomato flagellin receptor LeFLS2, an orthologue of *Arabidopsis* FLS2 exhibiting characteristically different perception specificities. Plant Mol Biol 64:539–547

Rockel P, Strube F, Rockel A, Wildt J, Kaiser WM (2002) Regulation of nitric oxide (NO) production by plant nitrate reductase in vivo and in vitro. J Exp Bot 53:103–110

Ronald PC, Beutler B (2010) Plant and animal sensors of conserved microbial signatures. Science 308:1783–17861

Rosli HG, Zheng Y, Pombo MA, Zhong S, Bombarely A, Fei Z, Collmer A, Martin GB (2013) Transcriptomics-based screen for genes induced by flagellin and repressed by pathogen effectors identifies a cell wall-associated kinase involved in plant immunity. Genome Biol 14:R139

Rotblat B, Enshel-Seijffers D, Gershoni MJ, Schuster S, Avni A (2002) Identification of an essential component of the elicitation active site of the EIX protein elicitor. Plant J 32:1–7

Ryan CA, Huffaker A, Yamaguchi Y (2007) New insights into innate immunity in Arabidopsis. Cell Microbiol 9:1902–1908

Sagi M, Fluhr R (2006) Production of reactive oxygen species by plant NADPH oxidases. Plant Physiol 336–340

Sahana N, Kaur H, Basavaraj TF, Jain RK, Palukaitis P, Canto T, Praveen S (2012) Inhibition of the host proteasome facilitates papaya ringspot virus accumulation and proteasomal catalytic activity is modulated by viral factor HcPro. PLoS One 7:e52546

Saijo Y (2010) ER quality control of immune receptors and regulators in plants. Cell Microbiol 12:716–724

Sänchez G, Gerhardt N, Siciliano F, Vojnov A, Malcuit I, Marono MR (2010) Salicylic acid is involved in the *Nb*-mediated defense responses to *Potato virus X* in *Solanum tuberosum*. Mol Plant-Microbe Interact 23:394–405

Sánchez-Vallet A, López C, Ramos B, Delgado-Cerezo M, Riviere P, Llorente F, Fernández PV, Miedes E, Estevez JM, Grant M, Molina A (2012) Disruption of abscisic acid signaling constitutively activates Arabidopsis resistance to the necrotrophic fungus *Plectosphaerella cucumerina*. Plant Physiol 160:2109–2124

Sauer M, Robert S, Kleine-Vehn J (2013) Auxin: simply complicated. J Expt Bot 64:2565–2577

Schiene K, Puhler A, Niehaus K (2000) Transgenic tobacco plants that express an antisense construct derived from a *Medicago sativa* cDNA encoding a Rac-related small GTP-binding protein fail to develop necrotic lesions upon elicitor infiltration. Mol Gen Genet 263:761–770

Schikora A, Schenk ST, Stein E, Molitor A, Zuccaro A, Kogel K-H (2011) *N*-acyl-homoserine lactone confers resistance toward biotrophic and hemibiotrophic pathogens via altered activation of AtMPK6. Plant Physiol 157:1407–1418

Schilmiller AL, Koo AJK, Howe GA (2006) Functional diversification of acyl-coenzyme A oxidases in jasmonic acid biosynthesis and action. Plant Physiol 143:812–824

Schlessinger J (2000) Cell signaling by receptor tyrosine kinases. Cell 103:211–225

Schmidt K, Pflugmacher M, Klages S, Maser A, Mock A, Stahl DJ (2008) Accumulation of the hormone abscisic acid (ABA) at the infection site of the fungus *Cercospora beticola* supports the role of ABA as a repressor of plant defence in sugar beet. Mol Plant Pathol 9:661–673

Schornack S, Huitema E, Cano LM, Bozkurt TO, Oliva R, Van Damme M, Schwizer S, Raffaele S, Chaparro-Garcia A, Farrer R, Segretin ME, Bos J, Haas BJ, Zody MC, Nusbaum C, Win J, Thines M, Kamoun S (2009) Ten things to know about oomycete effectors. Mol Plant Pathol 10:795–801

Schultheiss H, Dechert C, Kogel K-H, Hückelhoven R (2003) Functional analysis of barley RAC/ROP G-protein family members in susceptibility to the powdery mildew fungus. Plant J 36:589–601

Schulze B, Mentzel T, Jehle A, Mueller K, Beeler S, Boller T, Felix G, Chinchilla D (2010) Rapid heteromerization and phosphorylation of ligand-activated plant transmembrane receptors and their associated kinase BAK1. J Biol Chem 285:9444–9451

Schweizer P, Buchala A, Dudler RA, Metraux JP (1998) Induced systemic resistance in wounded rice plants. Plant J 14:475–481

Segonzac C, Zipfel C (2011) Activation of plant pattern-recognition receptors by bacteria. Curr Opin Microbiol 1(4):54–61

Seifert GJ, Blaukopf C (2010) Irritable walls: the plant extracellular matrix and signaling. Plant Physiol 153:467–478

Seo PJ, Park C-M (2010) MYB96-mediated abscisic acid signals induce pathogen resistance response by promoting salicylic acid biosynthesis in *Arabidopsis*. New Phytol 186:471–483

Shan L, He P, Sheen J (2007) Intercepting host MAPK signaling cascades by bacterial type III effectors. Cell Host Microbe 1:167–174

Shan L, He P, Li J, Heese A, Peck SC, Nümberger T, Martin GB, Sheen J (2008) Bacterial effectors target the common signaling partner BAK1 to disrupt multiple MAMP receptor-signaling complexes and impede plant immunity. Cell Host Microbe 4:17–27

Sheard LB, Zheng N (2009) Plant biology signal advance for abscisic acid. Nature 462:575–576

Shen X, Liu H, Yuan B, Li X, Xu C, Wang S (2011) OsEDR1 negatively regulates rice bacterial resistance via activation of ethylene biosynthesis. Plant Cell Environ 34:179–191

Shimizu T, Nakano T, Takamizawa D, Desaki Y, Ishi-Minami N, Nishizawa Y, Minami E, Okada K, Yamane H, Kaku H, Shibuya N (2010) Two LysM receptor molecules, CEBiP and OsCERK1, cooperatively regulate chitin elicitor signaling in rice. Plant J 64:204–214

Shiu S-H, Bleecker AB (2001) Plant receptor-like kinase gene family: diversity, function, and signaling. Sci STKE 113:RE22

Siemens J, Keller I, Sarx J, Kunz S, Schuller A, Nagel W, Schmülling T, Parniske M, Ludwig-Müller J (2006) Transcriptome analysis of *Arabidopsis* clubroots indicate a key role for cytokinins in disease development. Mol Plant Microbe Interact 19:480–494

Silipo A, Mollinaro A (2010) The diversity of the core oligosaccharide in lipopolysaccharides, In: X Wang and P J Quinn (eds) Endotoxins: structure, function and recognition, subcellular biochemistry, vol 53, Part 1. Springer Science, Business Media, BV, pp 69–99

Silipo A, Sturiale L, Garozzo D, Erbs G, Tandrup Poulsen T, Lanzetta R, Dow JM, Parrilli M, Newman M-A, Molinaro A (2008) The acylation and phosphorylation pattern of lipid A from *Xanthomonas campestris* strongly influence its ability to trigger the innate immune response in *Arabidopsis*. Chem BioChem 9:896–904

Sklodowska M, Gajewski E, Kuzniak E, Mikiciński A, Sobiczewski P (2010) BTH-mediated antioxidant system responses in apple leaf tissues. Sci Hortic 125:34–40

Slaughter A, Daniel X, Flors V, Luna E, Hohn B, Mauch-Mani B (2012) Descendants of primed Arabidopsis plants exhibit resistance to biotic stress. Plant Physiol 158:835–843

Slaymaker DH, Navarre DA, Clark D, del Pozo O, Martin GB, Klessig DF (2002) The tobacco salicylic acid-binding protein-3 (SABP3) is the chloroplast carbonic anhydrase, which exhibits antioxidant activity and plays a role in the hypersensitive defense response. Proc Natl Acad Sci U S A 99:11640–11645

Sohn KH, Lei R, Nemri A, Jones JDG (2007) The downy mildew effectors ATR1 and ATR13 promote disease susceptibility in *Arabidopsis thaliana*. Plant Cell 19:4077–4090

Son GH, Wan J, Kim HJ, Nguyen XC, ChungW-S HJC, Stacey G (2012) Ethylene-responsive element-binding factor 5, ERF5, is involved in chitin-induced innate immunity response. Mol Plant Microbe Interact 25:48–60

Song C, Yang B (2010) Mutagenesis of 18 type II effectors reveals virulence function of $XopZ_{Pxo99}$ in *Xanthomonas oryzae* pv *oryzae*. Mol Plant-Microbe Interact 23:893–902

Song JT, Lu H, McDowell JM, Greenberg JT (2004) A key role for *ALD1* in activation of local and systemic defenses in *Arabidopsis*. Plant J 40:200–212

Spoel SH, Dong X (2008) Making sense of hormone crosstalk during plant immune response. Cell Host Microbe 3:348–351

Spoel SH, Johnson JS, Dong X (2007) Regulation of tradeoffs between plant defenses against pathogens with different lifestyles. Proc Natl Acad Sci U S A 104:18842–18847

Spoel SH, Koornneef A, Claessens SMC, Korzelius JP, Van Pelt JA, Mueller MJ, Buchala AJ, Metraux J-P, Brown R, Kazan K, Van Loon LC, Dong X, Pieterse CMJ (2003) NPR1 modulates cross-talk between salicylate- and jasmonate-dependent defense pathways through a novel function in the cytosol. Plant Cell 15:760–770

Suharsono U, Fujisawa Y, Kawasaki T, Iwasaki Y, Satoh H, Shimamoto K (2002) The heterotrimeric G protein subunit acts upstream of the small GTPase Rac in disease resistance of rice. Proc Natl Acad Sci U S A 99:13307–13312

Sun F, Liu P, Xu J, Dong H (2010) Mutation in RAP26L, a transactivator of the ERF transcription factor family, enhances *Arabidopsis* resistance to *Pseudomonas syringae*. Physiol Mol Plant Pathol 74:295–302

Sun W, Cao Y, Labby KJ, Bittel P, Boller T, Bent AF (2012) Probing the *Arabidopsis* flagellin receptor: FLS2-FLS2 association and contributions of specific domains to signaling function. Plant Cell 24:1096–1113

Szczesny R, Buttner B, Escolar L, Schulze S, Seiferth A, Bonas U (2010) Suppression of the AvrBs1-specific hypersensitive response by the YopJ effector homolog AvrBsT from *Xanthomonas* depends on a SNF1-related kinase. New Phytol 187:1058–1074

Tada Y, Kusaka K, Betsuyaku S, Shinogi T, Sakamoto M, Ohura Y, Hata S, Mori T, Tosa Y, Mayama S (2005) Victorin triggers programmed cell death and the defense response via interaction with a cell surface mediator. Plant Cell Physiol 46:1787–1798

Taguchi F, Shimizu R, Inagaki Y, Toyoda K, Shiraishi T, Ichinose Y (2003a) Post-translational modification of flagellin determines the specificity of HR induction. Plant Cell Physiol 44:342–349

Taguchi F, Shimizu R, Nakagima R, Toyoda K, Shiraishi T, Ichinose Y (2003b) Differential effects of flagellins from *Pseudomonas syringae* pv *tabaci*, *tomato*, and *glycinea* on plant defense response. Plant Physiol Biochem 41:165–174

Taguchi F, Takeuchi K, Katoh E, Murata K, Suzuki T, Marutani M, Kawasaki T, Eguchi M, Katoh S, Kaku H, Yasuda C, Inagaki Y, Toyoda K, Shiraishi T, Ichinose Y (2006) Identification of glycosylation genes and glycosylated amino acids of flagellin in *Pseudomonas syringae* pv *tabaci*. Cell Microbiol 8:923–938

Takahashi F, Yoshida R, Ichimura K, Mizoguchi T, Seo S, Yonezawa M, Maruyama K, Yamaguchi-Shinozaki K, Shinozaki K (2007) The mitogen-activated protein kinase cascade MKK3-MPK6 is an important part of the jasmonate signal transduction pathway in Arabidopsis. Plant Cell 19:805–818

Takai R, Isogai A, Takayama S, Che F-S (2008) Analysis of flagellin perception mediated by flg22 receptor OsFLS2 in rice. Mol Plant Microbe Interact 21:1635–1642

Takeuchi K, Taguchi F, Inagaki Y, Toyoda K, Shiraishi T, Ichinose Y (2003) Flagellin glycosylation island in *Pseudomonas syringae* pv *glycinea* and its role in host specificity. J Bacteriol 185:6658–6665

Takeuchi K, Ono H, Yoshida M, Ishi T, Katch E, Taguchi F, Miki R, Murata K, Kaku H, Ichinose Y (2007) Flagellin glycans from two pathovars of *Pseudomonas syringae* contain rhamnose in D and L configurations in different ratios and modified 4-amino-4,6-dideoxyglucose. J Bacteriol 189:6945–6956

Takezawa D (2000) A rapid induction by elicitors of the mRNA encoding CCD-1, a 14 kDa Ca^{2+}-binding protein in wheat cultured cells. Plant Mol Biol 42:807–817

Talke IN, Blaudez D, Maathuis FJM, Sanders D (2003) CNGCs: prime targets of plant cyclic nucleotide signaling? Trend Plant Sci 8:286–293

Tatsuki M, Nakajima N, Fujii H, Shimada T, Nakano M, Hayashi K-I, Hayama H, Yoshioka H, Nakamura Y (2013) Increased levels of IAA are required for system 2 ethylene synthesis causing fruit softening in peach (*Prunus persica* L. Batsch). J Expt Bot 64:1049–1059

Tavernier E, Wendehenne J-P, Blein J-P, Pugin A (1995) Involvement of free calcium in action of cryptogein, a proteinaceous elicitor of a hypersensitive reaction in tobacco cells. Plant Physiol 109:1025–1031

Thatcher LF, Manners JM, Kazan K (2009) *Fusarium oxysporum* hijacks COI1-mediated jasmonate signaling to promote disease development in *Arabidopsis*. Plant J 58:927–939

Thomma BPHJ, Penninckx I, Broekaert WF, Cammue BPA (2001a) The complexity of disease signaling in *Arabidopsis*. Curr Opin Immunol 13:63–68

Thomma BPHJ, Tierens KFM, Penninckx I, Mauch-Mani B, Broekaert WF, Cammue BPA (2001b) Different microorganisms differentially induce Arabidopsis disease response pathways. Plant Physiol Biochem 39:673–680

Thomma BPHJ, Nürnberger T, Joosten M (2011) Of PAMPs and effectors: the blurred PTI-ETI dichotomy. Plant Cell 23:4–15

Tsuda K, Katagiri P (2010) Comparing signaling mechanisms engaged in pattern-triggered and effector-triggered immunity. Curr Opin Plant Biol 13:459–465

Tsuda K, Sato M, Glazebrook J, Cohen JD, Katagiri F (2008) Interplay between MAMP-triggered and SA-mediated defense responses. Plant J 53:763–775

Ueguchi-Tanaka M, Hirano K, Hasegawa Y, Kitano H, Matsuoka M (2008) Release of the rice DELLA protein SLR1 by gibberellin does not require SLR1 degradation in the *gid2* mutant. Plant Cell 20:2437–2446

Umezawa T, Sugiyama N, Mizoguchi M, Hayashi S, Myouga F, Yamaguchi-Shinozaki K, Ishihama Y, Hirayama T, Shinozaki K (2009) Type 2C protein phosphatases directly regulate abscisic acid-activated protein kinases in *Arabidopsis*. Proc Natl Acad Sci U S A 106:17588–17593

Uppalapati SR, Ishiga Y, Wangdi T, Kunkel BN, Anand A, Mysore KS, Bender CL (2007) Phytotoxin coronatine contributes to pathogen fitness and is required for suppression of salicylic acid accumulation in tomato inoculated with *Pseudomonas syringae* pv *tomato* DC3000. Mol Plant-Microbe Interact 20:955–965

Uraji M, Katagiri T, Okuma E, Te W, Hossain M-A, Masuda C, Miura A, Nakamura Y, Mori I-C, Shinozaki K, Murata Y (2012) Cooperative function of PLDδ and PLDα1in abscisic acid-induced stomatal closure in Arabidopsis. Plant Physiol 159:450–460

Valent B, Khang CH (2010) Recent advances in rice blast effector research. Curr Opin Plant Biol 13:434–441

van Damme M, Huibers RP, Elberse J, Van den Ackerveken G (2008) Arabidopsis DMR6 encodes a putative 2OG-Fe(II)oxygenase that is defense-associated but required for susceptibility to downy mildew. Plant J 54:785–793

Vandenabeele S, Van Der Kelen K, Dat J, Gadjev I, Boonefaes T, Morsa S, Rottiers P, Slooten L, Van Montagu M, Zabeau M, Inzé D, Van Breusegem F (2003) A comprehensive analysis of hydrogen peroxide-induced gene expression in tobacco. Proc Natl Acad Sci U S A 100:16113–16118

van den Burg HA, Tsitsigiannis DI, Rowland O, Lo J, Rallapalli G, Maclean D, Takken FL, Jones JD (2008) The F-box protein ACRE189/ACIF1 regulates cell death and defense responses activated during pathogen recognition in tobacco and tomsto. Plant Cell 20:697–719

Van der Does D, Leon-Reyes A, Koornneef A, Van Verk MC, Rodenburg N, Pauwels L, Goossens A, Körbes AP, Memelink J, Ritsema T, Van Wees SCM, Pieterse CMJ (2013) Salicylic acid suppresses jasmonic acid signaling downstream of SCF^{COI1}-JAZ by targeting GCC promoter motifs via transcription factor ORA59. Plant Cell 25:744–751

van Verk MC, Pappaaioannou D, Neeleman L, Bol JF, Linthorst HJM (2008) A novel WRKY transcription factor is required for induction of *PR-1a* gene expression by salicylic acid and bacterial elicitors. Plant Physiol 140:1983–1995

Vera-Estrella R, Barkla BJ, Higgins VJ, Blumwald E (1994) Plant defense response to fungal pathogens Activation of host-plasma membrane H^+-ATPase by elicitor-induced enzyme dephosphorylation. Plant Physiol 104:209–215

Vidhyasekaran P (2004) Concise encyclopedia of plant pathology. Haworth Press, Binghamton, p 619

Vidhyasekaran P (2007) Fungal pathogenesis in plants and crops: molecular biology and host defense mechanisms, IInd edn. CRC Press/Taylor Francis Group, Boca Raton, p 510

Vidhyasekaran P (2014) PAMP signals in plant innate immunity: signal perception and transduction. Springer, Dordrecht, p 442

Vlad F, Rubio S, Rodrigues A, Sirichandra C, Belin C, Robert N, Leung J, Rodriguez PL, Lauriere C, Meriot S (2009) Protein phosphatases 2C regulate the activation of the Snf1-related kinase OST1 by abscisic acid in *Arabidopsis*. Plant Cell 21:3170–3184

Vleeshouwers VGAA, Oliver RP (2014) Effectors as tools in disease resistance breeding against biotrophic, hemibiotrophic, and necrotrophic plant pathogens. Mol Plant Microbe Interact 27:196–206

Waller F, Müller A, Chung K-M, Yap Y-K, Nakamura K, Weiler E, Sano H (2006) Expression of a WIPK-activated transcription factor results in increase of endogenous salicylic acid and pathogen resistance in tobacco plants. Plant Cell Physiol 47:1169–1174

Wan J, Zhang S, Stacey G (2004) Activation of a mitogen-activated protein kinase pathway in Arabidopsis by chitin. Mol Plant Pathol 5:125–135

Wan J, Zhang X-C, Neece D, Ramonell KM, Clough S, Kim S-Y, Stacey MG, Stacey G (2008a) A LysM receptor-like kinase plays a critical role in chitin signaling and fungal resistance in *Arabidopsis*. Plant Cell 20:471–481

Wan J, Zhang X-C, Stacey G (2008b) Chitin signaling and plant disease resistance. Plant Signal Behav 3:831–833

Wang C, Zien C, Afitlhile M, Welti R, Hildebrand DF, Wang X (2000) Involvement of phospholipase D in wound-induced accumulation of jasmonic acid in Arabidopsis. Plant Cell 12:2237–2246

Wang XQ, Ullah H, Jones AM, Assmann SM (2001) G protein regulation of ion channels and abscisic acid signaling in *Arabidopsis* guard cells. Science 292:2070–2072

Wang X, Li X, Meisenhelder J, Hunter T, Yoshida S, Asami T, Chory J (2005) Autoregulation and homodimerization are involved in the activation of the plant steroid receptor BRI1. Dev Cell 8:855–865

Wang D, Eyles A, Mandich D, Bonello P (2006) Systemic aspects of host-pathogen interactions in Austrian pine (*Pinus nigra*): a proteomics approach. Physiol Mol Plant Pathol 68:149–157

Wang D, Pajerowska-Mukhtar K, Culler AH, Dong X (2007) Salicylic acid inhibits pathogen growth in plants through repression of the auxin signaling pathway. Curr Biol 17:1784–1790

Wang Y, Gao M, Li Q, Wang L, Jeon J-S, Qu N, Zhang Y, He Z (2008) *OsRAR1* and *OsSGT1* physically interact and function in rice basal disease resistance. Mol Plant-Microbe Interact 21:294–303

Wang Z, Mao H, Dong C, Ji R, Cai L, Fu H, Liu S (2009) Overexpression of *Brassica napus* MPK4 enhances resistance to *Sclerotinia sclerotiorum* in oilseed rape. Mol Plant Microbe Interact 22:235–244

Wang G, Fiers M, Ellendorff U, Wang Z, deWit PJGM, Angenent GC, Thomma BPHJ (2010a) The diverse roles of extracellular leucine-rich repeat-containing receptor-like proteins in plants. Critic Rev Plant Sci 29:285–299

Wang P, Du Y, Li Y, Ren D, Song C-P (2010b) Hydrogen peroxdide-mediated activation of MAP kinase 6 modulates nitric oxide biosynthesis and signal transduction in *Arabidopsis*. Plant Cell 22:2981–2998

Wang Y, Li J, Hou S, Wang X, Li Y, Ren D, Chen S, Tang X, Zhou JM (2010c) A *Pseudomonas syringae* ADP-ribosyltransferase inhibits Arabidopsis mitogen-activated kinase kinases. Plant Cell 22:2033–2044

Westfall CS, Zubieta C, Herrmann J, Kapp U, Nanao MH, Jez JM (2012) Structural basis for pre-receptor modulation of plant hormones by GH3 proteins. Science 336:1708–1711

Wiermer M, Feys BJ, Parker JE (2005) Plant immunity: the EDS1 regulatory node. Curr Opin Plant Biol 8:383–389

Wild M, Davière JM, Cheminant S, Regnault T, Baumberger N, Heintz D, Baltz R, Genschik P, Achard P (2012) The Arabidopsis DELLA RGA-LIKE3 is a direct target of MYC2 and modulates jasmonate signaling responses. Plant Cell 24:3307–3319

Wildermuth MC, Dewdney J, Wu G, Ausubel FM (2001) Isochorismate synthesis is required to synthesize salicylic acid for plant defense. Nature 414:562–565

Wilson ID, Neill SJ, Hancock JT (2008) Nitric oxide synthesis and signaling in plants. Plant Cell & Environ 31:622–631

Wolpert TJ, Dunkle LD, Ciuffetti LM (2002) Host-selective toxins and avirulence determinants: what's in a name? Annu Rev Phytopathol 40:251–285

Woodward AW, Bartel B (2005) Auxin: regulation, action, and interaction. Ann Bot (Lond) 95:707–735

Wu W-H, Assmann SM (1994) A membrane-delimited pathway of G-protein regulation of the guard-cell inward K^+ channel. Proc Natl Acad Sci U S A 91:6310–6314

Wu S, Lu D, Kabbage M, Wei HL, Swingle B, Records AR, Dickman M, He P, Shan L (2011) Bacterial effector HopF2 suppresses arabidopsis innate immunity of the plasma membrane. Mol Plant Microbe Interact 24:585–593

Xia Y, Yu K, Navarre D, Seebold K, Kachroo A, Kachroo P (2010) The *glabra1* mutation affects cuticle formation and plant responses to microbes. Plant Physiol 154:833–846

Xiang T, Zong N, Zou Y, Wu Y, Zhang J, Xing W, Li Y, Tang X, Zhu L, Chai J, Zhou J-M (2008) *Pseudomonas syringae* effector AvrPto blocks innate immunity by targeting receptor kinases. Curr Biol 18:74–80

Xiang T, Zong N, Zhang J, Chen J, Chen M, Zhou J-M (2011) BAK1 is not a target of the *Pseudomonas syringae* effector AvrPto. Mol Plant-Microbe Interact 24:100–107

Xin XF, He SY (2013) *Pseudomonas syringae* pv. *tomato* DC3000: a model pathogen for probing disease susceptibility and hormone signalling in plants. Annu Rev Phytopathol 51:1003–1011

Xing D, Chen Z (2006) Effects of mutations and constitutive overexpression of *EDS1* and *PAD4* on plant resistance to different types of microbial pathogens. Plant Sci 171:251–262

Xing DH, Lai ZB, Zheng ZY, Vinod KM, Fan BF, Chen ZX (2008) Stress- and pathogen-induced *Arabidopsis* WRKY48 is a transcriptional activator that represses plant basal defense. Mol Plant 1:459–470

Xu R, Song F, Zheng Z (2006) *OsBISAMT1*, a gene encoding *S*-adenosyl-L-methionine: salicylic acid carboxyl methyltransferase, is differentially expressed in rice defense responses. Mol Biol Rep 33:223–231

Xu J, Audenaert K, Hofte M, De Vleesschauwer D (2013) Abscisic acid promotes susceptibility to the rice leaf blight pathogen *Xanthomonas oryzae* pv. *oryzae* suppressing salicylic acid-mediated defenses. PLoS One 8:e67413

Yalowsky S, Baluska F, Jones A (2010) Integrated G proteins signaling in plants. Springer, Heidelberg

Yan Y, Christensen S, Isakeit T, Engelberth J, Meeley R, Hayward A, Emery RJ, Kolomiets MV (2012) Disruption of OPR7 and OPR8 reveals the versatile functions of jasmonic acid in maize development and defense. Plant Cell 24:1420–1436

Yang D-L, Yang Y, He Z (2013) Roles of plant hormones and their interplay in rice immunity. Mol Plant 6:675–685

Yao T, Ndoja A (2012) Regulation of gene expression by the ubiquitin-proteasome system. Semin Cell Dev Biol 23:523–529

Yao C, Wu Y, Nie H, Tang D (2012) RPN1a, a 26S proteasome subunit, is required for innate immunity in Arabidopsis. Plant J 71:1015–1028

Yin W, Dong S, Zhai L, Lin Y, Zheng X, Wang Y (2013) The *Phytophthora sojae Avr1d* gene encodes an RxLR-dEER effector with presence and absence polymorphisms among pathogen strains. Mol Plant Microbe Interact 26:958–968

Yong G, Tingting L, Yang L, Cabda R, Yun Z, Maolin W (2010) Isolation and characterization of gene encoding G protein α subunit protein responsive to plant hormones and abiotic stresses in *Brassicus napus*. Mol Biol Rep 37:3957–3965

Yoshioka K, Moeder W, Kang H-G, Kachroo P, Masmoudi K, Berkowitz G, Klessig D-F (2006) The chimeric *Arabidopsis* CYCLIC NUCLEOTIDE-GATED ION CHANNEL11/12 activates multiple pathogen responses. Plant Cell 18:747–763

Zamioudis C, Peterse CMJ (2012) Modulation of host immunity by beneficial microbes. Mol Plant-Microbe Interact 25:139–150

Zeidler D, Zahringer U, Gerber I, Dubery I, Hartung T, Bors W, Hutzler P (2004) Innate immunity in Arabidopsis thaliana: lipopolysaccharides activate nitric oxide synthase (NOS) and induce defense genes. Proc Natl Acad Sci U S A 101:15811–15816

Zeng Q, Wang X, Running MP (2007) Dual lipid modification of Arabidopsis Gγ-subunits is required for efficient plasma membrane targeting. Plant Physiol 143:1119–1131

Zeng L, Velasquez AC, Munkvold KR, Zhang J, Martin GB (2012) A tomato LysM receptor-like kinase promotes immunity and its kinase activity is inhibited by AvrPtoB. Plant J 69:92–103

Zhang J, Zhou J-M (2010) Plant immunity triggered by microbial molecular signatures. Mol Plant 3:783–793

Zhang DP, Wu ZY, Li XY, Zhao ZX (2002) Purification and identification of a 42-kilodalton abscisic acid-specific-binding protein from epidermis of broad bean leaves. Plant Physiol 128:714–725

Zhang Y, Tessaro MJ, Lassner M, Li X (2003) Knockout analysis of *Arabidopsis* transcription factors *TGA2*, *TGA5*, and *TGA6* reveals their redundant and essential roles in systemic acquired resistance. Plant Cell 15:2647–2653

Zhang Z-G, Wang C, Li J, Ji R, Shen G, Wang S-C, Zhou X, Zheng X-B (2004) The role of SA in the hypersensitive response and systemic acquired resistance induced by elicitor PB90 from *Phytophthora boehmeriae*. Physiol Mol Plant Pathol 65:31–38

Zhang AY, Jiang MY, Zhang JH, Tan M, Hu XL (2006) Mitogen-activated protein kinase is involved in abscisic acid-induced antioxidant defense and acts downstream of reactive oxygen species production in leaves of maize plants. Plant Physiol 141:475–487

Zhang A, Jiang MY, Zhang J, Ding H, Xu S, Hu X, Tan M (2007a) Nitric oxide induced by hydrogen peroxide mediates abscisic acid-induced activation of the mitogen-activated protein kinase cascade involved in antioxidant defense in maize leaves. New Phytol 175:36–50

Zhang J, Shao F, Li Y, Cui H, Chen L, Li H, Zou Y, Long C, Lan L, Chai J, Chen S, Tang X, Zhou J-M (2007b) A *Pseudomonas syringae* effector inactivates MAPKs to suppress PAMP-induced immunity in plants. Cell Host Microbe 1:175–185

Zhang W, He SY, Assmann SM (2008) The plant innate immunity response in stomatal guard cells invokes G-protein-dependent ion channel regulation. Plant J 56:984–996

Zhang J, Li W, Xiang T, Liu Z, Laluk K, Ding X, Zou Y, Gao M, Zhang X, Chen S, Mengiste T, Zhang Y, Zhou JM (2010) Receptor-like cytoplasmic kinases integrate signaling from multiple plant immune receptors and are targeted by a *Pseudomonas syringae* effector. Cell Host Microbe 7:290–301

Zhang W, Jeon BW, Assmann SM (2011) Heterotrimeric G-protein regulation of ROS signaling and calcium currents in *Arabidopsis* guard cells. J Expt Bot 62:2371–2379

Zhang H, Gao Z, Zheng X, Zhang Z (2012a) The role of G-proteins in plant immunity. Plant Signal Behav 7:1284–1288

Zhang L, Li Y, Lu W, Meng F, Wu C, Guo X (2012b) Cotton *GhMKK5* affects disease resistance, induces HR-like cell death, and reduces the tolerance to salt and drought stress in transgenic *Nicotiana benthamiana*. J Exp Bot 63:3935–3952

Zhang X, Wang C, Zhang Y, Sun Y, Mou Z (2012c) The Arabidopsis mediator complex subunit16 positively regulates salicylate-mediated systemic acquired resistance and jasmonate/ethylene-induced defense pathways. Plant Cell 24:4294–4309

Zhang X, Wu Q, Ren J, Qian W, He S, Huang K, Yu XC, Gao Y, Huang P, An C (2012d) Two novel RING-type ubiquitin ligases, RGLG4 and RGLG4, are essential for jasmonate-mediated responses in Arabidopsis. Plant Physiol 160:808–822

Zhang L, Kars I, Essenstam B, Liebrand TWH, Wagemakers L, Elberse J, Tagkalaki P, Tjoitang D, Ackerveken G, van Kan JAL (2014) Fungal endoploygalacturonases are recognized as microbe-associated molecular patterns by the Arabidopsis receptor-like protein RESPONSIVENESS TO BOTRYTIS POLYGALACTURONASES1. Plant Physiol 164:353–364

Zhao XM, She XP, Yu W, Liang XM, Du YG (2007) Effects of oligochitosans on tobacco cells and role of endogenous nitric oxide burst in resistance of tobacco to *Tobacco mosaic virus*. J Plant Pathol 89:55–65

Zheng Z, Qamar SA, Chen Z, Mengiste T (2006) Arabidopsis WRKY33 transcription factor is required for resistance to necrotrophic fungal pathogens. Plant J 48:596–605

Zheng X-y, Spivey NW, Zeng W, Liu P-P, Fu ZQ, Klessig DF, He SY, Dong X (2012) Coronatine promotes *Pseudomonas syringae* virulence by activating a signaling cascade that inhibits salicylic acid accumulation. Cell Host Microbe 11:587–596

Zheng X, McLellan H, Fraiture M, Liu X, Boevink PC, Gilroy EM, Chen Y, Kandel K, Sessa G, Birch PRJ, Brunner F (2014) Functionally redundant RXLR effectors from *Phytophthora infestans* act at different steps to suppress early flg22-triggered immunity. PLoS Pathog 10(4):e1004057

Zhu YJ, Qiu X, Moore PH, Borth W, Hu J, Ferreira S, Albert HH (2003) Systemic acquired resistance induced by BTH in papaya. Physiol Mol Plant Pathol 63:237–248

Zhu H, Li G-J, Ding L, Cui X, Berg H, Assmann SM, Xia Y (2009) *Arabidopsis* extra large G-protein 2 (XLG2) interacts with the Gβ subunit of heterotrimeric G protein and functions in disease resistance. Mol Plant 2:513–525

Zhu L, Li Y, Li L, Yang J, Zhang M (2011a) Ethylene is involved in leafy mustard systemic resistance to *Turnip mosaic virus* infection through the mitochondrial alternative oxidase pathway. Physiol Mol plant Pathol 76:166–172

Zhu Z, An F, Feng Y, Li P, Xue LAM, Jiang Z, Kim JM, To TK, Li W, Zhang X, Yu Q, Dong Z, Chen WQ, Seki M, Zhou JM, Guo H (2011b) Derepression of ethylene-stabilized transcription factors (EIN3/EIL1) mediates jasmonate and ethylene signaling synergy in *Arabidopsis*. Proc Natl Acad Sci U S A 108:12539–12544

Zipfel C (2009) Early molecular events in PAMP-triggered immunity. Curr Opin Plant Biol 12:414–420

Zipfel C, Rathjen P (2008) Plant immunity: AvrPto targets the frontline. Curr Biol 18:R218–R220

Zipfel C, Robatzek S, Navarro L, Oakeley EJ, Jones JDG, Felix G, Boller T (2004) Bacterial disease resistance in Arabidopsis through flagellin perception. Nature 428:764–767

Zipfel C, Kunze G, Chinchilla D, Caniard A, Jones JD, Boller T, Felix G (2006) Perception of the bacterial PAMP EF-Tu by the receptor EFR restricts *Agrobacterium*-mediated transformation. Cell 125:749–760

Zottini M, Costa A, De Michele R, Ruzzene M, Carimi F, Lo Schiavo F (2007) Salicylic acid activates nitric oxide synthesis in *Arabidopsis*. J Exp Bot 58:1397–1405

Chapter 3
Switching on Plant Immune Signaling Systems Using Microbe-Associated Molecular Patterns

Abstract Plants are endowed with innate immune system with ability to confer resistance against wide-range of oomycete, fungal, bacterial and viral pathogens. This basal resistance is not expressed in healthy unstressed plants. The pathogen associated molecular patterns (PAMPs)/microbe associated molecular patterns (MAMPs)/microbe-derived elicitors switch on various signaling systems activating the basal resistance. They can induce resistance against a wide range of biotrophic, hemibiotrophic, and necrotrophic pathogens. Thus engineering of PAMPs/MAMPs/elicitors may offer new opportunities for generating broad-spectrum disease resistance in various crops. Genetic engineering using genes encoding PAMPs may be highly effective in controlling diseases. Alternatively, formulations of the PAMPs can be developed and used as plant defense activators to manage wide-spectrum of diseases. The time of induction, intensity of induction, and duration of induction of the defense signals may vary depending on PAMPs. Amount of PAMP available in the plant-pathogen interaction site may determine the intensity of induced gene expression. Each PAMP may regulate distinctly different signaling pathway(s). Sometimes different PAMPs may induce the same signaling system, but the intensity of the defense signaling gene expression may differ. The same PAMP may behave differently in different plant system. A single PAMP may not be able to activate all the defense signaling-related genes and several PAMPs may be required to activate the complex signaling systems. PAMPs may act synergistically or antagonistically in inducing defense signaling. Some PAMPs have additive effect, while others show antagonistic effect between them. Selection of suitable PAMPs to manage different pathogens in different host plants is important in exploiting the PAMPs for disease management. Several commercial formulations of PAMPs/MAMPs/elicitors have been developed in different countries and widely used as foliar spray. The PAMP harpin formulation induces both local and systemic resistance in foliage and also induces resistance in fruits. Several factors such as environment, genotype, and crop nutrition determine the efficacy of harpin in controlling diseases under field conditions. The time of application is very critical in enhancing the efficacy of harpin in controlling diseases. Harpin should be applied before the pathogen invasion. The concentration of the harpin applied also determines the efficacy of the treatment in controlling diseases. Foliar application of harpin not only reduces disease incidence but also acts as a growth promoter. Bioengineering the genes

P. Vidhyasekaran, *Switching on Plant Innate Immunity Signaling Systems*,
Signaling and Communication in Plants, DOI 10.1007/978-3-319-26118-8_3

encoding proteinaceous PAMPs such as harpins, elicitins, and flagellins has been found to be effective tool to manage crop diseases. Levels of the PAMP harpin gene expression may vary among different transgenic plant lines. The transgenic plants which show high level of expression of harpin gene expression show very high level of resistance, while the transgenic plants which show low level of expression of the harpin gene show very low level of resistance against pathogens, suggesting that the transgenic lines should be carefully selected to generate highly useful disease-resistant cultivars. Proper selection of promoters for developing transgenic disease-resistant plants using PAMP genes is necessary. Expression of harpin genes can be enhanced resulting in higher accumulation of harpin by properly selecting the promoter for gene transcription. A promoter of the rice phenylalanine ammonia-lyase (PAL) gene was used to regulate the expression of cryptogein (*crypt*) gene in tobacco. The PAL promoter had a low level of constitutive expression and was strongly induced by pathogen infection. The transgenic tobacco plants expressing cryptogein with the inducible PAL promoter showed significantly enhanced resistance against various pathogens, suggesting that low-level constitutive expression of elicitin gene may have potential use in generating broad-spectrum disease-resistant plants. The constitutive expression of elicitin gene in transgenic plants will be ideal to induce resistance against wide-range of pathogens. Continuous recognition of the elicitin signal has been shown to be a prerequisite for prolonged activation of signaling events in tobacco cells. However, elicitin is known to induce cell necrosis and hence constitutive overexpression of the gene may affect the agronomic characters of the transgenic plants. To reduce the phytotoxicity of the elicitin, transgenic plants harboring a pathogen-inducible promoter were developed to express the elicitin at low level. The gene encoding the proteinaceous elicitor FsphDNase shows DNase activity, which can cause damage of DNA within nuclei of plant cells. Constitutive activation of *FsphDNase* within the plant cell may be destructive. However, use of pathogen-inducible promoters overcomes the adverse effect and the transgenic tobacco plants expressing the elicitor gene showed no detectable morphological differences from the wild-type plants. These transgenic plants also showed enhanced resistance against fungal, bacterial, and oomycete pathogens. Selection of suitable pathogen-inducible promoter for expressing the elicitor gene appears to be a perquisite for developing disease-resistant plants without any reduction in yield potential. Collectively these studies suggest that the PAMPs have high potential to engineer and manipulate defense signaling systems to intervene in pathogensesis of a wide-range of pathogens. Transgenic plants are often considered as poor yielder with adverse agronomic characters. The major drawback in developing transgenic plants for management of crop diseases is in their adverse effect on crop growth and yield potential. However, the transgenic plants expressing harpin gene did not affect crop growth and yield characters. Technologies have also been developed to reduce the adverse effect of some PAMPs on plant growth characters by using pathogen-inducible promoters instead of using constitutively expressing promoters.

3.1 PAMP-Triggered Immunity

3.1.1 PAMPs Detected in Bacterial, Fungal, Oomycete, and Viral Pathogens

Plants have built-in defense mechanisms to fight against fungal, bacterial, viral and oomycete pathogens (Vidhyasekaran 1997, 2007, 2014, 2015). Plant innate immunity is a potential basal defense system existing in plant kingdom (Iriti and Faoro 2007; Silipo et al. 2010; Drutskaya et al. 2011; Zamioudis and Peterse 2012; Macho and Zipfel 2014; Malinovsky et al. 2014; Zipfel 2014). This system provides an arsenal of deadly weapons to the host plants to fight against the pathogens (Kwon and Yun 2014). This immune system is in sleeping state in normal healthy plants, but when activated it can produce and secrete several antimicrobial compounds to ward off the pathogen attack (Segonzac and Zipfel 2011; Vidhyasekaran 2014). Pathogen-associated molecular patterns (PAMPs) are the potential tools to trigger the plant innate immune responses (Hwang and Hwang 2011; Alkan et al. 2012; Fu et al. 2014; Lloyd et al. 2014; Macho and Zipfel 2014; Trouvelot et al. 2014). PAMPs are the molecular fingerprints/signatures of pathogens and they are evolutionarily conserved building blocks of microbial surfaces that directly bind to the plant pattern recognition receptors (PRRs) residing in plant cell membrane and induce defense responses (Nicaise et al. 2009; Tsuda and Katagiri 2010; Thomma et al. 2011; Chen et al. 2014a; Vidhyasekaran 2014).

Several PAMPs have been isolated from various bacterial, fungal, and oomycete pathogens (Fig. 3.1) The PAMPs isolated and characterized from bacterial pathogens include flg22 (Gómez-Gómez and Boller 2002; Takai et al. 2008; Boller and Felix 2009; Clarke et al. 2013; Yu et al. 2014), flgII-28 (Cai et al. 2011; Clarke et al. 2013; Rosli et al. 2013; Veluchamy et al. 2014), liposaccharides (LPS) (Silipo et al. 2005; Newman et al. 2007; Silipo et al. 2008; Molinaro et al. 2009; Nicaise et al. 2009; Zeidler et al. 2010; Proietti et al. 2014), and peptidoglycans (Gust et al. 2007; Willmann et al. 2011; Erbs et al. 2008). These PAMPs have been purified from bacterial flagella structure and bacterial cell envelope lipopolysaccharides and peptidoglycans. PAMPs have also been detected in several bacterial proteins including elongation factor proteins (Felix and Boller 2003; Kunze et al. 2004; Zipfel et al. 2006; Zipfel 2008), cold-shock proteins (Felix and Boller 2003), harpin proteins (Engelhardt et al. 2009; Tampakaki et al. 2010; Boureau et al. 2011; Haapalainen et al. 2011; Pavli et al. 2011; Li et al. 2014a, b), sulfated proteins (Lee et al. 2006, 2009; Park et al. 2010a, b; Zhang and Zhou 2010), and the bacterial superoxide dismutase enzyme (Watt et al. 2006). A proteinaceous PAMP termed eMax (enigmatic MAMP of *Xanthomonas*) has been isolated from *Xanthomonas axonopodis* pv. *citri* (Jehle et al. 2013a, b). PAMPs have also been detected in bacterial rhamnolipids (Varnier et al. 2009; Vatsa et al. 2010; Sanchez et al. 2013) and bacterial nonmethylated CpG DNA (Yakushiji et al. 2009). Nep1-like proteins (NLPs) have been detected in some bacterial pathogens (Mattinen et al. 2004; Pemberton et al. 2005; Küfner et al. 2009) and these NLPs have been shown to trigger immune responses.

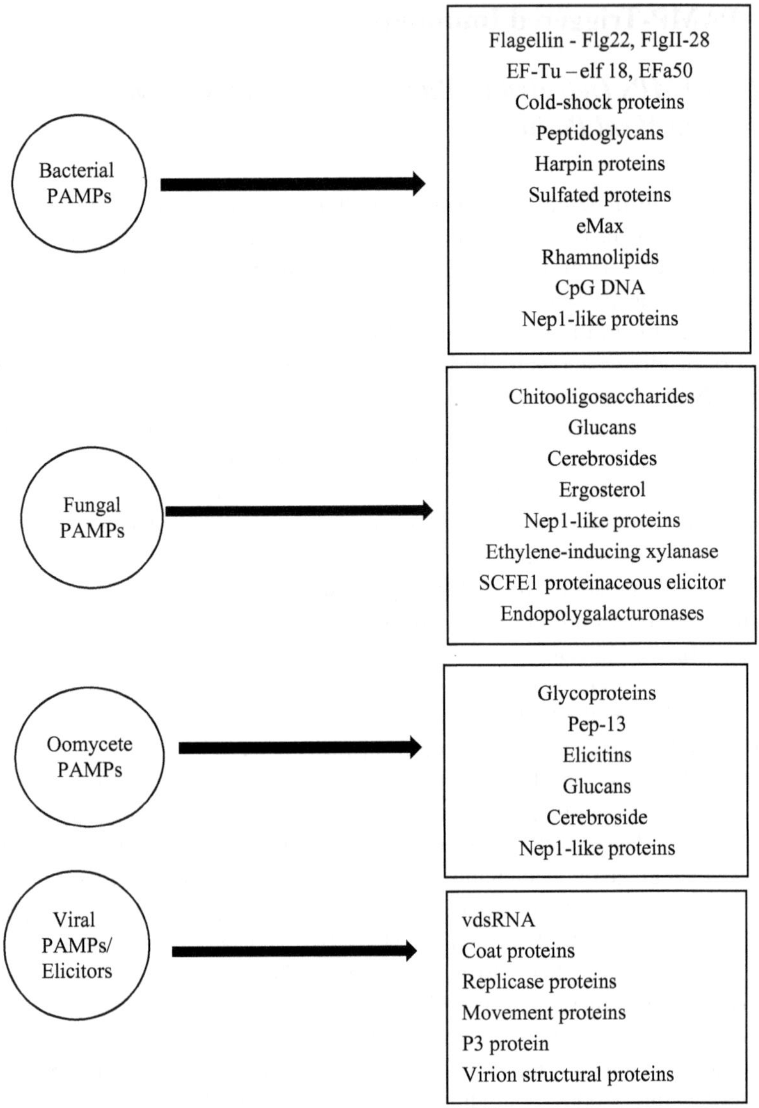

Fig. 3.1 PAMPs detected in bacterial, fungal, oomycete, and viral pathogens'

Several PAMPs have been isolated from fungal pathogens. Chitooligosaccharides are the important PAMPs detected in fungal pathogens (Miya et al. 2007; Hamel and Beaudoin 2010; Lizasa et al. 2010; Shimizu et al. 2010; Lehtonen et al. 2012; Son et al. 2012) and they are known to trigger the plant innate immune responses in a wide range of plants including both monocots and dicots (Shimizu et al. 2010; Son et al. 2012). Several chitooligosaccharides are generated from fungal cell walls at the site of infection through the action of apoplastic chitinases. Chitooctaose generated from fungal cell wall has been reported to be the most potential chitin fragment in eliciting defense responses in plants (Wan et al. 2008). *N*-acetylchitoheptaose generated from cell walls of the rice blast pathogen *Magnaporthe oryzae* has been identified as a potential elicitor triggering defense responses in rice cells (Yamaguchi et al. 2002). Polymeric chitin has also been reported to be a potential elicitor (Petutschnig et al. 2010). Glucans are important components of cell walls of various fungi. Several β-glucan poly- and oligo-saccharides seem to be generated from fungal cell walls at the site of infection through the action of plant β-1,3-glucanases (Silipo et al. 2010). Most of these β-glucan wall components have been recognized as PAMPs (Yamaguchi et al. 2000a, b; Shibuya and Minami 2001; Aziz et al. 2007; Silipo et al. 2010). Cerebrosides are sphingolipid PAMPs detected in a wide range of fungal pathogens (Umemura et al. 2000, 2004; Nakashima et al. 2008). Ergosterol has been detected in cell walls of most fungi (Kasparovsky et al. 2003, 2004; Laquitaine et al. 2006; Lochman and Mikes 2006) and it has been recognized as a fungal PAMP (Naito et al. 2008; Rossard et al. 2010). The fungal protein ethylene-inducing xylanase (EIX) is a PAMP inducing defense reactions in tobacco and tomato (Bailey et al. 1990; Avni et al. 1994; Bar et al. 2010). The necrosis- and ethylene-inducing protein (Nep1)-like proteins (NLPs) have been isolated from various fungal pathogens. NLPs show important characters of PAMPs: they trigger nonself immunity in plants, functional orthologs of NLPs occur in a wide range of pathogens, and they are absent in plant genome (Wang et al. 2004; Staats et al. 2007, Schouten et al. 2008; Motteram et al. 2009; Arenas et al. 2010). A proteinaceous elicitor called SCLEROTINIA CULTURE FILTRATE ELICITOR1 (SCFE1) from *Sclerotinia sclerotiorum* has been shown to induce typical PAMP-triggered immune responses in *Arabidopsis thaliana* (Zhang et al. 2013). Fungal endopolygalacturonases have been recognized as PAMPs by the arabidopsis receptor-like protein RESPONSIVENESS TO BOTRYTIS POLYGALACTURONASES1 (Zhang et al. 2014). An endopolygalacturonase from *Sclerotinia sclerotiorum* induces calcium-mediated signaling and programmed cell death in soybean cells (Zuppini et al. 2005). An endocellulase (EG1) isolated from *Rhizoctonia solani* shows elicitor activity (Ma et al. 2015).

Several different PAMPs have been detected in various oomycete pathogens. A peptide fragment (Pep-13), within an abundant cell wall glycoprotein (GP42) from *Phytophthora sojae*, has been identified as a PAMP in oomycete pathogens (Nürnberger et al. 1994; Hahlbrock et al. 1995). Elicitins are small lipid binding proteins secreted by the oomycetes *Phytophthora* and *Pythium* (Simon-Plas et al. 1997; Baillieul et al. 2003; Qutob et al. 2003; Kim et al. 2010) and they have been identified as PAMPs (Ponchet et al. 1999; Vidhyasekaran 2014). A glucan fragment, hexa-β-glucopyranosyl-D-glucitol isolated from the cell walls of *Phytophthora megasperma* f. sp. *glycinea,* acts as PAMP triggering defense responses (Sharp et al. 1984). A doubly-branched hepta-β-glucoside generated from *P. megasperma* f. sp. *glycinea* glucan is a highly active PAMP (Cheong et al. 1991). A glucan fragment β-1,6-1,3 heptaglucan from *Phytophthora sojae* induced defense responses in soybean (Cheong et al. 1991). *Pythium graminicola* contains a cerebroside elicitor (Umemura et al. 2004).

A cell wall glycoprotein named CBEL (for Cellulose-Binding Elicitor Lectin) isolated from *Phytophthora parasitica* var. *nicotianae* has been identified as an elicitor of defense responses in tobacco (Sèjalon-Delmas et al. 1997; Villalba Mateos et al. 1997). CBEL was found to be localized in the inner and outer layers of the cell walls of the oomycete, and it was present in close contact to the host cell wall during infection (Séjalon et al. 1995). The protein portion of CBEL is composed of two cysteine-rich domains, each containing subdomain homologous to cellulose-binding domains (CBDs) of fungal glucan hydrolases (Gaulin et al. 2006). CBDs of CBEL have been shown to be essential and sufficient to stimulate defense responses and CBDs are considered as PAMPs (Gaulin et al. 2006). CBEL homologues have been detected in various *Phytophthora* species and these homologues contain highly conserved CBD motifs (Khatib et al. 2004). Nep1-like proteins (NLPs) have been detected in a wide range of oomycete pathogens (*Pythium* spp., *Phytophthora* spp., and *Hyaloperonospora arabidopsis*) (Veit et al. 2001; Fellbrich et al. 2002; Qutob et al. 2002, 2006; Gijzen and Nürnberger 2006; Kanneganti et al. 2006; Küfner et al. 2009; Ottmann et al. 2009; Cabral et al. 2012). These NLPs have been identified as PAMPs (Fellbrich et al. 2002; Qutob et al. 2002; Kanneganti et al. 2006; Küfner et al. 2009).

No classical PAMPs have been detected in plant viruses. Viral double-stranded RNA (vdsRNA) has been identified as a PAMP in plant viruses (Ding 2010). The plant innate immune system detects the presence of dsRNA as aberrant RNA molecule (Wypijewski et al. 2009) and generates small RNAs. The generated small RNAs direct the antiviral machinery to cleave and destroy the invading viral genome (Alvarado and Scholthof 2009). Coat proteins (Culver and Dawson 1991; Bendahmane et al. 1999; Gilardi et al. 2004; Matsumoto et al. 2008; Lovato et al. 2008), replicase proteins (Padgett et al. 1997; Fellers et al. 2002; Erickson et al. 1999), movement proteins (MP) (Ghazala and Varrelmann 2007), P3 protein (Jenner et al. 2002, 2003), and virion structural proteins have been shown to act as elicitors inducing defense responses in plants (Love et al. 2005).

PAMPs are generally detected in a wide range of pathogens. The PAMP flg22 is widespread in bacterial species (Robatzek et al. 2007; Boller and Felix 2009; Nicaise et al. 2009). It has been detected in *Pseudomonas syringae* pv. *tomato* (Li et al. 2005b), *P. syringae* pv. *glycinea* (Taguchi et al. 2003b), *P. syringae* pv. *tabaci* (Taguchi et al. 2003a; Ishiga et al. 2005), *Acidovorax avenae* (Takai et al. 2008), *Xanthomonas campestris* pv. *campestris* (Sun et al. 2006), *Xanthomonas oryzae* pv. *oryzae* (Yu et al. 2014), *Ralstonia solanacearum* (Pfund et al. 2004), and *Agrobacterium* spp. (Felix et al. 1999), It has been detected in a phloem-limited alphaproteobacterium, *Candidatus* Liberibacter solanacearum (Hao et al. 2014). Some PAMPs are detected not only in plant pathogens, but also in human and animal pathogens (Felix and Boller 2003; Robatzek et al. 2007; Lee et al. 2009). The PAMP flg22 has been detected in *Escherichia coli*, a human and animal pathogen (Robatzek et al. 2007). The human pathogen *Salmonella enterica* PAMP flg22 has been shown to activate PAMP-triggered immunity in plants (Garcia and Hirt 2014). Flg22 has been detected in the root-colonizing bacterium *Pseudomonas fluorescens* (Leon-kloosterziel et al. 2005). Ax21 is a proteinaceous PAMP detected in

Xanthomonas spp. It is a sulfated protein and is present in all sequenced *Xanthomonas* species, in *Xyllella fastidiosa* and in the human pathogen *Stenotrophomonas maltophila* (Lee et al. 2006, 2009; Park et al. 2010a, b; Zhang and Zhou 2010).

PAMPs are detected not only in pathogens, but also in several nonpathogens, saprophytes, and biocontrol agents (Felix and Boller 2003; Denoux et al. 2008; Takakura et al. 2008; Silipo et al. 2008; Thuerig et al. 2006; Van Loon et al. 2008; Molinaro et al. 2009; Vallance et al. 2009; Varnier et al. 2009; Garcia and Hirt 2014; Quang et al. 2015). Since the PAMPs are detected in all microbes, the PAMPs are often called as microbe-associated molecular patterns (MAMPs) (Viterbo et al. 2007; Denoux et al. 2008; Aslam et al. 2009; Boller and Felix 2009; Jeworutzki et al. 2010; Thomma et al. 2011; Serrano et al. 2012; Jehle et al. 2013a, b; Sanchez et al. 2013; Burketova et al. 2015) and PAMP-triggered immunity is called MAMP-triggered immunity.

3.1.2 Variability in Structure and Function of PAMPs

Variability in structure and function has been reported among various PAMPs. *Acidovorax avenae* strains isolated from rice such as K1 can infect only rice plants, whereas the NI141 strain isolated from finger millet cannot infect rice plants. The rice avirulent NI141 strain induced several immune responses, whereas the rice-avirulent K1 strain of *A. avenae* did not induce these immune responses (Che et al. 2000; Tanaka et al. 2003; Fujiwara et al. 2004). The flagellins isolated from these two strains were structurally different (Che et al. 2000). Flagellin purified from an incompatible strain of *A. avenae* (NI141), induced rapid generation of H_2O_2 accompanying hypersensitive cell death and expression of various defense-related genes in cultured rice cells, whereas the flagellin from the compatible strain did not (Che et al. 2000; Tanaka et al. 2003). Introduction of NI141 flagellin gene into rice also triggered immune responses (Takakura et al. 2008). However, flg22 did not induce defense responses in rice. These results suggest that the recognition mechanism for flagellin may be different between rice and dicotyledonous plants, such as *Arabidopsis* and tomato. When expressed and purified from *Escherichia coli*, both types of flagellins induced H_2O_2 generation in rice. In addition, a deglycosylated flagellin from the compatible strain induced the same immune responses as the flagellin of incompatible strain did (Hirai et al. 2011). These studies suggest that post-translational modifications of flagellins might be associated with the specific induction of immune responses in rice.

Flagellins purified from *P. syringae* pv. *glycinea*, an incompatible pathogen for tobacco, induced immune responses in tobacco, whereas flagellin from *P. syringae* pv. *tabaci*, a compatible pathogen, did not, despite complete amino acid identity (Taguchi et al. 2003b). Flagellins derived from nonadapted bacteria but having identical protein sequences differentially induce strong defense responses in nonhost plants, suggesting that other domains and/or posttranslational modifications of flagellin may be involved in triggering immune responses (Taguchi et al. 2003a, b,

2006; Takeuchi et al. 2003, 2007). The major difference between various flagellins has been suggested to be in the glycosylation sites in flagellin (Ishiga et al. 2005; Taguchi et al. 2006; Takai et al. 2008). Two distinct EF-Tu epitopes acting as PAMPs have been detected in EF-Tu of *Acidovorax avenae*. The epitope elf18 induces defense responses in plants belonging to Brassicaceae family and does not induce PAMP-triggered immune responses (PTI) in rice. The other epitope in the EF-Tu of *A. avenae* is an EF-Tu middle region comprising Lys176 to Gly225, termed EFa50, which induces PTI in rice (Furukawa et al. 2014).

Pathogens contain a wide array of PAMPs of diverse chemical structures and every pathogen contains or secretes multiple PAMPs (Vidhyasekaran 2014). Each PAMP may regulate induction of different defense genes (Gust et al. 2007; Denoux et al. 2008; Aslam et al. 2009). The time of induction (Luna et al. 2011), intensity of induction (Lecourieux-Ouaked et al. 2000), and duration of induction (Aziz et al. 2007) of the defense signals may vary depending on PAMPs. Amount of PAMP available in the plant-pathogen interaction site may determine the intensity of induced gene expression (Peng et al. 2004b). Each PAMP may regulate distinctly different signaling pathway(s) (Kariola et al. 2003; Khatib et al. 2004; Qutob et al. 2006). Sometimes different PAMPs may induce the same signaling system, but the intensity of the defense signaling gene expression may differ (Kunze et al. 2004; Zipfel et al. 2006). The same PAMP may behave differently in different plant system (Reboutier et al. 2007). A single PAMP may not be able to activate all the defense signaling-related genes and several PAMPs may be required to activate the complex signaling systems (Fujiwara et al. 2004; Zipfel et al. 2004; Thilmony et al. 2006). PAMPs may act synergistically or antagonistically in inducing defense signaling (Yamaguchi et al. 2002; Kariola et al. 2003; Aslam et al. 2009). Some PAMPs have additive effect, while others show antagonistic effect between them (Aslam et al. 2009).

3.2 Harpin PAMPs as Molecular Tools to Manipulate PAMP-Triggered Immunity

3.2.1 Harpins Acting as PAMPs

Harpins are glycine-rich and heat-stable proteins that are secreted through type III secretion system (TTSS) in gram-negative plant pathogenic bacteria (Choi et al. 2013). The structural proteins in the TTSS are encoded by hypersensitive response and pathogenicity (*hrp*) genes (Tampakaki et al. 2010). The *hrp* genes have been detected in several phytopathogenic bacteria including members of the genera *Erwinia*, *Pantoea*, *Pseudomonas*, *Xanthomonas*, and *Ralstonia* (Table 3.1; Kvitko et al. 2007; Chen et al. 2008a, b; Engelhardt et al. 2009; Tampakaki et al. 2010; Boureau et al. 2011; Tsunemi et al. 2011; Choi et al. 2013). Harpins are structurally unrelated proteins that are produced and secreted by many bacterial pathogens

Table 3.1 Harpins detected in plant pathogenic bacterial pathogens

Harpins	Plant Pathogenic bacteria
$HrpN_{Ea}$	*Erwinia amylovora*
$HrpN_{Ecc}$	*Erwinia carotovora* subsp. *carotovora*
$HrpN_{Ech}$	*Erwinia chrysanthemi*
$HrpN_{Ep}$	*Erwinia pyrifoliae*
$HrpZ_{Psph}$	*Pseudomonas syringae* pv. *phaseolicola*
HrpW1, HopAK1	*Pseudomonas syringae* pv. *syringae*
HrpZ	*Pseudomonas syringae* pv. *tabaci*
HrpW1, HrpZ1	*Pseudomonas syringae* pv. *tomato*
$Hpa1_{Xag}$	*Xanthomonas axonopodis* pv. *glycines*
$Hpa1_{Xooc}$	*Xanthomonas oryzae pv. oryzicola*
$Hpa1_{Xoo}$	*Xanthomonas oryzae pv. oryzae*
Hpa1	*Xanthomonas axonopodis* pv. *citri*
Hpa_{Xm}	*Xanthomonas citri* subsp. *malvacearum*
PopA1, PopW	*Ralstonia solanacearum*

(Engelhardt et al. 2009; Sang et al. 2012; Choi et al. 2013; Ger et al. 2014; Li et al. 2014a, b). The Hpa1 proteins have been identified as harpins in *Xanthomonas* spp. The Hpa1 harpins include $Hpa1_{Xoo}$ produced by *Xanthomonas oryzae* pv. *oryzae*, the rice bacterial blight pathogen (Li et al. 2004b, 2005a; Wen and Wang 2001; Peng et al. 2004a, b; Fu et al. 2014; Li et al. 2014a, b), $Hpa1_{Xooc}$ (Chen et al. 2008a, b) produced by the rice bacterial leaf streak pathogen *X. oryzae* pv. *oryzicola* (Liu et al. 2006), and $Hpa1_{Xag}$ of *X. axonopodis* pv. *glycines* (Kim et al. 2003). *Pseudomonas syringae* pv. *tomato* DC3000 produces two harpins, HrpZ1 and HrpW1 (Kvitko et al. 2007). The harpins $HrpN_{Ea}$ (Wei et al. 1992; Reboutier et al. 2007), $HrpN_{Ech}$ (Bauer et al. 1995), $HrpZ_{Pss}$ (He et al. 1993), and $HrpZ_{Psph}$ (Tampakaki and Panopoulos 2000), are produced by *Erwinia amylovora*, *Erwinia chrysanthemi*, *Pseudomonas syringae* pv. *syringae*, and *P. syringae* pv. *phaseolicola*, respectively. The harpin HrpW orthologs include *E. amylovora* HrpW, *P. syringae* pv. *syringae* HrpW1 and HopAK1, and *Ralstonia solanacearum* PopW (Li et al. 2010; Choi et al. 2013). The harpin PopA1 has been detected in *R. solanacearum*, while Hpa_{Xm} has been detected in *X. citri* subsp. *malvacearum* (Choi et al. 2013).

Harpin protein enhances plant disease resistance by activating PAMP-induced immunity (Chuang et al. 2014). Harpins act as PAMPs triggering plant immune responses in several plants (Dong et al. 1999, 2004; Kariola et al. 2003; Alfano and Collmer 2004; Fontanilla et al. 2005a, b; Jang et al. 2006; Sohn et al. 2007; Chuang et al. 2010). The harpin HrpN, derived from *E. amylovora*, *E. pyrifoliae*, and *E. carotovora* was shown to trigger plant immune hormone signaling systems in *Arabidopsis*, tomato, and tobacco (Dong et al. 2004; Sohn et al. 2007; Chuang et al. 2010; Boureau et al. 2011). The *P. syringae* harpin HrpZ1 (Pavli et al. 2011), the *X. oryzae* pv. *oryzae* harpin Hpa1 (Peng et al. 2004a; Miao et al. 2010a, b), and the harpin protein PopA from *Ralstonia solanacearum* (Belbahri et al. 2001) elicit innate immune responses in various plants. The harpin HrpZ1 triggers several defense signaling systems and hypersensitive response (HR) in various plant species (Nürnberger et al. 2004; Grant et al. 2006).

The 24-amino-acid HrpZ fragment found in the C-terminal region showed the PAMP activity of HrpZ of *Pseudomonas syringae* (Haapalainen et al. 2011). A C-terminal fragment of the HrpZ1 protein of *P. syringae* pv. *phaseolicola* retained the ability of the harpin to trigger plant immunity. Random insertion mutagenesis of HrpZ1 further revealed that the C- terminus is important for the PAMP activity of the protein (Engelhardt et al. 2009). A fragment ($HpaG_{10–42}$) in the $HpaG_{Xooc}$ harpin triggered stronger defense responses in rice than the full-length parent protein $HpaG_{Xooc}$, suggesting that the Hpa fragment may have the PAMP elicitor activity (Chen et al. 2008a, b). The fragment comprising residues 10 through 42 of Hpa1, HpaG10-42, triggered defense responses in wheat (Wang et al. 2014). These studies suggest that a PAMP resides within the harpin structure.

The harpin HrpZ1 produced by *Pseudomonas syringae* is a reminiscent of a PAMP that triggers immunity-associated responses in a variety of plants (Engelhardt et al. 2009). HrpZ1 binds to plant membranes with high affinity and specificity, suggesting that the activation of plant immunity-associated responses by HrpZ1 is receptor-mediated. The binding site found in the microsomes was protease- and heat-resistant, suggesting that the binding site may therefore not be a protein at all (Engelhardt et al. 2009). HrpN harpin from *E. amylovora* has been shown to bind to a small 6.5-kDa plasma membrane-associated protein from apple (HrpN-interacting protein from *Malus*, HIPM). Orthologs of HIPM have been found in rice (OsHIPM) and in *Arabidopsis* (AtHIPM) (Oh and Beer 2007). These studies suggest that harpins possess all important characters of PAMPs: wide occurrence in various bacterial species, binding with PRRs and triggering innate immune responses.

Many bacterial pathogens use type-III protein secretion systems (TTSS) to infect plants. TTSS are molecular conduits that facilitate the injection of bacterial effectors into plant cells to manipulate host physiology. Harpins from various *P. syringae* pathovars form ion-conducting pores, suggesting a role of the harpin proteins in effector delivery during infection (Lee et al. 2001b; Fu et al. 2006; Tampakaki et al. 2010). The harpin HrpZ of *P. syringae* showed membrane-binding and pore-forming activities *in vitro*, suggesting that it could be targeted to the host cell plasma membrane (Haapalainen et al. 2011). HrpZ was found to interact with the lipid phosphatidic acid and pore-formation by HrpZ in artificial lipid vesicles was found to be dependent on the presence of phosphatidic acid. In addition, HrpZ was able to form pores in vesicles prepared from *Arabidopsis thaliana* plasma membrane (Haapalainen et al. 2011). HrpG forms dimers and higher order oligomers. The oligomerization was mainly mediated by a region near the C-terminus of the protein and the same region was also found to be essential for membrane pore formation. Phosphatidic acid binding appears to be mediated by two regions separate in the primary structure. A 24-amino-acid HrpG fragment found in the region was shown to be indispensable for the oligomerization and pore formation functions (Haapalainen et al. 2011). These studies suggest that the pore formation activity of harpins may facilitate translocation of the PAMP found within the harpin structure into plant cells and trigger the expression of genes involved in defense signaling systems.

3.2.2 Harpin-Induced Plant Immune Signal Transduction Systems

When harpin is applied to a plant, it binds with the specific receptors found in the plasma membranes of the plant surface. A nonproteinaceous receptor site for the harpin from the bean halo-blight pathogen *P. syringae* pv. *phaseolicola* has been identified in tobacco plasma membranes (Lee et al. 2001a). Most of the PRRs are receptor-like kinases (LRR-RLKs). The PRRs interact with several transmembrane LRR receptor kinases (Zipfel 2009). Expression of several LRR receptor kinases have been found to be upregulated in transgenic cotton plants expressing the harpin $Hpa1_{Xoo}$ gene (Miao et al. 2010b). On perception of harpin, several downstream signaling events are activated (Fig. 3.2). Hapins alter the Ca^{2+} signaling system and activate plant immune signal transduction system. $Harpin_{Psph}$ derived from *Pseudomonas syringae* pv. *phaseolicola* can make pores in artificial membranes and cause Ca^{2+}-channel responses (Lee et al. 2001a). $HrpN_{ea}$ induced Ca^{2+} influxes, K^+/H^+ exchange responses, and membrane depolarization in tobacco cell suspension cell cultures (Wei et al. 1992; Pike et al. 1998). Harpin treatment induced a rapid and sustained alkalinization of the *Arabidopsis* cell suspension media (Clarke et al. 2005). The K^+/H^+ exchange response may be mediated by Ca^{2+} influx (Atkinson et al. 1990). K^+/H^+ exchange response is an important component in the defense signaling system (Orlandi et al. 1992). Transient shifts of intracellular and apoplastic pH have been reported to be essential steps in several signal transduction processes (Felle et al. 2004). NAD kinase is an important enzyme involved in Ca^{2+} signaling and in ROS signaling systems. NAD kinase is activated over a pH range of 7.1–6.8. The cytosolic pH in unstimulated tobacco cells was 7.5, which changed to lower level after being stimulated. At the acidic pH, NAD kinase was activated resulting in HR-related defense responses (Karita et al. 2004). The Ca^{2+}-dependent protein kinase, activated by increased cytosolic Ca^{2+}, may evoke ion fluxes that lead to extracellular alkalinization and depolarization of the plasma membrane (Schaller and Oecking 1999). Harpin induces ion flux modulations in *Arabidopsis* cells (El-Maarouf et al. 2001).

More than threefold increase in expression of outward rectifying K^+ channel was observed in harpin- expressing cotton plants (Miao et al. 2010b). K^+ channels modulate the membrane potential and thereby activate Ca^{2+} signaling system (Vidhyasekaran 2014). The major function of K^+ channels is in regulation of membrane voltage control (Maathuis et al. 1997). Two types of K^+ channels have been identified, each with a characteristic voltage dependence, one opens in hyperpolarizing (inward rectifying K^+ channel, K^+_{in}) and another opens at depolarizing (outward-rectifying K^+ channel, K^+_{out}) conditions. Stimulation of H^+-ATPase will hyperpolarize the membrane and direct the K^+ gradient inward, while anion channel activation will depolarize the membrane and enhance the activity of K^+_{out} channel activity (Maathuis et al. 1997). Membrane depolarization (White 2000) or hyperpolarization (Pei et al. 2000) triggers Ca^{2+} channel activation as part of the Ca^{2+} signaling system. K^+ channels regulate both depolarization and hyperpolarization of cell

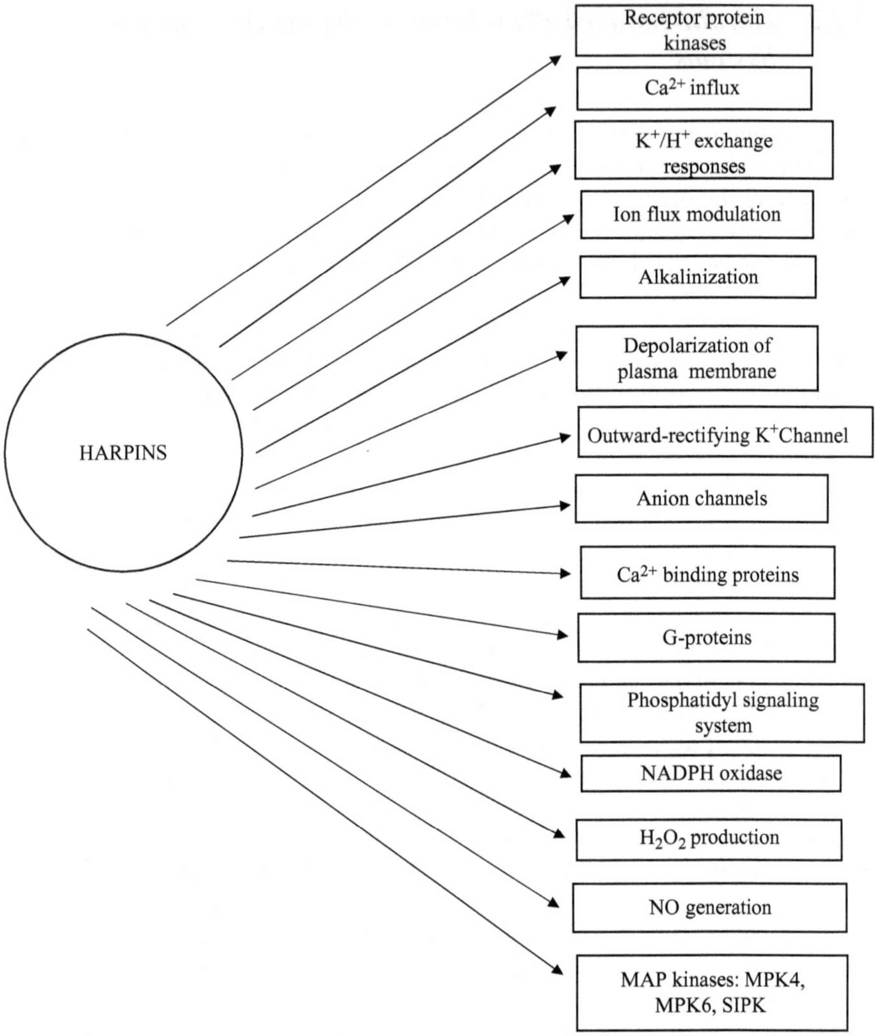

Fig. 3.2 Harpin-triggered early defense signaling systems

membrane (Maathuis et al. 1997) and hence K^+ channels play important role in modulation of Ca^{2+} signaling system.

Purified harpin from *Erwinia amylovora* ($HrpN_{ea}$) activated anion channels in apple cells (Reboutier et al. 2007). Anion efflux results in plasma membrane depolarization (Sanders et al. 2002). Membrane depolarization induces Ca^{2+} influx across the plasma membrane (Thuleau et al. 1994). It has also been shown that Ca^{2+} influx may activate anion channels and increases in cytoplasmic Ca^{2+} activate anion channels (Ward et al. 1995; Marten et al. 2007; Suh et al. 2007). Cytoplasmic Ca^{2+} elevation has been shown to result in activation of S-type anion channels via phosphorylation (Schmidt et al. 1995; Allen et al. 1995; Geiger et al. 2010). The activated anion channels further intensify Ca^{2+} influx which will pave the way for triggering Ca^{2+}-mediated signaling system.

Differentially expressed genes in transgenic cotton plants expressing $Hpa1_{Xoo}$ include calcium binding proteins (Miao et al. 2010b). The calcium signature is perceived by different Ca^{2+}-binding proteins (Kudla et al. 2010). These intracellular Ca^{2+}-binding proteins include calmodulin (CaM) and CaM-related proteins (Snedden and Fromm 2001). The complex Ca^{2+}/CaM has been shown to regulate NAD kinase, which generates NADPH for NADPH oxidase activity (Harding et al. 1997). The activated NADPH oxidase generates superoxide radicals in plant cells, which are converted to H_2O_2 (Sagi and Fluhr 2006).

G-proteins-mediated signaling was also altered in harpin HpaXoo transgenic plants (Miao et al. 2010b). G-proteins have been shown to trigger generation of ROS (Park et al. 2000; Suharsono et al. 2002; Wong et al. 2007; Vidhyasekaran 2014). More than fourfold increase in expression of phosphatidylinositol signaling system has been reported in harpin expressing transgenic plants (Miao et al. 2010b). Phosphatidylinositol is involved in triggering G-protein–mediated ROS production (Fig. 3.3).

Harpin is known to induce ROS generation in plant cells. Purified harpin from *Erwinia amylovora* ($HrpN_{Ea}$) induced ROS production in *Arabidopsis thaliana* (Reboutier et al. 2007) and tobacco cells (Baker et al. 1993). Harpin from *P. syringae* pv. *glycinea* induced ROS production in tobacco cells (Andi et al. 2001). An oxidative burst was detected in response to harpin in *Vitis* cells (Chang and Nick 2012). Harpin-derived from *Xanthomonas oryzae* pv. *oryzae* ($harpin_{Xoo}$) induces ROS production in tobacco leaves (Peng et al. 2004b). Transgenic cotton plants expressing a harpin-encoding gene hpa_{Xoo} showed increased generation of H_2O_2, when the transgenic plants were challenged with the wilt pathogen *Verticillium dahliae* (Miao et al. 2010a). Harpin proteins have been shown to activate the plant defense pathway, which involves transduction of a H_2O_2 signal generated in the apoplast. In the harpin $Hpa1_{Xoo}$-expressing transgenic *Arabidopsis* plants, H_2O_2 was generated in apoplasts in a NADPH oxidase-dependent manner, but accumulated to a greater extent in the cytoplasm than in the apoplast. Inhibiting apoplastic H_2O_2 generation abrogated both cytoplasmic H_2O_2 accumulation and plant resistance to bacterial pathogens (Sang et al. 2012). The results suggest that the apoplastic H_2O_2 is subjected to a cytoplasmic translocation for participation in the pathogen defense.

Both harpin and exogenous H_2O_2 act as signaling molecules to induce the expression of plant defense genes encoding enzymes such as phenylalanine ammonialyase (PAL), glutathione S-transferase (GST) and anthranilate synthase (ASA1) in *Arabidopsis*. H_2O_2 induces the expression of *PAL1* and *GST* but not that of *ASA1*. It is suggested that harpin induces two different signaling pathways, one leading to increased ROS generation and expression of *PAL1* and *GST* mRNA, and another leading to increased *GST* and *ASA1* expression, independent of H_2O_2 (Desikan et al. 1998).

Harpin induces nitric oxide (NO) generation in Arabidopsis cells (Krause and Durner 2004). Application of harpin activated the expression of genes involved in

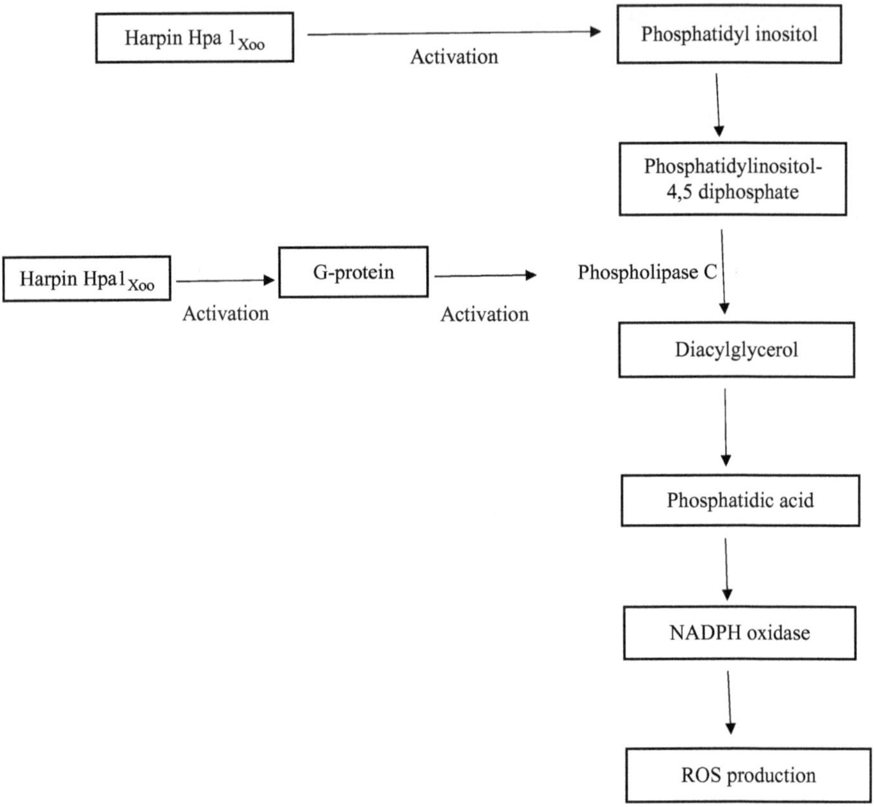

Fig. 3.3 Harpin-triggered phosphatidylinositol signaling system in triggering ROS production (Adapted from Miao et al. 2010b; Vidhyasekaran 2014)

NO generation in *Phalaenopsis* orchids (Chuang et al. 2014). It activates a MAPK in tobacco leaves (Adam et al. 1997). The harpin secreted by *P. syringae* pv. *syringae* induced activation of the *Arabidopsis* MAP kinases AtMPK4 and AtMPK6 (Desikan et al. 2001). $\text{Harpin}_{\text{Psph}}$ induced activation of the salicylic acid-responsive MAPK (SIPK), which was shown to be essential for the expression of *PR* genes in tobacco cells (Lee et al. 2001). Harpin treatment induced a rapid cytochrome C release from mitochondria into the cytosol, which is regarded as a hallmark of programmed cell death or apoptosis (Krause and Durner 2004). Together, these results suggest that harpins induce several early signaling events (Fig. 3.2). Harpin modulates plant gene expression, most likely through its effects on ROS and NO signaling pathways in *Phalaenopsis* orchids (Chuang et al. 2014).

Harpin treatment induced accumulation of SA in Arabidopsis (Fig. 3.4; Dong et al. 2004). Harpin appears to act upstream of SA signaling system. Harpin could not induce resistance in *Arabidopsis* plants that could not accumulate salicylic acid (Peng et al. 2003).

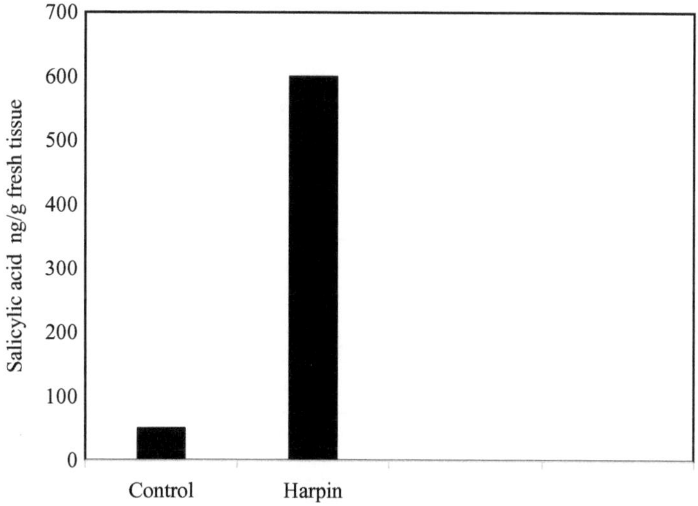

Fig. 3.4 Effect of harpin treatment on accumulation of salicylic acid in *Arabidopsis* (Adapted from Dong et al. 2004)

The regulatory protein EDS1 is required for SA production and it controls SA production (Eulgem et al. 2004; Song et al. 2004). EDS1 processes ROS-derived signals in the SA defense signaling system (Mateo et al. 2004). Harpin-elicited *EDS1*gene and induced disease resistance (Peng et al. 2003). Mutations in *eds1* lead to reduced SA levels in infected leaves (Feys et al. 2001). Collectively these results suggest that harpin acts upstream of SA signaling pathway and is involved in SA production in *Arabidopsis*. The *NPR1* (also known as *NIM1*) gene is a master regulator of the SA-mediated induction of defense genes and it functions downstream of SA (Chern et al. 2008). The *NPR1* gene had low constitutive expression in control plants. However, its expression was enhanced strikingly in transgenic *Arabidopsis* plants expressing harpin gene (Peng et al. 2004b). Harpin treatment induced expression of *NPR1* in *Arabidopsis* (Dong et al. 2004). Harpin induces disease resistance in *Arabidopsis* mediated by *NPR1* gene (Dong et al. 1999). TGA transcription factors play an important role in SA signaling system (Zhou et al. 2000). Several SA-responsive genes are regulated by bZIP transcription factors of TGA family (Ndamukong et al. 2007). Some of the bZIP transcription factors, such as TGA2 and TGA5 in *Arabidopsis*, interact with NPR1 and recognize the *as-1 cis* element found within the promoter of several *PR* genes (Kim and Delaney 2002). These results suggest that harpin induces SA signaling pathway mediated by NPR1-TGA (Fig. 3.5).

Harpin induces transient accumulation of ethylene in *Arabidopsis* (Dong et al. 2004). Harpin triggered several genes essential for ET signaling (Table 3.2). Downstream of ET synthesis, the ET is perceived by several receptors. The ET receptor gene *ETR1* is activated by harpin (Dong et al. 2004). ETR1 functions as an ROS sensor and has been shown to mediate H_2O_2 signaling in *Arabidopsis* (Desikan

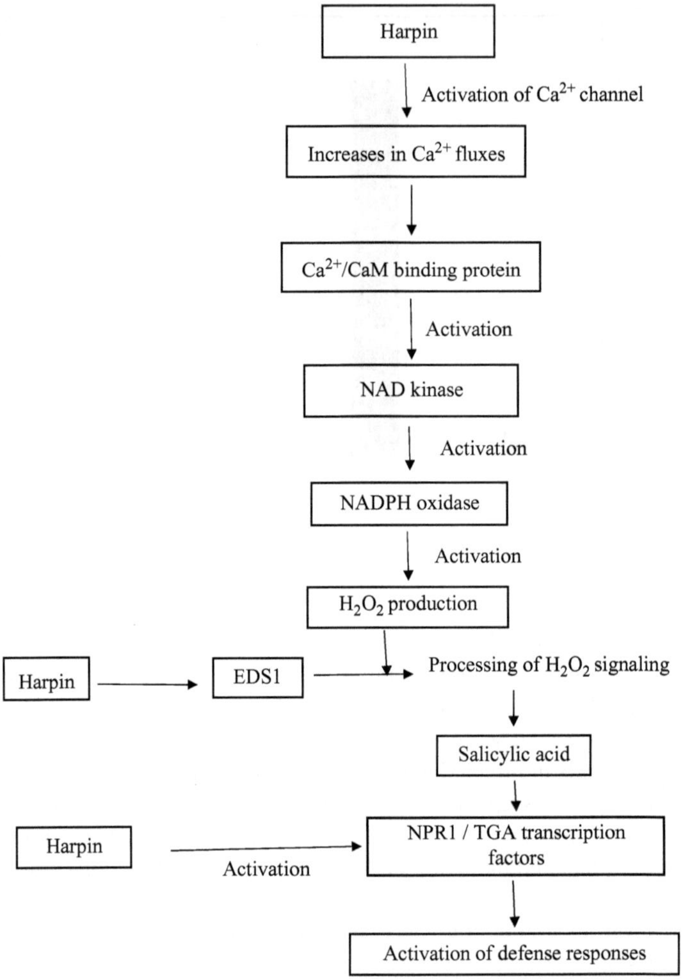

Fig. 3.5 Schematic diagram of possible role of harpin in SA-mediated activation of defense genes

Table 3.2 Effect of harpin treatment in induction of ethylene receptors/ethylene response factors in *Arabidopsis*

Ethylene receptors/ response factors	Intensity of induction of ethylene receptors and ethylene response factors at different hours after harpin treatment				
	0 h	6 h	12 h	24 h	72 h
ETR1	–	–	++	++	++
ERS1	+	+	+++	++	++
EIN2	–	–	+	+	++
CTR1	+	+	+	+	+
ERF1	–	–	++	++	++

Adapted from Dong et al. (2004)

et al. 2005). Another ET receptor gene *ERS1* is also activated by harpin. The ET receptor *CTR1* is not induced by harpin (Dong et al. 2004). However, the main function of ETR1 and ERS1 appears to maintain the ET receptor gene *CTR1* in an active state (Qu et al. 2007). Another ET receptor gene *EIN2* is also increasingly expressed in harpin-treated *Arabidopsis* plants (Table 3.2; Dong et al. 2004). These ethylene receptors also interact with each other and these interactions among the receptors may contribute to ethylene signal output (Gao et al. 2008). The family of ET receptors transmits the signal to downstream effectors (Grefen et al. 2008). The ethylene inducible transcription factor ERF1 has been shown to be a downstream component of ethylene signaling system (Berrocal-Lobo and Molina 2004). Harpin induces expression of the transcription factor ERF1 in *Arabidopsis* (Table 3.2; Dong et al. 2004). The harpin Hpa1 activates the ET signaling pathway (Chen et al. 2008a; Li et al. 2014a). Harpin induced the expression of tomato ethylene response factor5 (designated as SIERF5) in tomato. The amino acid sequence of SIERF5 was closely related to that of CaEREBP-C4, NtERF4, and NsERF4. Overexpression of SIERF5 in *Arabidopsis thaliana* activated a large number of genes involved in signaling pathways of disease resistance (Chuang et al. 2010).

SA mediates transcription of *Chia5*, *PR-1a*, and other genes encoding acidic PR proteins, whereas JA or ethylene mediates *PR-1b* and other basic PR genes. Harpin induces both acidic and basic *PR* genes in *Arabidopsis* and tobacco (Dong et al. 1999; Peng et al. 2003). Hrp_{Ea} induces both SA-dependent and JA/ET-dependent signaling pathways. The Hrp_{Ea} enhances the expression of the *PR1* gene, which is a marker gene for SA-dependent signaling system and it also enhances *PDF1.2*, a marker gene for JA/ET-dependent signaling system in *Arabidopsis* (Kariola et al. 2003). Ectopic expression of harpin *hrf1* gene activates expression of *PR1a* involved in SA signaling-dependent SAR (Systemic Acquired Resistance) pathway in rice and it also activates lipoxygenase (LOX) and allene oxide synthase2 (AOS2), which are involved in JA signaling–dependent ISR (Induced Systemic Resistance) pathway (Li et al. 2012). These results suggest that harpin functions through complex signaling systems involving SA, JA, and ET pathways. Harpin modulates plant gene expression through its effects on jasmonic acid signaling pathway in *Phalaenopsis* orchids (Chuang et al. 2014). However, Dong et al. (2004) suggested that harpin may induce only ET and SA signaling systems and JA signaling system may not be activated by harpin. Harpin does not induce *COI1*, the marker gene for JA signaling system in Arabidopsis (Dong et al. 2004). In fact, methyl jasmonate application inhibits harpin-induced defense response in tobacco (Andi et al. 2001).

3.2.3 Harpin-Induced Defense Response Genes

Harpins have been shown to trigger expression of several defense response genes. Some of the defense genes are expressed only in transgenic plants expressing harpin gene but not in control plants. The defense genes *Chi5*, *PR-1a*, and *GST1* were expressed only in transgenic *Arabidopsis* plants, not in untransformed plants

(Peng et al. 2004a). Harpin ($HrpZ_{Pss}$) induced several PR proteins including peroxidase (PR-9), β-1,3-glucanase (PR-2), and chitinase (PR-3) in cucumber (Strobel et al. 1996). The ectopic expression of harpin *hrf1* in rice significantly increased the expression of *PR-1a* gene, which is involved in the SA signal pathway (Fig. 3.6; Li et al. 2012). The *PR-1b* was expressed in both control and transgenic plants expressing harpin gene; however, its expression was very low in control plants (Peng et al. 2004b). Harpin induces peroxidase and chitinase in Hami melons (*Cucumis melo* L. var. *inodorus*) and protects against *Trichothecium roseum* causing postharvest decay (Yang et al. 2005). $Harpin_{Ea}$ induced activities of peroxidases and induced resistance in tomato against both fungal and bacterial diseases (Tosun et al. 2003). Harpin treatment induced expression of *PR-3b*, *PR-1*, *PDF1.2*, and *Hel* genes in *Arabidopsis* (Dong et al. 1999, 2004).

Harpin was able to induce alterations in cell walls such as papillae formation and wall thickening in apple fruits and conferred resistance against *Penicillium expansum* (de Capdeville et al. 2008). The HrpN contributes directly or indirectly to callose elicitation on apple leaves (Boureau et al. 2011). Harpin from *P. syringae* pv. *glycinea* induced *PAL* mRNA in tobacco cells (Andi et al. 2001). $HarpinG_{Xooc}$ treatment induced expression of the *CHS* gene encoding a tea chalcone synthase, which is involved in biosynthesis of catechols in tea (Wu et al. 2007). Harpin induced tannins in apple fruits (de Capdeville et al. 2008).

Ectopic expression of the *hrf1* gene encoding the $harpin_{Xoo}$ activated the phytoalexin biosynthetic pathway in rice plants (Li et al. 2012). A cytochrome p450 gene showed a 114.6-fold increase of expression levels in *hrf1*-transformed rice line. The p450-overexpressing rice lines showed increased phytoalexin accumulation (Li et al. 2012). The leaf silicon content was dramatically enhanced in *hrf1*-transgenic plants, inducing the accumulation of diterpenoids and flavonoid phytoalexins in rice (Rodrigues et al. 2004; Shao et al. 2008; Li et al. 2012).

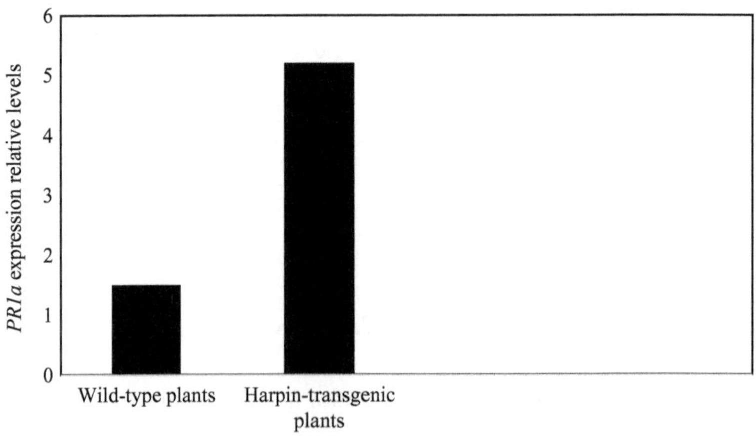

Fig. 3.6 Expression of *PR-1a* in wild-type and harpin *hrf1* gene-expressing transgenic rice plants at 24 h after inoculation with *Xanthomonas oryzae* pv. *oryzae* (Adapted from Li et al. 2012)

3.2.4 Development of Harpin Formulations for Management of Crop Diseases

The harpins have been developed as potential molecular tools to manage not only bacterial pathogens but also oomycete, fungal, and viral pathogens, nematodes and insect pests (Dong et al. 2004; de Capdeville et al. 2008; Shao et al. 2008). Commercial formulations of harpin have also been developed for management of crop diseases. A weakened nonpathogenic strain of *Escherichia coli* (K-12) was modified to produce harpin on a commercial scale. The harpin is concentrated from the growth medium of the genetically modified *E. coli* and the bacterial cells are killed and removed from the marketed product. Commercially produced harpin protein is identical to the harpin protein produced by *Erwinia amylovora*. Harpin has been produced commercially under two trade names "Messenger" and "Extend". It is marketed for the control of crop diseases as well as a plant growth enhancer (Wei et al. 1998; Zitter and Beer 1998; de Capdeville et al. 2008).

3.2.5 Foliar Application of Harpin Induces Plant Immune Responses against Wide Range of Pathogens

Harpin application induces resistance against various viral, bacterial, oomycete, and fungal pathogens in several crops including wheat, rice, apple, pear, citrus, cucumber, tobacco, tomato, pepper and strawberry (Strobel et al. 1996; Tosun et al. 2003; Li et al. 2005a; Pertot et al. 2007; Chen et al. 2008a, b; de Capdeville et al. 2008; Choi et al. 2013). Harpin induces both local and systemic resistance in foliage and also induces resistance in fruits (Dong et al. 1999; de Capdeville et al. 2002, 2003; Yang et al. 2005; Akbudak et al. 2006).

The harpin PopW induced resistance against *Tobacco mosaic virus* (TMV) in tobacco (Li et al. 2011). The efficacy of PopW in control of TMV in tobacco was demonstrated under field conditions. The harpin was sprayed at 25 μg/ml on tobacco seedlings in the transplanted field and this treatment significantly reduced the TMV disease incidence (Fig. 3.7; Li et al. 2011). Harpin$_{Xoo}$ induces resistance against TMV in tobacco (Li et al. 2005a). 'Messenger' (harpin) effectively controls TMV in tomato, *Cucumber mosaic virus* (CMV) in tomato and pepper and *Beet curly top virus* in pepper (Jones 2001). Harpin (HrpZ$_{Pss}$) treatment induced resistance against *Tobacco necrosis virus* (TNV) in tobacco (Strobel et al. 1996). Treatment of HrpN of *E. amylovora* increased resistance to aphids and several aphids have been shown to be the vectors of viral diseases (Dong et al. 2004; Vidhyasekaran 2004; Li et al. 2010, 2011).

Foliar application of Hpa1$_{10-42}$ fragment induced strong resistance in rice to *Xanthomonas oryzae* pv. *oryzicola* in greenhouse and field conditions (Chen et al. 2008a, b). HrpZ$_{Pss}$ induces resistance against the bacterial pathogen *Pseudomonas syringae* pv. *lachrymans* in tobacco (Strobel et al. 1996). Spray treatment with

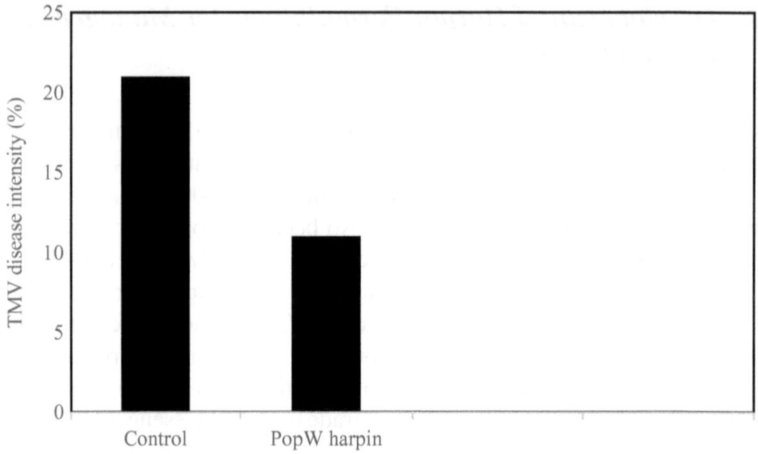

Fig. 3.7 Effect of PopW harpin treatment on *Tobacco mosaic virus* (TMV) disease incidence in tobacco

HrpN of *Erwinia amylovora* in *Arabidopsis* activated systemic acquired resistance (SAR), conferring disease resistance to the bacterial pathogen *Pseudomonas syringae* pv. *tomato* (Dong et al. 1999; Tosun et al. 2003). The harpin $HrpN_{Ea}$ induces systemic resistance against *Erwinia carotovora* subsp. *carotovora* in *Arabidopsis*. A clear reduction of the bacterial growth in $HrpN_{Ea}$-pretreated plants was observed (Kariola et al. 2003). Harpin conferred resistance against the bacterial blight pathogen *Xanthomonas oryzae* pv. *oryzae* in rice (Shao et al. 2008). Harpin (Messenger) reduced the incidence of grapevine Pierce's disease caused by the bacterium *Xylella fastidiosa*. In the field trials the disease incidence was 13, 7, and 6 % when 160, 320, and 420 g harpin a.i/ha were used. The disease incidence was 19 % in untreated control plants (Tubajika et al. 2007). Interestingly, harpin induced resistance also against insects (Zitter and Beer 1998) and it reduced *X. fastidiosa* transmission by the insect vector glassy-winged sharpshooter (*Homalodisca coagulata*) (Tubajika et al. 2007).

Harpin is effective in management of several fungal pathogens. Harpin is effective in inducing resistance against the biotrophic pathogen *Podosphaera aphanis*, the powdery mildew pathogen of strawberry (Pertot et al. 2007). $HrpZ_{Pss}$ controls the hemibiotrophic fungal pathogen *Colletotrichum lagenarium* in cucumber by inducing resistance (Strobel et al. 1996). Foliar application of $Hpa1_{10–42}$ fragment induced strong resistance in rice to the blast pathogen *Magnaporthe oryzae* in greenhouse and field conditions (Chen et al. 2008a, b). Harpin induced resistance to the rice blast fungus *M. oryzae* (Shao et al. 2008). Foliar application of $HrpN_{Ea}$ (the commercial product 'Messenger') reduced the damage caused by the gray mold pathogen *Botrytis cinerea* in tomato (Fontanilla et al. 2005b). Harpin induces resistance against *Penicillium expansum*, the apple blue mold pathogen (de Capdeville et al. 2008). Harpin induces local and systemic resistance against *Trichothecium roseum* in harvested Hami melons (Yang et al. 2005). Messenger (harpin) was

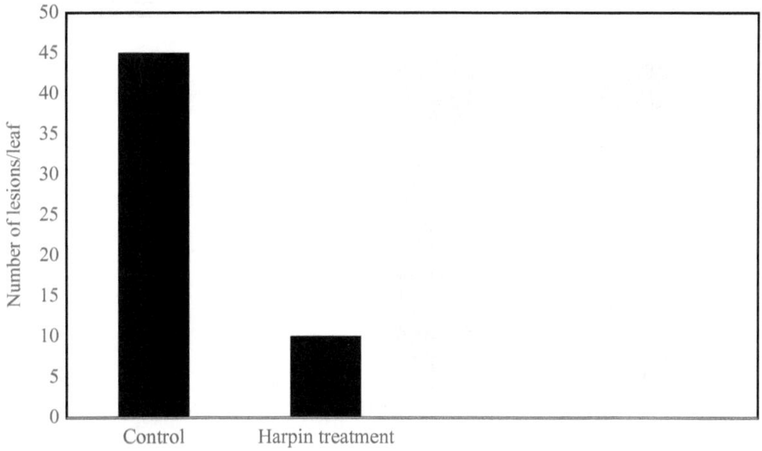

Fig. 3.8 Efficacy of harpin (Messenger) on the development of Citrus scab (*Elsinoe fawcettii*) on citrus seedlings (Adapted from Agostini et al. 2003)

effective in controlling citrus scab pathogen *Elsinoe fawcettii* in rough lemon (Fig. 3.8; Agostini et al. 2003).

Harpin has been shown to induce resistance against oomycete pathogens. Foliar application of $HrpN_{Ea}$ (Mesenger) in tomato reduced the late blight caused by the oomycete pathogen *Phytophthora infestans* (Fontanilla et al. 2005b). $Harpin_{Ea}$ induces resistance against the oomycete pathogen *Hyaloperonospora parasitica* in *Arabidopsis* (Dong et al. 1999). These results suggest that harpin can be a potential tool for management of viral, bacterial, fungal, and oomycete diseases. However, harpin-induced resistance may not be sufficient or suitable for controlling certain diseases. Messenger, the commercial product of harpin, effectively controlled citrus scab (*Elsinoe fawcettii*), but was ineffective in controlling melanose (*Diaporthe citri*) or brown spot (*Alternaria alternata*) in citrus (Agostini et al. 2003).

3.2.6 Harpin Treatment Triggers SA-Dependent Systemic Acquired Resistance

Systemic Acquired Resistance (SAR) is a heightened state of defense against a broad spectrum of pathogens activated throughout a plant following a local infection (Liu et al. 2011a). The induced resistance is detected in the uninoculated systemic tissue of a pathogen-infected plant (Liu et al. 2011b). Salicylic acid (SA) accumulation is required for activation of local defenses at the initial site of attack, and in the distant pathogen-free organs for the induction of SAR. SA signaling has been shown to be required for the manifestation of SAR (Du et al. 2012; Návarová et al. 2012; Shah and Zeier 2013). Spray application of the harpin PopW protein has

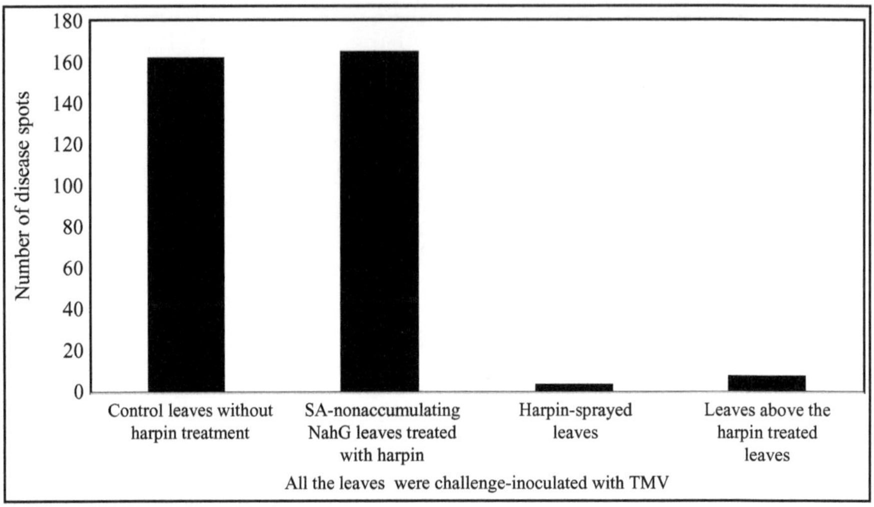

Fig. 3.9 Harpin (PopW) triggers systemic acquired resistance (SAR) against *Tobacco mosaic virus* (TMV) disease incidence (Adapted from Li et al. 2011)

been shown to induce systemic acquired resistance against TMV infection in tobacco (Fig. 3.9; Li et al. 2011). SAR induction by PopW was demonstrated by inoculating leaves above or below the PopW-treated leaves. Both leaves treated with the harpin PopW and inoculated with the virus and the leaves above or below the PopW-treated leaves (leaves without harpin treatment) showed reduced disease incidence compared to control plants (Fig. 3.9). The role of SA signaling in induction of SAR by the harpin PopW was assessed by developing transgenic plants expressing the SA degrading enzyme salicylate hydroxylase encoded by the *Pseudomonas putida nahG* gene. The disease incidence in SA-non-accumulating NahG plants were almost similar to the control plants inoculated with TMV (Li et al. 2011). The results suggest that the harpin induces SAR by activating SA signaling system.

The harpin $HrpN_{Ea}$ treatment induced systemic acquired resistance against the oomycete pathogen *Peronospora parasitica* and the bacterial pathogen *Pseudomonas syringae* pv. *tomato* in *Arabiodopsis* (Dong et al. 1999). The induction of disease resistance was accompanied by induction of the characteristic SAR genes *PR-1* and *PR-2*. The transgenic *Arabidopsis* plants containing the *nahG* gene which prevents accumulation of SA, the harpin neither elicited resistance nor activated SAR gene expression. Harpin also failed to activate SAR when applied to *nim1* (non-inducible immunity) mutants, which are defective in responding to SA and regulation of SAR. In contrast, mutants compromised in responsiveness to methyl jasmonate and ethylene developed the same resistance as did wild-type plants (Dong et al. 1999). The results suggest that the harpin $HrpN_{Ea}$ triggers defense responses through the NIM1-mediated SAR signal transduction pathway in SA-dependent pathway.

3.2.7 *Time of Application of Harpin Determines Its Efficacy in Induction of Defense Response*

Harpin has been shown to effectively control several diseases in various crop plants. However, when its efficacy was tested under greenhouse/field conditions, harpins have been found to be ineffective in controlling certain pathogens in some trials (Hopkins 2002; Keinath et al. 2007). Messenger (harpin formulation) failed to provide significant control of the sweet orange diseases caused by the bacterial pathogens *Xanthomonas axonopodis* pv. *citri* and *X. axonopodis* pv. *citrumelo* (Graham and Leite 2004). Messenger did not induce effective plant defense responses against *X. campestris* pv. *vesicatoria* in tomato (Obradovic et al. 2005). Several factors such as environment, genotype, and crop nutrition determine the efficacy of elicitors in controlling diseases under field conditions (Walters et al. 2005). The time of application is very critical in enhancing the efficacy of harpin in controlling diseases. Harpin induced resistance against blue mold of apple. When the harpin (commercial formulation Messenger) is applied only at the time of pathogen inoculation, it was ineffective in control of the pathogen *Penicillium expansum.* It should be applied well before the pathogen invasion. Harpin applied at 48 h before inoculation of the pathogen was highly effective in control of the blue mold disease in apple (Fig. 3.10; de Capdeville et al. 2002). Spraying apple trees with harpin a few days before harvest was found to be highly effective for the control of blue mold decay in storage (de Capdeville et al. 2003).

Application of the commercial product of harpin 'Messenger' at 5 days before inoculation with the citrus scab pathogen *Elsinoe fawcettii* was highly effective in

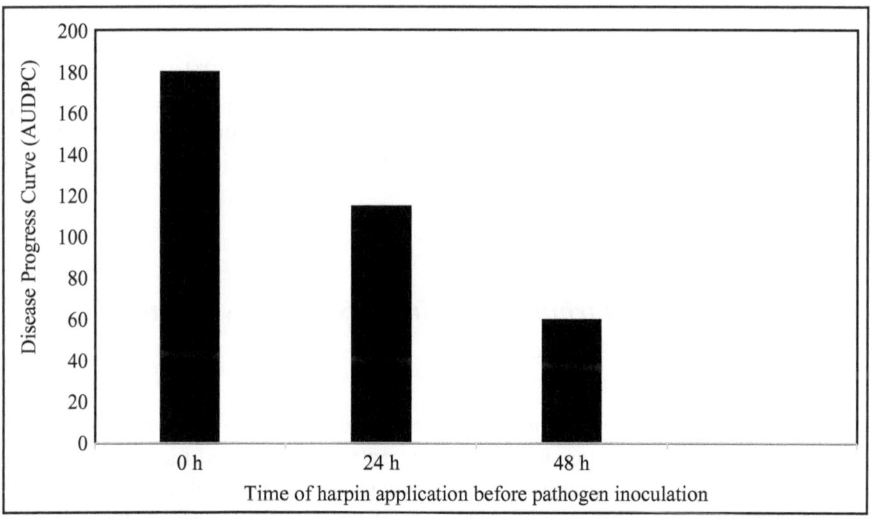

Fig. 3.10 Effect of time of application of harpin in the control of blue mold of apple fruit (Adapted from de Capdeville et al. 2002)

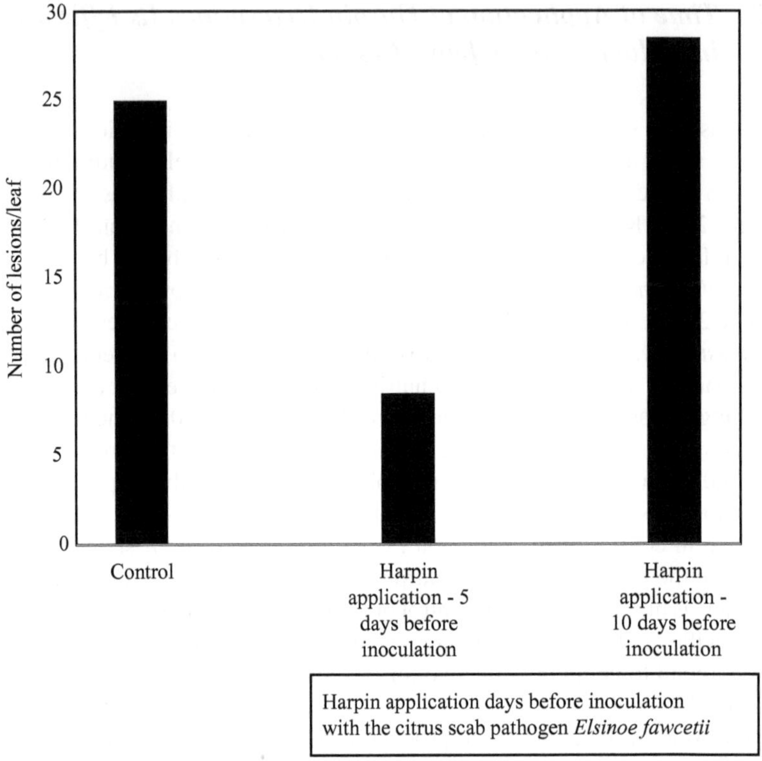

Fig. 3.11 Effect of time of application in determining the efficacy of harpin in the control of citrus scab (Adapted from Agostini et al. 2003)

controlling the scab disease in rough lemon (*Citrus jambhiri*). However, when Messenger was applied 10 days before inoculation with the pathogen, it failed to control the disease (Fig. 3.11; Agostini et al. 2003). In geraniums, resistance against *Botrytis* and *Xanthomonas* was induced 3 days after application of Messenger (Qiu et al. 2001).

3.2.8 Amount of Harpin Determines Its Efficacy in Inducing Disease Resistance

The concentration of the harpin applied also determines the efficacy of the treatment in controlling diseases. Increased concentrations showed increased efficacy of harpin in controlling *Penicillium expansum* in apple (de Capdeville et al. 2002). However, application of low concentrations of $HpaG_{10-42}$ was more effective than application of high concentrations in controlling rice bacterial blight (Chen et al. 2008b).

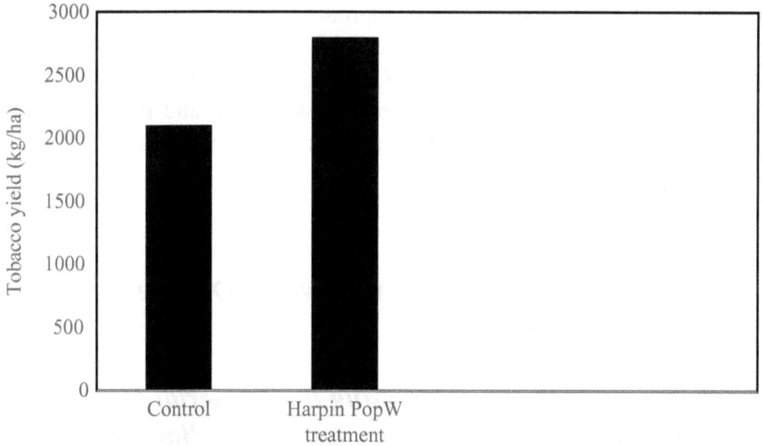

Fig. 3.12 Effect of harpin (PopW) treatment in increasing tobacco yield under field conditions (Adapted from Li et al. 2011)

3.2.9 *Harpin Increases Crop Growth and Crop Yield*

Foliar application of harpin not only reduces disease resistance but also acts as a growth promoter. It increases photosynthesis, nutrient uptake, seed germination, root development and plant growth (Jones 2001; Qiu et al. 2001; Dong et al. 2004; Li et al. 2005a; Li et al. 2014a, b). Harpin treatment increased peach fruit yield primarily by increasing its fruit size (Taylor 2006). It also improves fruit development and induces earlier fruit maturation (Jones 2001). Harpin actually increases crop yield (Jones 2001; Keinath et al. 2007; Chen et al. 2008a, b). Green tea (*Cammelia sinensis*) yield increased by more than 55 % over untreated control when the tea shrubs were treated with harpin$_{Xooc}$ (Wu et al. 2007). When the harpin PopW was applied as spray treatment, it increased the yield of tobacco leaves, which was 30.4 % more than the yield of control plants (Fig. 3.12; Li et al. 2011). The harpin PopW also improved the quality of tobacco leaves, with an increase in the number of first-class tobacco leaves of 50.2 % compared with control plants (Li et al. 2011). The harpin protein Hpa1 induces several growth-promoting responses in plants, activating the ethylene signaling system, increasing photosynthesis rates and *EXPANSIN* (*EXP*) gene expression (Li et al. 2014a, b). Harpin protein has been shown to act as a growth promoter in *Phalaenopsis* orchids (Chuang et al. 2014).

3.2.10 *Foliar Spray Application of* Bacillus thuringiensis *Expressing Harpin Gene*

Foliar application of commercial formulations of harpin protein has been found to be useful in management of various crop diseases. Genetically engineered *Escherichia coli* K-12 is used for commercial production of harpin$_{Ea}$ of *Erwinia amylovora*

(Fontanilla et al. 2005a). In another approach, genetically engineered *Bacillus thuringiensis* strain expressing *harpin*$_{Pss}$ gene was spray-applied on groundnut leaves. This treatment reduced the incidence of late leaf spot caused by *Phaeoisariopsis personata* in groundnut and promoted crop growth (Anil and Podile 2012). These results suggest that the genetically-engineered bacteria will be highly useful in crop disease management.

3.3 Engineering Harpin Gene to Develop Disease Resistant Plants

3.3.1 Transgenic Plants Expressing the Bacterial Harpin Gene Show Enhanced Resistance against the Bacterial Pathogen

Several harpin genes have been cloned from various bacterial pathogens (Wei et al. 1992; He et al. 1993; Li and Fan 1999; Li et al. 2004b; Peng et al. 2004b; Takakura et al. 2004; Li et al. 2005; Malnoy et al. 2005; Jang et al. 2006; Ren et al. 2006b; Sohn et al. 2007; Shao et al. 2008; Huo et al. 2010; Miao et al. 2010a, b; Xu et al. 2010; Pavli et al. 2011; Zhang et al. 2011; Pavli et al. 2012; Miao and Wang 2013; Fu et al. 2014; Wang et al. 2014). These harpin genes have been used to engineer transgenic plants with constitutive expression of the genes. Constitutive expression of the harpin genes conferred resistance against pathogens. Transgenic rice plants that express a harpin-encoding gene (*hrf1*), derived from *Xanthomonas oryzae* pv. *oryzae* were developed (Shao et al. 2008). Transgenic rice plants expressing *hrf1* gene encoding the harpin of the rice bacterial blight pathogen *X. oryzae* pv, *oryzae* showed enhanced disease resistance against the bacterial blight pathogen *X. oryzae* pv. *oryzae* strain PXO79 (Fig. 3.13; Li et al. 2012). These transgenic plants also showed broad-spectrum resistance to the main *X. oryzae* pv. *oryzae* strains (Li et al. 2012). Ectopic expression of *hrf1* gene in rice remarkably induced the expression of genes related to the SA and JA signaling pathways. The enhanced disease resistance induced by the *hrf1* transgene was accompanied by high and rapid induction of phytoalexin production and enhanced silicon accumulation (Li et al. 2012). Collectively these results suggest that the harpin gene conferred broad-spectrum non-specific disease resistance in rice by inducing multiple defense responses, Transgenic apple plants expressing *hrpN* encoding harpin$_{Ea}$ isolated from the apple bacterial pathogen *Erwinia amylovora* show increased resistance against *E.amylovora* (Borejsza-Wysocka et al. 2000). Transgenic pear plants expressing *HrpN*$_{Ea}$ gene, isolated from *E. amylovora*, under the control of the constitutive promoter *CaMV35S* were developed. Stable constitutive expression of the harpin gene was found in most of the transgenic clones. These transgenic clones showed resistance against the fire blight pathogen *E. amylovora*, which was correlated to their degree of expression of the transgene *hrpN*$_{Ea}$ (Malnoy et al. 2005).

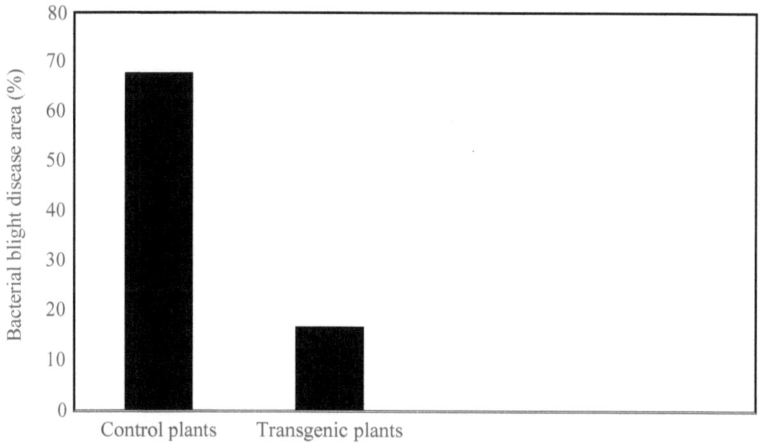

Fig. 3.13 Transgeic rice plants expressing *hrf1* gene show resistance against *Xanthomonas oryzae* pv. *oryzae* (Adapted from Li et al. 2012)

3.3.2 Harpin Gene from a Bacterial Pathogen Triggers Defense Responses against Viral, Fungal, Oomycete and Also Bacterial Pathogens in Different Host Plants

Transgenic plants expressing a bacterial harpin gene may show resistance to a wide range of fungal, oomycete and viral pathogens, besides bacterial pathogens. Transgenic tobacco plants expressing the harpin gene isolated from the rice pathogen *Xanthomonas oryzae* pv. *oryzae*, $hpaG_{Xoo}$ gene, produced $\text{harpin}_{\text{Xoo}}$ intracellularly. These plants showed resistance to the fungal pathogen *Alternaria alternata*, the bacterial pathogen *Ralstonia solanacearum* and the viral pathogen *Tobacco mosaic virus* (Fig. 3.14; Peng et al. 2004b).

Transgenic tobacco plants expressing the $hrpZ_{Pss}$ gene encoding a harpin from *Pseudomonas syringae* pv. *syringae* strain LOB2-1, the causal agent of bacterial blight of lilac showed enhanced resistance to the tobacco powdery mildew fungus *Erysiphe cichoracearum* (Takakura et al. 2004). Transgenic tobacco lines expressing $hrpN_{Ea}$ showed enhanced resistance to *Botrytis cinerea* (Jang et al. 2006). Transgenic chrysanthemum plants expressing $hpaG_{Xoo}$ gene from *X. oryzae* pv. *oryzae* showed enhanced resistance against the alternaria leaf spot disease (Xu et al. 2010).

The rapeseed was genetically engineered via *Agrobacterium* –mediated transformation to express the *hrf2* gene from the rice bacterial pathogen *X. oryzae* pv. *oryzicola.* The resistance to *Sclerotinia sclerotiorum* was greatly enhanced in the T_1 generation of transgenic rapeseed (Huo et al. 2010). Introduction of the *hrf2* gene into rapeseed may be an effective strategy for inducing resistance to *S. sclerotiorum* in rapeseed. Transgenic tobacco plants expressing HrpN_{EP} from *Erwinia pyrifoliae* were developed by introducing the harpin gene driven by the OsCc1 promoter into

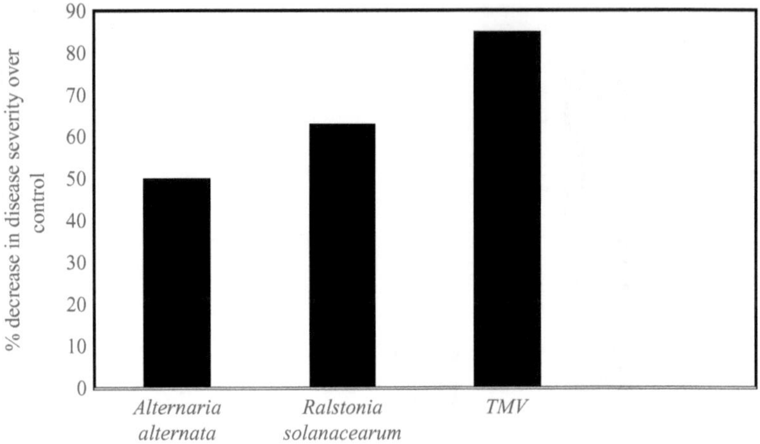

Fig. 3.14 Transgenic tobacco plants expressing *hpaG*$_{Xoo}$ gene, show resistance against fungal, bacterial, and viral pathogens (Adapted from Peng et al. 2004b)

tobacco plants (Sohn et al. 2007). These transgenic tobacco plants showed enhanced resistance to *Botrytis cinerea*. Resistance to *B. cinerea* was correlated with enhanced induction of SA-dependent genes such as *PR-1a*, *PR-2*, and *Chia5*, of JA-dependent genes such as PR-1b, and of genes related to ethylene production, such as *NT-EFR26*, *NT-1A1C*, *DS321*, *NTACS1* and *NT-ACS2* (Sohn et al. 2007).

Transgenic potato plants expressing the *hrpN* gene encoding harpin produced by the bacterial pathogen *Erwinia amylovora* show increased resistance against the oomycete potato pathogen *Phytophthora infestans* (Li and Fan 1999). The bacterial pathogen *Ralstonia solanacearum* secretes the harpin protein, PopA and the gene (*popA*) encoding PopA protein has been cloned (Belbahri et al. 2001). The harpin PopA has been exploited to develop transgenic plants producing PopA early in the infection process in sufficient concentration to trigger resistance to pathogens. Transgenic tobacco plants in which the pathogen-inducible promoter of the *hsr203J* gene from tobacco controls the expression of the *popA* gene from *R. solanacearum* were developed (Belbahri et al. 2001). PopA itself induced the *hsr203J* promoter and hence the transgenic plants rapidly accumulated the PopA protein in the pathogen infection sites. All the regenerated plants showed normal phenotypes. The transgenic plants expressing the bacterial elicitor showed high resistance to the oomycete pathogen *Phytophthora parasitica* var. *nicotianae* (Fig. 3.15; Belbahri et al. 2001). When compared to the untransformed plants, the invasion of stems by the pathogen was reduced by 80–90 % in all transgenic lines (Belbahri et al. 2001). However, this strategy had the risk of uncontrolled cell-death proliferation. Among all transgenic tobacco lines tested, one line occasionally initiated a spreading necrosis upon local inoculation with *P. parasitica* var. *nicotianae* (Belbahri et al. 2001). Hence careful selection of the transgenic plants should be made and the selected lines should be tested under field conditions for their agronomic characters.

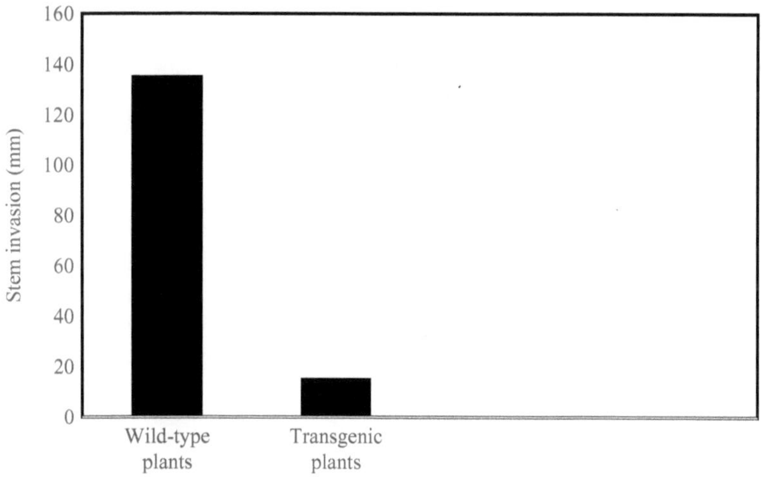

Fig. 3.15 Transgenic tobacco plants expressing the bacterial harpin *PopA* gene show enhanced resistance to the oomycete pathogen *Phytophthora parasitica* var. *nicotianae* (Adapted from Belbahri et al. 2001)

Transgenic rice plants that express a harpin-encoding gene (*hrf1*), derived from *Xanthomonas oryzae* pv. *oryzae* showed enhanced expression of defense-related genes and exhibited nonspecific resistance to the blast fungus *Magnaporthe oryzae* (formerly called as *M. grisea*). Transgenic plants and their T1 – T7 progenies were highly resistant to all major races of the pathogen in China. In the transgenic plants, formation of melonized appressoria, which is essential for infection, was inhibited on plant leaves (Shao et al. 2008). Transgenic rice plants expressing *hpa1* gene of *Xanthomonas oryzae* pv. *oryzae* encoding a bacterial harpin protein show enhanced resistance against the rice bacterial blight pathogen *X. oryzae* pv. *oryzae* and against the rice blast fungal pathogen *M. oryzae* (Choi et al. 2012). The transgenic Arabidopsis plants expressing the *hpa1* gene of *X. oryzae* pv. *oryzae* showed resistance against the necrotrophic fungal pathogen *Botrytis cinerea* and the bacterial pathogen *Xanthomonas campestris* pv. *campestris* (Choi et al. 2012). Transgenic Arabidopsis plants expressing *hrp* gene encoding $\text{harpin}_{\text{Pss}}$ constitutively showed an enhanced disease resistance against the soft rot pathogen *Erwinia carotovora* subsp. *carotovora* (Pandey et al. 2005).

Transgenic sugar beet and *Nicotiana benthamiana* plants expressing $hrpZ_{Psph}$ gene encoding harpin of *Pseudomonas syringae* pv. *phaseolicola* showed enhanced resistance against rhizomania disease caused by *Beet necrotic yellow vein virus* (BNYVV) (Pavli et al. 2011, 2012). Hpa1_{Xoo} is a harpin protein from *X. oryzae* pv. *oryzae*. When $hpa1_{Xoo}$ was transformed into the susceptible cotton line Z35 through *Agrobacterium*-mediated transformation, the transgenic cotton line showed improved resistance to the wilt fungal pathogen *Verticillium dahliae* (Fig. 3.16; Miao et al. 2010a; Miao and Wang 2013). The rice plants expressing the $hpaG_{Xoo}$ gene from *X. oryzae* pv. *oryzae* showed resistance to the rice sheath blight pathogen

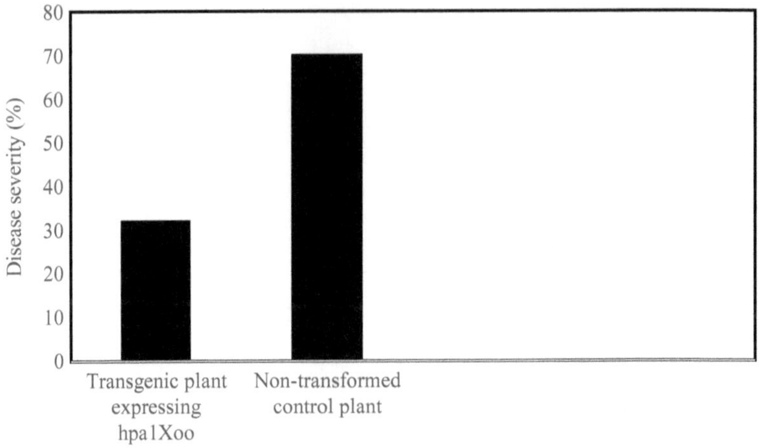

Fig. 3.16 Transgenic cotton plants expressing *hpa1Xoo* gene show enhanced *Verticillium* wilt disease resistance in the field (Adapted from Miao et al. 2010a)

Rhizoctonia solani (Ren et al. 2006a, b). These results suggest that the harpin gene from a bacterial pathogen may be a potential tool to develop crop plants resistant to soil-borne pathogens.

3.3.3 Variation in Levels of Harpin Gene Expression Resulting in Variation in Levels of Expression of Disease Resistance

Levels of harpin gene expression may vary among different transgenic plant lines (Peng et al. 2004b). Transgenic tobacco plants expressing the $hpaG_{XOO}$ gene encoding the harpin protein isolated from the rice bacterial blight pathogen *Xanthomonas oryzae* pv. *oryzae* were developed (Peng et al. 2004b). The $hpaG_{Xoo}$ was found to be expressed in both T1 and T2 plants of 20 $hpaG_{Xoo}$-transgenic lines tested. However, levels of the gene expression varied in the $hpaG_{Xoo}$ -expressing tobacco (HARTOB) lines. The $hpaG_{Xoo}$ gene expression was low in the line15, intermediate in the line 6, and high in the line 7. The defense response genes *GST1*, *NPR1*, *Chia5*, *PR-1a*, and *PR-1b* were found to be expressed at the lowest, intermediate, and the greater levels in lines 15, 6, and 7, respectively (Peng et al. 2004b). The results suggest that levels of the defense-related gene expression are consistent with levels of $hpaG_{Xoo}$ expression. The varied levels of $hpaG_{Xoo}$ gene expression observed in the lines 7, 6, and 15 reflected in their ability to trigger defense responses against the fungal pathogen (*Alternaria alternata*), bacterial pathogen (*Ralstonia solanacearum*) and viral pathogen (*Tobacco mosaic virus*). The line 7, which showed high level of $hpaG_{Xoo}$ gene expression showed very high level of resistance against TMV infection, while

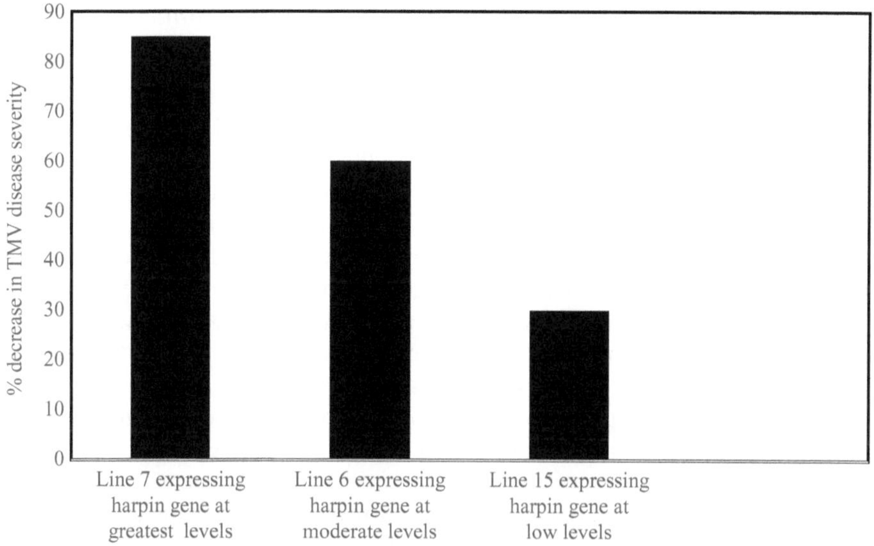

Fig. 3.17 Levels of *Tobacco mosaic virus* disease resistance in transgenic tobacco lines expressing $hpaG_{Xoo}$ gene at low (*line 15*), moderate (*line 6*), and greatest (*line 7*) levels (Adapted from Peng et al. 2004b)

the line 15, which showed low level of expression of the harpin gene showed very low level of resistance against the viral infection (Fig. 3.17; Peng et al. 2004b). These studies suggest that the transgenic lines should be carefully selected to generate highly useful disease-resistant cultivars.

It has been shown that expression of harpin genes can be enhanced resulting in higher accumulation of harpin by properly selecting the promoter for gene transcription. Takakura et al. (2004) developed transgenic tobacco plants using both an inducible promoter and a constitutive expression promoter to compare their efficacy in inducing disease resistance. The $HrpZ_{Pss}$ gene was fused to a *Cauliflower mosaic virus* 35S promoter for constitutive expression and a pea phenylalanine ammonia-lyase (PAL) promoter for pathogen-inducible expression and the gene constructs were introduced into tobacco plants. The transgenic plants expressing CaMV 35S promoter showed higher accumulation of harpin (0.1–0.5 % of the total soluble protein) while PAL promoter induced only 0.03–0.1 % of the total soluble protein after inoculation with the pathogen *Erysiphe cichoracearum* (Takakura et al. 2004). The results suggest that the *Cauliflower mosaic virus* 35S promoter, which enhances the accumulation of harpin, may be more useful in engineering disease resistance utilizing harpin gene. No morphological changes were observed in the leaves, stems, and flowers in the transgenic tobacco plants expressing $HrpZ_{Pss}$ gene plants even with high constitutive or induced expression of harpin. Seed fertility was normal in almost all transgenic plants. These results suggest that harpin protein even at high concentrations does not adversely affect the growth of plants (Takakura et al. 2004).

3.3.4 Growth and Yield Potential of Transgenic Plants Expressing Bacterial Harpin Gene

Transgenic plants are often considered as poor yielder with adverse agronomic characters (Belbahri et al. 2001; Century et al. 2008; Takakura et al. 2008; Takatsuji 2014). However transgenic plants expressing harpin gene show good agronomic characters. Harpin is known to promote crop growth (Li et al. 2011). Transgenic *Nicotiana benthamiana* plants expressing *hrpZ*$_{Psph}$ gene showed increased vigor and grew faster in comparison with non-transgenic control plants (Pavli et al. 2011). Transgenic chrysanthemum plants expressing *hpaG*$_{Xoo}$ gene from *X. oryzae* pv. *oryzae* flowered earlier than the wild-type plants did, suggesting that the harpin gene might have accelerated chrysanthemum development (Xu et al. 2010). Overexpression of a harpin-encoding gene *hrf1* in rice enhances drought tolerance (Zhang et al. 2011). Transgenic expression of a functional fragment of harpin protein Hpa1 in wheat induces defense against insect pests (Fu et al. 2014). These results suggest that harpin may enhance the expression of agronomic traits and it may be a potential tool to develop disease resistant crop plants.

3.4 Molecular Manipulation of Plant Innate Immune Signaling Systems Using Flagellin

3.4.1 Activation of Plant Immune Signaling System by Flg22

Bacterial flagella have been found to be potential sources for the PAMPs. Flagellin is the structural protein that forms the major portion of flagellar filaments. A peptide flg22, the stretch of 22 amino acids in the N terminus of bacterial flagellin has been identified as the bacterial PAMP epitope (Boller and Felix 2009). Flg22 is the PAMP detected in flagella of several bacterial pathogens (Fujiwara et al. 2004; Denoux et al. 2008; Takakura et al. 2008; Bardoel et al. 2011; Cai et al. 2011; Sun et al. 2013; Clarke et al. 2013; Rosli et al. 2013; Hao et al. 2014; Veluchamy et al. 2014). Flg22 has been shown to be perceived by the pattern recognition receptor (PRR) FLS2 in *Arabidopsis* (Geldner and Robatzek 2008; Göhre et al. 2008; Bardoel et al. 2011). Similar PRRs have been detected in rice, tomato, barley, and tobacco (Meindl et al. 2000; Chinchilla et al. 2006, 2007a, b; Hann and Rathjen 2007; Robatzek et al. 2007; Shen et al. 2007; Takai et al. 2008; Chen et al. 2014b; Hao et al. 2014). On perception by the receptors, flg22 activates several signaling events. A receptor-like kinase (SERK3/BAK1) appears to integrate diverse perception events into downstream PAMP responses (Sun et al. 2013). Another PAMP epitope derived from flagellin, flgII-28, has been identified in *Pseudomonas syringae* flagellin (Cai et al. 2011). The recognition of flgII-28 appears to be restricted to a number of Solanaceae species (Cai et al. 2011; Clarke et al. 2013). It has been predicted that there is an additional and yet to be identified receptor, FLS3 (Flagellin sensing 3), which is

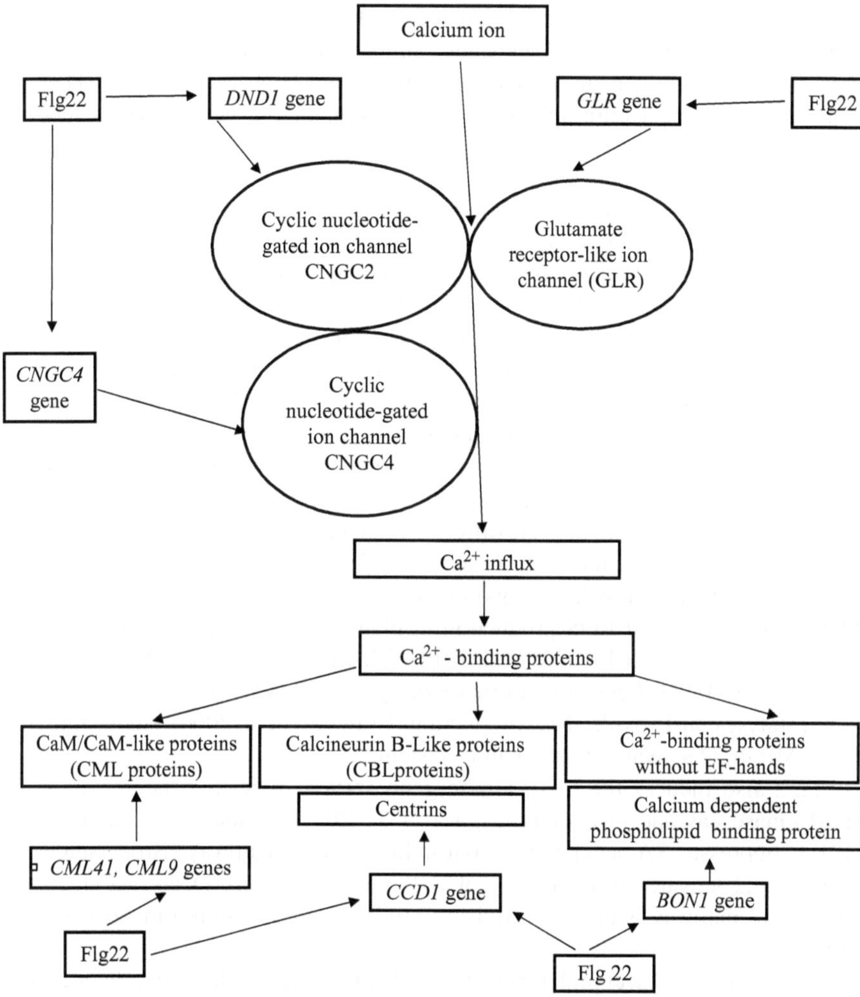

Fig. 3.18 Flg22-triggered Ca^{2+} signaling system

involved in the perception of flgII-28 in the Solanaceous species (Clarke et al. 2013; Veluchamy et al. 2014).

The PAMP flg22 has been shown to trigger Ca^{2+} signaling system (Fig. 3.18). Flg22 triggers Ca^{2+} influx in plants (Kwaaitaal et al. 2012). Cyclic nucleotide-gated ion channels (CNGCs) are involved in Ca^{2+}-dependent signaling pathways (Talke et al. 2003; Yoshioka et al. 2006). *Arabidopsis DND1* codes for a cyclic nucleotide-gated channel 2 (CNGC2) (Clough et al. 2000) and Flg22 induces the expression of *DND1* in *Arabidopsis* (Denoux et al. 2008). Flg22 also activates cyclic nucleotide gated channel 4 encoding gene *CNGC4* (Denoux et al. 2008). These channels have been found to be calmodulin (CaM)-binding proteins (Borsics et al. 2007).

Flg22 triggers glutamate receptor-like channels functioning in the Ca^{2+} influx (Kwaaitaal et al. 2012). Flg22 activates calmodulin-like gene *CML41* within 1 h after treatment and the peak expression of the gene was observed at 12 h (Denoux et al. 2008). The flagellin-upregulated expression of the gene *CML9* encoding CML9, a calmodulin-like protein in *Arabidopsis* (Leba et al. 2012). Flg22 triggered the expression of Ca^{2+}-dependent protein kinase in rice cells (Fujiwara et al. 2004). It also activated *CCD1* in cultured rice cells (Fujiwara et al. 2004). *CCD-1* encodes a Ca^{2+}-binding protein that shares homology with the C-terminal half domain of centrin and centrins are involved in Ca^{2+} signaling (Takezawa 2000). Flg22 also induces the expression of *BON1* gene which encodes a calcium-dependent phospholipid binding protein (Denoux et al. 2008). Some Ca^{2+}-binding proteins do not contain EF-hand structural motifs. These proteins contain other Ca^{2+}-binding domains such as the C2 domain (Reddy 2001). The C2 domain is a Ca^{2+}-phospholipid-binding site, and Ca^{2+} binding is coordinated by four to five amino acid residues provided by bipartite loops (Rizo and Südhof 1998). These domains often mediate Ca^{2+}-dependent phospholipid binding.

The flagellin appears to activate different branches of defense signaling network. It triggered the expression of *RbohD* and *RbohC* genes encoding NADPH oxidases (Respiratory burst oxidase homologs) involved in generation of ROS (Denoux et al. 2008). Flg22 triggered ROS production in plants (Takakura et al. 2008; Serrano et al. 2012; Hao et al. 2014; Trdă et al. 2014; Veluchamy et al. 2014). It triggered the expression of *AtNOS1* encoding nitric oxide synthase (NOS) (Denoux et al. 2008). NOS catalyses nitric oxide (NO) production (Crawford et al. 2006), suggesting that the PAMP flg22 activates NO signaling system.

Flg22 also induces the MAP kinase gene *MPK4*, indicating the activation of MAP kinase cascade (Suarez-Rodriguez et al. 2007; Denoux et al. 2008). The MPK4 suppresses SA accumulation but induces the JA pathway (Brodersen et al. 2006). Flg22 activates the expression of *MPK3* in Arabidopsis (Denoux et al. 2008). The MPK3 pathway is also involved in JA/ET signaling system (Takahashi et al. 2007b).

Flagellin induced accumulation of SA in *Arabidopsis* (Mishina and Zeier 2007). Flg22 up-regulated the expression of genes involved in SA signaling system and these include *PAD4*, *EDS1*, *EDS5*, *SID2* and *NPR1* (Fig. 3.19; Denoux et al. 2008). The NI141 flagellin isolated from *Acidovorax avenae* induced the expression of *OsEDS1* in rice cells (Fujiwara et al. 2004). The major function of PAD4 and EDS1 is to process ROS-derived signals in the defense signaling (Rustérucci et al. 2001; Mateo et al. 2004). They transduce ROS-derived signals leading to SA production (Wiermer et al. 2005). *EDS5* encodes a protein, which transports precursors for SA biosynthesis (Nawrath et al. 2002). SID2 is an isochorismate synthase that is involved in biosynthesis of SA (Wildermuth et al. 2001). SA signaling system consists of a linear pathway in which *PAD4* and *EDS1* activate *EDS5* and *SID2*, which produce SA (Glazebrook et al. 2003). The *NPR1* is a master regulator of the SA-mediated induction of defense genes (Cao et al. 1997; Chern et al. 2008). While PAD4, EDS1, EDS5, and SID2 act upstream of SA, NPR1 acts downstream of SA (Nawrath et al. 2002; Glazebrook et al. 2003; Lippok et al. 2007; Chern et al. 2008; Spoel and Dong 2012; Wu et al. 2012).

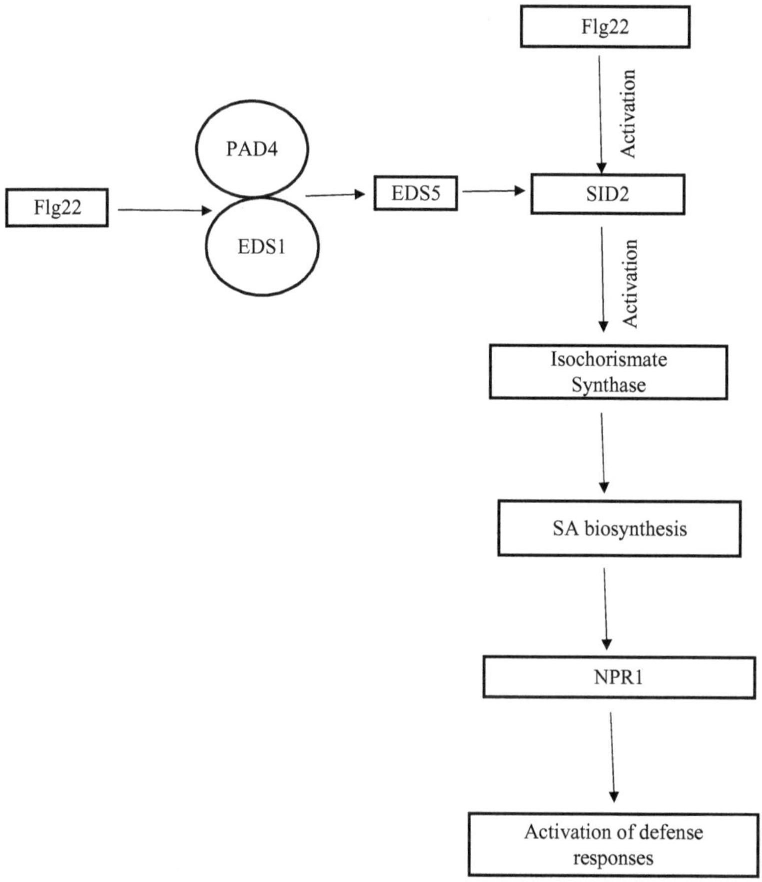

Fig. 3.19 Activation of SA biosynthesis and signaling pathway by the PAMP flg22 (Adapted from Glazebrook et al. 2003; Chern et al. 2008; Denoux et al. 2008)

Flg22 activates the transcription factor WRKY7, which is a negative regulator of SA-mediated responses in Arabidopsis (Denoux et al. 2008). Probably the transcription factor would have suppressed the SA- induced defense response. The transgenic *Arabidopsis* plants overexpressing *WRKY7* showed reduced expression of defense-related genes, including *PR1* (Kim et al. 2006). Ferrari et al. (2007) showed that resistance to *Botrytis cinerea* induced in *Arabidopsis* by flg22 was independent of SA signaling.

Flg22 induced the activation of JA signaling system (Fig. 3.20). It enhanced the expression of *LOX3* and *LOX4* genes encoding lipoxygenases (LOX), *OPR3* gene encoding 12-oxophytodienoate reductase (OPR) and *ACX1* gene encoding acyl-CoA oxidase (ACX), (Denoux et al. 2008). LOX, OPR, and ACX are involved in the biosynthesis of JA (Wasternack 2007; Balbi and Devoto 2008).

Flg22 up-regulated *ACS* genes encoding 1-amino-cyclopropane-1-carboxylase synthase (ACS) and ACS is the key enzyme involved in biosynthesis of ethylene.

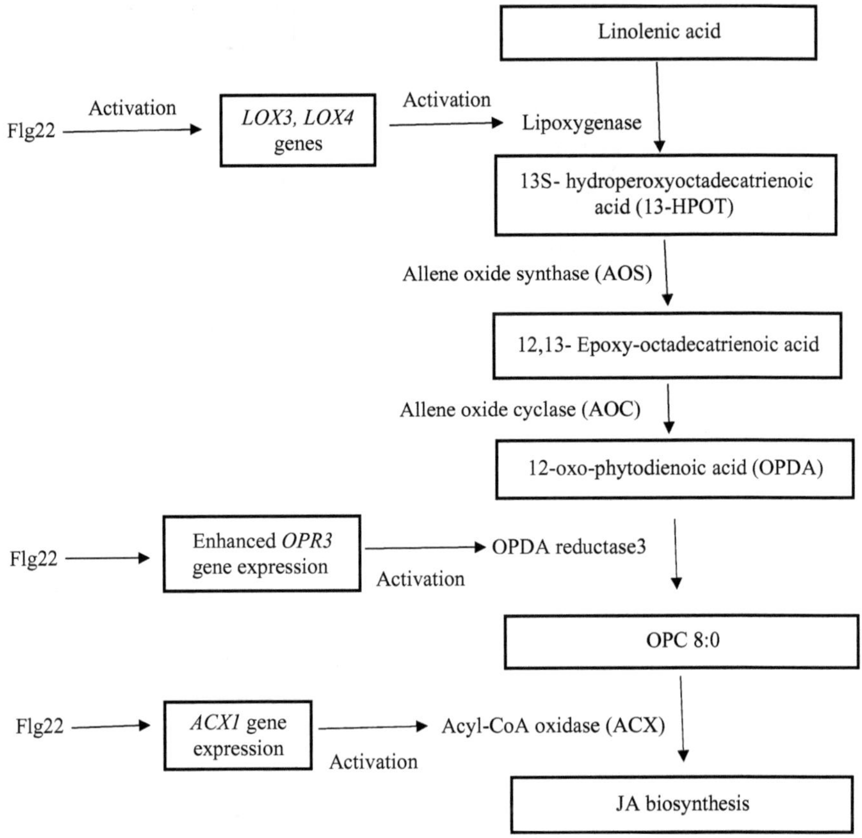

Fig. 3.20 Activation of genes involved in JA biosynthesis pathway by the PAMP flg22 (Adapted from Denoux et al. 2008; Wasternack and Hause 2013)

It also induced expression of the ethylene receptors ETR1 and EIN4 (Denoux et al. 2008). In *Arabidopsis*, ethylene is perceived by membrane-bound receptors such as ETR1 and EIN4, which transmit the signal to downstream effectors (Qu et al. 2007; Grefen et al. 2008). The ethylene receptor ETR1 functions as an ROS sensor and mediates H_2O_2 signaling in *Arabidopsis* (Desikan et al. 2005). Flg22 also induced the negative regulator of ethylene responses *CTR1* (Denoux et al. 2008). CTR1is a MAPKKK and acts through MAPK cascade. A MAPK cascade, MAPKKK (CTR1) – MKK9 – MPK3/MPK6, has been shown to be an important downstream component in ethylene signaling system. Both the inhibition of CTR1 and activation of MKK9 are induced by ethylene signaling (Yoo et al. 2008). Flg22 activates the ethylene-regulated transcription factor ERF1. ERF1 is a downstream component of ethylene signaling system (Berrocal-Lobo and Molina 2004). These results suggest that flg22 is involved in activation of ethylene signaling system.

Flg22 induces activation of several transcription factors. It activates WRKY33 and WRKY40 transcription factors (Denoux et al. 2008). These transcription factors function as activators of JA signaling system and repressors of SA signaling system (Zheng et al. 2006). WRKY33 positively regulates JA-mediated responses, while it retards the SA-mediated PR genes expression (Zheng et al. 2006). After flg22 treatment WRKY70 was consistently induced around 20-fold in *A. thaliana*. WRKY70 induces expression of SA-responsive PR genes (Li et al. 2004a). WRKY70 has been shown to function downstream of ROS and SA (Knoth et al. 2007). The WRKY70 transcription factor acts as a node of convergence for jasmonate-mediated and salicylate-mediated signals in plant defense (Li et al. 2004a). It modulates the selection of signaling pathways in plant defense (Li et al. 2006).

3.4.2 Flg22 Triggers Host Defense Responses

Flg22 induces expression of several defense-related genes. The pathogenesis-related protein-1 encoding gene *PR-1* was induced in *Arabidopsis* seedlings that were infiltrated with the flagellin flg22 elicitor solution (Denoux et al. 2008). Flg22 activates *THI2.2* gene encoding thionin, and *DIR1* gene encoding lipid transfer protein (Denoux et al. 2008). Thionin belongs to PR-13 family and lipid transfer protein belongs to PR-14 family of PR proteins (Sels et al. 2008; Maghuly et al. 2009). Both of them have been shown to play important role in disease resistance (Iwai et al. 2002; Jayaraj and Punja 2007; Sun et al. 2008). Another PR protein PR-4 is also induced by flg22 (Denoux et al. 2008) and *PR-4* genes are involved in conferring resistance to diseases (Flocchetti et al. 2008). Fujiwara et al. (2004) showed the induction of thaumatin-like protein (PR-5) in rice cells. Flg22 induces callose deposition in *Arabidopsis* (Denoux et al. 2008).

3.4.3 Foliar Application of Flg22 Induces Disease Resistance

The PAMP flg22 can be developed as a molecular tool for management of crop diseases. To demonstrate this possibility flg22 was sprayed on 4-week-old *Arabidopsis* plamts and the plants were inoculated with the fungal pathogen *Botrytis cinerea* 24 h after the spray. The flg22 treatment significantly reduced the disease development caused by the pathogen (Fig. 3.21; Ferrari et al. 2007). The results suggest that the bacterial flagellin peptide elicitor flg22 enhances resistance to *B. cinerea* in *Arabidopsis* (Ferrari et al. 2007). Treatment of *Arabidopsis* plants with flg22 induced the expression of numerous defense-related genes and triggered resistance to the bacterial pathogen *Pseudomonas syringae* pv. *tomato* in *Arabidopsis* (Zipfel et al. 2004). In grapevine, flg22 triggered immune responses and the treatment led to partial resistance against *B, cinerea* (Trdă et al. 2014). These studies suggest that the PAMP flg22 can be developed as a potential tool for the management of a wide range of crop diseases.

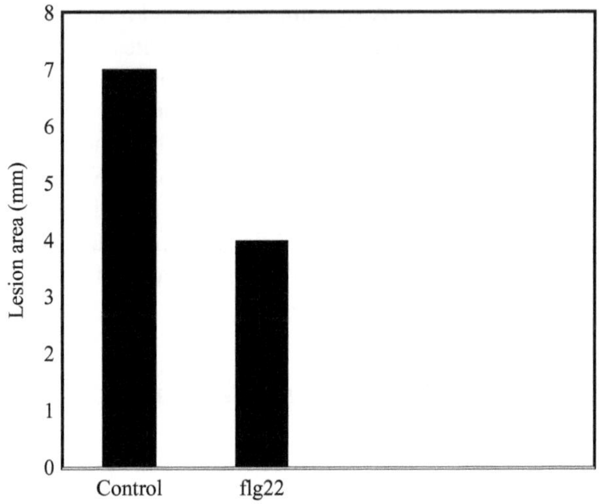

Fig. 3.21 Induction of disease resistance against *Botrytis cinerea* by flg22 application in *Arabidopsis* (Adapted from Ferrari et al. 2007)

3.4.4 Genetic Engineering to Develop Disease Resistant Plants Using Flagellin

The PAMP flagellin has been exploited to develop disease resistant plants. A flagellin gene from the bacterial pathogen *Acidovorax avenae* strain N1141 was fused to the *Cauliflower mosaic virus* 35S promoter and introduced to rice (Takakura et al. 2008). The transgenic rice plants showed accumulation of flagellin and increased H_2O_2 production. These plants exhibited increased expression of the defense-related phenylalanine ammonia-lyase (*PAL*) gene. They showed enhanced resistance to the rice blast pathogen *Magnaporthe grisea* (Takakura et al. 2008). These results suggest that the PAMP flagellin production in plants confers disease resistance and the PAMP may be a potential tool to trigger the defense signaling system and develop disease resistant plants.

3.5 Molecular Manipulation of Plant Immune Systems Using the PAMP Elicitins

3.5.1 Oomycetes-Secreted Elicitins

Elicitins are low molecular weight proteinaceous elicitors secreted by some species of the oomycetes *Phytophthora* and *Pythium* (Table 3.3; Panabières et al. 1995; Suty et al. 1995; Keller et al. 1999; Ponchet et al. 1999; Churngchow and Rattarasan

Table 3.3 Elicitins detected in *Phytophthora* and *Pythium* spp.

Oomycete	Elicitin
Phytophthora infestans	Infestin (INF1, INF2A, INF2B, INF4, INF5, INF6)
Phytophthora megasperma	α-Megaspermin, β-Megaspermin, γ- Megaspermin
Phytophthora cryptogea	α-Cryptogein, β-Cryptogein
Phytophthora parasitica	Parasiticein
Phytophthora sojae	Soj1, Soj2
Phytophthora capsici	α-Capsicein
Phytophthora cinnamomi	α-Cinnamomin, β-Cinnamomin
Phytophthora drechsleri	Dre- α, Dre- β
Pythium vexans	Vex1, Vex2
Pythium oligandrum	Oligandrin (Oli-d1, Oli-d2), POD1, POD2, CWPs, Elicitin-like glycoproteins

2000; Picard et al. 2000; Brummer et al. 2002; Koehl et al. 2003; Lherminier et al. 2003; Qutob et al. 2003; Donghua et al. 2004; Mohamed et al. 2007; Vidhyasekaran 2007; Chirapongsatonkul et al. 2008; Quang et al. 2015). These are structurally similar to lipid-transfer proteins of plant cells (Blein et al. 2002). They behave like sterol carrier proteins (Mikes et al. 1998) and they are able to pick up sterols from plasma membranes (Vauthrin et al. 1999).

Elicitin-like proteins have been identified in the biocontrol agent *Pythium oligandrum* (Takenaka et al. 2006). Cell wall protein fractions of *P. oligandrum* show elicitor properties in sugar beet and wheat. The fractions contained two major forms (POD-1 and POD-2). The POD proteins were identified as glycoproteins; however, only their peptide portions showed elicitor activity (Takenaka et al. 2006). POS1 is another elicitin-like protein detected in *P. oligandrum* (Masunaka et al. 2010). Oligandrin is an extracellular protein detected in *P. oligandrum*. *P. oligandrum* isolates can be divided into two groups based on the cell wall elicitor proteins. Several genes encoding the elicitin-like proteins have been isolated from these isolates. The D-type isolates carried *pod-1*, *pod-2* and two oligandrin genes, termed *oli-d1* and *oli-d2*, while the S-type isolates carried *pos-1*and one oligandrin gene termed *oli-s1* (Masunaka et al. 2010).

3.5.2 Elicitin-Induced Early Plant Immune Signaling Events

3.5.2.1 Protein Phosphorylation

Elicitins are perceived by specific receptor sites in the plasma membrane of plant cells (Wendehenne et al. 1995; Bourque et al. 1999; Buhot et al. 2001). A lectin-like receptor kinase has been identified in *Nicotiana tabacum* and this protein was encoded by *NtlecRK1* gene (Sasabe et al. 2007). *NtlecRK1* forms a gene family with three members in tobacco. NtlecRK1 protein was located on the plasma membrane and it was responsive to the elicitin infestin INF1 (Sasabe et al. 2007).

The elicitin cryptogein is initially recognized by a plasma membrane receptor in tobacco (Wendehenne et al. 1995; Buhot et al. 2001). The plasma membrane perception step is followed by activation of various protein kinases, such as the wound-induced protein kinase (WIPK), the salicylic acid-induced protein kinase (SIPK), and the mitogen-activated protein kinase (MAPK) Ntf4 (Tavernier et al. 1995; Lebrun-Garcia et al. 1998; Hoeberichts et al. 2013). The very early response to elicitins appears to be protein phosphorylation (Lecourieux-Ouaked et al. 2000), The elicitin-induced plant defense responses were inhibited by the protein kinase inhibitors K252a and staurosporine and it indicates that induced defense responses require protein phosphorylation (Sasabe et al. 2007). The protein phosphorylation induced by the kinases triggers calcium influx (Amelot et al. 2012). Protein phosphorylation precedes Ca^{2+} influx in tobacco cells treated with cryptogein (Fig. 3.22; Tavernier et al. 1995).

3.5.2.2 Plant Plasma Membrane Depolarization

Cryptogein induces plasma membrane depolarization in tobacco cells (Pugin et al. 1997). Depolarization-activated Ca^{2+}-permeable channels are common in plasma membrane of plant cells. These voltage-gated Ca^{2+} channels mediate Ca^{2+} influx across the plasma membrane of cells (Huang et al. 1994). Ca^{2+} channel opens upon depolarization of the membrane electrical potential (Huang et al. 1994; White 2000, 2004). Membrane depolarization may be due to the activation of anion channels (Sanders et al. 2002; Jeworutzki et al. 2010).

3.5.2.3 Calcium Ion Influx

Cryptogein induces anion efflux in tobacco cells (Pugin et al. 1997; Kadota et al. 2004). Phamacological analyses with anion channel blockers including DIDS, 4,49-diisothiocyanostilbene-2,29-disulfonic acid, suggest the importance of anion efflux mediated by the plasma membrane anion channels in induction of cryptogein-induced responses in tobacco BY-2 cells (Kadota et al. 2004). An S-type anion channel SLAC1 has been shown to be involved in cryptogein-induced ion fluxes (Kurusu et al. 2013). Overexpression of *SLAC1* in cultured tobacco BY2 cells enhanced anion efflux and extracellular alkalinization (Kurusu et al. 2013). The results suggest that SLAC1 expressed in BY2 cells mediates cryptogein-induced plasma membrane anion efflux to mediate the cryptogein-triggered activation of other ion fluxes. It has also been reported that Ca^{2+} influx is a prerequisite for the activation of plasma membrane anion channels in several systems (Ward et al. 1995; Jabs et al. 1997; Wendehenne et al. 2002). In cryptogein-treated tobacco cells, the major calcium influx did not result from plasma membrane depolarization. Instead, the Ca^{2+} influx occurred upstream and it triggered anion efflux and plasma

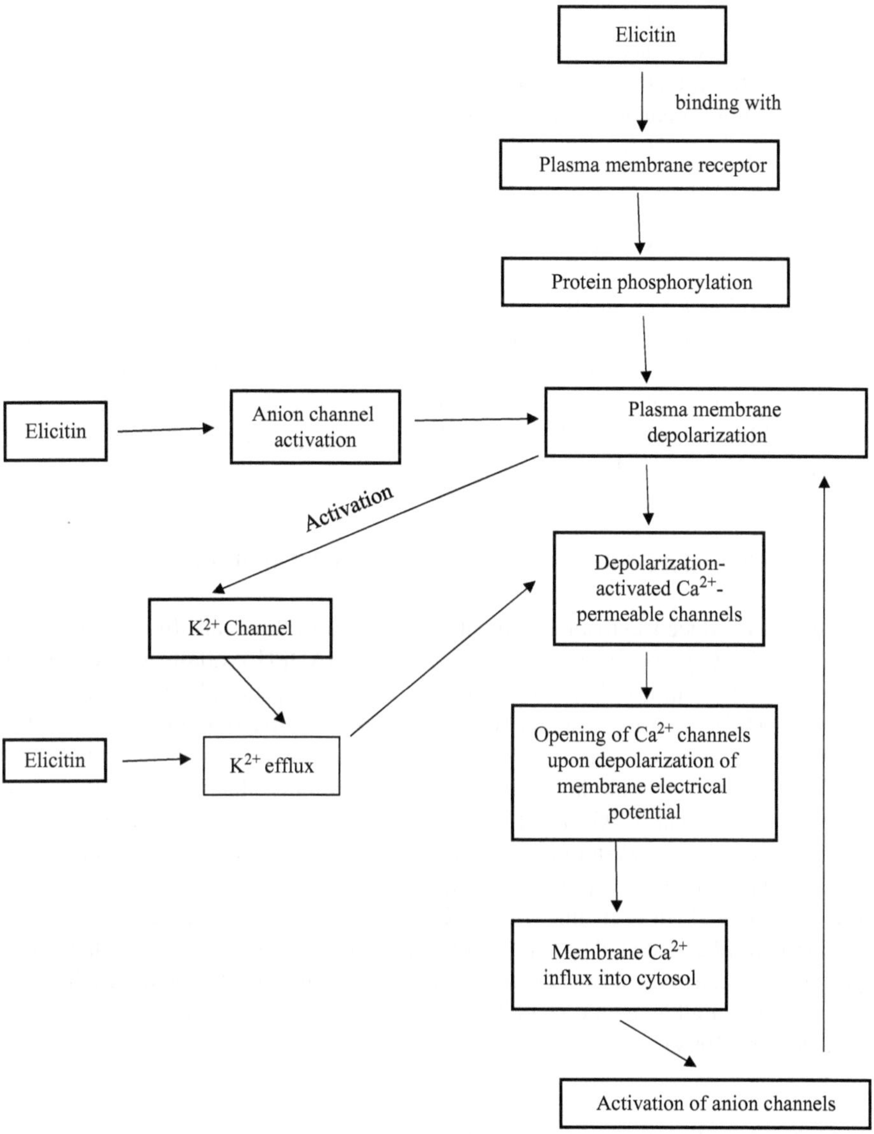

Fig. 3.22 Elicitins-triggered Ca^{2+} signaling system

membrane depolarization, which in turn may mobilize some Ca^{2+} voltage-dependent channels (Pugin et al. 1997; Wendehenne et al. 2002). Collectively these results suggest that initial Ca^{2+} -influx may be through some voltage-dependent channels, independent of anion efflux and subsequent activation of Ca^{2+} influx may be through the anion efflux-activated voltage-dependent channels.

It has also been shown that Ca^{2+} influx may activate anion channels and increases in cytoplasmic Ca^{2+} activate anion channels (Ward et al. 1995; Marten et al. 2007; Suh et al. 2007). Cytoplasmic Ca^{2+} elevation has been shown to result in activation of S-type anion channels via phosphorylation (Schmidt et al. 1995; Allen et al. 1999; Geiger et al. 2010). The activated anion channels further intensify Ca^{2+} influx which will pave the way for triggering Ca^{2+}-mediated signaling system.

The PAMP cryptogein induced K^+ efflux in tobacco cells (Viard et al. 1994). The outward-rectifying K^+ channels found in the plasma membrane of plant cells are also Ca^{2+}-permeable depolarization-activated channels. These Ca^{2+}-permeable outward rectifying K^+ channels activate significantly at voltages more positive than about -50 mV under most physiological conditions and catalyze a large K^+ efflux simultaneously with a small Ca^{2+} influx (White and Broadley 2003). Anion channel activation will depolarize the membrane and enhance the activity of K^+ channel activity (Maathuis et al. 1997). These channels may play a role in the initial Ca^{2+} influx into the cytosol (Fig. 3.22).

3.5.2.4 Cytosol Acidification and Extracellular Alkalinization

Cryptogein induced cytosol acidification and extracellular alkalinization (Pugin et al. 1997). Cryptogein activates a membrane NADPH oxidase whose activity depends on Ca^{2+} influx (Pugin et al. 1997). The cryptogein-induced plasma membrane NADPH oxidase is responsible for cytoplasm acidification (Pugin et al. 1997). Cryptogein triggers Ca^{2+} influx, which induces extracellular alkalinization (Tavernier et al. 1995). A Ca^{2+} ionophore triggered an extracellular alkalinization in tobacco cells (Tavernier et al. 1995). K^+/H^+ exchange response is an important component in the defense signaling system (Orlandi et al. 1992). The enzyme H^+/K^+-ATPase is a proton pump which is responsible for the acidification of cytoplasm. The enzyme transports one H^+ in exchange of one K^+ (Kuhlbrandt 2004). The K^+/H^+ exchange response may be mediated by Ca^{2+} influx (Atkinson et al. 1990). The Ca^{2+}-dependent protein kinase, activated by increased cytosolic Ca^{2+}, may evoke ion fluxes that lead to extracellular alkalinization and depolarization of the plasma membrane (Schaller and Oecking 1999).

3.5.2.5 Reactive Oxygen Species Production

Cryptogein induced plasma membrane NADPH oxidase (respiratory burst oxidase homologue [RBOH]) activity resulting in production of ROS (Fig. 3.23; Bottin et al. 1994; Sasabe et al. 2000; Kadota et al. 2004; Leborgne-Castel et al. 2008; Kurusu et al. 2013). RBOHs are membrane proteins composed of six transmembrane domains associated with two haem groups, a C-terminal region with two calcium-binding EF-hands and phosphorylation domains (Suzuki et al. 2011). RBOHs catalyse the formation of the superoxide anion (O_2^-) by transferring an electron from intracellular NADPH to apoplastic molecule of oxygen (Sagi and

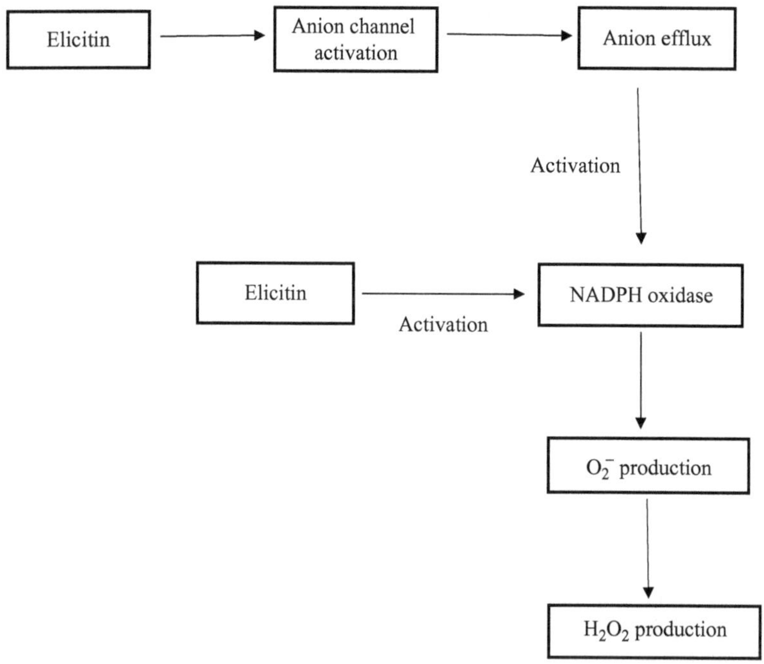

Fig. 3.23 Elicitin-triggered anion efflux in ROS production (Adapted from Kurusu et al. 2013)

Fluhr 2006). Superoxide anions are short-lived and the half life of O_2^- is less than a second. It is usually dismutated either nonenzymatically or via superoxide dismutase (SOD) and then converted to H_2O_2, which is relatively stable (Grant and Loake 2000; Mori and Schroeder 2004; Vidhyasekaran 2014). Several isoforms of RBOHs have been reported in different plants (Zhang et al. 2009). RBOHD is the sole isoform responsible for the ROS burst triggered by the elicitin cryptogein in tobacco (Simon-Plas et al. 2002; Lherminier et al. 2009).

RBOHs are intrinsic plasma membrane (PM) proteins of plant cells (Takeda et al. 2008; Lee et al. 2013; Coursol et al. 2015; Noirot et al. 2014). Cryptogein induced, within 1 h, a 1.5 fold increase in RBOHD abundance at the PM and a concomitant decrease in the internal compartments. Most of the proteins targeted to the PM upon elicitation by cryptogein were not newly synthesized but may originate from the Golgi Pool. ROS accumulation preceded RBOH transcript- and protein–upregulation, indicating that ROS resulted from the activation of a PM-resident pool of enzymes, and that enzymes newly addressed to the PM were inactive (Noirot et al. 2014).

ROS production was very early event in elicitin-treated tobacco cells and ROS production was observed within the first 15 min after elicitation (Leborgne-Castel et al. 2008). The cryptogein-induced ROS production mediated by NADPH oxidase was enhanced in the S-type anion channel *SLAC1*-overexpressing tobacco cells (Fig. 3.23; Kurusu et al. 2013). The basal expression level of *NtRbohD*, the gene encoding NADPH oxidase and the basal ROS level before the elicitation in the *SLAC1* overexpressors were comparable to those in the control, suggesting that overexpression of *SLAC1* may lead to enhanced activation of NADPH oxidases by its posttranslational regulation (Kurusu et al. 2013).

The oomycete PAMP cryptogein has been shown to stimulate endocytosis in tobacco cells (Leborgne-Castel et al. 2008). PAMP is known to activate PRRs by inducing its phosphorylation (Segonzac and Zipfel 2011). Plasma membrane resident PRRs are translocated into endosomes (Geldner and Robatzek 2008). PAMP-induced PRR-mediated endocytosis has been shown to be dependent on phosphorylation of the PRR (Robatzek et al. 2006). PAMP-induced internalization of PRRs from the plasma membrane (PM) is closely correlated with their immune function (Robatzek et al. 2006; Bar and Avni 2008). Translocation of PRRs from plasma membrane to endocytic compartments has been widely reported (Fliegmann et al. 2004; Gross et al. 2005; Robatzek et al. 2006; Leborgne-Castel et al. 2008). Receptor endocytosis appears to be a common phenomenon in plant defense signaling system (Groen et al. 2008; Chen et al. 2010).

The PRRs are translocated to endosomes mostly through clathrin-mediated pathway. (Robatzek et al. 2006). Clathrin is a basket-like network of protein molecules that forms on the cell membrane in response to the attachment of ligands (PAMPs) to receptors (PRRs). It coats the endocytotic vesicles which bud off from the membrane and it becomes inside surface of the coated vesicle during endocytosis (Vidhyasekaran 2014). The oomycete PAMP cryptogein has been shown to stimulate endocytosis in tobacco cells, which is correlated with ROS production (Leborgne-Castel et al. 2008). Internalization of the lipophilic dye FM4-64, which is a marker of endocytosis, was stimulated a few minutes after addition of cryptogein to tobacco cells. The cryptogein increased clathrin-mediated endocytosis. In the presence of cryptogein, increases in both FM4-64 internalization and clathrin-mediated endocytosis are specifically blocked upon treatment with the receptor-mediated endocytosis inhibitor tyrphostin A23 (Leborgne-Castel et al. 2008). The kinetics of the transient increase in clathrin-coated pits at the plasma membrane coincided with that of transitory reactive oxygen species (ROS) production occurring within the first 15 min after elicitation. In tobacco cells expressing the NADPH oxidase gene *NtrbohD* antisense cDNA, which are unable to produce ROS when treated with cryptogein, the clathrin-coated pits stimulation was inhibited. These results indicate that the very early endocytic process induced by cryptogein in tobacco is due to clathrin-mediated endocytosis and is dependent on ROS production induced by the NADPH oxidase (Leborgne-Castel et al. 2008).

3.5.2.6 Induction of Nitric Oxide Burst

The oomycete PAMP cryptogein triggers a nitric oxide (NO) burst within minutes in tobacco cells (Foissner et al. 2000; Lamotte et al. 2004). NO is a key mediator for rapid induction of plant immune signaling systems (Saito et al. 2006; Neill et al. 2008; Perchepied et al. 2010; Gaupels et al. 2011; Vandelle and Delledonne 2011;

Chun et al. 2012; Bellin et al. 2013). Rapid NO production has been shown to be dependent on Ca^{2+} signaling system (Lamotte et al. 2004; Ali et al. 2007; Courtois et al. 2008; Choi et al. 2009; Vatsa et al. 2011; Ma et al. 2012). It has been demonstrated that different calcium channels involved in calcium influx are also involved in NO production. Glutamate receptor (GLR)-like channels are another type of ion channels involved in Ca^{2+} influx (Ma et al. 2012; Forde and Roberts 2014). The oomycete PAMP elicitor signal cryptogein activates GLR calcium channels, which are also involved in triggering NO production (Vatsa et al. 2011).

Treatment of protoplasts prepared from *Nicotiana benthamiana* leaves with the PAMP elicitin INF1 elevated NO production. INF1-induced NO generation was suppressed by an NO-specific scavenger. Silencing of nitrate reductase (NR) genes significantly decreased INF1-induced NO production (Yamamoto-Katou et al. 2006). These results suggest that nitrate reductase is involved in the elicitin-triggered NO generation. NO may be formed from nitrite by the action of nitrate reductase, which catalyzes the reduction of nitrate to nitrite using NAD(P)H as an electron donor and also generates NO from nitrite (Fig. 3.24; Yamamoto-Katou et al. 2006). NO generation was reduced in ammonium –fed tobacco plants compared to NO_3^- -fed plants and NO_3^- -fed plants showed enhanced disease resistance against *Pseudomonas syringae* pv. *phaseolicola* (Gupta et al. 2013). NO production was

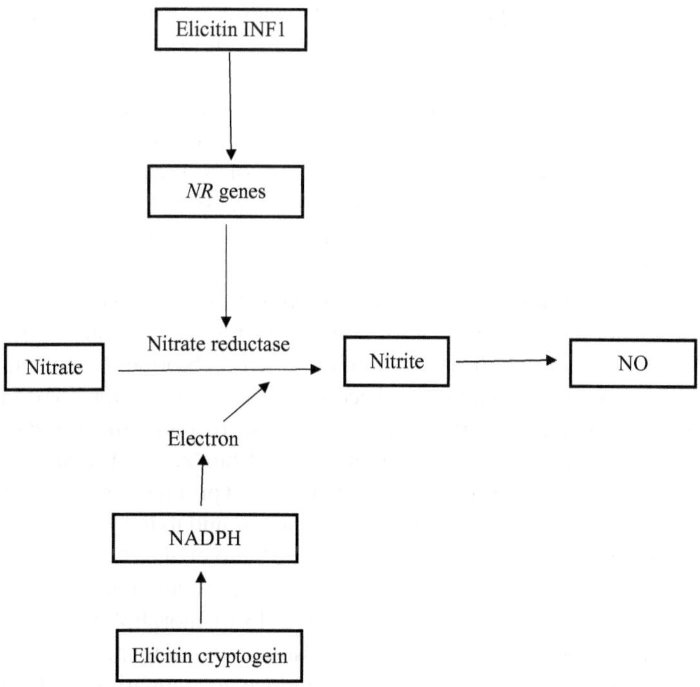

Fig. 3.24 Elicitin-induced NO production through nitrate reductase pathway (Adapted from Yamamoto-Katou et al. 2006)

completely absent in ammonium-grown tobacco cell suspensions totally devoid of nitrate (Planchet et al. 2005, 2006). It indicates that NO synthesis is catalyzed through a nitrate/nitrite-dependent pathway in these tobacco cells.

Cryptogein has also been shown to trigger NO production by another NO biosynthesis pathway, nitric oxide synthase (NOS) pathway, in tobacco cells. NO production was observed in tobacco cells treated with cryptogein. Cryptogein-induced NO production in tobacco cell suspensions was sensitive to nitric oxide synthase inhibitors and it was catalyzed by a variant P, a pathogen-inducible plant nitric oxide synthase (Lamotte et al. 2004). The results suggest that the elicitin elicits NO production by activating nitric oxide synthase.

3.5.2.7 Peroxynitrite ($ONOO^-$) Generation

Peroxynitrite ($ONOO^-$) is a reactive nitrogen species generated by the interaction of ROS and NO during oxidative burst (Romero-Puertas et al. 2007). Peroxynitrite is generated by the reaction of NO with superoxide anion (O_2^-) (Fig. 3.25; Bryk et al. 2000; Neill et al. 2008; Vandelle and Delledonne 2011). Peroxynitrite generation was observed in the elicitin INF1-treated tobacco cells (Saito et al. 2006). $ONOO^-$ generation in tobacco cells treated with the elicitin occurred within 1 h and reached a maximum level at 6–12 h after INF1 treatment (Saito et al. 2006). Urate, an $ONOO^-$ scavenger, abolished INF1-induced $ONOO^-$ generation (Saito et al. 2006). $ONOO^-$ reacts with tyrosine residues in proteins to form nitrotyrosine in a nitration reaction as an $ONOO^-$ specific reaction. NO can also perform posttranslational protein modifications through nitration, besides through S-nitrosylation (Zaninotto et al. 2006). During the hypersensitive reaction (HR), the formation of $ONOO^-$ is promoted by the rate of NO reaction with O_2^-, which is approximately three times faster than the reaction of O_2^- with superoxide dismutase (SOD) forming H_2O_2 during the oxidative burst (Ichiropoulos and al-Mehdi 1995).

$ONOO^-$ causes protein tyrosine nitration through the nitration of Tyr residues in Tyr kinase (Schopfer et al. 2002; Radi 2004; Romero-Puertas et al. 2008). $ONOO^-$ donor SIN-1 [3-(4-morpholinyl) sydnonimine hydrochloride) treatment induced nitrotyrosine-containing proteins in tobacco cells (Saito et al. 2006). The number of nitrated proteins increased during disease resistance response in *Arabidopsis thaliana* (Romero-Puertas et al. 2008). Protein extracts from leaves of *A. thaliana* treated with $ONOO^-$ showed a significant increase in nitrated proteins when pretreated with GSNO (Romero-Puertas et al. 2008). $ONOO^-$ was found to induce protein nitration in soybean and tobacco (Delledonne et al. 2001; Saito et al. 2006).

$ONOO^-$is an important signal molecule triggering accumulation of PR proteins (Durner et al. 1998). $ONOO^-$ donor SIN-1 [3-(4-morpholinyl) sydnonimine hydrochloride) treatment induced nitrotyrosine-containing proteins in tobacco cells (Saito et al. 2006). SIN was found to induce accumulation of the transcript encoding PR-1 in tobacco leaves (Durner et al. 1998). $ONOO^-$ causes protein tyrosine nitration through the nitration of tyrosine (Tyr) residues in Tyr kinase (Schopfer

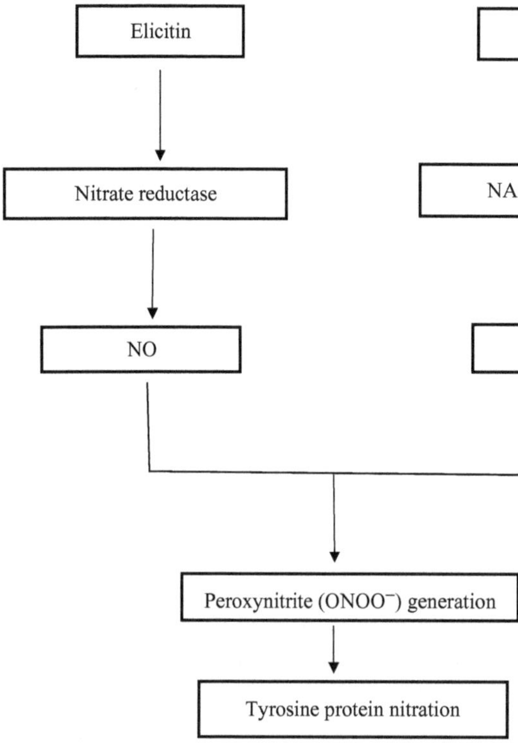

Fig. 3.25 Peroxynitrite (ONOO) generation in INF1-treated tobacco cells (Adapted from Saito et al. 2006; Vandelle and Delledonne 2011)

et al. 2002; Radi 2004; Romero-Puertas et al. 2008). The number of nitrated proteins increased during disease resistance response in *Arabidopsis thaliana* (Romero-Puertas et al. 2008). $ONOO^-$ was found to induce protein nitration in soybean and tobacco (Delledonne et al. 2001; Saito et al. 2006). Protein nitration alters catalytic activity and interferes with cellular signaling processes (Schopfer et al. 2002).

Protein extracts from leaves of *A. thaliana* treated with $ONOO^-$ showed a significant increase in nitrated proteins when pretreated with S-nitrosoglutathione (GSNO) (Romero-Puertas et al. 2008). NO reacts rapidly with glutathione (GSH), the major intracellular low-molecular-mass antioxidant to yield GSNO. GSNO is a bioactive, stable, and mobile reservoir of NO and it is an important player in plant defense responses against pathogens (Espunya et al. 2012). GSNO is considered to represent a functionally relevant signaling molecule that might act both as NO reservoir and NO donor (Lindermayr et al. 2005) or independently of homolytic cleavage to NO (Gaston 1999). GSNO acts synergistically with SA in systemic acquired resistance (SAR) (Espunya et al. 2012). It has been suggested that GSNO would act as a long- distance phloematic signal in SAR (Durner and Klessig 1999).

3.5.2.8 Activation of MAPK Cascade

Cryptogein activates mitogen-activated protein kinase (MAPK) cascade (Yang et al. 2001; Lebrun-Garcia et al. 2002; Takemoto et al. 2005). MAPK activation is compromised when the extracellular calcium influx is blocked by Ca^{2+} channel blockers or Ca^{2+} chelators in tobacco cells treated with cryptogein (Garcia-Brugger et al. 2006). Mitogen-activated protein kinase cascades are major signal transduction systems functioning downstream of PRRs upon perception of PAMP elicitor signals (Hettenhausen et al. 2012; Zhang et al. 2012). MAPKKKs receive PAMP signals from upstream receptors/sensors to activate the MAPK signaling system (Teige et al. 2004; Li et al. 2012).

Two separate MAPK pathways including MAPKKK – NtMEK2 – WIPK and MAPKKK – NtMEK2 – SIPK have been shown to be involved in defense response signaling in tobacco (Liu et al. 2003). Both these pathways are interconnected in induction of defense signals. Elicitins from *Phytophthora* species activated the 44-kDa MAP kinase encoded by WIPK and the activation of this kinase occurred 2–4 h after elicitin treatment in tobacco cells (Zhang et al. 2000). It was preceded by dramatic increases in WIPK mRNA and de novo protein synthesis in tobacco cells (Liu et al. 2003). The tobacco MAP kinase SIPK is activated by the oomycete PAMP, β-megaspermin (Hall et al. 2007). The SIPK activation induced by the PAMP required external calcium influx, suggesting that SIPK activation occurs downstream of Ca^{2+} influx (Hall et al. 2007). SIPK, NtMEK2 and the upstream MAPKKK pre-exist in cells. Upon recognition of elicitor signals, NtMEK2 is activated by its upstream MAPKKK, which in turn activates the pre-existing SIPK. Activation of SIPK turns on the transcription of *WIPK* gene, which leads to the accumulation of WIPK protein. The newly synthesized WIPK protein is then activated by the NtMEK2 which is activated by a putative MAPKKK (Liu et al. 2003). WIPK triggers HR-related cell death probably by the production of H_2O_2. SIPK besides activating synthesis of WIPK, activates transcription of various defense genes, particularly 3-hydroxy-3-methylglutaryl CoA reductase (*HMGR*) and phenylalanine ammonia lyase (*PAL*) genes encoding key enzymes in the phytoalexin and salicylic acid biosynthesis pathways (Yang et al. 2001; Liu et al. 2003).

A MAPKK, NbMKK1, has been shown to be involved in the elicitin INF1-induced cell death signaling in *Nicotiana benthamiana* (Takahashi et al. 2007a). NbMKK1 functions upstream of NbSIPK (Takahashi et al. 2007b). It has been reported that NbMKK1 and NbSIPK physically interact (Takahashi et al. 2007a). NbMKK1 is localized to the nucleus, and the N-terminal putative MAPK docking site of NbMKK1 is required for its function as a cell-death inducer. The elicitin INF1-mediated hypersensitive response (HR) was delayed in NbMKK1-silenced plants (Takahashi et al. 2007a). These results suggest that the elicitin-induced NbMKK1 is involved in the defense pathway.

3.5.3 Induction of Salicylic Acid Biosynthesis and SA-Dependent Signaling Pathway by Elicitins

The elicitins megaspermins triggered the activity of phenylalanine ammonia-lyase (PAL), the key enzyme involved in the SA biosynthesis pathway in cultured tobacco cells. The elicitins also caused SA accumulation (Dorey et al. 1999). When α-aminooxy-β-phenylpropionate (AOPP) is applied along with the elicitin, it suppressed the induction of SA-responsive acidic isoforms of PR-1, PR-2, PR-3 and PR-5 in tobacco plants. AOPP is a specific inhibitor of PAL (Cordelier et al. 2003). The results suggest that the elicitin activates SA biosynthesis and downstream signaling pathway. The elicitin-induced resistance against *Tobacco mosaic virus* breaksdown in tobacco plants expressing the bacterial *nahG* gene which are unable to accumulate SA (Keller et al. 1994). It suggests that SA signaling system takes part in elicitin-induced defense responses. Salicylic acid mediates elicitin-induced systemic acquired resistance in tobacco (Keller et al. 1996; Cordelier et al. 2003).

3.5.4 Elicitin-Induced Jasmonic Acid Biosynthesis and JA-dependent Signaling Pathway

When the cultured tobacco cells were treated with the elicitin cryptogein, an increase in the activity of lipoxygenase, the key enzyme in the JA biosynthesis pathway was observed (Bottin et al. 1994). There was also an accumulation of JA-responsive proteinase inhibitors in the tobacco cells treated with the elicitin (Bottin et al. 1994). The elicitin preparations (homogenate of mycelia) from *P. oligandrum* induced JA-dependent defense responses in tomato. The expression of the JA-responsive gene for the basic PR-6 protein was induced in the tomato cultivars treated with the mycelial homogenate (Hase et al. 2008). The induced resistance against *Ralstonia solanacearum* was compromised in the elicitin-treated *jai1-1* mutant plants defective in JA signaling. These results suggest the JA-dependent signaling pathway is required for the elicitin-induced defense responses. By contrast, SA did not accumulate in the elicitin-treated plants. The SA-inducible *PR-1(P6)* gene was also not upregulated in these plants. Induced resistance against *R. solanacearum* was not compromised in SA-non-accumulating *NahG* transgenic plants treated with the elicitor (Hase et al. 2008). These results suggest that SA signaling pathway is not involved in the elicitin-induced defense response. These studies also suggest that the elicitin triggers JA biosynthesis and activates downstream JA signaling pathway.

The cell wall protein fraction (CWP) containing elicitin-like proteins induced JA-responsive genes *PDF1.2* and *JR2* and induced resistance against *Ralstonia solanacearum* and *Pseudomonas syringae* pv. *tomato* in *Arabidopsis thaliana* (Kawamura et al. 2009). Such CWP-induced defense responses were completely compromised in CWP-treated *coi1-1* and *jar1-1* mutants with an impaired JA signalling (Kawamura et al. 2009). These results suggest the CWP triggers JA signalling system and induces disease resistance.

3.5.5 Induction of Ethylene Biosynthesis and Ethylene-Dependent Signaling Pathway

Cryptogein applied on the petiole section of excised tobacco leaves induced production of ethylene (Milat et al. 1991). The elicitin induced production of ethylene in tobacco cultured cells (Blein et al. 1991). Treatment of tomato roots with mycelial homogenates (containing the elicitin oligandrin) of *Pythium oligandrum* resulted in transient elevation of ethylene. It was followed by high expression of three basic ethylene-inducible defense-related genes (*PR-2b, PR-3b and PR-5b*). The expression of *LeERF2* and *LeETR4* genes, which are involved in the ethylene-dependent signaling pathway, was also accelerated by the elicitin treatment (Hase et al. 2006). *Phytophthora megasperma* elicitin triggered induction of the ethylene-responsive basic isoforms of the PR proteins, PR-1, PR-2, PR-3, and PR-5 in tobacco (Cordelier et al. 2003). Accumulation of the basic isoforms of the PR proteins was suppressed when the inhibitor of ACC synthase CoCl2 was applied along with the elicitin (Cordelier et al. 2003). ACC synthase is the key enzyme involved in biosynthesis of ethylene (Brisson et al. 2012; Vidhyasekaran 2015). These results suggest that elicitin triggers ethylene biosynthesis and activates downstream ethylene signaling pathway. The elicitin-like proteins, oligogandrin and cell wall protein fractions, trigger early signaling events preceding activation of the JA- and ethylene signalling pathways (Benhamou et al. 2012).

3.5.6 Elicitin-Induced Defense Responses

The late reactions induced by elicitin include stimulation of phenylalanine ammonia-lyase (Dorey et al. 1999) and chalcone synthase activities (Keller et al. 1994), accumulation of phytoalexins (Tavernier et al. 1995) and the PR proteins PR-1, PR-2, PR-3, and PR-5 (Bonnet et al. 1986; Keller et al. 1994) and induction of programmed cell death (Binet et al. 2001). It induced callose deposition in tobacco plants (Lherminier et al. 2003). The elicitin cryptogein induced the key enzyme in the sesquiterpenoid phytoalexin pathway, sesquiterpene cyclase (Keller et al. 1998). The elicitin megaspermin induced PR-1, PR-2, PR-3, and PR-5 proteins and it induced *O*-methyltransferases, the enzymes of phenylpropanoid pathway in tobacco (Cordelier et al. 2003).

Oligandrin isolated from *Pythium oligandrum* induced accumulation of phenolic compounds and cell wall thickening in tomato (Mohamed et al. 2007). It induced expression of three defense-related genes encoding lipid transfer protein-1 (LTP-1), β-1,3-glucanase, and stilbene synthase in tomato leaves (Mohamed et al. 2007). The elicitin purified from *Phytophthora palmivora* induced increased activitiy of peroxidase and the biosynthesis of the phytoalexin scopoletin in calli of the rubber tree, *Hevea brasiliensis* (Chirapongsatonkul et al. 2008). Elicitins isolated from *Pythium oligandrum* induced phenylalanine ammonia-lyase in sugarbeet (Takenaka et al. 2006).

3.5.7 Elicitins Trigger Systemic Acquired Resistance

Tobacco plants pretreated with the elicitin cryptogein showed systemic acquired resistance (SAR) against the tobacco black shank pathogen *Phytophthora parasitica* (Ricci et al. 1989). Elicitin isolated from *Phytophthora megasperma* induced systemic resistance against *Tobacco mosaic virus* in tobacco (Cordelier et al. 2003). Another elicitin, oligandrin, secreted by *Pythium oligandrum* was found to be translocated through the vascular system and it induced systemic resistance against the oomycete pathogen *Phytophthora parasitica* in tomato plants (Picard et al. 2000). Oligandrin induced systemic resistance to Fusarium crown and root rot caused by the fungal pathogen *Fusarium oxysporum* in tomato plants (Benhamou et al. 2001). When oligandrin was applied to roots of tomato plants, it induced resistance against foliar infection by *Botrytis cinerea* (Mohamed et al. 2007). In contrast, application of cryptogein or capsicein to the aerial parts of tobacco plants was found to inhibit *Phytophthora cryptogea* or *P. capsici* infection through roots (Keller et al. 1996). Systemic induction induced by elicitins was persistent and the acquired resistance in tobacco treated with elicitin was observed against several fungal pathogens including *Sclerotinia sclerotiorum*, *Botrytis cinerea* and *Rhizoctonia solani* (Keller et al. 1996).

3.5.8 Management of Crop Diseases Using Elicitin

An elicitin-like protein isolated from *P. oligandrum* induced resistance against the bacterial pathogen *Ralstonia solanacearum* in tomato (Hase et al. 2006, 2008). Elicitin-like proteins isolated from *P. oligandrum* induced resistance against sugar beet seedling disease caused by the oomycete *Aphanomyces cochlioides* (Takenaka et al. 2006). The two oligandrins, Oli-D1 and Oli-D2, isolated from *P. oligandrum* induced disease resistance responses against *Botrytis cinerea* in tomato (Quang et al. 2015). Oligandrin induced resistance to *Phytophthora parasitica* infection in tomato plants (Picard et al. 2000). P. *oligandrum* has been developed as a commercial product by Vyzkummy ustov rastlinnej, Slovak Republic as "Polygandron" to induce disease resistance (Madsen and de Neergaard 1999). *P. oligandrum* is also marketed as "Polyversum" (Rekanovic et al. 2007; Kurzawińska and Mazur 2009).

Polyversum effectively controlled Verticillium wilt in pepper (Rekanovic et al. 2007). Cryptogein induced resistance against *Phytophthora parasitica* var. *nicotianae* and *Sclerotinia sclerotiorum* in tobacco (Bonnet et al. 1996).

Elicitins have been shown to be effective in controlling phytoplasma diseases (Lherminier et al. 2003). Tobacco plants were treated with cryptogein or oligandrin at 1 mg/ml and the treated plants showed inhibitory effects on the development of phytoplasma in sieve elements for at least 5 months. Inoculated elicited plants had normal development, flowering, and seed production, whereas diseased nonelicited plants showed typical symptom expression of sterility of floral organs (Lherminier et al. 2003).

Foliar spray of cell wall protein fraction (CWP) containing elicitin-like proteins of *P. oligandrum* induced defense-related genes (acidic class III chitinase, β-1,3-glucanase, 5-enol-pyruvylshikimate-phosphate synthase and oxalate oxidase-like germin) and increased resistance against *Cercospora* leaf spot in sugar beet in growth chamber trials (Takenaka and Tamagake 2009). In a field experiment, CWP treatment reduced the severity of the sugar beet Cercospora leaf spot disease caused by *Cercospora beticola* (Takenaka and Tamagake 2009). CWP treatment induced resistance against the bacterial pathogens *Ralstonia solanacearum* and *Pseudomonas syringae* pv. *tomato* in Arabidopsis (Kawamura et al. 2009).

Exogenous, foliar applications of oligandrin have been shown to sensitize susceptible tomato plants to react more rapidly and more efficiently to *Fusarium oxysporum* f. sp. *radicis-lycopersici* attack, mainly through the massive accumulation of fungitoxic compounds at sites of attempted pathogen penetration (Benhamou et al. 2001). The results suggest that oligandrin is involved in priming of defense responses against the pathogen. Oligandrin induced systemic resistance to Fusarium crown and root rot in tomato plants (Benhamou et al. 2001).

3.5.9 Genetic Engineering to Develop Disease Resistant Plants Using Elicitin Gene

Elicitins induce several defense signaling genes, resulting in activation of defense responses against various pathogens. The constitutive expression of elicitin gene in transgenic plants will be ideal to induce resistance against wide-range of pathogens. Continuous recognition of the elicitin signal has been shown to be a prerequisite for prolonged activation of signaling events in tobacco cells (Kadota et al. 2006). However, elicitin is known to induce cell necrosis and hence constitutive overexpression of the gene may affect the agronomic characters of the transgenic plants. To reduce the phytotoxicity of the elicitin, transgenic plants harboring a pathogen-inducible promoter were developed to express the elicitin at low level. Keller et al. (1999) generated transgenic tobacco plants harboring a fusion between the pathogen-inducible tobacco *hsr203J* gene promoter and a *Phytophthora cryptogea* gene encoding the highly active elicitin cryptogein. Under noninduced conditions, the

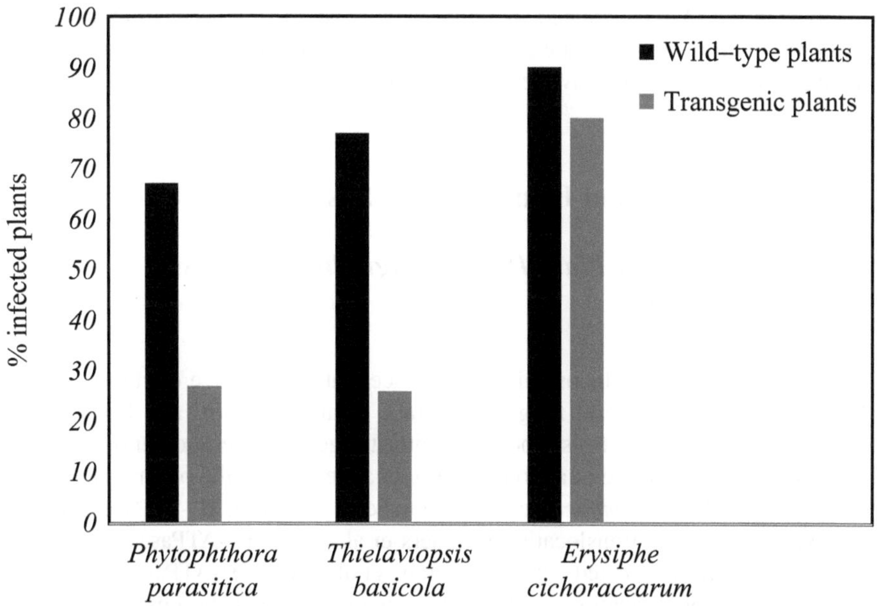

Fig. 3.26 Transgenic tobacco plants expressing elicitin gene show enhanced resistance to *Phytophthora parasitica* var. *nicotianae*, *Thielaviopsis basicola*, and *Erysiphe cichoracearum* (Adapted from Keller et al. 1999)

transgene was silent, and no elicitin could be detected in transgenic plants. When the pathogen was inoculated, the elicitin started to accumulate in the transgenic plants. The induced elicitin triggered the disease resistance response without damaging the plants. The transgenic plants showed resistance to several pathogens including *Phytophthora parasitica* var. *nicotianae*, *Thielaviopsis basicola*, and *Botrytis cinerea* (Fig. 3.26; Keller et al. 1999). The results suggest that elicitin is a powerful tool to develop disease-resistant plants.

Donghua et al. (2004) from China Agricultural University used a promoter of the rice phenylalanine ammonia-lyase (PAL) gene to regulate the expression of cryptogein (*crypt*) gene in tobacco. The PAL promoter had a low level of constitutive expression and was strongly induced by pathogen infection. The transgenic tobacco plants expressing cryptogein with the inducible PAL promoter showed significantly enhanced resistance against black shank oomycete pathogen (*Phytophthora parasitica* var. *nicotianae*), brown spot fungal pathogen (*Alternaria alternata*), and wild fire bacterial pathogen (*Pseudomonas syringae* pv. *tabaci*) (Donghua et al. 2004). These studies indicate that low-level constitutive expression of elicitin gene may have potential use in generating broad-spectrum disease-resistant plants.

Two oligandrins, Oli-D1 and Oli-D2 induced an immune response in *Nicotiana benthamiana* and tomato (Quang et al. 2015). When Oli-D1 and Oli-D2 were transiently expressed in tomato, they induced resistance against *Botrytis cinerea* (Quang et al. 2015). The elicitins triggered the expression of a subset of genes involved in the JA/ET-mediated signaling pathway in the transgenic tomato plants, while the elicitins triggered the expression of NPR1 and SGT1 involved in SA signaling system in transgenic *N. benthamiana* plants (Quang et al. 2015),

3.6 Manipulation of Plant Immune System Using Chitosan

3.6.1 Induction of Plant Defense Signaling Systems by Chitosan

Chitosan (β-1,4-linked glucosamine) is a deacetylated derivative of chitin and is a MAMP detected in several fungi (Amborabé et al. 2008; Iriti and Faoro 2009; Bueter et al. 2013). It acts as a potent elicitor triggering several defense signaling systems. The primary site of action of chitosan appears to be the plasma membrane H^+-ATPase in *Mimosa pudica* cells (Amborabé et al. 2008). H^+-ATPases directly use ATP to drive ion translocation (Sanders et al. 2002). H^+-ATPase activity was inhibited in chitosan-treated cells (Amborabé et al. 2008). H^+-ATPase inactivation results in the depolarization of the plasma membrane and the chitosan-induced membrane depolarization is known to be involved in increasing Ca^{2+} influx (Wendehenne et al. 2002; Amborabé et al. 2008). Increases in cytosolic Ca^{2+} concentration ($[Ca^{2+}]_{cyt}$) induced by chitosan have been reported in soybean cells (Zuppini et al. 2004). These events result in activation of Ca^{2+}-mediated signaling pathway (Karita et al. 2004). Chitosan treatment triggers activation of NADPH oxidase leading to the production of H_2O_2 (Rossard et al. 2006). It suggests that chitosan may activate the ROS signaling system. Chitosan strongly induced NO production in tobacco cells (Zhao et al. 2007). Chitosan may also trigger JA signaling system. Chitosan triggers activity of lipoxygenase, the key enzyme in JA biosynthetic pathway, in wheat (Bohland et al. 1997). The fungal elicitor chitosan treatment induces octadecanoid signaling system, resulting in increased accumulation of JA in rice (Rakwal et al. 2002) (Fig. 3.27).

3.6.2 Induction of Host Defense Responses by Chitosan

Chitosan is effective in inducing several defense responses (Fig. 3.28). The host defense responses induced by chitosan include synthesis of phenolics (Liu et al. 2007), lignin (Pearce and Ride 1982), phytoalexins (Jayaraj et al. 2009), and saponins (Hu et al. 2004), induction of callose deposition (Reddy et al. 2000; Benhamou 2004), and production of several pathogenesis-related (PR) proteins (Romanazzi et al. 2002; Amborabé et al. 2004). Accumulation of phenolics in plants induced by chitosan has been reported (Reddy et al. 1999; Rossard et al. 2006). Chitosan treatment enhanced the content of phenolic compounds in tomato fruit (Liu et al. 2007).

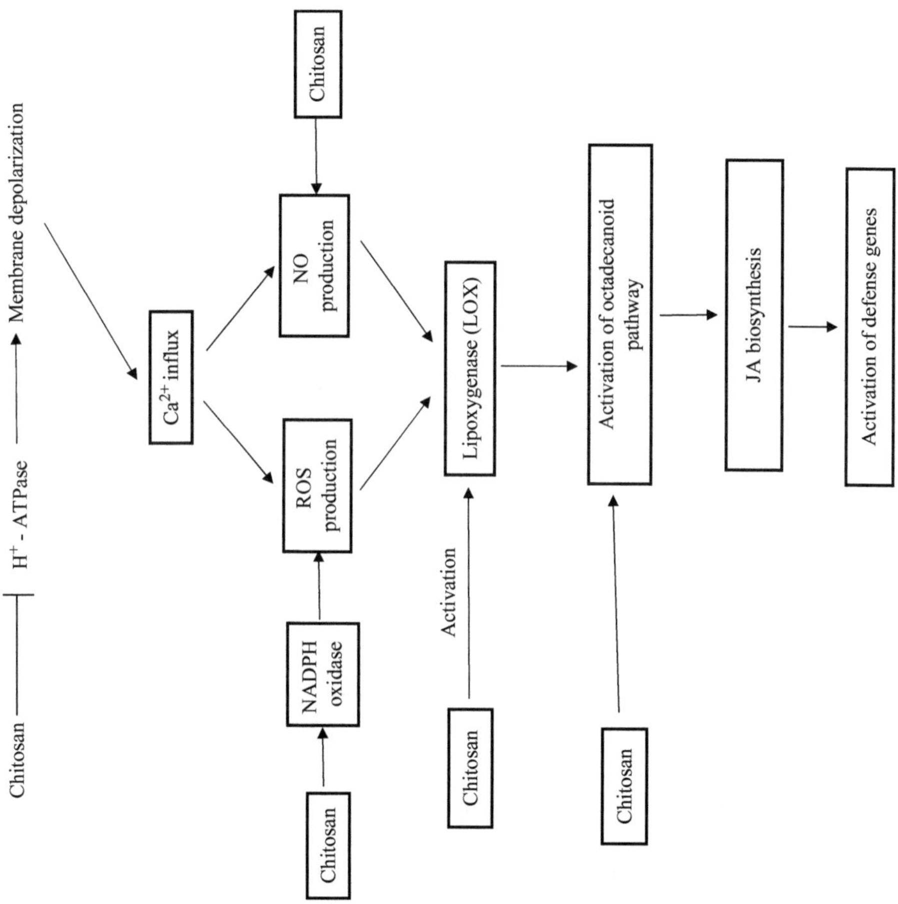

Fig. 3.27 Chitosan-induced signaling systems (Adapted from Vranová et al. 2002; Zago et al. 2006; Courtois et al. 2008; Palmieri et al. 2008)

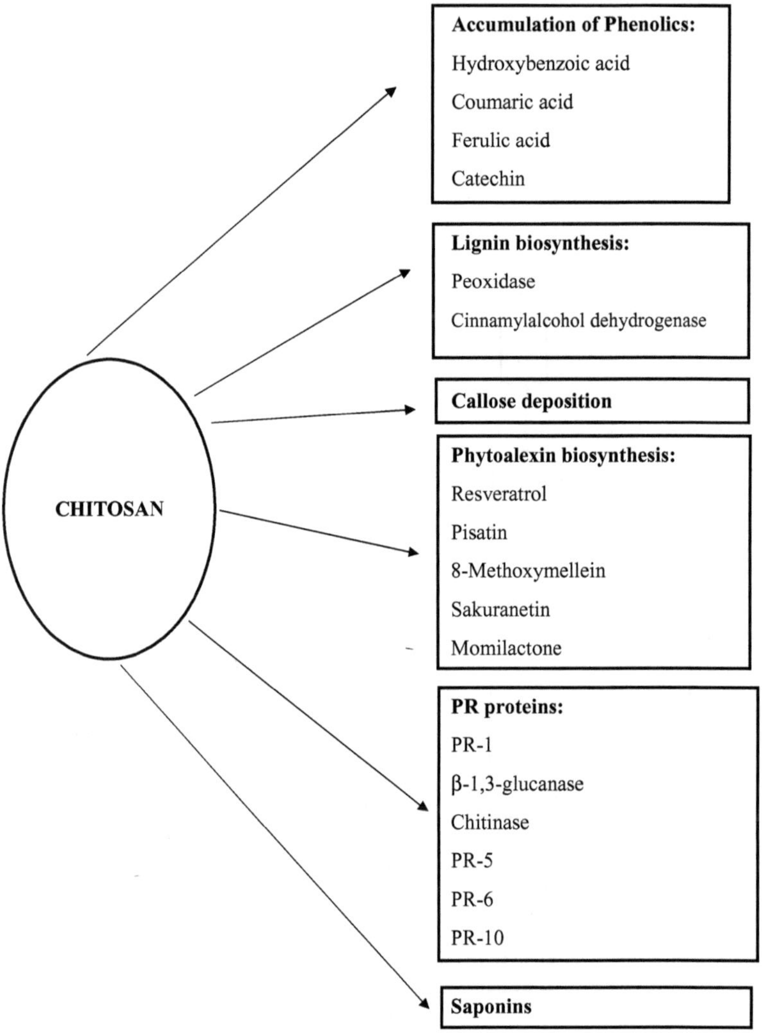

Fig. 3.28 Chitosan-induced defense responses in plants

Chitosan induced synthesis of several different phenolic compounds in roots of tomato plants. At 24 h after chitosan treatment, the increase in ferulic acid content of root cell wall was 3.3 times. The increase of 4-hydroxybenzoic acid was 2.2 times and the level of 4-coumaric acid was more than double in roots of tomato plants treated with chitosan treatment (Mandal and Mitra 2007). An enhanced accumulation of *p*-hydroxybenzoic acid, *p*-coumaric acid, and ferulic acid was observed in coconut cell cultures treated with chitosan treatment (Chakraborty et al. 2009). Chitosan treatment elicited production of catechin and resveratrol in grapes (Romanazzi et al. 2006). Increased total polyphenol content was observed in grapevine treated with the chitosan formulation Kendal Cops (Iriti et al. 2011). Chitosan treatment induced phenolic compound deposition in barley leaves, creating a hostile environment that slowed down the spreading of the pathogen *Blumeria graminis* f. sp. *hordei* by impairing haustorium development (Faoro et al. 2008).

Phenylalanine ammonia-lyase (PAL) is the first enzyme in the phenylpropanoid pathway leading to synthesis of phenolic compounds (Lin et al. 2005; Zhao et al. 2007; Vidhyasekaran 2007). Chitosan activates PAL in potato and *Mimosa pudica* (Dornenburg and Knorr 1997; Rossard et al. 2006). It induced increases in PAL activity and elevated *PAL* mRNA level in tobacco cells (Zhao et al. 2007). The activities of the early phenylpropanoid pathway enzymes such as, PAL, *p*-coumaroyl-CoA ligase (4CL) and *p*-hydroxybenaldehyde dehydrogenase increased in coconut cell cultures treated with chitosan (Chakraborty et al. 2009). The activities of PAL, peroxidase (PO), and polyphenol oxidase (PPO) significantly showed significant increase in chitosan-treated carrot plants (Jayaraj et al. 2009). Chitosan treatment induced activities of PAL, PPO, and PO in tomato roots resulting in reinforcement of cell wall and conferring resistance against *Fusarium oxysporum* f. sp. *lycopersici* (Mandal and Mitra 2007).

Chitosan treatment induces polyphenol oxidase (PPO) in carrot plants (Jayaraj et al. 2009). Increase in PPO activity has been reported in 'Valencia' orange fruits treated with chitosan (Rappussi et al. 2009). PPO catalyzes the oxidation of phenolics to quinones, which are highly toxic to pathogens and are involved in inducing resistance to pathogens (Thipyapong et al. 2007). The quinones activate accumulation of ROS and it is also involved in the phenylpropanoid pathway (Thipyapong et al. 2007).

Chitosan treatment induced cinnamyl alcohol dehydrogenase (CAD) in tomato roots (Mandal and Mitra 2007). CAD is a molecular marker specific for lignin synthesis (Walter et al. 1988). CAD catalyzes the final step in a branch of phenylpropanoid synthesis specific for production of lignin monomers (Walter et al. 1988; Kim et al. 2004). Chitosan enhanced the activity of peroxidase in tomato plants and peroxidase activity showed a sharp increase at 24 h after treatment with chitosan (Mandal and Mitra 2007). Increases in peroxidase activities in citrus fruits treated with chitosan have been reported (Rappussi et al. 2009). Peroxidases are involved in biosynthesis of lignins (Vidhyasekaran 2007; Uzal et al. 2009). Chitosan treatment increased lignin deposition by 6.8 times in roots of tomato plants (Mandal and Mitra 2007). Chitosan treatment induced callose deposition in bean (Faoro and Iriti 2007; Iriti and Faoro 2008), soybean (Köhle et al. 1985), and tobacco (Iriti et al. 2006) and conferred resistance against virus diseases (Iriti et al. 2006; Faoro and Iriti 2007; Iriti and Faoro 2008). Chitosan triggers callose formation in several plants (Köhle et al. 1985; Conrath et al. 1989; Kauss et al. 1989; Reddy et al. 2000; Benhamou 2004).

Chitosan is an effective inducer of phytoalexin synthesis (Köhle et al. 1984). Chitosan induced the expression of chalcone synthase (*chs*) gene in soybean cells and CHS is an important enzyme in phytoalexin synthesis (Zuppini et al. 2004). Chitosan treated carrot plants accumulated high amounts of the phytoalexin 8-methoxymellin (Jayaraj et al. 2009). Chitosan triggered the production of the flavonoid phytoalexin sakuranetin and the diterpenoid lactone phytoalexin momilactone A (Agrawal et al. 2002). It induced accumulation of the phytoalexin pisatin in pea (Hadwiger et al. 1994).

Chitosan triggers the transcription of several pathogenesis-related genes. Chitosan-treated carrot plants had elevated transcript levels of *PR-1*, chitinase, and *PR-5* genes (Jayaraj et al. 2009). Chitosan treatment induced accumulation of two major classes of PR proteins, namely OsPR5 and OsPR10 in rice leaves. Accumulation of the OsPR5 and OsPR10 mRNAs was also observed in the chitosn-treated rice plants (Agrawal et al. 2002). Chitosan activates the lytic enzymes chitinase and β-1,3-glucanase (El Ghaouth 1997; Fajardo et al. 1998; Romanazzi et al. 2002; Amborabé et al. 2004; Lin et al. 2005; Jayaraj et al. 2009). Chitinase and β-1,3-glucanase activities increased in chitosan-treated Valencia orange fruits (Rappussi et al. 2009). Chitosan treatment induced chitinase and β-1,3-glucanase activities in papaya fruits (Hewajulige et al. 2009). Chitosan induces the production of proteinase inhibitors (Doares et al. 1995). It induced the accumulation of phytocystatins (cysteine proteinase inhibitors) in rice leaves (Agrawal et al. 2002). Some of the proteinase inhibitors are considered as PR-6 proteins (Vidhyasekaran 2007).

Saponins are preformed antimicrobial triterpenoid compounds found in several plants (Hu et al. 2004; Yokosuka and Mimaki 2009) They play an important role in disease resistance (Osbourn 1996; Papadoulou et al. 1999; Bouarab et al. 2002). Chitosan treatment induces synthesis of saponins in plants (Hu et al. 2004). Chitosan would have induced biosynthesis of saponins by activating Ca^{2+} influx, MAPK cascade, and ROS signaling systems. These signaling systems activate squalene synthase, squalene epoxidase, and β-amyrin synthase, resulting in synthesis of saponins (Fig. 3.29; Hu et al. 2003, 2004).

Cell death program is involved in activating plant immunity system. It activates systemic acquired resistance (SAR) in plants (Vidhyasekaran 2007). Chitosan elicited cell death program showing some typical morphological features of apoptosis in tobacco (Wang et al. 2008) and induced resistance against *Tobacco necrosis virus* (TNV) (Iriti et al. 2006; Faoro and Iriti 2007). Chitosan induces programmed cell death in soybean cells and activates several defense responses (Zuppini et al. 2004). Chitosan-induced programmed cell death in pea leaves involves ROS generated by the NADPH oxidase of the plasma membrane (Vasil'ev et al. 2009).

3.6.3 Chitosan Induces Resistance against Wide-Range of Pathogens

A commercial formulation of chitosan has been developed by Glycogene Sys Inc. (Boston) under the trade name 'Elexa'. It protects a range of crops against several pathogens. Ecobulle, France developed chitosan as a commercial product under the trade name Chitogel (Elmer and Reglinski 2006). Another formulation containing 4 % chitosan has been developed as "Elexa 4 Plant Defense Booster" by Plant Defense Boosters Inc. USA (Elmer and Reglinski 2006). ChiPro GmbH, Bremen,

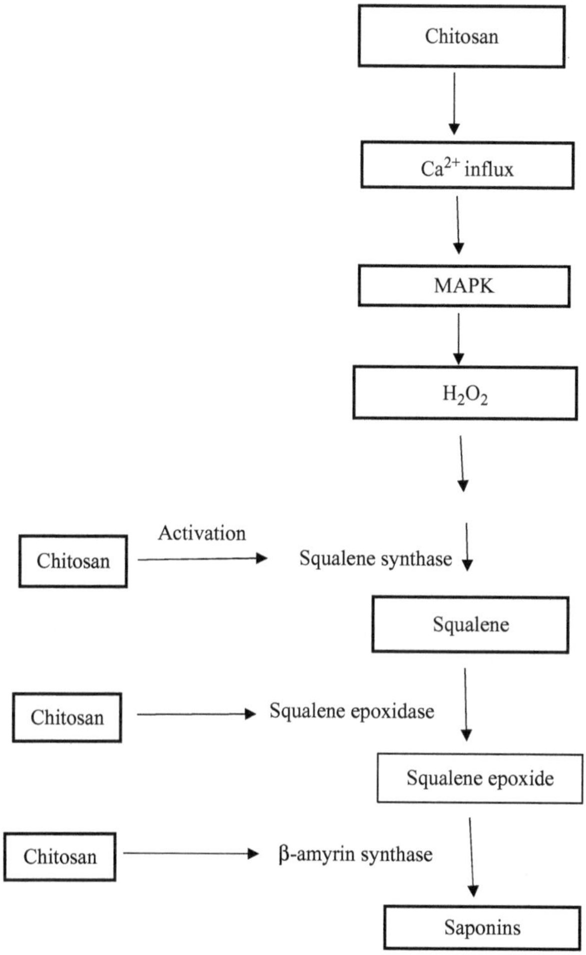

Fig. 3.29 Chitosan triggers biosynthesis of saponins (Adapted from Hu et al. 2003, 2004)

Germany developed the chitosan formulation under the trade name ChitaPlant®. Biochikol is another commercial formulation of chitosan (Kurzawińska and Mazur 2009). "Chitoplant" is the commercial product sold as a plant strengthener in Germany (Dafermos et al. 2012). A new chitosan formulation based on a 4 % chitosan with 1.5 % Cu^{2+} and 0.5 % Mn under the trade name Kendal Cops® has been developed (Iriti et al. 2011).

Chitosan is effective in controlling various *Fusarium* diseases. It enhances resistance to the wheat crown rot pathogen *Fusarium graminearum* (Reddy et al. 1999), the tomato crown and root rot pathogen *F. oxysporum* f. sp. *radicis-lycopersici* (Benhamou 1992; Lafontaine and Benhamou 1996), the soybean sudden death syndrome pathogen *F. solani* f. sp. *glycines* (Prapagdee et al. 2007), and the celery yellows pathogen *F. oxysporum* (Bell et al. 1998). Chitosan treatment effectively controlled the dry rot of potato caused by *Fusarium sulphureum* (Li et al. 2009). Chitosan treatment of wheat seeds induces resistance to *Fusarium graminearum*, the crown rot pathogen (Reddy et al. 1999). Chitosan seed treatment protects tomato plants from Fusarium crown rot and root rot (Benhamou et al. 1994). Chitosan treatment enhances resistance of tomato plants to infection by the crown and root rot pathogen *F. oxysporum* f. sp. *radicis-lycopersici* (Benhamou 1992; Benhamou and Thériault 1992; Lafontaine and Benhamou 1996). When applied as a foliar spray, it protects tomato seedlings against *Fusarium oxysporum* (Benhamou et al. 1998).

Chitosan triggers defense responses against powdery mildew pathogens (Borkowski and Szwonek 2004; Moret et al. 2009; Dafermos et al. 2012). Chitosan treatment has been shown to induce resistance against powdery mildew fungal pathogen *Oidium lycopersici* in field-grown tomato (Gorbatenko et al. 1996). Elexa treatment reduced the powdery mildew (*O. lycopersici*) disease symptom severity in tomato (Isebaert et al. 2002). Foliar application of chitosan reduced incidence of powdery mildew in green house-grown tomato caused by *Leveillula taurica* (*Oidiopsis taurica*) (Dafermos et al. 2012). Chitosan was effective in controlling powdery mildew on cucumber cotyledons caused by *Spherotheca fuliginea* and *Erysiphe cichoracearum* (Moret et al. 2009). Chitosan induces defense responses against the grapevine powdery mildew pathogen *Uncinula necator*. Foliar spray of chitosan applied shortly after flowering and then again 14 days later induced resistance to powdery mildew in grapes under field conditions (Gorbatenko et al. 1996). The chitosan formulation Kendal Cops controlled grapevine powdery mildew effectively. The disease severity was very much reduced in Kendal Cops-treated grapes, and the reduction was as good as that observed in conventional fungicide-treated plants (Fig. 3.30; Iriti et al. 2011). Foliar application of chitosan reduced significantly the infection of the powdery mildew pathogen *Blumeria graminis* f. sp.

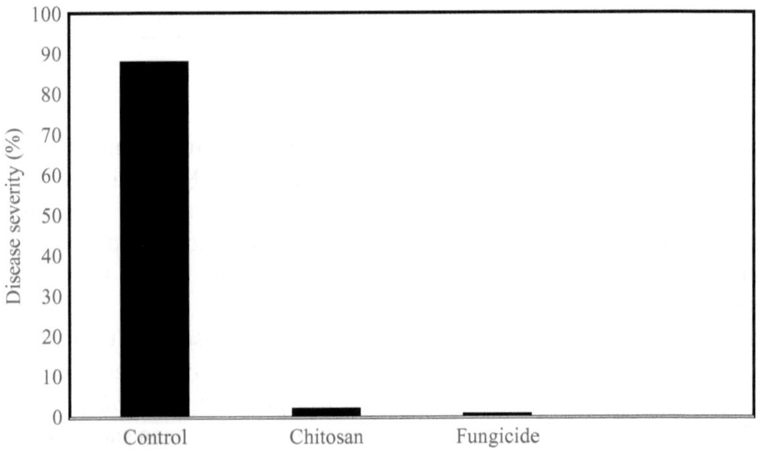

Fig. 3.30 Efficacy of the chitosan formulation Kendal Cops in controlling grapevine powdery mildew (Adapted from Iriti et al. 2011)

hordei in barley plants by about 56 %, showing the induction of a good level of local acquired resistance resistance (LAR). Chitosan also induced systemic acquired resistance (SAR), being the infection in the second non-treated leaves reduced by 57 % (Faoro et al. 2008).

Chitosan improves development, and protects grapevine against *Botrytis cinerea* (Ait Barka et al. 2004). Application of chitosan 21 days and/or 5 days before harvest reduced botrytis bunch rot severity on table grapes (Romanazzi et al. 2002). Chitosan provided both protective and curative control of *B. cinerea* in grapevine (Amborabé et al. 2004). Foliar spray of chitosan at 0.01 % concentration applied at 10-day interval starting from flowering stage very effectively controlled grape botrytis bunch rot (Elmer and Reglinski 2006). Chitosan controls *Botrytis cinerea* infection in tomato fruit (Liu et al. 2007). Chitosan treatment reduced *Botrytis cinerea* infection in greenhouse-grown carrot plants (Jayaraj et al. 2009). Chitosan confers resistance against gray mold (*Botrytis cinerea*) and blue mold (*Penicillium expansum*) in tomato fruit (Liu et al. 2007). Chitosan treatment reduced *Alternaria radicina* infection in greenhouse-grown carrot plants (Jayaraj et al. 2009). Chitosan treatment enhanced the chitinase Chitosan induced chitinase activity and also induced new isoforms of chitinase, resulting in the reduction of early blight (*Alternaria solani*) disease severity in tomato leaves (Sathiyabama et al. 2014). It induces resistance against *Sclerotinia sclerotorum* infection in carrot (Molloy et al. 2004). Chitosan protects citrus fruit against *Penicillium digitatum*, the causal agent of green mold (Benhamou 2004).

Chitosan effectively controls several oomycete pathogens. It controlled cucumber root rot caused by the oomycete *Pythium aphanidermatum* (El Ghaouth et al. 1994). It is also effective in controlling the oomycete downy mildew pathogens *Sclerospora graminicola* in pearl millet (Sharathchandra et al. 2004) and *Plasmopara viticola* in grapevine (Aziz et al. 2006). Elexa applied as seed treatment and/or foliar spray induces resistance against the oomycete downy mildew pathogen *S. graminicola* in pearl millet. The chitosan formulation treatment enhanced plant growth, besides controlling the disease (Sharathchandra et al. 2004).

Chitosan also controls virus diseases and induces resistance against *Tobacco mosaic virus* (TMV) infection in tobacco (Zhao et al. 2007). Chitosan treatment induces high level of resistance against *Tobacco necrosis virus* (TNV) in bean leaves (Hu et al. 2008; Iriti and Faoro 2008). A formulated product of chitosan, guanidinylated chitosan hydrochloride (GCH), induced resistance against TMV in tobacco. It was more effective in controlling the virus disease than chitosan hydrochloride (Hu et al. 2009).

Chitosan treatment induces resistance and controls diseases only partially in most cases. Chitosan partially protects soybean against sudden death syndrome (SDS) caused by *Fusarium solani* f. sp. *glycines* (Prapagdee et al. 2007). Chitosan treatment activated chitinase activity in soybean and retarded SDS development in soybean leaves. However, the SDS symptoms gradually appeared and were associated with the reduction of chitinase activity level after five days of infection period. The results suggest that chitosan alone may not be able to offer complete protection against pathogens (Prapagdee et al. 2007). Chitosan applied as a root dip did not reduce *Fusarium* yellows of celery but significantly reduced disease severity when used with a tolerant celery cultivar (Bell et al. 1998), suggesting that chitosan may require some additional induction of defense mechanisms to confer effective control of diseases. Benhamou et al. (1998) showed that a combination of chitosan with an endophytic bacterial strain induced a substantial increase in extent and magnitude of the cellular changes, when compared to individual applications. This combination induced significantly higher callose accumulation and induced enhanced disease resistance response against *Fusarium oxysporum* f. sp. *radicis-lycopersici* (Benhamou et al. 1998).

Chitosan alone induced resistance against the gray mold pathogen *Botrytis cinerea* and the downy mildew pathogen *Plasmopara viticola* in grapevine. However, in combination with CuSO4, chitosan induced much higher percentage of protection against both the diseases. Chitosan alone induced accumulation of phytoalexins in grapevine leaves; however in combination with $CuSO_4$, it much strongly induced the accumulation of the phytoalexins (Aziz et al. 2006). Collectively these studies suggest that chitosan in combination with other defense inducers may be highly useful in controlling diseases.

Chitosan has been widely used to control postharvest diseases. Chitosan controls *B. cinerea* infection of table grapes (Romanazzi et al. 2006). It also protects citrus fruit against the green mold pathogen *Penicillium digitatum* (Benhamou 2004) and blue mold pathogen *Penicillium expansum* in tomato fruit (Liu et al. 2007). Chitosan reduces infection by *Guignardia citricarpa* in postharvest oranges. Chitosan inhibited the development of new lesions in oranges and effectively reduced the black spot disease incidence in the orange fruits during storage (Rappussi et al. 2009). Chitosan in combination with the yeast *Candida saitoana* effectively controlled the postharvest diseases of apple and citrus fruit (El-Ghaouth et al. 2000). Chitosan treatment reduced the anthracnose disease incidence in papaya fruits (Hewajulige et al. 2009). Chitosan in combination with *Burkholderia cepacia* and calcium chloride gave complete control of the anthracnose disease caused by *Colletotrichum gloeosporioides* in papaya fruits during storage. This combination offered a greater control by reducing 99 % disease severity in naturally infected fruits at the end of 14 days storage, which was superior to that obtained with the standard fungicide benomyl (Table 3.4; Rahman et al. 2009).

Table 3.4 Efficacy of chitosan in the control of anthracnose disease development in papaya fruit during storage

Treatment	Area Under Disease Progress Curve (AUDPC) (Unit/square)
Control	32.1
Chitosan	9.5
Chitosan + *Burkholderia cepacia* + $CaCl_2$	0.3
Benomyl	12.8

Adapted from Rahman et al. (2009)

Chitosan treatments have plant growth promoting effects, resulting in improved yields and plant health in numerous crops and fruits (Kowlaski et al. 2006; Boonreung and Boonlertnirun 2013). Seed treatment with chitosan improves seed germination and seedling growth of maize plants (Guan et al. 2009). Application of chitosan by seed soaking and soil application four times throughout cropping season significantly increased rice yield (Boonlertnirun et al. 2008). Chitosan appears to be a useful tool to manage several crop diseases without any sacrifice in the crop yield potential.

3.7 Manipulation of Plant Immune System Using Cerebrosides

Cerebrosides A, B, and C, categorized as sphingolipids, are the elicitors derived from a wide range of fungi, including *Pythium graminicola*, *Botrytis allii*, *Sclerotium cepivorum*, *Fusrium oxysporum* f. sp. *lycopersici*, *F. o.* f. sp. *melonis*, *F. o.* f. sp. *cucumerinum*, *F. o.* f. sp. *lactucae*, and *Magnaporthe grisea* (Umemura et al. 2000, 2004). Cerebroside elicitor induced ROS signaling system in tomato roots. The elicitor treatment induced rapid production of H_2O_2 (Umemura et al. 2004). The elicitor induced the activation of several defense responses in rice and tomato. It induced accumulation of phytoalexins and PR proteins in rice (Umemura et al. 2000, 2002). *PR-1* mRNA accumulated at 48 h after treatment with the elicitor in tomato roots. Expression of *PR-5* mRNA was also induced by cerebroside treatment (Umemura et al. 2004).

Application of cerebroside effectively controlled tomato wilt caused by *F. oxysporum* f. sp. *lycopersici* (Table 3.5; Umemura et al. 2004). The elicitor was applied at 5–10 μg/ml as seedling root treatment and transplanted in pots under greenhouse conditions. The plants were challenge-inoculated with the pathogen. The elicitor treatment reduced the disease incidence. It also reduced the disease severity (Umemura et al. 2004). The elicitor treatment controlled *Fusarium* wilt of melon (*F. oxysporum* f. sp. *melonis*) and root rot of lettuce caused by *F. oxysporum* f. sp. *lactucae* under greenhouse conditions (Table 3.5; Umemura et al. 2004). The efficacy of cerebroside elicitor in the management of diseases has also been demonstrated under field conditions. The elicitor treatment reduced the stem rot (*F. oxysporum* f. sp. *batatas*) disease incidence in sweet potato (*Ipomoea batatas*) under field conditions (Table 3.5). It effectively protected rice plants against the blast pathogen *Magnaporthe oryzae* under field conditions (Umemura et al. 2000). Cerebroside elicited systemic resistance against pearl millet (*Pennisetum glacum*) downy mildew disease caused by *Sclerospora graminicola*). The elicitor treatment enhanced yield of the crop due to disease suppression (Deepak et al. 2003).

Table 3.5 Efficacy of cerebroside in the control of *Fusarium* diseases of various crops

Treatment	% disease incidence			
	Fusarium wilt of tomato	*Fusarium* wilt of melon	*Fusarium* wilt of sweet potato	*Fusarium* wilt of lettuce
Cerebroside	68	17	23	30
Control	93	57	72	100

Adapted from Umemura et al. (2004)

3.8 Manipulation of Plant Immune System Using CfHNNI1 Elicitor

A proteinaceous elicitor, CfHNNI1 (for *Cladosporium fulvum* host and nonhost plant necrosis inducer 1), has been isolated from *Cladosporium fulvum*, the tomato leaf mold pathogen and has been exploited to develop genetically engineered disease resistant crop plants (Cai et al. 2007). Stable transgenic tobacco plants overexpressing *CfHNNI1* were developed and the transgene was expressed under the control of the pathogen-inducible hsr203J promoter. The transgenic plants induced the expression of several defense-related *PR* genes. They showed enhanced resistance to the oomycete pathogen *Phytophthora parasitica* var. *nicotianae*, the bacterial pathogen *Pseudomonas syringae* pv. *tabaci* and the viral pathogen *Tobacco mosaic virus* (Cai et al. 2007). These studies demonstrated that the proteinaceous elicitor is another potential tool to develop broad-spectrum disease resistant crop plants.

3.9 Bioengineering *FsphDNase* Elicitor Gene to Trigger Plant Immune Responses against Wide Range of Pathogens

Another proteinaceous elicitor, FsphDNase (for *Fusarium solani* f. sp. *phaseoli* DNase), is secreted by *Fusarium solani* f. sp. *phaseoli* (Klosterman et al. 2001; Hadwiger 2008). The elicitor induced the accumulation of several PR genes and induced phytoalexin synthesis in pea tissue (Hadwiger et al. 1995; Choi et al. 2001). The gene encoding the elicitor was cloned and transgenic tobacco plants expressing the elicitor gene under the control of pathogen-inducible DRR206 promoter were developed. These transgenic plants showed enhanced resistance against the fungal pathogen *Alternaria alternata* (Fig. 3.31; Choi et al. 2004). These plants also showed resistance to the bacterial pathogen *Pseudomonas syringae* pv. *tabaci* (Choi et al. 2004). Transgenic potato plants expressing *FsphDNase* elicitor gene showed resistance to the oomycete pathogen *Phytophthora infestans* (Choi et al. 2004).

Since this elicitor gene shows DNase activity, it can cause damage of DNA within nuclei of plant cells (Klosterman et al. 2001). Constitutive activation of

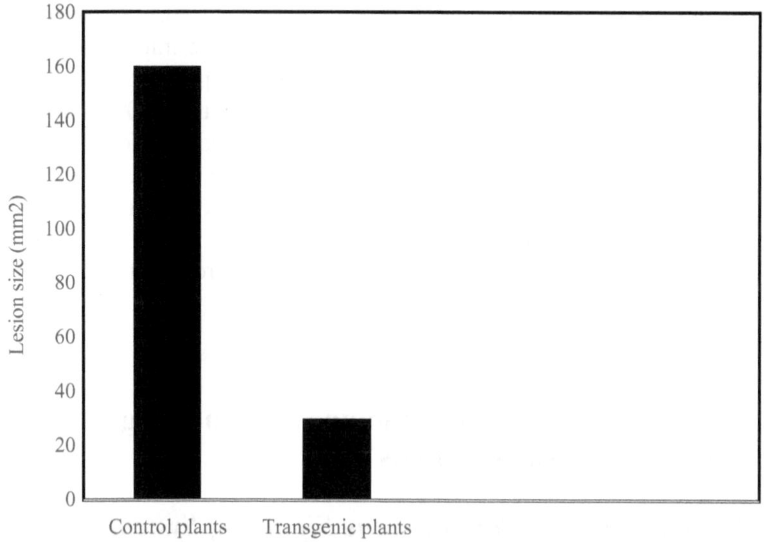

Fig. 3.31 Efficacy of *FsphDNase* elicitor gene in conferring resistance to *Alternaria alternata* in transgenic tobacco plants (Adapted from Choi et al. 2004)

FsphDNase within the plant cell may be destructive. However, use of pathogen-inducible promoters overcomes the adverse effect and the transgenic tobacco plants expressing the elicitor gene showed no detectable morphological differences from the wild type (Choi et al. 2004). These studies revealed that the elicitor gene can be transferred to a wide-range of plants and the transgenic plants expressing the *FsphDNase* elicitor gene show resistance against broad-spectrum of pathogens. Selection of suitable pathogen-inducible promoter for expressing the elicitor gene appears to be a perquisite for developing disease-resistant plants without any reduction in yield potential.

3.10 Engineering the Elicitor-Encoding *pemG1* Gene for Crop Disease Management

PemG1 (Protein Elicitor from M. grisea1) is a proteinaceous elicitor isolated from the rice blast pathogen *Magnaporthe oryzae* (previously called as *M. grisea*). It is a 40 kDa heat-stable protein. Its exogenous application triggered H_2O_2 production in plants (Xu et al. 2006). The elicitor application increased resistance against the blast pathogen *M. oryzae* in rice (Xu et al. 2006). The elicitor-encoding gene, *pemG1*, was isolated from the rice blast pathogen *M. oryzae* and used to develop blast disease-resistant rice plants (Qiu et al. 2009). The *pemG1* gene was cloned from *M. oryzae* and a plant expression vector carrying the synthetic pemG1 under the control of maize ubiquitin 1 (ubi-1) promoter was constructed. Transgenic rice plants constitutively expressing the *pemG1* gene were developed. The phenylalanine ammonia-lyase activity was about two times higher in the transgenic plants than that in wild-type plants. The transgenic rice plants expressing *pemG1* showed increased resistance against the rice blast pathogen (Qiu et al. 2009). The wild-type plants developed clear blast lesions at 8 days after pathogen inoculation and the lesions clearly increased in size and the symptoms were severe. In contrast, only small lesions were observed on part of the leaves of *pemG1*-expressing plants and most leaves remained green and healthy in the transgenic plants (Qiu et al. 2009). These studies suggest a feasibility of developing disease resistant plants using proteinaceous elicitor genes.

3.11 Manipulation of Plant Immune System Using the MAMP Rhamnolipids

3.11.1 Activation of Plant Immune Signaling System

Rhamnolipids are glycolipid MAMPs produced by various bacterial species including some *Pseudomonas* and *Burkholderia* species (Varnier et al. 2009; Vatsa et al. 2010; Chrzanowski et al. 2012; Sanchez et al. 2012, 2013). Rhamnolipids produced by *Pseudomonas aeruginosa* are amphiphilic molecules typically composed of 3-hydroxyfatty acids linked through a beta-glycosidic bond to mono- or di-rhamnoses (Soberon-Chavez et al. 2005; Vatsa et al. 2010). Rhamnolipids triggered various signaling systems involved in plant immunity (Fig. 3.32). Rhamnolipids treatment induced a rapid and transient burst of $[Ca^{2+}]_{cyt}$ within 2 min after treatment in grapevine cells (Varnier et al. 2009). When grapevine cells were treated with rhamnolipids, a biphasic elevation of $[Ca^{2+}]_{cyt}$ was observed with a first peak at 2 min followed by a second peak at 5 min (Varnier et al. 2009). Rhamnolipids activated mitogen-activated protein kinases (MAPKs) with apparent molecular mass of 46 and 50 kDa. The activation of both MAPKs was detected as soon as 5 min. The MAMP rhamnolipids also induced the activity of NADPH oxidase, which is involved in H_2O_2 production. Strong accumulation of H_2O_2 was observed after rhamnolipid treatment in grapevine cells (Varnier et al. 2009).

The rhamnolipids have been shown to induce the activation of SA, JA and ethylene signaling systems (Sanchez et al. 2012). The MAMP triggered the expression of *pal* gene which encodes phenylalanine ammonia-lyase (PAL) in grapevine (Varnier et al. 2009). PAL is the key enzyme involved in SA biosynthesis (Vidhyasekaran 2015). Rhamnolipids treatment induced accumulation of salicylic acid and a two-fold increase in salicylic acid content was observed at 24 h after treatment in *Arabidopsis* (Sanchez et al. 2012). PR-1 is a SA-dependent defense gene marker (Hondo et al. 2007; Xing et al. 2008; Vlot et al. 2009; Vidhyasekaran 2015) and rhamnolipids triggered *PR1* gene expression within 24 h after rhamnolipids treatment (Sanchez et al. 2012).

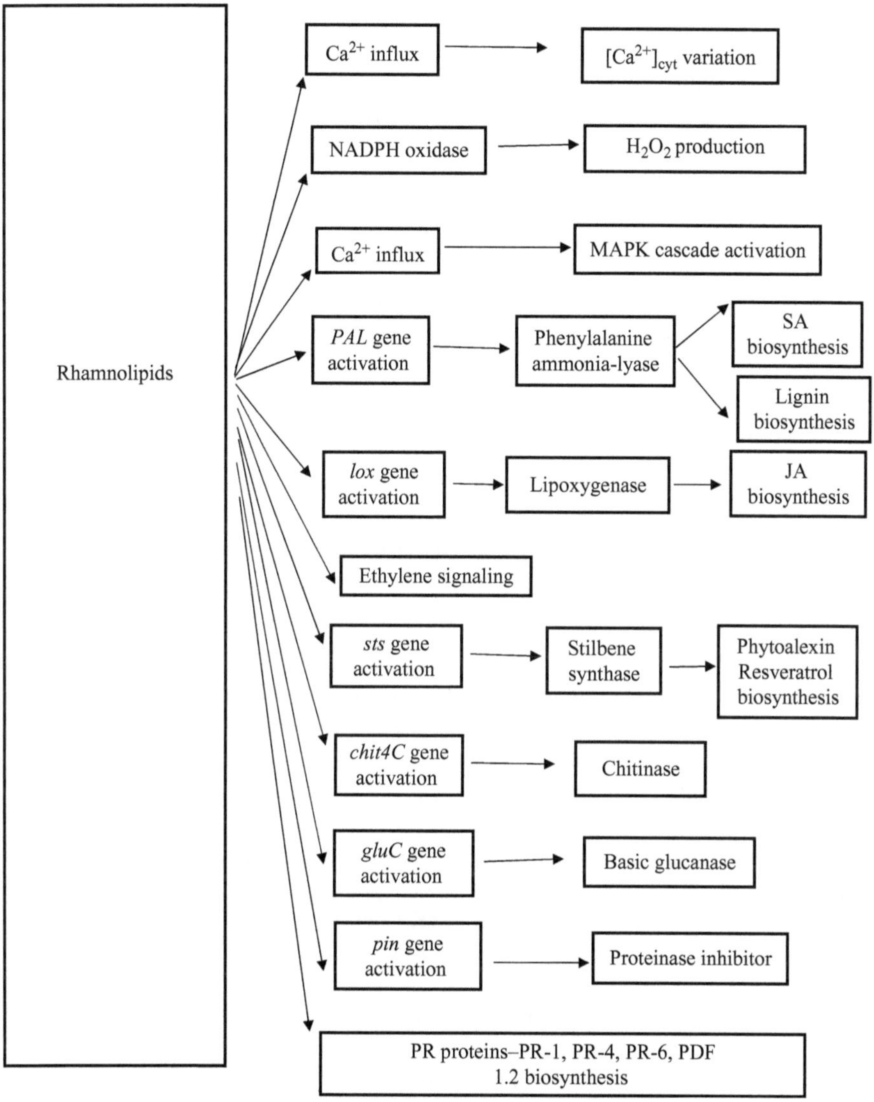

Fig. 3.32 Rhamnolipids-triggered plant immune systems in grapevine and *Arabidopsis* (Adapted from Varnier et al. 2009; Sanchez et al. 2012)

Rhamnolipids triggered the expression of *lox* gene encoding lipoxygenase in grapevine (Varnier et al. 2009). Lipoxygenase is an important enzyme involved in JA biosynthesis (Bertoni 2012). Rhamnolipid treatment induced a 200-fold increase in JA level at 6 h after treatment in Arabidopsis (Sanchez et al. 2012). *PDF1.2* is a marker gene in JA/ET signaling pathways (Vidhyasekaran 2007; Zander et al. 2010; Kravchuk et al. 2011; Cevik et al. 2012; Wathugala et al. 2012; Po-Wen et al. 2013; Vidhyasekaran 2014, 2015). Rhamnolipids treatment induced the expression of

PDF1.2 gene in Arabidopsis and a sharp increase in the gene expression was observed at 6 h after rhamnolipids treatment (Sanchez et al. 2012). Expression of *PR4* gene encoding pathogenesis-related protein PR-4 (Vidhyasekaran 2007) has been shown to be dependent on ethylene signaling (Lawton et al. 1994; Penninckx et al. 1998; Vidhyasekaran 2015). The PAMP rhamnolipids triggered the expression of *PR4* gene by several fold (Sanchez et al. 2012). The results suggest that rhamnolipids can trigger the expression of SA, JA, and ethylene signaling pathways.

Rhamnolipids triggered the expression of several defense-related genes in grapevine (Fig. 3.32). The chitinase gene *chit4c* expression was strongly induced as soon as 6 h after treatment. Rhamnolipids-induced enhanced expression of the other PR protein genes, a basic glucanase (*gluc*) and a protease inhibitor (*pin*) in grapevine has also been reported (Varnier et al. 2009). The MAMP induced the expression of *sts* gene encoding stilbene synthase involved in phytoalexin biosynthesis and *pal* gene encoding PAL involved in lignin biosynthesis in grapevine (Varnier et al. 2009).

3.11.2 Potential of the MAMP Rhamnolipids for Management of Crop Diseases

Arabidopsis plants sprayed with rhamnolipids at 1 mg/ml showed high resistance to the necrotrophic fungal pathogen *Botrytis cinerea*, hemibiotrophic bacterial pathogen *Pseudomonas syringae* pv. *tomato*, and the biotrophic oomycete pathogen *Hyaloperonospora arabidopsidis* (Sanchez et al. 2012). Rhamnolipids have been shown to efficiently protect grapevine leaves and berries against *B. cinerea* by eliciting plant defense responses (Sanchez et al. 2013). Rhamnolipids produced by *Pseudomonas aeruginosa* efficiently protected grapevine against *B. cinerea* (Varnier et al. 2009). Rhamnolipids treatment protect pepper plants from Phytophthora blight disease caused by *Phytophthora capsici* (Kim et al. 2000). The MAMP treatment also prevents the development of *Colletotrichum orbiculare* infection on leaves of cucumber plants (Kim et al. 2000).

Rhamnolipids have been formulated to be effective in the field for management of crop diseases. The rhamnolipids are used as biosurfactants, wetter, emulsifier and adjuvant for herbicidal and pesticidal systems in agriculture, and have the potential to be effective in the field (Varnier et al. 2009).

3.12 Manipulation of Plant Immune System Using the Proteinaceous Elicitor Sm1 Derived from *Trichoderma virens*

A proteinaceous elicitor, Sm1 (small protein1), secreted by *Trichoderma virens* induces defense responses and systemic resistance in plants (Djonovic et al. 2006; 2007; Vargas et al. 2008). Application of Sm1 elicitor resulted in the production of

ROS (H_2O_2) in rice leaves and cotton cotyledons (Djonovic et al. 2006). It induced the peroxidase gene *POD6*, which is involved in the ROS signaling system in cotton. The elicitor treatment also induced the *LOX1* (*lipoxygenase 1*) gene involved in JA signaling system. It triggered the induction of several defense-related genes. Expression of both *CAD* and *HMG* genes was upregulated in Sm1-treated tissues. The *CAD1-C* gene encodes (+) δ-cardine synthase and the *HMG* encodes 3-hydroxy-3-methylglutaryl-CoA reductase and these enzymes are involved in sesquiterpenoid phytoalexin synthesis. The Sm1 elicitor induced the expression of the PR genes, such as *GLU* encoding β-1,3-glucanase and *CHT* encoding chitinase (Djonovic et al. 2006). The *Sm1* overexpressing *T. virens* upregulated *PAL* gene in maize (Djonovic et al. 2007). These results suggest that *Sm1* induces defense signaling systems resulting in activation of several defense-related genes.

Sm1 treatment was able to protect cotton cotyledons from infection by *Colletotrichum* sp. (Djonovic et al. 2006). Instead of application of purified Sm1 elicitor for the control of diseases, the Sm1 overproducing *Trichoderma virens* can be employed to manage crop diseases. *T. virens* strains overexpressing *Sm1* gene were developed and these *Sm1*-overexpression strains effectively controlled maize anthracnose leaf blight disease caused by *Colletotrichum graminicola* (Djonovic et al. 2007). Maize plants grown with *Sm1*-deletion strains failed to induce resistance against *C. graminicola*, suggesting that the Sm1 elicitor is essential to induce disease resistance. *Sm1* gene expression and secretion of the protein in *T. virens* was significantly higher when the fungus was associated with cotton plants (Djonovic et al. 2006). *Sm1* was indeed expressed in planta after the root colonization by *T. virens* strains (Djonovic et al. 2007). These results suggest that the application of *T. virens* may result in production of Sm1 elicitor in vivo and the secreted Sm1 may induce systemic resistance against pathogens. *T. virens* seems to be a potential tool to trigger defense signaling systems and manage crop diseases. Several commercial formulations of the fungus have been developed under the trade names of GlioGard™ and SoilGard™ (Lumsden et al. 1996; Vidhyasekaran 2004).

3.13 Manipulation of Plant Immune Responses Using Yeast-Derived Elicitors

Elicitor preparations from the cell wall of the yeast *Saccharomyces cerevisiae* were found to be effective in inducing resistance against various diseases. Commercial formulations of the yeast elicitors are also available. A yeast-derived elicitor formulation under the trade name ISR2000™ is supplied by Improcrop, USA. The yeast elicitor triggered the genes encoding glutathione-S-transferase (*Gst2*, *Gst11*) in *Arabidopsis* (Raacke et al. 2006) and GST is involved in redox signaling system. The elicitor induced both SA-signaling (Raacke et al. 2006) and JA-signaling (Blechert et al. 1995; Parchmann et al. 1997; Suzuki et al. 2005) systems. The yeast elicitor induced accumulation of PR-1, PR-2, and PR-5 proteins in *Arabidopsis*

(Raacke et al. 2006). The yeast elicitor induced PAL activity in barley (Reglinski et al. 1994b) and induced phytoalexin synthesis in *Arabidopsis* (Raacke et al. 2006).

ISR2000 was found to be effective in reducing grapevine powdery mildew (*Uncinula necator*), late blight of potato and tomato (*Phytophthora infestans*), and tomato bacterial speck (*Pseudomonas syringae* pv. *tomato*) disease intensity (Tosun 2007). Elicitor derived from the yeast *S. cerevisiae* effectively controlled powdery mildew infection in barley in field trials (Reglinski et al. 1994a, b). ISR2000 treatment reduced the incidence of bacterial canker (*Clavibacter michiganensis* subsp. *michiganensis*) in tomato from 87 % (water control) to 44 % (Ustun et al. 2009). Yeast elicitor induces resistance against the bacterial pathogen *Pseudomonas syringae* and the fungal pathogen *Botrytis cinerea* in *Arabidopsis* (Raacke et al. 2006). Although yeast elicitor reduced the incidence of the diseases caused by oomycete, fungal and bacterial pathogens, its efficacy was not as high as the standard fungicides or bactericides. The yeast extract can be used as one of the treatments in the integrated disease management program to reduce the use of fungicides and avoid the development of fungicide resistance in field pathogen population. ISR2000™ treatment generally increases crop yield (Türküsay et al. 2009) and it is an added advantage in using the elicitor.

References

Adam A, Pike S, Hoyos ME, Stone JM, Walker JC, Novacky A (1997) Rapid and transient activation of a myelin basic protein kinase in tobacco leaves treated with harpin from *Erwinia amylovora*. Plant Physiol 115:853–861

Agostini JP, Bushong PM, Timmer LW (2003) Greenhouse evaluation of products that induce host resistance for control of scab, melanose, and Alternaria brown spot of citrus. Plant Dis 87:69–74

Agrawal GK, Rakwal R, Tamogami S, Yonekura M, Kubo A, Saji H (2002) Chitosan activates defense/stress response(s) in the leaves of *Oryza sativa* seedlings. Plant Physiol Biochem 40:1061–1089

Ait Barka E, Eullaffroy P, Clement C, Vernet G (2004) Chitosan improves development, and protects *Vitis vinifera* L against *Botrytis cinerea*. Plant Cell Rep 22:608–614

Akbudak N, Akbudak B, Seniz V, Eris A (2006) Prehavest application of harpin on the cool storage life of pepper. Acta Hortic (ISHS) 712:517–522

Alfano JR, Collmer A (2004) Type III secretion system effector proteins: double agents in bacterial disease and plant defense. Annu Rev Phytopathol 42:385–414

Ali R, Ma W, Lemtiri-Chlieh F, Tsaltas D, Leng Q, von Bodman S, Berkowitz GA (2007) Death don't have no mercy and neither does calcium: Arabidopsis CYCLIC NUCLEOTIDE GATED CHANNEL2 and innate immunity. Plant Cell 19:1081–1095

Alkan N, Fluhr R, Prusky D (2012) Ammonium secretion during *Colletotrichum coccodes* infection modulates salicylic and jasmonic acid pathways of ripe and unripe tomato fruit. Mol Plant-Microbe Interact 25:85–96

Allen GJ, Muir SR, Sanders D (1995) Release of Ca^{2+} from individual plant vacuoles by both InsP3 and cyclic ADP-ribose. Science 268:735–737

Alvarado VY, Scholthof HB (2009) Plant responses against invasive nucleic acids: RNA silencing and its suppression by plant viral pathogens. Semin Cell Dev Biol 20:1032–1040

Amborabé B-E, Aziz A, Trotel-Aziz P, Quantinet D, Dhuicq L, Vernet G (2004) Stimulation des defenses naturelles de la vigne Essais d'emploi du chitosan contre *Botrytis cinerea*. Phytoma 571:26–29

Amborabé B-E, Bonmort J, Fleurat-Lessard P, Roblin G (2008) Early events induced by chitosan on plant cells. J Expt Bot 59:2317–2324

Amelot N, de Dorthac B, San Clemente H, Mazars C, Grima-Pettenati J, Brière C (2012) Transcriptome analysis of tobacco BY-2 cells elicited by cryptogein reveals new potential actors of calcium-dependent and calcium-independent plant defense pathways. Cell Calcium 51:117–130

Andi S, Taguchi F, Toyoda K, Shiraishi T, Ichinose Y (2001) Effect of methyl jasmonate on harpin-induced hypersensitive cell death, generation of hydrogen peroxide and expression of *PAL* mRNA in tobacco suspension cultured BY-2 cells. Plant Cell Physiol 42:446–449

Anil K, Podile AR (2012) Harpin$_{Pss}$-mediated enhancement in growth and biological control of late leaf spot in groundnut by a chlorothalonil-tolerant *Bacillus thuringiensis* SFC24. Microbiol Res 167:194–198

Arenas YC, Kalkman ERIC, Schouten A, Dieho M, Vredenbregt P, Uwumukiza B, Ruiz MO, van Kan JAL (2010) Functional analysis and mode of action of phytotoxic Nep1-like proteins of *Botrytis cinerea*. Physiol Mol Plant Pathol 74:376–386

Aslam SN, Erbs G, Morrissey KL, Newman M-A, Chinchilla D, Boller T, Molinaro A, Jackson RW, Cooper RM (2009) Microbe-associated molecular pattern (MAMP) signatures, synergy, size, and charge: influences on perception or mobility and host defense responses. Mol Plant Pathol 10:375–387

Atkinson MM, Keppler LD, Orlandi EW, Baker CJ, Mischke CF (1990) Involvement of plasma membrane calcium influx in bacterial induction of the K^+/H^+ and hypersensitive responses in tobacco. Plant Physiol 92:215–221

Avni A, Bailey BA, Mattoo AK, Anderson JD (1994) Induction of ethylene biosynthesis in *Nicotiana tabacum* by a *Trichoderma viride* xylanase is correlated to the accumulation of 1-aminocyclopropane-1-carboxylic acid (ACC) synthase and ACC oxidase transcripts. Plant Physiol 106:1049–1055

Aziz A, Trotel-Aziz P, Dhuicq L, Jeandet P, Couderchet M, Vernet G (2006) Chitosan oligomers and copper sulfate induce grapevine defense reactions and resistance to gray mold and downy mildew. Phytopathology 96:1188–1194

Aziz A, Gauthier A, Bezier A, Poinssot B, Joubert JM, Pugin A, Heyraud A, Baillieu F (2007) Elicitor and resistance-inducing activities of β-1,4 cellodextrins in grapevine, comparison with β-1,3 glucans and α-1,4 oligogalacturonides. J Expt Bot 58:1463–1472

Bailey BA, Dean JFD, Anderson AJ (1990) An ethylene biosynthesis-inducing endoxylanase elicits electrolyte leakage and necrosis in *Nicotiana tabacum* cv. Xanthi leaves. Plant Physiol 94:1849–1854

Baillieul F, de Ruffray P, Kauffmann S (2003) Molecular cloning and biological activity of alpha-, beta-, and gamma-megaspermin, three elicitins secreted by *Phytophthora megasperma* H20. Plant Physiol 131:155–166

Baker CJ, Orlandi EW, Mock NM (1993) Harpin, an elicitor of the hypersensitive response in tobacco caused by *Erwinia amylovora*, elicits active oxygen production in suspension cells. Plant Physiol 102:1341–1344

Balbi V, Devoto A (2008) Jasmonate signalling network in *Arabidopsis thaliana*: Crucial regulatory nodes and new physiological scenarios. New Phytol 177:301–318

Bar M, Avni A (2008) EHD2 inhibits ligand-induced endocytosis and signaling of the leucine-rich repeat receptor-like protein LeEix2. Plant J 59:600–611

Bar M, Sharfman M, Ron M, Avni A (2010) BAK1 is required for the attenuation of ethylene-inducing xylanase (Eix)-induced defense responses by the decoy receptor LeEix1. Plant J 63:791–800

Bardoel BW, van der Ent S, Pel MJC, Tommassen J, Pieterse CMJ, van Kessel KPM, van Strijp JAG (2011) *Pseudomonas syringae* evades immune recognition of flagellin in both mammals and plants. PLoS Pathog 7:e1002206

Bauer DW, Wei Z-M, Beer SV, Coolmer A (1995) *Erwinia chrysanthemi* harpin$_{Ech}$, an elicitor of the hypersensitive response that contributes to soft rot pathogenesis. Mol Plant-Microbe Interact 8:484–491

Belbahri L, Boucher C, Candresse T, Nicole M, Ricci P, Keller HA (2001) A local accumulation of the *Ralstonia solanacearum* PopA protein in transgenic tobacco renders a compatible interaction incompatible. Plant J 28:419–430

Bell AA, Hubbard JC, Liu L, Davis RM, Subbarao KV (1998) Effects of chitin and chitosan on the incidence and severity of Fusarium yellows in celery. Plant Dis 82:322–328

Bellin D, Asai S, Delledonne M, Yoshioka H (2013) Nitric oxide as a mediator for defense responses. Mol Plant Microbe Interact 26:271–277

Bendahmane A, Kanyuka K, Baulcombe DC (1999) The *Rx* gene from potato controls separate virus resistance and cell death responses. Plant Cell 11:781–791

Benhamou N (1992) Ultrstructural and cytochemical aspects of chitosan on *Fusarium oxysporum* f. sp. *radicis-lycopersici*, agent of tomato crown and root rot. Phytopathology 82:1185–1193

Benhamou N (2004) Potential of mycoparasite, *Verticillium lecanii*, to protect citrus fruit against *Penicillium digitatum*, the causal agent of green mold: a comparison with the effect of chitosan. Phytopathology 94:693–705

Benhamou N, Thériault G (1992) Treatment with chitosan enhances resistance of tomato plants to the crown and root rot pathogen *Fusarium oxysporum* f sp *radicis* f sp *lycopersici*. Physiol Mol Plant Pathol 41:33–52

Benhamou N, Kloepper JW, Tuzun S (1994) Induction of systemic resistance to Fusarium crown rot and root rot in tomato plants by seed treatment with chitosan. Phytopathology 84:1432–1444

Benhamou N, Kloepper JW, Tuzun S (1998) Induction of resistance to *Fusarium* wilt of tomato by combination of chitosan and endophytic bacterial strain: Ultrastructurte and cytochemistry of host response. Planta 204:153–168

Benhamou N, Bélanger RR, Rey P, Tirilly Y (2001) Oligandrin, the elicitin-like protein produced by the mycoparasite *Pythium oligandrum*, induces systemic resistance to crown and root rot in tomato plants. Plant Physiol Biochem 39:681–698

Benhamou N, Le Floch G, Vallance J, Gebore J, Grizard D, Rey P (2012) *Pythium oligandrum*: an example of opportunistic success. Microbiology 158:2679–2694

Berrocal-Lobo M, Molina A (2004) Ethylene response factor 1 mediates *Arabidopsis* resistance to the soilborne fungus *Fusarium oxysporum*. Mol Plant-Microbe Interact 17:763–770

Bertoni G (2012) Oxylipins and plant palability. Plant Cell 24:1305

Binet M, Humbert C, Lecourieux D, Vantard M, Pugin A (2001) Disruption of microtubular cytoskeleton induced by cryptogein, an elicitor of hypersensitive response in tobacco cells. Plant Physiol 125:564–572

Blechert S, Brodschelm W, Hoelder S, Kammerer L, Kutchan T, Mueller MJ, Xia Z-Q, Zenk MH (1995) The octadecanoid pathway: signal molecules for the regulation of secondary pathways. Proc Natl Acad Sci USA 92:4099–4105

Blein JP, Milat ML, Ricci P (1991) Responses of cultured tobacco cells to cryptogein, a proteinaceous elicitor from *Phytophthora cryptogea*: Possible plasmalemma involvement. Plant Physiol 95:486–491

Blein JP, Coutos-Thevenot P, Marion D, Ponchet M (2002) From elicitins to lipid-transfer proteins: a new insight in cell signalling involved in plant defence mechanisms. Trends Plant Sci 7:293–296

Bohland C, Balkenhohl T, Loers G, Feussner I, Grambow HJ (1997) Differential induction of lipoxygenase isoforms in wheat upon treatment with rust fungus elicitor, chitin oligosaccharides, chitosan, and methyl jasmonate. Plant Physiol 114:679–685

Boller T, Felix G (2009) A renaissance of elicitors: perception of miceobe-associated molecular patterns and danger signals by pattern-recognition receptors. Annu Rev Plant Biol 60:379–406

Bonnet P, Poupet A, Abad P, Venard P, Cardin L (1986) Induction de nècroses foliaires, de protéines b et de resistance dans les interactions tabac-*Phytophthora*. Agronomie 6:829–837

Bonnet P, Bourdon E, Ponchet M, Blein J-P, Ricci P (1996) Acquired resistance triggered by elicitins in tobacco and other plants. Eur J Plant Pathol 102:181–192

Boonlertnirun S, Boonraung C, Suvannasara R (2008) Application of chitosan in rice production. J Metals Mater Miner 18:47–52

Boonreung C, Boonlertnirun S (2013) Efficiency of chitosan for controlling dirty panicle disease in rice plants. ARPN J Agric Biol Sci 8:380–384

Borejsza-Wysocka E, Kader AA, Norelli JL, Bauer DW, Garr ER, Beer SV, Aldwinckle HS (2000) Effect of expressing *hrpN* in apple on resistance to *Erwinia amylovora*. In Vitro Cell Dev Biol Anim 34:1023

Borkowski J, Szwonek E (2004) Powdery mildew control on green-house tomatoes by chitosan and other selected substances. Acta Hortic 633:435–438

Borsics T, Webb D, Andeme-Ondzihi C, Staehelin A, Christopher DA (2007) The cyclic nucleotide-gated calmodulin-binding channel AtCNGC10 localizes to the plasma membrane and influences numerous growth responses and starch accumulation in *Arabidopsis thaliana*. Planta 225:563–573

Bottin A, Véronési C, Pontier D, Esquerré-Tugayé MT, Blein JP, Rustérucci C, Ricci P (1994) Differential responses of tobacco cells to elicitors from two *Phytophthora* species. Plant Physiol Biochem 32:373–378

Bouarab K, Melton R, Peart J, Baulcombe D, Osbourn A (2002) A saponin-detoxifying enzyme mediates suppression of plant defences. Nature 418:889–892

Boureau T, Siamer S, Perino C, Gaubert S, Patrit O, Degrave A, Fagard M, Chevreau E, Barny MA (2011) The HrpN effector of *Erwinia amylovora*, which is involved in type III translocation, contributes directly or indirectly to callose elicitation on apple leaves. Mol Plant-Microbe Interact 24:577–584

Bourque S, Binet MN, Ponchet M, Pugin A, Lebrun-Garcia A (1999) Characterization of the cryptogein binding sites on plant plasma membranes. J Biol Chem 274:34699–34705

Brisson L, El-Bakkali-Taheri N, Giorgi M, Fadel A, Kaizer J, Rěglier M, Tron T, el Ajandouz H, Simaan AJ (2012) 1-Aminocyclopropane-1-carboxylic acid oxidase: insight into cofactor binding from experimental and theoretical studies. J Biol Inorg Chem 17:939–949

Brodersen P, Petersen M, Bjorn Nielsen H, Zhu S, Newman MA, Shokat KM, Rietz S, Parker J, Mundy J (2006) *Arabidopsis* MAP kinase 4 regulates salicylic acid- and jasmonic acid/ethylene-dependent responses via EDS1 and PAD4. Plant J 47:532–546

Brummer M, Arend M, Fromm J, Schlenzig A, Oßwald WF (2002) Ultrastructural changes and immunocytochemical localization of the elicitin quercinin in *Quercus robur* L roots infected with *Phytophthora quercina*. Physiol Mol Plant Pathol 61:109–120

Bryk R, Griffin P, Nathan C (2000) Peroxynitrite reductase activity of bacterial peroxiredoxins. Nature 407:211–215

Bueter CL, Specht CA, Levitz SM (2013) Innate sensing of chitin and chitosan. PLoS Pathog 9(1):e1003080

Buhot N, Douliez J-P, Jacquemard A, Marion D, Tran V, Maume BF, Milat M-L, Ponchet M, Mikès V, Kader J-C, Blein J-P (2001) A lipid transfer protein binds to a receptor involved in the control of plant defence responses. FEBS Lett 509:27–30

Burketova L, Trda L, Ott P, Valentova O (2015) Bio-based resistance inducers for sustainable plant protection against pathogens. Biotech Adv 33:994–1004. doi:10.1016/j.biotechadv.2015

Cabral A, Oome S, Sander N, Kūfner I, Nūrnberger T, van den Ackerveken G (2012) Nontoxic Nep-1 like proteins of the downy mildew pathogen *Hyaloperonospora arabidopsis*: Repression of necrosis-inducing activity by a surface-exposed region. Mol Plant-Microbe Interact 25:697–708

Cai X-Z, Zhou X, Xu Y-P, Joosten MHAJ, deWit PJGM (2007) *Cladosporium fulvum* CfHNNI1 induces hypersensitive necrosis, defence gene expression and disease resistance in both host and nonhost plants. Plant Mol Biol 64:89–101

Cai R, Lewis J, Yan S, Liu H, Clarke CR, Campanile F, Almeida NF, Studholme DJ, Lindeberg M, Schneider D, Zaccardelli M, Setubal JC, Morales-Lizcano NP, Bernal A, Coaker G, Baker C,

Bender CL, Leman S, Vinatzer BA (2011) The plant pathogen *Pseudomonas syringae* pv. *tomato* is genetically monomorphic and under strong selection to evade tomato immunity. PLoS Pathog 7:e1002130

Cao H, Glazebrook J, Clarke J, Volko S, Dong X (1997) The Arabidopsis *npr1* gene that controls systemic acquired resistance encodes a novel protein containing ankyrin repeats. Cell 88:57–63

Century K, Reuber TL, Ratcliffe OJ (2008) Regulating the regulators: The future prospects for transcription-factor-based agricultural biotechnology products. Plant Physiol 147:20–29

Cevik V, Kidd BN, Zhang P, Hill C, Kiddle S, Denby KJ, Holub EB, Cahill DM, Manners JM, Schenk PM, Beynon J, Kazan K (2012) MEDIATOR25 acts as an integrative hub for the regulation of jasmonate-responsive gene expression in Arabidopsis. Plant Physiol 160:541–555

Chakraborty M, Karun A, Mitra A (2009) Accumulation of phenylpropanoid derivatives in chitosan-induced cell suspension culture of *Cocos nucifera*. J Plant Physiol 166:63–71

Chang X, Nick P (2012) Defense signaling triggered by flg22 and harpin is integrated into a different stilbene output in *Vitis* cells. PLoS ONE 7(7):e40446

Che FS, Nakajima Y, Tanaka N, Iwano M, Yoshida T, Takayama S, Kadota I, Isogai A (2000) Flagellin from an incompatible strain of *Pseudomonas avenae* induces a resistance response in cultured rice cells. J Biol Chem 275:32347–32356

Chen L, Qian J, Qu S, Long J, Yin Q, Zhang C, Wu X, Sun F, Wu T, Hayes M, Beer SV, Dong H (2008a) Identification of specific fragments of $HpaG_{Xooc}$, a harpin from *Xanthomonas oryzae* pv. *oryzicola*, that induce disease resistance and enhance growth in plants. Phytopathology 98:781–791

Chen L, Zhang S-J, Zhang S-S, Qu S, Ren X, Long J, Yin Q, Qian J, Sun F, Zhang C, Wang L, Wu X, Wu T, Zhang Z, Cheng Z, Hayes M, Beer SV, Dong H (2008b) A fragment of the *Xanthomonas oryzae* pv. *oryzicola* harpin $HpaG_{Xooc}$ reduces disease and increases yield of rice in extensive grower plantings. Phytopathology 98:792–802

Chen F, Gao J, Miao S, Yuan Y-X, Wang M-Y, Li Q, Mao B-Z, Jiang L-W, He ZH (2010) Plasma membrane localization and potential endocytosis of constitutively expressed XA21 proteins in transgenic rice. Mol Plant 3:917–926

Chen C-W, Panzeri D, Yeh Y-H, Kadota Y, Huang P-Y, Tao C-N, Roux M, Chien SC, Chin T-C, Chu P-W, Zipfel C, Zimmerli L (2014a) The *Arabidopsis* malectin-like leucine-rich repeat receptor-like kinase IOS1 associates with the pattern recognition receptors FLS2 and EFR and is critical for priming of pattern-triggered immunity. Plant Cell 26:3201–3219

Chen X, Zuo S, Schwessinger B, Chern M, Canlas PE, Ruan D, Zhou X, Wang J, Daudi A, Petzold CJ (2014b) An XA21-associated kinase (OsSERK2) regulates immunity mediated by the XA21 and XA3 immune receptors. Mol Plant 7:874–892

Cheong J-J, Birberg W, Fugedi P, Pilotti Å, Garegg PJ, Hong N, Ogawa T, Hahn MG (1991) Structure-activity relationships of oligo-β-glucoside elicitors of phytoalexin accumulation in soybean. Plant Cell 3:127–136

Chern M, Canlas PE, Ronald PC (2008) Strong suppression of systemic acquired resistance in *Arabidopsis* by NRR is dependent on its ability to interact with NPR1 and its putative repressive domain. Mol Plant 1:552–559

Chinchilla D, Bauer Z, Regenass M, Boller T, Felix G (2006) The *Arabidopsis* receptor kinase FLS2 binds flg22 and determines the specificity of flagellin perception. Plant Cell 18:465–476

Chinchilla D, Boller T, Robatzek S (2007a) Flagellin signaling in plant immunity. Adv Exp Med Biol 598:358–371

Chinchilla D, Zipfel C, Robatzek S, Kemmerling B, Nürnberger T, Jones JDG, Felix G, Boller T (2007b) A flagellin-induced complex of the receptor FLS2 and BAK1 initiates plant defence. Nature 448:497–500

Chirapongsatonkul N, Nak-udom P, Churngchow N (2008) Defense responses of calli and seeds of *Hevea brasiliensis* to zoospores and the elicitin of *Phytophthora palmivora*. J Phytopathol 156:732–741

Choi JJ, Klosterman SJ, Hadwiger LA (2001) A comparison of the effects of DNA-damaging agents and biotic elicitors on the induction of plant defense genes, nuclear distortion, and cell death. Plant Physiol 125:752–762

Choi JJ, Klosterman SJ, Hadwiger LA (2004) A promoter from pea gene DRR206 is suitable to regulate an elicitor-coding gene and develop disease resistance. Phytopathology 94:651–660

Choi HW, Lee DH, Hwang BK (2009) The pepper calmodulin gene CaCaM1 is involved in reactive oxygen species and nitric oxide generation required for cell death and the defense response. Mol Plant Microbe Interact 22:1389–1400

Choi M-S, Heu S, Paek N-C, Koh H-J, Lee J-S, Oh C-S (2012) Expression of *hpa1* gene encoding a bacterial harpin protein on *Xanthomonas oryzae* pv. *oryzae* enhances disease resistance to both fungal and bacterial pathogens in rice and *Arabidopsis*. Plant Pathol J 28:364–372

Choi M-S, Kim W, Lee C, Oh C-S (2013) Harpins, multifunctional proteins secreted by Gram-negative plant-pathogenic bacteria. Mol Plant-Microbe Interact 26:1115–1122

Chrzanowski L, Lawniczak L, Czaczyk K (2012) Why do microorganisms produce rhamnolipids? World J Microbiol Biotechnol 28:401–419

Chuang H-W, Hamrak A, Chen Y-C, Hsu C-M (2010) A harpin-induced ethylene–responsive factor regulates plant growth and responses to biotic and abiotic stresses. Biochem Biophys Res Commun 402:414–420

Chuang H-W, Chang P-Y, Syu Y (2014) Harpin protein, an elicitor of disease resistance, acts as a growth promoter in *Phalaenopsis* orchids. J Growth Regul 33:788–797

Chun HJ, Park HC, Koo SC, Lee JH, Park CY, Choi MS, Kang CH, Baek D, Cheong YH, Yun DJ, Chung WS, Cho MJ, Kim MC (2012) Constitutive expression of mammalian nitric oxide synthase in tobacco plants triggers disease resistance to pathogens. Mol Cells 34:463–471

Churngchow N, Rattarasan M (2000) The elicitin secreted by *Phytophthora palmivora*, a rubber tree pathogen. Phytochemistry 54:33–38

Clarke A, Mur LAJ, Darby RM, Kenton P (2005) Harpin modulates the accumulation of salicylic acid by *Arabidopsis* cells via apoplastic alkalization. J Expt Bot 56:3129–3136

Clarke CR, Chinchilla D, Hind SR, Taguchi F, Miki R, Ichinose Y, Martin GB, Leman S, Felix G, Vinatzer BA (2013) Allelic variation in two distinct *Pseudomonas syringae* flagellin epitopes modulates the strength of plant immune responses but not bacterial mobility. New Phytol 200:847–860

Clough SJ, Fengler KA, Yu IC, Lippok B, Smith RK Jr, Bent AF (2000) The Arabidopsis *dnd1* "defense, no death" gene encodes a mutated cyclic nucleotide-gated ion channel. Proc Natl Acad Sci U S A 97:9323–9328

Conrath U, Domard A, Kauss H (1989) Chitosan-elicited synthesis of callose and of coumarin derivatives in parsley cell suspension cultures. Plant Cell Reptr 8:152–154

Cordelier S, de Ruffray P, Fritig B, Kauffmann S (2003) Biological and molecular comparison between localized and systemic acquired resistance induced in tobacco by a *Phytophthora megasperma* glycoprotein elicitin. Plant Mol Biol 51:109–118

Coursol S, Fromentin J, Noirot E, Brière C, Robert F, Morel J, Liang YK, Lherminier J, Simon-Plas F (2015) Long-chain bases and their phosphorylated derivatives differentially regulate cryptogein-induced production of reactive oxygen species in tobacco (*Nicotiana tabacum*) BY-2 cells. New Phytol 205:1239–1249. doi:10.1111/nph.13094

Courtois C, Besson A, Dahan J, Bourque S, Dobrowolska G, Pugin A, Wendehenne D (2008) Nitric oxide signaling in plants: interplays with Ca^{2+} and protein kinases. J Expt Bot 59:155–163

Crawford NM, Galli M, Tischner R, Heimer YM, Okamoto M, Mack A (2006) Response to Zemojtel *et al*: plant nitric oxide synthase: back to square one. Trends Plant Sci 11:526–527

Culver JN, Dawson WO (1991) Tobacco mosaic virus elicitor coat protein genes produce a hypersensitive phenotype in transgenic *Nicotiana sylvestris* plants. Mol Plant-Microbe Interact 4:458–463

Dafermos NG, Kasselaki AM, Goumas DE, Spantidakis K, Eyre MD, Leifert C (2012) Integration of elicitors and less-susceptible hybrids for the control of powdery mildew in organic tomato crops. Plant Dis 96:1506–1512

de Capdeville G, Beer SV, Wilson CL, Aist JR (2002) Alternative disease control agents induced resistance to blue mold in harvested 'Red Delicious' apple fruit. Phytopathology 92:900–908
de Capdeville G, Beer SV, Watkins CB, Wilson CL, Tedeschi LO, Aist JR (2003) Pre- and post-harvest harpin treatments of apples induce resistance to blue mold. Plant Dis 87:39–44
de Capdeville G, Beer SV, Wilson CL, Aist JR (2008) Some cellular correlates of harpin-induced resistance to blue mold of apples, Trop. Plant Pathol 33:103–113
Deepak SA, Niranjan Raj S, Umemura K, Kono T, Shekar Shetty H (2003) Cerebroside as an elicitor for induced resistance against the downy mildew pathogen in pearl millet. Ann Appl Biol 143:169–173
Delledonne M, Zeier J, Marocco A, Lamb C (2001) Signal interactions between nitric oxide and reactive oxygen intermediates in the plant hypersensitive disease resistance response. Proc Natl Acad Sci U S A 98:13454–13459
Denoux C, Galletti R, Mammarella N, Gopalan S, Werck D, De Lorenzo G, Ferrari S, Ausubel FM, Dewdney J (2008) Activation of defense response pathways by OGs and Flg22 elicitors in Arabidopsis seedlings. Mol Plant 1:423–445
Desikan R, Reynolds A, Hancock JT, Neill SJ (1998) Harpin and hydrogen peroxide both initiate programmed cell death but have differential effects on defense gene expression in *Arabidopsis* suspension cell cultures. Biochem J 330:115–120
Desikan R, Hancock JT, Ichimura K, Shinozaki K, Neill SJ (2001) Harpin induces activation of the *Arabidopsis* mitogen-activated protein kinases AtMPK4 and AtMPK6. Plant Physiol 126:1579–1587
Desikan R, Hancock JT, Bright J, Harrison J, Weir I, Hooley R, Neill SJ (2005) A role for ETR1 in hydrogen peroxide signaling in stomatal guard cells. Plant Physiol 137:831–834
Ding S-W (2010) RNA-based antiviral immunity. Nat Rev Immunol 10:632–644
Djonovic S, Pozo MJ, Dangott LJ, Howell CR, Kenerley CM (2006) Sm1, a proteinaceous elicitor s'ecreted by the biocontrol fungus *Trichoderma virens* induces plant defense responses and systemic resistance. Mol Plant-Microbe Interact 19:838–853
Djonovic S, Vargas WA, Kolomiets MV, Horndeski M, Wiest A, Kenerley CM (2007) A proteinaceous elicitor Sm1 from the beneficial fungus *Trichoderma virens* is required for induced systemic resistance in maize. Plant Physiol 145:875–889
Doares SH, Syrovets T, Weiler EW, Ryan CA (1995) Oligogalacturonides and chitosan activate plant defensive genes through the octadecanoid pathway. Proc Natl Acad Sci U S A 92:4095–4098
Dong H, Delaney TP, Bauer DW, Beer SV (1999) Harpin induces disease resistance in *Arabidopsis* through the systemic acquired resistance pathway mediated by salicylic acid and *NIM1* gene. Plant J 20:207–215
Dong H-P, Peng J, Bao Z, Meng X, Bonasera JM, Chen G, Beer SV, Dong H (2004) Downstream divergence of the ethylene signaling pathway for harpin-stimulated Arabidopsis growth and insect resistance. Plant Physiol 136:3628–3638
Donghua J, Zejian G, Xujun C, Zhiqiang C, Zhong Z (2004) Study on tobacco disease resistances mediated by the elicitor gene cryptogein from *Phytophthora cryptogea*. Chin J Agric Biotechnol 1:125–131
Dorey S, Kopp M, Geoffroy P, Fritig B, Kauffmann S (1999) Hydrogen peroxide from the oxidative burst is neither necessary nor sufficient for hypersensitive cell death induction, phenylalanine ammonia lyase stimulation, salicylic acid accumulation or scopoletin consumption in elicitin-treated cultured tobacco cells. Plant Physiol 121:163–172
Dornenburg H, Knorr D (1997) Evaluation of elicitor and high pressure induced enzymatic browning utilizing potato (*Solanum tuberosum*) suspension culture as a model for plant tissues. J Agric Food Chem 45:4173–4177
Drutskaya MS, Belousov PV, Nedopasov SA (2011) Innate mechanisms of viral recognition. Mol Biol 45:5–15
Du Q, Zhu W, Zhao Z, Qian X, Xu Y (2012) Novel benzo-1,2,3-thiadiazole-7-carboxylate derivatives as plant activators and the development of their agricultural applications. J Agric Food Chem 60:346–353

Durner J, Klessig DF (1999) Nitric oxide as a signal in plants. Curr Opin Plant Biol 2:369–374
Durner J, Wendehenne D, Klessig DF (1998) Defense gene induction in tobacco by nitric oxide, cyclic GMP, and cyclic ADP ribose. Proc Natl Acad Sci U S A 95:10328–10333
El Ghaouth A (1997) Biologically based alternatives to synthetic fungicides for the control of postharvest diseases. J Ind Microbiol Biotechnol 19:160–162
El Ghaouth A, Arul J, Benhamou N, Asselin A, Belanger RR (1994) Effect of chitosan on cucumber plants: suppression of *Pythium aphanidematum* and induction of defense reactions. Phytopathology 84:313–320
El-Ghaouth A, Smilanaick JL, Brown GE, Ippolito A, Wisniewski M, Wilson CL (2000) Application of *Candida saitoana* and glycolchitosan for the control of postharvest diseases of apple and citrus fruit under semi-commercial conditions. Plant Dis 84:243–248
El-Maarouf H, Barny MA, Rona JP, Bouteau F (2001) Harpin, a hypersensitive response elicitor from *Erwinia amylovora*, regulates ion channel activities in *Arabidopsis thaliana* suspension cells. FEBS Lett 497:82–84
Elmer PAG, Reglinski T (2006) Biosuppression of *Botrytis cinerea* in grapes. Plant Pathol 55:155–177
Engelhardt S, Lee J, Gabler Y, Kemmerling B, Haapalainen ML, Li C-M, Wei Z, Keller H, Joosten M, Taira S, Nurnberger T (2009) Separable roles of the *Pseudomonas syringae* pv. *phaseolicola* accessary protein HrpZ1 in ion-conducting pore formation and activation of plant immunity. Plant J 57:706–717
Erbs G, Silipo A, Aslam S, De Castro C, Liparoti V, Flagiello A, Pucci P, Lanzetta R, Parrilli M, Molinaro A, Newman M-A, Cooper RM (2008) Peptidoglycan and muropeptides from pathogens *Agrobacterium* and *Xanthomonas* elicit plant innate immunity: Structure and activity. Chem Biol 15:438–448
Erickson FL, Holzberg S, Calderon-Urrea A, Handley V, Aktell M, Corr C, Baker B (1999) The helicase domain of the TMV replicase proteins induces the N-mediated defence response in tobacco. Plant J 18:67–75
Espunya C, De Michele R, Gomez-Cadenas AA, Martinez MC (2012) S-Nitrosoglutathione is a component of wound- and salicylic acid-induced systemic responses in *Arabidopsis thaliana*. J Expt Bot 63:3219–3227
Eulgem T, Weighman VJ, Chang H-S, McDowell JM, Holub EB, Glazebrook J, Zhu T, Dangl JL (2004) Gene expression signatures from three genetically separable resistance gene signaling pathways for downy mildew resistance. Plant Physiol 135:1129–1144
Fajardo JE, McCollum TG, McDonald RE, Mayer RT (1998) Differential induction of proteins in orange flavedo by biologically based elicitors and challenged by *Penicillium digitatum* Sacc. Biol Control 13:143–151
Faoro F, Iriti M (2007) Callose synthesis as a tool to screen chitosan efficacy in inducing plant resistance to pathogens. Caryolgia 60:121–124
Faoro F, Maffi D, Cantu D, Iriti M (2008) Chemical-induced resistance against powdery mildew in barley: the effects of chitosan and benzothiadiazole. Biocontrol 53:387–401
Felix G, Boller T (2003) Molecular sensing of bacteria in plants. The highly conserved RNA-binding motif RNP-1 of bacterial cold shock proteins is recognized as an elicitor signal in tobacco. J Biol Chem 278:6201–6208
Felix G, Duran JD, Volko S, Boller T (1999) Plants have a sensitive perception system for the most conserved domain of bacterial flagellin. Plant J 18:265–276
Fellbrich G, Romanski A, Varet A, Blume B, Brunner F, Stefan Engelhardt S, Felix G, Kemmerling B, Krzymowska M, Nürnberger T (2002) NPP1, a *Phytophthora*-associated trigger of plant defence in parsley and *Arabidopsis*. Plant J 32:375–390
Felle HH, Hermann A, Hanstein S, Hückelhoven R, Kogel K-H (2004) Apoplasic pH signaling in barley leaves attacked by the powdery mildew fungus *Blumeria graminis* f. sp. *hordei*. Mol Plant-Microbe Interact 17:118–123
Fellers JP, Tremblay D, Handest MF, Lommel SA (2002) The *Potato virus Y* M^SN^R Nlb-replicase is the elicitor of a veinal necrosis-hypersensitive response in root knot nematode resistant tobacco. Mol Plant Pathol 3:145–152

Ferrari S, Galletti R, Denoux C, De Lorenzo G, Ausubel FM, Dewdney J (2007) Resistance to *Botrytis cinerea* induced in Arabidopsis by elicitors is independent of salicylic acid, ethylene, or jasmonate signaling but requires PHYTOALEXIN DEFICIENT3. Plant Physiol 144:367–379

Feys BJ, Moisan LJ, Newman MA, Parker JE (2001) Direct interaction between the Arabidopsis disease resistance signaling proteinsm EDS1 and PAD4. EMBO J 20:5400–5411

Fliegmann J, Mithöfer A, Wanner G, Ebel J (2004) An ancient enzyme domain hidden in the putative β-glucan elicitor of soybean may play an active part in the perception of pathogen-associated molecular patterns during broad host resistance. J Biol Chem 279:1132–1140

Flocchetti F, D'Amore R, De Palma M, Bertini L, Caruso C, Caporale C, Testa A, Cristinzio G, Saccardo F, Tucci M (2008) Constitutive over-expression of two wheat pathogenesis-related genes enhances resistance of tobacco plants to *Phytophthora nicotianae*. Plant Cell Tissue Organ Cult 92:73–84

Foissner L, Wendehenne D, Langebartels C, Durner J (2000) In vivo imaging of an elicitor-induced nitric oxide burst in tobacco. Plant J 23:817–824

Fontanilla M, Montes M, De Prado R (2005a) Effects of the foliar-applied protein "$\text{Harpin}_{\text{Ea}}$" (messenger) on tomatoes infected with *Phytophthora infestans*. Commun Agric Appl Biol Sci 70:41–45

Fontanilla M, Montes M, De Prado R (2005b) Induction of resistance to the pathogenic agent *Botrytis cinerea* in the cultivation of the tomato by means of the application of the protein "Harpin" (Messenger). Commun Agric Appl Biol Sci 70:35–40

Forde BG, Roberts MR (2014) Glutamate receptor-like channels in plants; a role as amino acid sensors in plant defence? F1000 Prime Rep 6:37. doi:10.12703/P6-37

Fu ZQ, Guo M, Alfano JR (2006) *Pseudomonas syringae* HrpJ is a type III secreted protein that is required for plant pathogenesis, injection of effectors, and secretion of the HrpZ1 harpin. J Bacteriol 188:6060–6069

Fu M, Xu M, Zhou T, Wang D, Tian S, Han L, Dong H, Zhang C (2014) Transgenic expression of a functional fragment of harpin protein Hpa1 in wheat induces the phloem-based defence against English grain aphid. J Exp Bot 65:1439–1453

Fujiwara S, Tanaka N, Kaneda T, Takayama S, Isogai A, Che F-S (2004) Rice cDNA microarray-based gene expression profiling of the response to flagellin perception in cultured rice cells. Mol Plant-Microbe Interact 17:986–998

Furukawa T, Inagaki H, Takai R, Hirai H, Che F-S (2014) Two distinct EF-TU epitopes induce immune responses in rice and *Arabidopsis*. Mol Plant Microbe Interact 27:113–124

Gao Z, Wen C-K, Binder BM, Chen Y-F, Chang J, Chiang Y-H, Kerris RJ III, Chang C, Schaller GE (2008) Heteromeric interactions among ethylene receptors mediate signaling in *Arabidopsis*. J Biol Chem 283:23801–23810

Garcia AV, Hirt H (2014) *Salmonella enterica* induces and subverts the plant immune system. Front Microbiol 5:141. doi:10.3389/fmicb.2014.00141

Garcia-Brugger A, Lamotte O, Vandelle E, Bourque S, Lecourieux D, Poinssot B, Wendehenne D, Pugin A (2006) Early signaling events induced by elicitors of plant defenses. Mol Plant-Microbe Interact 19:711–724

Gaston B (1999) Nitric oxide and thiol groups. Biochim Biophys Acta 1411:323–333

Gaulin E, Drame N, Lafitte C, Torto-Alalibo T, Martinez Y, Ameline-Torregrosa C, Khatib M, Mazarguil H, Villalba-Mateos F, Kamoun S, Mazars C, Dumas B, Bottin A, Esquerré-Tugayé M-T, Rickauer M (2006) Cellulose binding domains of a *Phytophthora* cell wall protein are novel pathogen-associated molecular patterns. Plant Cell 18:1766–1777

Gaupels F, Kuruthukulangarakoola GT, Durner J (2011) Upstream and downstream signals of nitric oxide in pathogen defence. Curr Opin Plant Biol 14:704–714

Geiger D, Scherzer S, Mumm P, Marten I, Ache P, Matschi S, Liese A, Wellmann C, Al-Rasheid KAS, Grill E, Romeis T, Hedrich R (2010) Guard cell anion channel SLAC1 is regulated by CDPK protein kinases with distinct Ca^{2+} affinities. Proc Natl Acad Sci U S A 107:8023–8028

Geldner N, Robatzek S (2008) Plant receptors go endosomal: a moving view on signal transduction. Plant Physiol 147:1565–1574

Ger M-J, Louh G-Y, Lin Y-H, Feng T-Y, Huang H-E (2014) Ectopically expressed sweet pepper ferredoxin PFLP enhances disease resistance to Pectobacterium carotovorum subsp. carotovorum affected by harpin and protease-mediated hypersensitive response in Arabidopsis. Mol Plant Pathol 15:892–906. doi:10.1111/mpp.12150

Ghazala W, Varrelmann M (2007) *Tobacco rattle virus* 29 K movement protein is the elicitor of extreme and hypersensitive-like resistance in two cultivars of *Solanum tuberosum*. Mol Plant-Microbe Interact 20:1396–1405

Gijzen M, Nürnberger T (2006) Nep1-like proteins from plant pathogens: recruitment and diversification of the NPP1 domain across taxa. Phytochemistry 67:1800–1897

Gilardi P, Garcia-Luque I, Serra MT (2004) The coat protein of tobamovirus acts as elicitor of both *L2* and *L4* gene-mediated resistance in *Capsicum*. J Gen Virol 85:2077–2085

Glazebrook J, Chen W, Estes B, Chang HS, Nawrath C, Metraux JP, Zhu T, Katagiri F (2003) Topology of the network integrating salicylate and jasmonate signal transduction derived from global expression phenotyping. Plant J 34:217–228

Göhre V, Spallek T, Häweker H, Mersmann S, Mentzel T, Boller T, de Torres M, Mansfield JW, Robatzek S (2008) Plant pattern-recognition receptor FLS2 is directed for degradation by the bacterial ubiquitin ligase AvrPtoB. Curr Biol 18:1824–1832

Gómez-Gómez L, Boller T (2002) Flagellin perception: a paradigm for innate immunity. Trends Plant Sci 7:251–256

Gorbatenko IY, Onischuk JA, Krivtsov GG, Vanyushin BF (1996) Eliciting and growth-regulating effects of chitosan on plants. Biol Bull Russ Acad Sci 23:327–330

Graham JH, Leite RP (2004) Lack of control of citrus canker by induced systemic resistance compounds. Plant Dis 88:745–750

Grant JJ, Loake GJ (2000) Role of reactive oxygen intermediates and cognate redox signaling in disease resistance. Plant Physiol 8:629–643

Grant SR, Fisher EJ, Chang JH, Mole BM, Dangl JL (2006) Subterfuge and manipulation: type III effector proteins of phytopathogenic bacteria. Annu Rev Microbiol 60:425–449

Grefen C, Städele K, Ruzicka K, Obrdlik P, Harter K, Horák J (2008) Subcellular localization and in vivo interactions of the *Arabidopsis thaliana* ethylene receptor family members. Mol Plant 1:308–320

Groen AJ, de Vries SC, Lilley KS (2008) A proteomics approach to membrane trafficking. Plant Physiol 147:1584–1589

Gross A, Kapp D, Nielsen T, Niehaus K (2005) Endocytosis of *Xanthomonas campestris* pathovar *campestris* lipopolysaccharides in non-host plant cells of *Nicotiana tabacum*. New Phytol 165:215–226

Guan Y-J, Hu J, Wang X-J, Shao C-X (2009) Seed priming with chitosan improves maize germination and seedling growth in relation to physiological changes under low temperature stress. J Zhejang Univ Sci B 10:427–433

Gupta KJ, Brotman Y, Segu S, Zeier T, Zeier J, Persijn ST, Cristescu SM, Harren FJM, Bauwe H, Fernie AR, Kaiser WM, Mur LAJ (2013) The form of nitrogen nutrition affects resistance against *Pseudomonas syringae* pv. *phaseolicola* in tobacco. J Exp Bot 64:553–568

Gust AA, Biswas R, Lenz HD, Rauhut T, Ranf S, Kemmerling B, Götz F, Glawischnig E, Lee J, Felix G, Nurnberger T (2007) Bacteria-derived peptidoglycans constitute pathogen-associated molecular patterns triggering innate immunity in *Arabidopsis*. J Biol Chem 282:32338–32348

Haapalainen M, Engelhardt S, Küfner I, Li C-M, Nürnberger T, Lee J, Romantschuk M, Taira S (2011) Functional mapping of harpin HrpZ of *Pseudomonas syringae* reveals the sites responsible for protein oligomerization, lipid interactions and plant defence induction. Mol Plant Pathol 12:151–166

Hadwiger LA (2008) Pea-*Fusarium solani* interactions contributions of a system toward understanding disease resistance. Phytopathology 98:372–379

Hadwiger LA, Ogawa T, Kuyama H (1994) Chitosan polymer sizes effective in inducing phytoalexin accumulation and fungal suppression are verified with synthesized oligomers. Mol Plant-Microbe Interact 7:531–533

Hadwiger LA, Chang MM, Parsons A (1995) *Fusarium solani* DNase is a signal for increasing expression of nonhost disease resistance response genes, hypersensitivity, and pisatin production. Mol Plant-Microbe Interact 8:871–879

Hahlbrock K, Scheel D, Logemann E, Nürnberger T, Parniske M, Reinhold S, Sacks WR, Schmelzer E (1995) Oligopeptide elicitor-mediated defense gene activation in cultured parsley cells. Proc Natl Acad Sci U S A 92:4150–4157

Hall HC, Samuel MA, Ellis BE (2007) SIPK conditions transcriptional responses unique to either bacterial or oomycete elicitation in tobacco. Mol Plant Pathol 8:581–594

Hamel L-P, Beaudoin N (2010) Chitooligosaccharide sensing and downstream signaling contrasted outcomes in pathogenic and beneficial plant-microbe interactions. Planta 232:787–806

Hann DR, Rathjen JP (2007) Early events in the pathogenecity of *Pseudomonas syringae* on *Nicotiana benthamiana*. Plant J 49:607–618

Hao G, Pitino M, Ding F, Lin H, Stover E, Duan Y (2014) Induction of innate immune responses by flagellin from the intracellular bacterium from the intracellular bacterium, '*Candidatus* Liberibacter solanacearum'. BMC Plant Biol 14:14211

Harding SA, Oh SH, Roberts DM (1997) Transgenic tobacco expressing a foreign calmodulin gene shows an enhanced production of active oxygen species. EMBO J 16:1137–1144

Hase S, Shimizu A, Nakaho K, Takenaka S, Takahashi H (2006) Induction of transient ethylene and reduction in severity of tomato bacterial wilt by *Pythium oligandrum*. Plant Pathol 55:537–543

Hase S, Takahashi S, Takenaka S, Nakaho K, Arie T, Seo S, Ohashi Y, Takahashi H (2008) Involvement of jasmonic acid signalling in bacterial wilt disease resistance induced by biocontrol agent *Pythium oligandrum* in tomato. Plant Pathol 57:870–876

He SY, Huang HC, Collmer A (1993) *Pseudomonas syringae* pv. *syringae* harpin$_{Pss}$: A protein that is secreted via the Hrp pathway and elicits the hypersensitive response in plant cells. Cell 73:1255–1266

Hettenhausen C, Baldwin IT, Wu J (2012) Silencing *MPK4* in *Nicotiana attenuata* enhances photosynthesis and seed production but compromises abscisic acid-induced stomatal closure and guard cell-mediated resistance to *Pseudomonas syringae* pv. *tomato* DC3000. Plant Physiol 158:759–776

Hewajulige IGN, Sultanbawa Y, Wijeratnam RSW, Wijesundara RLC (2009) Mode of action of chitosan coating on anthracnose disease control in papaya. Phytoparasitica 37:437–444

Hirai H, Takai R, Iwano M, Nakai M, Kondo M, Takayama S, Isogai A, Che FS (2011) Glycosylation regulates specific induction of rice immune responses by *Acidovorax avenae* flagellin. J Biol Chem 286:25519–25530

Hoeberichts FA, Davoine C, Vandorpe M, Morsa S, Ksas B, Stassen C, Triantaphylidès C, Van Breusegem F (2013) Cryptogein-induced transcriptional reprogramming in tobacco is light dependent. Plant Physiol 163:263–275

Hondo D, Hase S, Kanayama Y, Yoshikawa N, Takenaka S, Takahashi H (2007) The *LeATL6*-associated ubiquitin/proteasome system may contribute to fungal elicitor-activated defense response via the jasmonic acid-dependent signaling pathway in tomato. Mol Plant-Microbe Interact 20:72–81

Hopkins DL (2002) Control of gummy blight of watermelon with plant defense activators combined with fungicides. Proc Fla State Hortic Soc 115:183–186

Hu X, Neill S, Cai W, Tang Z (2003) Hydrogen peroxide and jasmonic acid mediate oligogalacturonic acid-induced saponin accumulation in suspension-cultured cells of *Panax ginseng*. Physiol Plantarum 118:414–421

Hu X-Y, Neill SJ, Cai WM, Tang Z-C (2004) Induction of defence gene expression by oligogalacturonic acid requires increases in both cytosolic calcium and hydrogen peroxide in *Arabidopsis thaliana*. Cell Res 14:234–240

Hu Y, Iriti M, Faoro F (2008) Abscisic acid is involved in chitosan-induced resistance to tobacco necrosis virus (TNV). Plant Physiol Biochem 46:1106–1111

Hu Y, Cai J, Du Y, Lin J, Wang C, Xiong K (2009) Preparation and anti-TMV activity of guanidinylated chitosan hydrochloride. J Appl Polym Sci 112:3522–3528

Huang JW, Grunes DL, Kochian LV (1994) Voltage-dependent Ca^{2+} influx into right-side-out plasma membrane vesicles isolated from wheat roots. Proc Natl Acad Sci U S A 91:3473–3477

Huo R, Wang Y, Ma L, Qiao J, Shao M, Gao X (2010) Assessment of inheritance pattern and agronomic performance of transgenic rapeseed having harpinXooC-encoding *hrf2* gene. Transgenic Res 19:841–847

Hwang IS, Hwang BK (2011) The pepper mannose-binding lectin gene *CaMBL1* is required to regulate cell death and defense responses to microbial pathogens. Plant Physiol 155:447–463

Ichiropoulos H, al-Mehdi AB (1995) Peroxynitrite-mediated oxidative protein modifications. FEBS Lett 364:279–282

Iriti M, Faoro F (2007) Review of innate and specific immunity in plants and animals. Mycopathologia 164:57–64

Iriti M, Faoro F (2008) Abscisic acid is involved in chitosan-induced resistance to *Tobacco necrosis virus* (TNV). Plant Physiol Biochem 46:1106–1111

Iriti M, Faoro F (2009) Chitosan as a MAMP, searching for a PRR. Plant Signal Behav 4:66–68

Iriti M, Sironi M, Gomarasca S, Casazza AP, Soave C, Faoro F (2006) Cell death-mediated antiviral effect of chitosan in tobacco. Plant Physiol Biochem 44:893–900

Iriti M, Vitalini S, Di Tommaso G, D'Amico S, Borgo M, Faoro F (2011) New chitosan formulation prevents grapevine powdery mildew infection and improves polyphenol content and free radical scavenging activity of grape and wine. Aust J Grape Wine Res 17:263–269

Isebaert S, Verhoeven R, Haesaert G (2002) Disease control by means of induced resistance. Meded Rijksuniv Gent Fak Landbouwkd Toegep Biol Wet 67:159–164

Ishiga Y, Takeuchi K, Inagaki Y, Toyoda K, Shiraishi T, Ichinose Y (2005) Defense responses of Arabidopsis thaliana inoculated with *Pseudomonas syringae* pv. *tabaci* wild type and defective mutants for flagellin (Δ*fliC*) and flagellin-glycosylation (Δ*orf1*). J Gen Plant Pathol 71:302–307

Iwai T, Kaku H, Honkura R, Nakamura S, Ochiai H, Sasaki T, Ohashi Y (2002) Enhanced resistance to seed-transmitted bacterial diseases in transgenic rice plants overproducing an oat cell-wall-bound thionin. Mol Plant-Microbe Interact 15:515–521

Jabs T, Tschöpe M, Colling C, Hahlbrock K, Scheel D (1997) Elicitor-stimulated ion fluxes and O_2^- from the oxidative burst are essential components in triggering defense gene activation and phytoalexin synthesis in parsley. Proc Natl Acad Sci U S A 94:4800–4805

Jang YS, Sohn SI, Wang MH (2006) The *hrpN* gene of *Erwinia amylovora* stimulates tobacco growth and enhances resistance to *Botrytis cinerea*. Planta 223:449–456

Jayaraj J, Punja ZK (2007) Combined expression of chitinase and lipid transfer protein genes in transgenic carrot plants enhances resistance to foliar fungal pathogens. Plant Cell Rep 26:1539–1546

Jayaraj J, Rahman M, Wan A, Punja ZK (2009) Enhanced resistance to foliar fungal pathogens in carrot by application of elicitors. Ann Appl Biol 155:71–80

Jehle AK, Fürst U, Lipschis M, Albert M, Felix G (2013a) Perception of the novel MAMP eMax from different *Xanthomonas* species requires the *Arabidopsis* receptor-like protein ReMAX and the receptor-kinase SOBIR. Plant Signal Behav 8e27408:PMID 24384530

Jehle AK, Lipschis M, Albert M, Fallahzadeh-Mannaghani V, Fürst U, Mueller K, Felix G (2013b) The receptor-like protein ReMAX of *Arabidopsis* detects the microbe-associated molecular pattern eMax from *Xanthomonas*. Plant Cell 25:2330–2340

Jenner CE, Tomimura K, Ohshima K, Hughes SL, Walsh JA (2002) Mutations in *Turnip mosaic virus* P3 and cylindrical inclusion bodies are separately required to overcome two *Brassica napus* resistance genes. Virology 300:50–59

Jenner CE, Wang X, Tomimura K, Ohshima F, Ponz F, Walsh JA (2003) The dual role of the potyvirus P3 protein of *Turnip mosaic virus* as a symptom and avirulence determinant in brassicas. Mol Plant-Microbe Interact 16:777–784

Jeworutzki E, Roelfsema MR, Anschütz U, Krol E, Elzenga JT, Felix G, Boller T, Hedrich R, Becker D (2010) Early signaling through the *Arabidopsis* pattern recognition receptors FLS2 and EFR involves Ca^{2+}-associated opening of plasma membrane anion channels. Plant J 62:367–378

Jones J (2001) Harpin. Pestic Outlook 12(August):134–135

Kadota Y, Goh T, Tomatsu H, Tamauchi R, Higashi K, Muto S, Kuchitsu K (2004) Cryptogein-induced initial events in tobacco BY-2 cells: pharmacological characterization of molecular relationship among cytosolic Ca^{2+} transients, anion efflux and production of reactive oxygen species. Plant Cell Physiol 45:160–170

Kadota Y, Fujii S, Ogasawara Y, Maeda Y, Higashi K, Kuchitsu K (2006) Continuous recognition of the elicitor signal for several hours is prerequisite for induction of cell death and prolonged activation of signaling events in tobacco BY-2 cells. Plant Cell Physiol 47:1337–1342

Kanneganti T-D, Huitema E, Cakir C, Kamoun S (2006) Synergistic interactions of the plant cell death pathways induced by *Phytophthora infestans* Nep1-like protein PinPP1.1 and INF1 elicitin. Mol Plant-Microbe Interact 19:854–863

Kariola T, Palomäki TA, Brader G, Palva ET (2003) *Erwinia carotovora* subsp. *carotovora* and *Erwinia*-derived elicitors HrpN and PehA trigger distinct but interacting defense responses and cell death in *Arabidopsis*, Mol. Plant-Microbe Interact 16:179–187

Karita E, Yamakawa H, Mitsuhara I, Kuchitsu K, Ohashi Y (2004) Three types of tobacco calmodulins characteristically activate plant NAD kinase at different Ca^{2+} concentrations and pHs. Plant Cell Physiol 45:1371–1379

Kasparovsky T, Milat M-L, Humbert C, Blein J-P, Havel L, Mikes V (2003) Elicitation of tobacco cells with ergosterol activates a signal pathway including mobilization of internal calcium. Plant Physiol Biochem 41:495–501

Kasparovsky T, Blein JP, Mikes V (2004) Ergosterol elicits burst in tobacco cells via phospholipase A2 and protein kinase C signal pathway. Plant Physiol Biochem 42:429–435

Kauss H, Jeblick W, Domard A (1989) The degrees of polymerization and N-acetylation of chitosan determine its ability to elicit callose formation in suspension cells and protoplasts of *Catharanthus roseus*. Planta 184:8–13

Kawamura Y, Takenaka S, Hase S, Kubota M, Ichinose Y, Kanayama Y, Nakaho K, Klessig DF, Takahashi H (2009) Enhanced defense responses in Arabidopsis induced by the cell wall protein fractions from *Pythium oligandrum* require *SGT1*, *RAR1*, *NPR1* and *JAR1*. Plant Cell Physiol 50:924–934

Keinath AP, Holmes GJ, Everts KL, Egel DS, Langston JDB (2007) Evaluation of combinations of chlorothalonil with azoxystrobin, harpin, and disease forecasting for control of downy mildew and gummy stem blight on melon. Crop Protect 26:83–88

Keller H, Blein J-P, Bonnet P, Bourdon E, Panabières F, Ricci P (1994) Responses of tobacco to elicitins, proteins from *Phytophthora* spp eliciting acquired resistance. In: Daniels MJ, Downie JA, Osbourn AE (eds) Advances in molecular genetics of plant-microbe interactions, vol 3. Springer, New York, pp 327–332

Keller H, Bonnet P, Galiana E, Pruvot L, Friedrich L, Ryals J, Ricci P (1996) Salicylic acid mediates elicitin-induced systemic acquired resistance, but not necrosis in tobacco. Mol Plant-Microbe Interact 9:696–703

Keller H, Czernic P, Ponchet M, Ducrot PH, Back K, Chappell J, Ricci P, Marco Y (1998) Sesquiterpene cyclase is not a determining factor for elicitor- and pathogen-induced capsidiol accumulation in tobacco. Planta 205:467–476

Keller H, Pamboukdjian N, Ponchet M, Poupet A, Delon R, Verrier J-L, Roby D, Ricci P (1999) Pathogen-induced elicitin production in transgenic tobacco generates a hypersensitive response and nonspecific disease resistance. Plant Cell 11:223–236

Khatib M, Lafitte C, Esquerré-Tugayé M-T, Bottin A, Rickauer M (2004) The CBEL elicitor of *Phytophthora parasitica* var. *nicotianae* activates defence in *Arabidopsis thaliana* via three different signalling pathways. New Phytol 162:501–510

Kim HS, Delaney TP (2002) Overexpression of *TGA5*, which encodes a bZIP transcription factor that interacts with NIM1/NPR1, confers SAR-independent resistance in *Arabidopsis thaliana* to *Peronospora parasitica*. Plant J 32:151–163

Kim BS, Lec JY, Hwang BK (2000) *In vivo* control and *in vitro* antifungal activity of rhamnolipid B, a glycolipid antibiotic, against *Phytophthora capsici* and *Colletotrichum orbiculare*. Pest Manage Sci 56:1029–1035

Kim JG, Park BK, Yoo CH, Jeon E, Oh J, Hwang I (2003) Characterization of the *Xanthomonas axonopodis* pv *glycines* HpaG pathogenicity island. J Bacteriol 185:3155–3166

Kim S-J, Kim M-R, Bedgar DL, Moinuddin SGA, Cardenas CL, Davin LB, Kang C, Lewis NG (2004) Functional reclassification of the putative cinnamyl alcohol dehydrogenase multigene family in *Arabidopsis*. Proc Natl Acad Sci U S A 101:1455–1460

Kim KC, Fan B, Chen Z (2006) Pathogen-induced Arabidopsis WRKY7 is a transcriptional repressor and enhances susceptibility to *Pseudomonas syringae*. Plant Physiol 142:1180–1192

Kim TH, Böhmer M, Hu H, Nishimura N, Schroeder JI (2010) Guard cell signal transduction network: advances in understanding abscisic acid, CO_2, and Ca^{2+} signaling. Annu Rev Plant Biol 61:561–591

Klosterman SJ, Chen J, Choi JJ, Chinn EE, Hadwiger LA (2001) Characterization of a 20 kDa DNase elicitor from *Fusarium solani* f. sp. *phaseoli* and its expression at the onset of induced resistance in *Pisum sativum*. Mol Plant Pathol 2:147–158

Knoth C, Ringler J, Dangl JL, Eulgem T (2007) *Arabidopsis* WRKY70 is required for full *RPP4*-mediated disease resistance and basal defense against *Hyaloperonospora parasitica*. Mol Plant-Microbe Interact 20:120–128

Koehl J, Oszwald W, Kohn H, Elstner EF, Heiser I (2003) Different responses of two tobacco cultivars and their cell suspension cultures to quercinin, a novel elicitin from *Phytophthora quercina*. Plant Physiol Biochem 41:261–269

Köhle H, Young DH, Kauss H (1984) Physiological changes in suspension-cultured soybean cells elicited by treatment with chitosan. Plant Sci Lett 33:221–230

Köhle H, Jeblick W, Poten F, Blaschek W, Kauss H (1985) Chitosan elicited callose synthesis in soybean cells as a Ca^{2+}dependent process. Plant Physiol 77(544):551

Kowlaski B, Terry FJ, Herrera L, Peñalver DA (2006) Application of soluble chitosan *in vitro* and in the greenhouse to increase yield and seed quality of potato minitubers. Potato Res 49:167–176

Krause M, Durner J (2004) Harpin inactivates mitochondria in Arabidopsis suspension cells. Mol Plant-Microbe Interact 17:131–139

Kravchuk Z, Vicedo B, Flors V, Camañes G, González-Bosch C, Garcia-Agustin P (2011) Priming for JA-dependent defenses using hexanoic acid is an effective mechanism to protect Arabidopsis against *B. cinerea*. J Plant Physiol 186:359–366

Kudla J, Batistić O, Hashimoto K (2010) Calcium signals: the lead currency of plant information processing. Plant Cell 22:541–563

Küfner I, Ottmann C, Oecking C, Nürnberger T (2009) Cytolytic toxins as triggers of plant immune response. Plant Signal Behav 10:977–979

Kuhlbrandt W (2004) Biology, structure, and mechanism of P-type ATPases. Nat Rev Mol Cell Biol 5:282–295

Kunze G, Zipfel C, Robatzek S, Niehaus K, Boller T, Felix G (2004) The N terminus of bacterial elongation factor Tu elicits innate immunity in *Arabidopsis* plants. Plant Cell 16:3496–3507

Kurusu T, Saito K, Horikoshi S, Hanamata S, Negi J, Yagi C, Kitahata N, Iba K, Kuchitsu K (2013) An S-type anion channel SLAC1 is involved in cryptogein-induced ion fluxes and modulates hypersensitive responses in tobacco BY-2 cells. PLoS ONE 8(8):e70623. doi:10.1371/journal pone.0070623

Kurzawiηska H, Mazur S (2009) The evaluation of *Pythium oligandrum* and chitosan in control of *Phytophthora infestans* (Mont.) de Bary on potato plants. Folia Hortic 21(2):13–23

Kvitko BH, Ramos AR, Morello JE, Oh HS, Collmer A (2007) Identification of harpins in *Pseudomonas syringae* pv. *tomato* DC3000, which are functionally similar to HrpK1 in promoting translocation of type III secretion system effectors. J Bacteriol 189:8059–8072

Kwaaitaal M, Maintz J, Cavdar M, Panstruga R (2012) On the ligand binding profile and desensitization of plant ionotropic glutamate receptor (iGluR)-like channels functioning in MAMP-triggered Ca^{2+} influx. Plant Signal Behav 7:1373–1377

Kwon C, Yun HS (2014) Plant exocytic secretion of toxic compounds for defense. Toxical Res 30:77–81

Lafontaine PJ, Benhamou N (1996) Chitosan treatment: an emerging strategy for enhancing resistance of greenhouse tomato plants to infection by *Fusarium oxysporum* f. sp. *lycopersici*. Biocontrol Sci Technol 6:111–124

Lamotte O, Gould K, Lecourieux D, Sequeira-Legrand A, Lebrun-Garcia A, Durner J, Pugin A, Wendehenne D (2004) Analysis of nitric oxide signaling functions in tobacco cells challenged by the elicitor cryptogein. Plant Physiol 135:516–529

Laquitaine L, Gomès E, Francois J, Marchive C, Pascal S, Hamdi S, Atanassova R, Delrot S, Coutos-Thévenot P (2006) Molecular basis of ergosterol-induced protection of grape against *Botrytis cinerea*: Induction of type 1 LTP promoter activity, WRKY, and stilbene synthase gene expression. Mol Plant-Microbe Interact 19:1103–1112

Lawton KA, Potter SL, Uknes S, Ryals J (1994) Acquired resistance signal transduction in *Arabidopsis* is ethylene independent. Plant Cell 6:581–588

Leba L-J, Cheval C, Ortiz-Martin I, Ranty B, Beuzón CR, Galaud J-P, Aldon D (2012) CML9, an Arabidopsis calmodulin-like protein, contributes to plant innate immunity through a flagellin-dependent signalling pathway. Plant J 71:976–989

Leborgne-Castel N, Lherminier J, Der C, Fromentin J, Houot V, Simon-Plas F (2008) The plant defense elicitor cryptogein stimulates clathrin-mediated endocytosis correlated with reactive oxygen species production in Bright Yellow-2 tobacco cells. Plant Physiol 146:1255–1266

Lebrun-Garcia A, Ouaked F, Chiltz A, Pugin A (1998) Activation of MAPK homologues by elicitors in tobacco cells. Plant J 15:773–781

Lebrun-Garcia A, Chiltz A, Gout E, Bligny R, Pugin A (2002) Questioning the role of salicylic acid and cytosolic acidification in mitogen activated protein kinase (MAPK) activation induced by cryptogein in tobacco cells. Planta 214:792–797

Lecourieux-Ouaked D, Pugin A, Lebrun-Garcia A (2000) Phosphoproteins involved in the signal transduction of cryptogein, an elicitor of defense reactions in tobacco. Mol Plant-Microbe Interact 13:821–829

Lee J, Klessig DF, Nurnberger T (2001a) A harpin binding site in tobacco plasma membranes mediates activation of the pathogenesis-related gene *HIN1* independent of extracellular calcium but dependent on mitogen-activated protein kinase activity. Plant Cell 13:1079–1093

Lee J, Klusener B, Tsiamis G, Stevens C, Neyt C, Tampakaki AP, Panopoulos NJ, Nöller J, Weiler EW, Cornelis GR, Mansfield JW, Nürnberger T (2001b) $HrpZ_{Psph}$ from the plant pathogen *Pseudomonas syringae* pv. *phaseolicola* binds to lipid bilayers and forms an ion-conducting pore in vitro. Proc Natl Acad Sci U S A 98:289–294

Lee S-W, Han S-W, Bartley LE, Ronald PC (2006) Unique characteristics of *Xanthomonas oryzae* pv. *oryzae* AvrXa21 and implications for plant innate immunity. Proc Natl Acad Sci U S A 103:18395–18400

Lee SW, Han S, Sririyanum M, Park C, Seo Y, Ronald PC (2009) A type 1-secreted, sulfated peptide triggers XA21-mediated innate immunity. Science 326:850–853

Lee Y, Rubio MC, Alassimone J, Geldner N (2013) A mechanism for localized lignin deposition in the endodermis. Cell 153:402–412

Lehtonen NT, Akita M, Frank W, Reski R, Valkonen JPT (2012) Involvement of a class III peroxidase and the mitochondrial protein TSPO in oxidative burst upon treatment of moss plants with a fungal elicitor. Mol Plant Microbe Interact 25:363–371

Leon-kloosterziel KM, Verhagen BWM, Keurentjes JJB, VanPelt JA, Rep M, VanLoon LC, Pieterse CMJ (2005) Colonization of the Arabidopsis rhizosphere by fluorescent *Pseudomonas* spp. activates a root-specific, ethylene-responsive *PR-5* gene in the vascular bundle. Plant Mol Biol 57:731–748

Lherminier J, Benhamou N, Larrue, Milat ML, Boudon-Padieu E, Nicole M, Blein J-P (2003) Cytological characterization of elicitin-induced protection in tobacco plants infected by *Phytophthora parasitica* or phytoplasma. Phytopathology 93:1308–1319

Lherminier J, Elmayan T, Fromentin J, Elaraqui KT, Vesa S, Morel J, Verrier J-L, Cailleteau B, Blein J-P, Simon-plas F (2009) NADPH oxidase-mediated reactive oxygen species production; subcellular localization and reassessment of its role in plant defense. Mol Plant Microbe Interact 22:868–881

Li R, Fan Y (1999) Reduction of lesion growth rate of late blight plant disease in transgenic potato expressing harpin protein. Sci China (Ser.C) 40:96–101

Li J, Brader G, Palva ET (2004a) The WRKY70 transcription factor: a node of convergence for jasmonate-mediated and salicylate-mediated signals in plant defense. Plant Cell 16:319–331

Li P, Lu X, Shao M, Long J, Wang J (2004b) Genetic diversity of harpins from *Xanthomonas oryzae* and their activity to induce hypersensitive response and disease resistance in tobacco. Sci China C Life Sci 47:461–469

Li M, Shao M, Lu X-Z, Wang J-S (2005a) Biological activities of purified $harpin_{Xoo}$ and $harpin_{Xoo}$ detection in transgenic plants using its polyclonal antibody. Acta Biochim Biophys Sinica 37:713–718

Li X, Lin H, Zhang W, Zou Y, Zhang J, Tang X, Zhou JM (2005b) Flagellin induces innate immunity in nonhost interactions that is suppressed by *Pseudomonas syringae* effectors. Proc Natl Acad Sci U S A 102:12990–12995

Li J, Brader G, Kariola T, Tapio Palva E (2006) WRKY70 modulates the selection of signaling pathways in plant defense. Plant J 46:477–491

Li Y-C, Sun X-J, Bi Y, Ge Y-H, Wang Y (2009) Antifungal activity of chitosan on *Fusarium sulphureum* in relation to dry rot of potato tuber. Agr Sci China 8:597–604

Li JG, Liu HX, Cao J, Chen LF, Gu C, Allen C, Guo JH (2010) PopW of *Ralstonia solanacearum*, a new two-domain harpin targeting the plant cell wall. Mol Plant Pathol 11:371–381

Li JG, Cao J, Sun FF, Niu DD, Yan F, Liu HX, Guo JH (2011) Control of *Tobacco mosaic virus* by PopW as a result of induced resistance in tobacco under greenhouse and field conditions. Phytopathology 101:1202–1208

Li W, Shao M, Zhong W, Yang J, Okada K, Yamane H, Zhang L, Wang G, Wang D, Xiao S, Chang S, Qian G, Liu F (2012) Ectopic expression of *Hrf1* enhances bacterial resistance via regulation of diterpene phytoalexins, silicon and reactive oxygen species burst in rice. PLoS ONE 7(9):e43914

Li X, Han B, Xu M, Han L, Zhao Y, Liu Z, Dong H, Zhang C (2014a) Plant growth enhancement and associated physiological responses are coregulated by ethylene and gibberellin in response to harpin protein Hpa1. Planta 239:831–846

Li X, Han B, Zhao Y, You Z, Dong H, Zhang C (2014b) Hpa1 harpin needs nitroxyl terminus to promote vegetative growth and leaf photosynthesis in *Arabidopsis*. J Biosci 39:127–137

Lin W, Hu X, Zhang W, Rogers WJ, Cai W (2005) Hydrogen peroxide mediates defence responses induced by chitosans of different molecular weights in rice. J Plant Physiol 162:937–944

Lindermayr C, Saalbach G, Durner J (2005) Proteomic identification of *S*-nitrosylated proteins in *Arabidopsis*. Plant Physiol 137:921–930

Lippok B, Birkenbihl RP, Rivory G, Brummer J, Schmelzer EL, Somssich IE (2007) Expression of *AtWRKY33* encoding a pathogen- or PAMP-responsive WRKY transcription factor is regulated by a composite DNA motif containing W box elements. Mol Plant-Microbe Interact 20:420–429

Liu Y, Jin H, Yang K, Kim C, Baker B, Zhang S (2003) Interaction between two mitogen-activated protein kinases during tobacco defense signaling. Plant J 34:149–160

Liu F, Liu H, Jia Q, Wu X, Guo X, Zhang S, Song F, Dong H (2006) The internal glycine-rich motif and cysteine suppress several effects of $HpaG_{Xooc}$ in plants. Phytopathology 96:1052–1059

Liu J, Tian S, Meng X, Yu Y (2007) Effects of chitosan on control of postharvest diseases and physiological responses of tomato fruit. Postharvest Biol Technol 44:300–306

Liu P-P, von Dahl CC, Klessig DF (2011a) The extent to which methyl salicylate is required for signaling systemic acquired resistance is dependent on exposure to light after infection. Plant Physiol 157:2216–2226

Liu P-P, von Dahl CC, Park S-W, Klessig DF (2011b) Interconnection between methyl salicylate and lipid-based long-distance signaling during the development of systemic acquired resistance in *Arabidopsis* and tobacco. Plant Physiol 155:1762–1768

Lizasa E, Mitsutomi M, Nagano Y (2010) Direct binding of a plant LysM receptor-like kinase LysM RLK1/CERK1 to chitin in vitro. J Biol Chem 285:2996–3004

Lloyd SR, Schoonbeek H, Trick M, Zipfel C, Ridout CJ (2014) Methods to study PAMP-triggered immunity in *Brassica* species. Mol Plant Microbe Interact 27:286–295

Lochman J, Mikes V (2006) Ergosterol treatment leads to the expression of a specific set of defence-related genes in tobacco. Plant Mol Biol 62:43–51

Lovato FA, Inoue-Nagata AK, Nagata T, de Avila AC, Pereira LAR, Resende RO (2008) The N protein of *Tomato spotted wilt virus* (TSWV) is associated with the induction of programmed cell death (PCD) in *Capsicum chinense* plants, a hypersensitive host to TSWV infection. Virus Res 137:245–252

Love AJ, Yun B-W, Laval V, Loake GJ, Milner JL (2005) *Cauliflower mosaic virus*, a compatible pathogen of Arabidopsis, engages three distinct defense-signaling pathways and activates rapid systemic generation of reactive oxygen species. Plant Physiol 139:935–948

Lumsden RD, Walter JF, Baker CP (1996) Development of *Gliocladium virens* for damping-off disease control. Can J Plant Pathol 18:463–468

Luna E, Bruce TJA, Roberts MR, Flors V, Ton J (2011) Next generation systemic acquired resistance. Plant Physiol 158:844–853

Ma Y, Walker RK, Zhao Y, Berkowitz GA (2012) Linking ligand perception by PEPR pattern recognition receptors to cytosolic Ca^{2+} elevation and downstream immune signaling in plants. Proc Nat Acad Sci USA 109:195852–19857

Ma Y, Han C, Chen J, Li H, He K, Liu A, Li D (2015) Fungal cellulose is an elicitor but its enzymatic activity is not required for its elicitor activity. Mol Plant Pathol 16:14–26. doi:10.1111/mpp.12156

Maathuis FJM, Ichida AM, Sanders D, Schroeder JI (1997) Roles of higher K^+ channels. Plant Physiol 114:1141–1149

Macho AP, Zipfel C (2014) Plant PRRs and the activation of innate immune signalling. Mol Cell 54:263–272

Madsen AM, de Neergaard E (1999) Interaction between the mycoparasite *Pythium oligandrum and sclerotia of the plant pathogen Sclerotinia sclerotiorum.* Eur J Plant Pathol 105:761–768

Maghuly F, Borroto-Fernandez EG, Khan MA, Herndl A, Marzban G, Laimer M (2009) Expression of calmodulin and lipid transfer protein genes in *Prunus incisa* × *serrula* under different stress conditions. Tree Physiol 29:437–444

Malinovsky FG, Fangel JU, Willats GT (2014) The role of the cell wall in plant immunity. Front Plant Sci 5:178. doi:10.3369/fpls.2014.00178

Malnoy M, Venisse JS, Chevreau E (2005) Expression of a bacterial effector, harpin N, causes increased resistance to fire blight in *Pyrus communis*. Tree Genet Genomes 1:41–49

Mandal S, Mitra A (2007) Reinforcement of cell wall in roots of *Lycopersicon esculentum* through induction of phenolic compounds and lignin by elicitors. Physiol Mol Plant Pathol 71:201–209

Marten H, Konrad KR, Dietrich P, Roelfsema RG, Hedrich R (2007) Ca^{2+}-dependent and –independent abscisic acid activation of plasma membrane anion channels in guard cells of *Nicotiana tabacum*. Plant Physiol 143:28–37

Masunaka A, Sekiguchi H, Takahashi H, Takenaka S (2010) Distribution and expression of elicitin-like protein genes of the biocontrol agent *Pythium oligandrum*. J Phytopathol 158:417–426

Mateo A, Muhlenbock P, Rustérucci C, Chang CC, Miszalski Z, Karpinska B, Parker JE, Mullineaux PM, Karpinski S (2004) *LESION SIMULATING DISEASE 1* is required for acclimation to conditions that promote excess excitation energy. Plant Physiol 136:2818–2830

Matsumoto K, Sawada H, Matsumoto K, Hamada H, Yoshimoto E, Ito T, Takeuchi S, Tsuda S, Suzuki K, Kobayashi K, Kiba A, Okuno T, Hikichi Y (2008) The coat protein gene of tobamovirus P_0 pathotype is a determinant for activation of temperature-insensitive L^{1a}-gene-mediated resistance in *Capsicum* plants. Arch Virol 155:645–650

Mattinen L, Tshulkina M, Mae A, Pihonen M (2004) Identification and characterization of Nip necrosis-inducing virulence protein of *Erwinia carotovora* subsp *carotovora*. Mol Plant-Microbe Interact 17:1366–1375

Meindl T, Boller T, Felix G (2000) The bacterial elicitor flagellin activates its receptor in tomato cells according to the address-message concept. Plant Cell 12:1783–1794

Miao W, Wang J (2013) Genetic transformation of cotton with a harpin-encoding gene hpa_{Xoo} confers defense response against *Verticillium dahliae* Kieb. Methods Mol Biol 958:223–246

Miao W, Wang X, Li M, Song C, Wang Y, Hu D, Wang J (2010a) Genetic transformation of cotton with a harpin-encoding gene hpa_{Xoo} confers an enhanced defense response against different pathogens through a priming mechanism. BMC Plant Biol 10:67

Miao W, Wang X, Song C, Wang Y, Ren Y, Wang J (2010b) Transcriptome analysis of $Hpa1_{Xoo}$ transformed cotton revealed constitutive expression of genes in multiple signalling pathways related to disease resistance. J Expt Bot 61:4263–4275

Mikes V, Milat ML, Ponchet M, Panabieres F, Ricci P, Blein JP (1998) Elicitins, proteinaceous elicitors of plant defense, are a new class of sterol carrier proteins. Biochem Biophys Res Commun 245:133–139

Milat ML, Ricci P, Blein JP (1991) Capsidiol and ethylene production by tobacco cells in response to cryptogein, an elicitor from *Phytophthora cryptogea*. Phytochemistry 30:2171–2173

Mishina TE, Zeier J (2007) Pathogen-associated molecular pattern recognition rather than development of tissue necrosis contributes to bacterial induction of systemic acquired resistance in Arabidopsis. Plant J 50:500–513

Miya A, Albert P, Shinya T, Desaki Y, Ichimura K, Shirasu K, Narusaka Y, Kawakami N, Kaku H, Shibuya N (2007) CERK1 a LysM receptor kinase is essential for chitin elicitor signaling in *Arabidopsis*. Proc Natl Acad Sci U S A 104:19613–19618

Mohamed N, Lherminier J, Farmer M-J, Fromentin J, Béno V, Houot M-L, Milat M-L, Blein J-P (2007) Defense responses in grapevine leaves against *Botrytis cinerea* induced by application of a *Pythium oligandrum* strain or its elicitin, oligandrin, to roots. Phytopathology 97:611–620

Molinaro A, Newman M-A, Lanzetta R, Parrilli M (2009) The structures of lipopolysaccharides from plant-associated gram-negative bacteria. Eur J Org Chem 34:5887–5896

Molloy C, Cheah L-H, Koolaard JP (2004) Induced resistance against *Sclerotinia sclerotiorum* in carrots treated with enzymatically hydrolysed chitosan. Postharvest Biol Technol 33:61–65

Moret A, Muñoz Z, Garcės S (2009) Control of powdery mildew on cucumber cotyledons by chitosan. J Plant Pathol:91. doi 10.4454/pp.v-9112.967

Mori IC, Schroeder JI (2004) Reactive oxygen species activation of plant Ca^{2+} channels. A signaling mechanism in polar growth, hormone transduction, stress signaling, and hypothetically mechanotransduction. Plant Physiol 135:702–708

Motteram J, Kūfner I, Deller S, Brunner F, Hammond-Kosack KE, Nurnberger T, Rudd JJ (2009) Molecular characterization and functional analysis of *MgNLP*, the sole NPP1 domain-containing protein, from the fungal wheat leaf pathogen *Mycosphaerella graminicola*. Mol Plant-Microbe Interact 22:790–799

Naito K, Taguchi F, Suzuki T, Inagaki Y, Toyoda K, Shiraishi TT, Ichinose Y (2008) Amino acid sequence of bacterial microbe-associated molecular pattern flg22 is required for virulence. Mol Plant-Microbe Interact 21:1165–1174

Nakashima A, Chen L, Thao NP, Fujiwara M, Wong HL, Kuwano M, Umemura K, Shirasu K, Kawasaki T, Shimamoto K (2008) RACK1 functions in rice immunity by interacting with the Rac1 immune complex. Plant Cell 20:2265–2279

Návarová H, Bernsdorff F, Döring A-C, Zeier J (2012) Pipecolic acid, an endogenous mediator of defense amplification and priming, is a critical regulator of inducible plant immunity. Plant Cell 24:5123–5141

Nawrath C, Heck S, Parinthawong N, Metraux JP (2002) EDS5, an essential component of salicylic acid-dependent signaling for disease resistance in Arabidopsis, is a member of MATE transporter family. Plant Cell 14:275–286

Ndamukong I, Abdallat AA, Thurow C, Fode B, Zander M, Weigel R, Gatz C (2007) SA-inducible *Arabidopsis* glutaredoxin interacts with TGA factors and suppresses JA-responsive *PDF1.2* transcription. Plant J 50:128–139

Neill S, Bright J, Desikan R, Hancock J, Harrison J, Wilson I (2008) Nitric oxide evolution and perception. J Expt Bot 59:25–35

Newman M-A, Dow JM, Molinaro A, Parrilli M (2007) Priming, induction and modulation of plant defense responses by bacterial lipopysaccharides. Plant Immun 13:69–84

Nicaise V, Roux M, Zipfel C (2009) Recent advances in PAMP-triggered immunity against bacteria: Pattern recognition receptors watch over and raise the alarm. Plant Physiol 150:1638–1647

Noirot E, Der C, Lherminier J, Robert F, Moricova P. Kiêu K, Leborgne-Castel N, Simon-Plas F, Bouhidel K (2014) Dynamic changes in the subcellular distribution of the tobacco ROS-producing enzyme RBOHD in response to the oomycete elicitor cryptogein. J Exp Bot 65:5011–5022

Nürnberger T, Nennstiel D, Jabs T, Sacks WR, Hahlbrock K, Scheel D (1994) High affinity binding of a fungal oligopeptide elicitor to parsley plasma membranes triggers multiple defense responses. Cell 78:449–460

Nürnberger T, Brunner F, Kemmerling B, Piater L (2004) Innate immunity in plants and animals: striking similarities and obvious differences. Immunol Rev 198:249–266

Obradovic A, Jones JB, Momol MT, Olson SM, Jackson LE, Balogh B, Guven K, Iriarte FB (2005) Integration of biological control agents and systemic acquired resistance inducers against bacterial spot on tomato. Plant Dis 89:712–716

Oh CS, Beer SV (2007) AtHIPM, an ortholog of the apple HrpN-interacting protein, is a negative regulator of plant growth and mediates the growth-enhancing effect of HrpN in Arabidopsis. Plant Physiol 145:426–436

Orlandi WE, Hutcheson SW, Baker CJ (1992) Early physiological responses associated with race-specific recognition in soybean leaf tissue and cell suspensions treated with *Pseudomonas syringae* pv. *glycinea*. Physiol Mol Plant Pathol 40:173–180

Osbourn A (1996) Saponins and plant defence – a soap story. Trends Plant Sci 1:4–9

Ottmann C, Luberacki B, Kūfner I, Koch W, Brunner F, Weyand M, Mattinen L, Pirhonen M, Anderluh G, Ulrich Seitz H, Nurnberger T, Oecking C (2009) A common toxin fold mediates microbial attack and plant defense. Proc Natl Acad Sci U S A 106:10359–10364

Padgett HS, Watanabe Y, Beachey RN (1997) Identification of the TMV replicase sequence that activates the *N* gene-mediated hypersensitive response. Mol Plant-Microbe Interact 10:709–715

Palmieri MC, Sell S, Huang X, Scherf M, Werner T, Durner J, Lindermayr C (2008) Nitric oxide-responsive genes and promoters in *Arabidopsis thaliana*: a bioinformatics approach. J Exp Bot 59:177–180

Panabières F, Marais A, Le Berre J-Y, Penot I, Fournier D, Ricci P (1995) Characterization of a gene cluster of *Phytophthora cryptogea* which codes for elicitins, proteins inducing a hypersensitive-like response in tobacco. Mol Plant-Microbe Interact 8:996–1003

Pandey A-K, Huang GM-J, H-E YM-K, Zeng J, Feng T-Y (2005) Expression of the hypersensitive response-assisting protein in *Arabidopsis* results in harpin-dependent cell death in response to *Erwinia carotovora*. Plant Mol Biol 59:771–780

Papadoulou K, Melton RE, Leggertt M, Daniels MJ, Osbourn AE (1999) Compromised disease resistance in saponin-deficient plants. Proc Natl Acad Sci U S A 96:12923–12928

Parchmann S, Gundlach H, Mueller MJ (1997) Induction of 12-oxo-phytodienoic acid in wounded plants and elicited plant cell cultures. Plant Physiol 115:1057–1064

Park J, Choi HJ, Lee S, Lee T, Yang Z, Lee Y (2000) Rac-related GTP-binding protein in elicitor-induced reactive oxygen generation by suspension-cultured soybean cells. Plant Physiol 124:725–732

Park C-J, Bart R, Chern M, Canlas PE, Bai W, Ronald PC (2010a) Overexpression of the endoplasmic reticulum chaperone BiP3 regulates XA21-mediated innate immunity in rice. PLoS ONE 5(2):e9262

Park C-J, Han S-W, Chen X, Ronald PC (2010b) Elucidation of XA21-mediated innate immunity. Cell Microbiol 12:1017–1025

Pavli OI, Kelaidi GI, Tampakaki AP, Skaracis GN (2011) The *hrpZ* gene of *Pseudomonas syringae* pv. *phaseolicola* enhances resistance to rhizomania disease in transgenic *Nicotiana benthamiana* and sugar beet. PLoS ONE 6:e17306

Pavli OI, Tampakaki AP, Skaracis GN (2012) High level resistance against rhizomania disease by simultaneously integrating two distinct defense mechanisms. PLoS ONE 7(12):e51414

Pearce RB, Ride JP (1982) Chitin and related compounds as elicitors of lignification response in wounded wheat leaves. Physiol Plant Pathol 20:119–123

Pei Z-M, Murata Y, Benning G, Thomine S, Klüsener B, Allen GJ, Grill E, Schroeder JI (2000) Calcium channels activated by hydrogen peroxide mediate abscisic acid signaling in guard cells. Nature 406:731–734

Pemberton CL, Whitehead NA, Sebaihia M, Bell KS, Hyman IJ, Harris SJ, Matlin AJ, Robson ND, Birch PRJ, Carr JP, Toth IK, Salmond GP (2005) Novel quorum-sensing-controlled genes in *Erwinia carotovora* subsp. *carotovora*: Identification of a fungal elicitor homologue in a soft-rotting bacterium. Mol Plant-Microbe Interact 18:343–353

Peng J-L, Dong H-S, Dong H-P, Delaney TP, Bonasera BM, Beer SV (2003) Harpin-elicited hypersensitive cell death, and pathogen resistance requires the *NDR1* and *EDS1* genes. Physiol Mol Plant Pathol 62:317–326

Peng J-L, Bao Z, Ren H, Wang J, Dong H (2004a) HarpinXoo and its functional domains activate pathogen-inducible plant promoters in *Arabidopsis*. Acta Bot Sin 46:1083–1090

Peng J-L, Bao Z-L, Ren H-Y, Wang J-S, Dong HS (2004b) Expression of $harpin_{Xoo}$ in transgenic tobacco induces pathogen defense in the absence of hypersensitive cell death. Phytopathology 94:1048–1055

Penninckx IAMA, Thomma BPHJ, Buchala A, Metraux JP, Broekaert WF (1998) Concomitant activation of jasmonate and ethylene response pathways is required for induction of a plant defensin gene in Arabidopsis. Plant Cell 10:2103–2113

Perchepied L, Balagué C, Riou C, Claudel-Renard C, Riviére N, Grezes-Besset B, Roby D (2010) Nitric oxide participates in the complex interplay of defense-related signaling pathways controlling disease resistance to *Sclerotinia sclerotiorum* in *Arabidopsis thaliana*. Mol Plant-Microbe Interact 23:846–860

Pertot I, Fiamingo F, Amsalem L, Maymom M, Freeman S, Gobbin D, Elad Y (2007) Sensitivity of two *Podosphaera aphanis* populations to disease control agents. J Plant Pathol 89:85–96

Petutschnig EK, Jones AME, Serazetdinova L, Lipka U, Lipka V (2010) The Lysin motif receptor-like kinase (LysM-RLK) is a major chitin binding protein in *Arabidopsis thaliana* and subject to chitin-induced phosphorylation. J Biol Chem 285:28902–28911

Pfund C, Tans-Kersten J, Dunning FM, Alonso JM, Ecker JR, Allen C, Bent AF (2004) Flagellin is not a major defense elicitor in *Ralstonia solanacearum* cells or extracts applied to *Arabidopsis thaliana*. Mol Plant-Microbe Interact 17:696–706

Picard K, Ponchet M, Blein J-P, Rey P, Tirillt Y, Benhamou N (2000) Oligandrin, a proteinaceous molecule produced by the mycoparasite, *Pythium oligandrum*, induces resistance to *Phytophthora parasitica* infection in tomato plants. Plant Physiol 124:379–395

Pike SM, Adam AL, Pu X-A, Hoyos ME, Laby R, Beer SV, Novacky A (1998) Effects of *Erwinia amylovora* harpin on tobacco leaf cell membranes are related to leaf necrosis and electrolyte leakage and distinct from perturbations caused by *E. amylovora*. Physiol Mol Plant Pathol 53:39–60

Planchet E, Gupta KJ, Sonoda M, Zeier J, Kaiser WM (2005) Nitric oxide emission from tobacco leaves and cell suspensions: rate limiting factors and evidence for the involvement of mitochondrial electron transport. Plant J 41:732–743

Planchet E, Sonoda M, Zeier J, Kaiser WM (2006) Nitric oxide (NO) as an intermediate in cryptogein-induced hypersensitive response: a critical evaluation. Plant Cell Environ 29:59–69

Ponchet M, Panabieres F, Milat ML, Mikes V, Montillet JL, Suty L, Triantaphylides C, Tirilly Y, Blein JP (1999) Are elicitins cryptograms in plant-oomycete communications? Cell Mol Life Sci 56:1020–1047

Po-Wen C, Singh P, Zimmerli L (2013) Priming of the Arabidopsis pattern-triggered immunity response upon infection by nerotrophic *Pectobacterium carotovorum* bacteria. Mol Plant Pathol 14:58–70

Prapagdee B, Kotchadat K, Kumsopa A, Visarathnonth N (2007) The role of chitosan in protection of soybean from sudden death syndrome caused by *Fusarium solani* f. sp. *glycines*. Bioresour Technol 98:1353–1358

Proietti S, Giangrande C, Amoresano A, Pucci P, Molarinaro A, Bertini L, Caporale C, Caruso C (2014) *Xanthomonas campestris* lipopolysaccharides trigger innate immunity and oxidative burst in *Arabidopsis*. Plant Physiol Biochem 85C:51–62

Pugin A, Frachisse J-M, Tavernier E, Bligny R, Gout E, Douce R, Guern J (1997) Early events induced by the elicitor cryptogein in tobacco cells: involvement of a plasma membrane NADPH oxidase and activation of glycolysis and the pentose phosphate pathway. Plant Cell 9:2077–2091

Qiu D, Clayton K, Wei Z-M (2001) Bioassay development for Messenger® activity. Phytopathology 91:S74

Qiu D, Mao J, Yang X, Zeng H (2009) Expression of elicitor-encoding gene from *Magnaporthe grisea* enhances resistance against blast disease in transgenic rice. Plant Cell Rep 28:925–933

Qu X, Hall BP, Gao Z, Schaller GE (2007) A strong constitutive ethylene-response phenotype conferred on Arabidopsis plants containing null mutations in the ethylene receptors *ETR1* and *ERS1*. BMC Plant Biol 7:3. doi:10.1186/1471-2229-7-3

Quang Z, Li X, Huang L, Hong Y, Zhang Y, Zhang H, Li D, Song F (2015) Elicitin-like proteins Oli-D1 and Oli-D2 from *Pythium oligandrum* trigger hypersensitive response in *Nicotiana benthamiana* and induce resistance against *Botrytis cinerea* in tomato. Mol Plant Pathol 16:238–250. doi:10.1111/mpp.12176

Qutob D, Kamoun S, Gijzen M (2002) Expression of a *Phytophthora sojae* necrosis-inducing protein occurs during transition from biotrophy to necrotrophy. Plant J 32:361–373

Qutob D, Huitema E, Gijzen M, Kamoun S (2003) Variation in structure and activity among elicitins from *Phytophthora sojae*. Mol Plant Pathol 4:119–124

Qutob D, Kemmerling B, Brunner F, Küfner I, Engelhardt S, Gust AA, Luberacki B, Seitz HU, Stahl D, Rauhut T, Glawischnig E, Schween G, Lacombe B, Watanabe N, Lam E, Schlichting R, Scheel D, Nau K, Dodt G, Hubert D, Gijzen M, Nürnberger T (2006) Phytotoxicity and innate immune responses induced by NEP1-like proteins. Plant Cell 18:3721–3744

Raacke IC, von Rad U, Mueller MJ, Berger S (2006) Yeast increases resistance in *Arabidopsis* against *Pseudomonas syringae* and *Botrytis cinerea* by salicylic acid-dependent as well as-independent mechanisms, Mol Plant Microbe Interact 19:1138–1146

Radi R (2004) Nitric oxide, oxidants, and protein tyrosine nitration. Proc Natl Acad Sci USA 101:4003–4008

Rahman MA, Mahmud TMM, Kadir J, Rahman RA, Begum MM (2009) Enhancing the efficacy of *Burkholderia cepacia* B23 with calcium chloride and chitosan to control anthracnose of papaya during storage. Plant Pathol J 25:361–368

Rakwal R, Tamogani S, Agrawal GK, Iwahashi H (2002) Octadecanoid signaling component 'burst' in rice (*Oryza sativa* L.) seedlings leaves upon wounding by cut and treatment with fungal elicitor chitosan. Biochem Biophys Res Commun 295:1041–1045

Rappussi MCC, Pascholati SF, Benato EA, Cia P (2009) Chitosan reduces infection by *Guignardia citricarpa* in postharvest 'Valencia' oranges. Brazilian Arch Biol Technol 52:513–521

Reboutier D, Frankart C, Briand J, Biligui B, Laroche S, Rona J-P, Barny M-A, Bouteau F (2007) The $HrpN_{ea}$ harpin from *Erwinia amylovora* triggers differential responses on the nonhost *Arabidopsis thaliana* cells and on the host apple cells. Mol Plant-Microbe Interact 20:94–100

Reddy AS (2001) Calcium: silver bullet in signaling. Plant Sci 160:381–404

Reddy MV, Arul J, Angers P, Couture L (1999) Chitosan treatment of wheat seeds induces resistance to *Fusarium graminearum* and improves seed quality. J Agric Food Chem 47:1208–1216

Reddy MVB, Angers P, Castaigne F, Arul J (2000) Chitosan effects on blackmold rot and pathogenic factors produced by *Alternaria alternata* in postharvest tomatoes. J Am Soc Hortic Sci 125:742–747

Reglinski T, Newton AC, Lyon GD (1994a) Assessment of the ability of yeast-derived elicitors to control powdery mildew in the field. J Plant Dis Prot 101:1–10

Reglinski T, Newton AC, Lyon GD (1994b) Induction of resistance mechanisms in barley by yeast-derived elicitors. Ann Appl Biol 124:509–517

Rekanovic E, Milijasevic S, Todorovic B, Potocnik I (2007) Possibilities of biological and chemical control of Verticillium wilt in pepper. Phytoparasitica 35:436–441

Ren H, Gu G, Long J, Yin Q, Wu T, Song T, Zhang S, Chen Z, Dong H (2006a) Combinative effects of a bacterial type-III effector and a biocontrol bacterium on rice growth and disease resistance. J BioSci 31:617–627

Ren H, Song T, Wu T, Sun L, Liu Y, Yang F, Chen Z, Dong H (2006b) Effects of a biocontrol bacterium on growth and defence of transgenic rice plants expressing a bacterial type-III effector. Ann Microbiol 56:281–287

Ricci P, Bonnet P, Huet JC, Sallantin M, BeauvaisCante F, Bruneteau M, Billard V, Michel G, Pernollet JC (1989) Structure and activity of proteins from pathogenic fungi *Phytophthora* eliciting necrosis and acquired resistance in tobacco. Eur J Biochem 183(555):563

Rizo J, Südhof TC (1998) C2-domains, structure and function of a universal Ca^{2+}-binding domain. J Biol Chem 273:15879–15882

Robatzek S, Chinchilla D, Boller T (2006) Ligand-induced endocytosis of the pattern recognition receptor FLS2 in *Arabidopsis*. Genes Dev 20:537–542

Robatzek S, Bittel P, Chinchilla D, Köchner P, Felix G, Shiu SH, Boller T (2007) Molecular identification and characterization of the tomato flagellin receptor LeFLS2, an orthologue of Arabidopsis FLS2 exhibiting characteristically different perception specificities. Plant Mol Biol 64:539–547

Rodrigues FA, McNally DJ, Datnoff LE, Jones JB, Labbe C, Benhamou N, Menzies JG, Bélanger C (2004) Silicon enhances the accumulation of diterpenoid phytoalexins in rice: a potential mechanism for blast resistance. Phytopathology 94:177–183

Romanazzi G, Nigro F, Ippolito AD, Venere D, Salerno M (2002) Effects of pre- and post-harvest chitosan treatments to control storage grey mold of table grapes. J Food Sci 67:1862–1867

Romanazzi G, Mlikota Gabler T, Smilanick JL (2006) Preharvest chitosan and postharvest UV irradiation treatments suppress gray mold of table grapes. Plant Dis 90:445–450

Romero-Puertas MC, Laxa M, Matté A, Zaninotto F, Finkemeier I, Jones AME, Perazzolli M, Vandelle E, Dietz K-J, Delledonne M (2007) *S*-nitrosylation of peroxiredoxin II E promotes peroxynitrite-mediated tyrosine nitration. Plant Cell 19:4120–4130

Romero-Puertas MC, Campostrini NN, Matté A, Righetti PG, Perazzolli M, Zolla L, Roepstorff P, Delledonne M (2008) Proteomic analysis of *S*-nitrosylated proteins in *Arabidopsis thaliana* undergoing hypersensitive response. Proteomics 8:1459–1469

Rosli HG, Zheng Y, Pombo MA, Zhong S, Bombarely A, Fei Z, Collmer A, Martin GB (2013) Transcriptomics-based screen for genes induced by flagellin and repressed by pathogen effectors identifies a cell wall-associated kinase involved in plant immunity. Genome Biol 14:R139

Rossard S, Luini E, Pérault JM, Bonmort J, Roblin G (2006) Early changes in membrane permeability, production of oxidative burst and modification of PAL activity induced by ergosterol in cotyledons of *Mimosa pudica*. J Exp Bot 57:1245–1252

Rossard S, Roblin G, Atanassova R (2010) Ergosterol triggers characteristic elicitation steps in *Beta vulgaris* leaf tissues. J Exp Bot 61:1807–1816

Rustérucci C, Aviv DH, Holt BF, Dangl JL, Parker JE (2001) The disease resistance signaling components *EDS1* and *PAD4* are essential regulators of the cell death pathway controlled by *LSD1* in *Arabidopsis*. Plant Cell 13:2211–2224

Sagi M, Fluhr R (2006) Production of reactive oxygen species by plant NADPH oxidases. Plant Physiol 141:336–340

Saito S, Yamamoto-Katou A, Yoshioka H, Doke N, Kawakita K (2006) Peroxynitrite generation and tyrosine nitration in defense responses in tobacco BY-2 cells. Plant Cell Physiol 47:689–697

Sanchez L, Courteaux B, Hubert J, Kauffmann S, Renault J-H, Clément C, Baillieul F, Dorey S (2012) Rhamnolipids elicit defense responses and induce disease resistance against biotrophic, hemibiotrophic, and necrotrophic pathogens that require different signaling pathways in Arabidopsis and highlight a central role for salicylic acid. Plant Physiol 160:1530–1641

Sanchez L, Mazeyrat-Gourbeyre F, Clèment C, Baillieul F, Dorey S (2013) Rhamnolipids elicit plant defence responses and enhance resistance against *Botrytis cinerea*. Acta Hortic (ISHS) 1009:105–109

Sanders D, Pelloux J, Brownlee C, Harper JF (2002) Calcium at the cross-roads of signalling. Plant Cell 14:S401–S417

Sang S, Li X, Gao R, You Z, Lü B, Liu P, Ma Q, Dong H (2012) Apoplastic and cytoplasmic location of harpin protein $Hpa1_{Xoo}$ plays different roles in H_2O_2 generation and pathogen resistance in *Arabidopsis*. Plant Mol Biol 79:375–391

Sasabe M, Takeuchi K, Kamoun S, Ichinose Y, Govers F, Toyoda K, Shiraishi T, Yamada T (2000) Independent pathways leading to apoptotic cell death, oxidative burst and defense gene expression in response to elicitin in tobacco cell suspension culture. Eur J Biochem 267:5005–5013

Sasabe M, Naito K, Suenaga H, Ikeda T, Toyoda K, Inagaki Y, Shiraishi T, Ichinose Y (2007) Elicitin-responsive lectin-like receptor kinase genes in BY-2 cells. DNA Seq J Seq Map 18:152–159

Sathiyabama M, Akila G, Einstein Charles R (2014) Chitosan-induced defense responses in tomato plants against early blight disease caused by *Alternaria solani* (Ellis and Martin) Sorauer. Arch Phytopathol Plant Protect 47:1963–1973

Schaller A, Oecking C (1999) Modulation of plasma membrane H^+-ATPase activity differentially activates wound and pathogen defense responses in tomato plants. Plant Cell 11:263–272

Schmidt C, Schelle I, Liao YJ, Schroeder JI (1995) Strong regulation of slow anion channels and abscisic acid signaling in guard cells by phosphorylation and dephosphorylation events. Proc Natl Acad Sci USA 92:9535–9539

Schopfer FJ, Baker PR, Freeman BA (2002) NO-dependent protein nitration: a cell signaling event or an oxidative inflammatory response? Trends Biochem Sci 28:646–654

Schouten A, van Baarlen P, van Kan JAL (2008) Phytotoxic Nep1-like proteins from the necrotrophic fungus *Botrytis cinerea* associated with membranes and the nucleus of plant cells. New Phytol 177:493–505

Segonzac C, Zipfel C (2011) Activation of plant pattern-recognition receptors by bacteria. Curr Opin Microbiol 14:54–61

Séjalon N, Villaba F, Bottin A, Rickauer M, Dargent R, Esquerré-Tugayé MT (1995) Characterization of a cell surface antigen isolated from the plant pathogen *Phytophthora parasitica* var *nicotianae*. Can J Bot 73:1104–1108

Sèjalon-Delmas N, Mateos FV, Bottin A, Rickauer M, Dargent R, Esquerrè-Tugayè MT (1997) Purification, elicitor activity, and cell wall localization of a glycoprotein from *Phytophthora parasitica* var *nicotianae*, a fungal pathogen of tobacco. Phytopathology 87:899–909

Sels J, Mathys J, De Coninck BMA, Cammue BPA, De Bolle MFC (2008) Plant pathogenesis-related (PR) proteins: a focus on PR peptides. Plant Physiol Biochem 46:941–950

Serrano M, Kanehara K, Torres M, Yamada K, Tintor N, Kombrink E, Schulze-Lefert P, Saijo Y (2012) Repression of sucrose/ultraviolet B light-induced flavonoid accumulation in microbe-associated molecular pattern-triggered immunity in Arabidopsis. Plant Physiol 158:408–422

Shah J, Zeier J (2013) Long-distance communication and signal amplification in systemic acquired resistance. Front Plant Sci 4:Article 30. doi:10.3389/fpls.2013.00030

Shao M, Wang J, Dean RA, Lin Y, Gao X, Hu S (2008) Expression of a harpin-encoding gene in rice confers durable nonspecific resistance to *Magnaporthe grisea*. Plant Biotechnol J 6:73–81

Sharathchandra RG, Niranjan Raj S, Shetty NP, Amruthesh KN, Shekar Shetty H (2004) A chitosan formulation Elexa induces downy mildew disease resistance and growth promotion in pearl millet. Crop Protect 23:881–888

Sharp JK, McNeil M, Albersheim P (1984) The primary structures of one elicitor active and seven elicitor inactive hexa-β-glucopyranosyl-D-glucitol isolated from the mycelial cell walls of *Phytophthora megasperma* f sp *glycinea*. J Biol Chem 259:11321–11336

Shen QH, Saijo Y, Mauch S, Biskup C, Bieri S, Keller B, Seki H, Ulker B, Somssich IE, Schulze-Lefert P (2007) Nuclear activity of MLA immune receptors links isolate-specific and basal disease resistance responses. Science 315:1098–1103

Shibuya N, Minami E (2001) Oligosaccharide signalling for defence responses in plant. Physiol Mol Plant Pathol 59:223–233

Shimizu T, Nakano T, Takamizawa D, Desaki Y, Ishi-Minami N, Nishizawa Y, Minami E, Okada K, Yamane H, Kaku H, Shibuya N (2010) Two LysM receptor molecules, CEBiP and OsCERK1, cooperatively regulate chitin elicitor signaling in rice. Plant J 64:204–214

Silipo A, Molinaro A, Sturiale L, Dow JM, Erbs JM, Lanzetta R, Newman M-A, Parrilli M (2005) The elicitation of plant innate immunity by lipooligosaccharide of *Xanthomonas campestris*. J Biol Chem 280:33660–33668

Silipo A, Sturiale L, Garozzo D, Erbs G, Tandrup Poulsen T, Lanzetta R, Dow JM, Parrilli M, Newman M-A, Molinaro A (2008) The acylation and phosphorylation pattern of lipid A from *Xanthomonas campestris* strongly influence its ability to trigger the innate immune response in *Arabidopsis*. Chem BioChem 9:896–904

Silipo A, Erbs G, Shinya T, Dow JM, Parrilli MT, Lanzetta Shibuya N, Newman M-A, Molinaro A (2010) Glyco-conjugates as elicitors or suppressors of plant innate immunity. Glycobiology 20:406–419

Simon-Plas F, Rusterucci C, Milat ML, Humbert C, Montillet JL, Blein JP (1997) Active oxygen species production in tobacco cells elicited by cryptogein. Plant Cell Environ 20:1573–1579

Simon-Plas F, Elmayan T, Blein J-P (2002) The plasma membrane oxidase NtrbohD is responsible for AOS production in elicited tobacco cells. Plant J 31:137–148

Snedden WA, Fromm H (2001) Calmodulin as a versatile calcium signal transducer in plants. New Phytol 151:35–66

Soberon-Chavez G, Lépine F, Déziel E (2005) Production of rhamnolipids by *Pseudomonas aeruginosa*. Appl Microbiol Biotechnol 68:718–725

Sohn SI, Kim YH, Kim BR, Lee SY, Lim CK, Hur JH, Lee JY (2007) Transgenic tobacco expressing the $hrpN_{EP}$ gene from *Erwinia pyrifoliae* triggers defense responses against *Botrytis cinerea*. Mol Cells 24:232–239

Son GH, Wan J, Kim HJ, Nguyen XC, Chung W-S, Hong JC, Stacey G (2012) Ethylene-responsive element-binding factor 5, ERF5, is involved in chitin-induced innate immunity response. Mol Plant Microbe Interact 25:48–60

Song JT, Lu H, McDowell JM, Greenberg JT (2004) A key role for *ALD1* in activation of local and systemic defenses in *Arabidopsis*. Plant J 40:200–212

Spoel SH, Dong X (2012) How do plants achieve immunity? Defence without specialized immune cells. Nat Rev Immunol 12:89–100

Staats M, Van Baarlen P, Schouten A, Van Kan JAL (2007) Functional analysis of NLP genes from *Botrytis elliptica*. Mol Plant Pathol 8:209–214

Strobel NE, Ji C, Gopalan S, Kuc JA, He SY (1996) Induction of systemic acquired resistance in cucumber by *Pseudomonas syringae* pv *syringae* 61 $HrpZ_{Pss}$ protein. Plant J 9:431–439

Suarez-Rodriguez MC, Adams-Phillips L, Liu Y, Wang H, Su S-H, Jester PJ, Zhang S, Bent AF, Krysan PJ (2007) MEKK1 is required for flg22-induced MPK4 activation in Arabidopsis plants. Plant Physiol 143:661–669

Suh SJ, Wang Y-F, Frelet A, Leonhardt A, Klein M, Forestier C, Mueller-Roeber B, Cho MH, Martinoia E, Schroeder JI (2007) The ATP binding cassette transporter AtMRP5 modulates anion and calcium channel activities in *Arabidopsis* guard cells. J Biol Chem 282:1916–1924

Suharsono U, Fujisawa Y, Kawasaki T, Iwasaki Y, Satoh H, Shimamoto K (2002) The heterotrimeric G protein subunit acts upstream of the small GTPase Rac in disease resistance of rice. Proc Nat Acad Sci USA 99:13307–13312

Sun W, Dunning FM, Pfund C, Weingarten R, Bent AF (2006) Within-species flagellin polymorphism in *Xanthomonas campestris* pv *campestris* and its impact on elicitation of *Arabidopsis FLAGELLIN SENSING2*-dependent defenses. Plant Cell 18:764–779

Sun J-Y, Gaudet DA, Lu Z-X, Frick M, Puchalski B, Laroche A (2008) Characterization and antifungal properties of wheat nonspecific lipid transfer proteins. Mol Plant-Microbe Interact 21:346–360

Sun Y, Li L, Macho AP, Han Z, Hu Z, Zipfel C, Zhou JM, Chai J (2013) Structural basis for flg22-induced activation of the Arabidopsis FLS2-BAK1 immune complex. Science 342:624–628

Suty L, Blein JP, Ricci P, Pugin A (1995) Early changes in gene expression in tobacco cells elicited with cryptogein. Mol Plant-Microbe Interact 8:644–651

Suzuki H, Reddy MSS, Naoumkina M, Aziz N, May GC, Huhman DV, Summer LW, Blount JW, Mendes P, Dixon RA (2005) Methyl jasmonate and yeast elicitor induce differential transcriptional and metabolic re-programming in cell suspension cultures of the model legume *Medicago truncatula*. Planta 220:696–707

Suzuki N, Miller G, Morales J, Shulaev V, Torres MA, Mittler R (2011) Respiratory burst oxidases: the engines of ROS signaling. Curr Opin Plant Biol 14:691–699

Taguchi F, Shimizu R, Inagaki Y, Toyoda K, Shiraishi T, Ichinose Y (2003a) Post-translational modification of flagellin determines the specificity of HR induction. Plant Cell Physiol 44:342–349

Taguchi F, Shimizu R, Nakagima R, Toyoda K, Shiraishi T, Ichinose Y (2003b) Differential effects of flagellins from *Pseudomonas syringae* pv *tabaci*, *tomato*, and *glycinea* on plant defense response. Plant Physiol Biochem 41:165–174

Taguchi F, Takeuchi K, Katoh E, Murata K, Suzuki T, Marutani M, Kawasaki T, Eguchi M, Katoh S, Kaku H, Yasuda C, Inagaki Y, Toyoda K, Shiraishi T, Ichinose Y (2006) Identification of glycosylation genes and glycosylated amino acids of flagellin in *Pseudomonas syringae* pv *tabaci*. Cell Microbiol 8:923–938

Takahashi Y, Nasir KHB, Ito A, Kanzaki H, Matsumura H, Saitoh H, Fujisawa S, Kamoun S, Terauchi R (2007a) A high-throughput screen of cell-death-inducing factors in *Nicotiana benthamiana* identifies a novel MAPKK that mediates INF1-induced cell death signaling and non-host resistance to *Pseudomonas cichorii*. Plant J 49:1030–1040

Takahashi F, Yoshida R, Ichimura K, Mizoguchi T, Seo S, Yonezawa M, Maruyama K, Yamaguchi-Shinozaki K, Shinozaki K (2007b) The mitogen-activated protein kinase cascade MKK3-MPK6 is an important part of the jasmonate signal transduction pathway in Arabidopsis. Plant Cell 19:805–818

Takai R, Isogai A, Takayama S, Che F-S (2008) Analysis of flagellin perception mediated by flg22 receptor OsFLS2 in rice. Mol Plant Microbe Interact 21:1635–1642

Takakura Y, Ishida Y, Inoue Y, Tsutsumi F, Kuwata S (2004) Induction of hypersensitine response-like reaction by powdery mildew in transgenic tobacco expressing harpin $_{pss}$. Physiol Mol Plant Pathol 64:83–89

Takakura Y, Che F-S, Ishida Y, Tsutsumi F, Kuotani K-I, Usami S, Isogai A, Imaseki H (2008) Expression of a bacterial flagellin gene triggers plant immune responses and confers disease resistance in transgenic rice plants. Mol Plant Pathol 9:525–529

Takatsuji H (2014) Development of disease-resistant rice using regulatory components of induced disease resistance. Front Plant Sci 5:Art 630. doi:10.3389/fpls.2014.00630

Takeda S, Gapper C, Kaya H, Bell E, Kuchitsu K, Dolan L (2008) Local positive feedback regulation determines cell shape in root hair cells. Science 319:1241–1244

Takemoto D, Hardham AR, Jones DA (2005) Differences in cell death induction by *Phytophthora* elicitins are determined by signal components downstream of MAP kinase kinase in different species of *Nicotiana* and cultivars of *Brassica rapa* and *Raphanus sativus*. Plant Physiol 138:1491–1504

Takenaka S, Nakamura Y, Kono T, Sekiguchi H, Masunaka A, Takahashi H (2006) Novel elicitin-like proteins isolated from the cell wall of the biocontrol agent *Pythium oligandrum* induce defence-related genes in sugar beet. Mol Plant Pathol 7:325–339

Takenaka S, Tamagake H (2009) Foliar spray of a cell wall protein fraction from the biocontrol agent *Pythium oligandrum* induces defence-related genes and increases resistance against Cercospora leaf spot. J Gen Plant Pathol 75:340–348

Takeuchi K, Taguchi F, Inagaki Y, Toyoda K, Shiraishi T, Ichinose Y (2003) Flagellin glycosylation island in *Pseudomonas syringae* pv *glycinea* and its role in host specificity. J Bacteriol 185:6658–6665

Takeuchi K, Ono H, Yoshida M, Ishi T, Katch E, Taguchi F, Miki R, Murata K, Kaku H, Ichinose Y (2007) Flagellin glycans from two pathovars of *Pseudomonas syringae* contain rhamnose in D and L configurations in different ratios and modified 4-amino-4,6-dideoxyglucose. J Bacteriol 189:6945–6956

Takezawa D (2000) A rapid induction by elicitors of the mRNA encoding CCD-1, a 14 kDa Ca^{2+}-binding protein in wheat cultured cells. Plant Mol Biol 42:807–817

Talke IN, Blaudez D, Maathuis FJ, Sanders D (2003) CNGCs: prime targets of plant cyclic nucleotide signaling? Trend Plant Sci 8:286–293

Tampakaki AP, Panopoulos NJ (2000) Elicitation of hypersensitive cell death by extracellularly targeted $HrpZ_{Pph}$ produced in planta. Mol Plant-Microbe Interact 13:1366–1374

Tampakaki AP, Skandalis N, Gazi AD, Bastaki MN, Sarris PF, Charova SN, Kokkinidis M, Panapoulos NJ (2010) Playing the "Harp" Evolution of our understanding of hrp/hrc genes. Annu Rev Phytopathol 48:347–370

Tanaka N, Che F-S, Watanabe N, Fujiwara S, Takayama S, Isogai A (2003) Flagellin from an incompatible strain of *Acidovorax avenae* mediates H_2O_2 generation accompanying hypersensitive cell death and expression of *PAL Cht-1* and *PBZ1* but not of *LOX* in rice. Mol Plant-Microbe Interact 16:422–428

Tavernier E, Wendehenne D, Blein J-P, Pugin A (1995) Involvement of free calcium in action of cryptogein, a proteinaceous elicitor of a hypersensitive reaction in tobacco cells. Plant Physiol 109:1025–1031

Taylor KC (2006) Harpin protein application impacts fruit yield, size and retention of peach fruit. Acta Hortic (ISHS) 713:237–242

Teige M, Scheikl E, Eulgem T, Doczi R, Ichimura K, Shinozaki K, Dangl JL, Hirt H (2004) The MKK2 pathway mediates cold and salt stress signaling in Arabidopsis. Mol Cell 15:141–152

Thilmony R, Underwood W, He SY (2006) Genome-wide transcriptional analysis of the *Arabidopsis thaliana* interaction with the plant pathogen *Pseudomonas syringae* pv. *tomato* DC3000 and the human pathogen *Escherichia coli* O157:H7 46:34–53

Thipyapong P, Stout MJ, Attajarusit J (2007) Functional analysis of polyphenol oxidases by antisense/sense technology. Molecules 12:1569–1595

Thomma BPHJ, Nürnberger T, Joosten MHAJ (2011) Of PAMPs and effectors: the blurred PTI-ETI dichotomy. Plant Cell 23:4–15

Thuerig B, Felix G, Binder A, Boller T, Tamm L (2006) An extract of *Penicillium chrysogenum* elicits early defense-related responses and induces resistance in *Arabidopsis thaliana* independently of known signaling pathways. Physiol Mol Plant Pathol 67:180–193

Thuleau P, Ward JM, Ranjeva R, Schroeder JI (1994) Voltage-dependent calcium-permeable channels in the plasma membrane of a higher plant cell. EMBO J 13:2970–2975

Tosun N (2007) Disease control with ISR2000™ (elicitor) in conjunction with fungicides. http://www.engormix.com/diseasecontrol/eart 658

Tosun N, Karabay NU, Turkusay H, Aki C, Turkan I, Schading RL, Onsekiz C (2003) The effect of $harpin_{Ea}$ as plant activator in control of bacterial and fungal diseases of tomato. Acta Hortic 613:251–254

Trdă L, Fernandez O, Boutrot F, Hėloir MC, Kelloniemi J, Daire X, Adrian M, Clėment C, Zipfel C, Dorey S, Poinssot B (2014) The grapevine flagellin receptor VvFLS2 differentially recognizes flagellin-derived epitopes from the endophytic growth-promoting bacterium *Burkholderia phytofirmans* and plant pathogenic bacteria. New Phytol 201:1371–1384

Trouvelot S, Hėloir M-C, Poinssot B, Gauthier A, Paris F, Guillier C, Combier M, Trdá L, Daire X, Adrian M (2014) Carbohydrates in plant immunity and plant protection: roles and potential application as foliar sprays. Front Plant Sci 5:article 592, p14. doi:10.3389/fpls.2014.00592

Tsuda K, Katagiri P (2010) Comparing signaling mechanisms engaged in pattern-triggered and effector-triggered immunity. Curr Opin Plant Biol 13:459–465

Tsunemi K, Taguchi F, Marutani M, Watanabe-Sugimoto M, Inagaki Y, Toyoda K, Shiraishi T, Ichinose Y (2011) Degeneration of *hrpZ* gene in *Pseudomonas syringae* pv. *tabaci* to evade tobacco defence: an arms race between tobacco and its bacterial pathogens. Mol Plant Pathol 12:709–714

Tubajika KM, Civerolo EL, Luvisis DA, Hashim JM (2007) The effects of kaolin, harpin, and imidacloprid on development of Pierce's disease in grape. Crop Protect 26:92–99

Türküsay H, Tosun N, Yildiz S, Saygili H (2009) Effects of plant activators on physiological and morphological parameters of processing tomato. Acta Hortic (ISHS) 808:431–436

Umemura K, Ogawa N, Yamauchi T, Iwata M, Shimura M, Koga J (2000) Cerebroside elicitors found in diverse phytopathogens activate defense responses in rice plants. Plant Cell Physiol 41:676–683

Umemura K, Ogawa N, Koga J, Iwata M, Usami H (2002) Elicitor activity of cerebroside, a sphingolipid elicitor, in cell suspension cultures of rice. Plant Cell Physiol 43:778–784

Umemura K, Tanino S, Nagatsuka T, Koga J, Iwata M, Nagashima K, Amemiya Y (2004) Cerebroside elicitor confers resistance to Fusarium disease in various plant species. Phytopathology 94:813–818

Ustun N, Ulutas E, Yasarakinci N, Kilic T (2009) Efficacy of some plant activators on bacterial canker of tomato in Aegean region of Turkey. Acta Hortic (ISHS) 808:405–408

Uzal EN, Gómez Ros LV, Pomar F, Bernal MA, Paradela A, Albar JP, Barceló A (2009) The presence of sinapyl lignin in *Ginkgo biloba* cell cultures changes our views of the evolution of lignin biosynthesis. Physiol Plant 135:196–213

Vallance J, Le Floch G, Déniel F, Barbier G, Levesque CA, Rey P (2009) Influence of *Pythium oligandrum* biocontrol on fungal and oomycete population dynamics in the rhizosphere. Appl Environ Microbiol 75:4790–4800

Van Loon LC, Bakker PAHM, Van der Heijdt WHW, Wendehenne D, Pugin A (2008) Early responses of tobacco suspension cells to rhizobacterial elicitors of induced systemic resistance. Mol Plant Microbe Interact 21:1609–1621

Vandelle E, Delledonne M (2011) Peroxynitrite formation and function in plants. Plant Sci 181:534–539

Vargas WA, Djonovic S, Sukno SA, Kenerley CM (2008) Dimerization controls the activity of fungal elicitors that trigger systemic resistance in plants. J Biol Chem 283:19804–19815

Varnier A-L, Sanchez L, Vatsa P, Boudesocque L, Garcia-Brugger A, Rabenoelina F, Sorokin A, Renault J-H, Kauffmann S, Pugin A, Clement C, Baillieul F, Dorey S (2009) Bacterial rhamnolipids are novel MAMPs conferring resistance to *Botrytis cinerea* in grapevine. Plant Cell Environ 32:176–193

Vasil'ev LA, Dzyubinskaya EV, Zinovkin RA, Kiselevsky DB, Lobysheva NV, Samuilov VD (2009) Chitosan-induced programmed cell death in plants. Biochem Mosc 74:1035–1043

Vatsa P, Sanchez L, Clement C, Baillieul F, Dorey S (2010) Rhamnolipid biosurfactants as new players in animal and plant defense against microbes. Int J Mol Sci 11:5095–5108

Vatsa P, Chiltz A, Bourgue S, Wendehenne D, Garcia-Brugger A, Pugin A (2011) Involvement of putative glutamate receptors in plant defence signaling and NO production. Biochimie 93:2095–2101

Vauthrin S, Mikes V, Milat M-L, Ponchet M, Maume B, Osman H, Blein J-P (1999) Elicitins trap and transfer sterols from micelles, liposomes and plasma membranes. Biochim Biophys Acta 1419:335–342

Veit S, Worle JM, Nürnberger T, Koch W, Seitz HU (2001) A novel protein elicitor (PaNie) from *Pythium aphanidermatum* induces multiple defense responses in carrot, Arabidopsis, and tobacco. Plant Physiol 127:832–841

Veluchamy S, Hind SR, Dunham DM, Martin GB, Panthee DR (2014) Natural variation for responsiveness to flg22, flgII-28, and CSP22 and *Pseudomonas syringae* pv. *tomato* in heirloom tomatoes. PLoS ONE 9(9):e105119

Viard MP, Martin F, Pugin A, Blein J-P (1994) Protein phosphorylation is induced in tobacco cells by the elicitor cryptogein. Plant Physiol 104:1245–1249

Vidhyasekaran P (1997) Fungal pathogenesis in plants and crops: molecular biology and host defense mechanisms. Marcel Dekker, New York, p 553

Vidhyasekaran P (2004) Concise Encyclopedia of plant pathology. Haworth Press, Binghamton/New York, p 619

Vidhyasekaran P (2007) Fungal pathogenesis in plants and crops: molecular biology and host defense mechanisms, IIth edn. CRC Press/Taylor Francis Group, Boca Raton, p 510

Vidhyasekaran P (2014) PAMP signals in plant innate immunity: signal perception and transduction. Springer, Dordrecht, p 442

Vidhyasekaran P (2015) Plant hormone signaling systems in plant innate immunity. Springer, Dordrecht, p 458

Villalba Mateos F, Rickauer M, Esquerré-Tugayé MT (1997) Cloning and characterization of a cDNA encoding an elicitor of *Phytophthora parasitica* var *nicotianae* that shows cellulose-binding and lectin-like activities. Mol Plant-Microbe Interact 10:1045–1053

Viterbo A, Wiest A, Brotman Y, Chet I, Kenerley C (2007) The 18mer peptaibols from *Trichoderma virens* elicit plant defense responses. Mol Plant Pathol 8:737–746

Vlot AC, Dempsey DA, Klessig DF (2009) Salicylic acid, a multifaceted hormone to combat disease. Annu Rev Phytopathol 47:177–206

Vranová E, Inzé D, Van Brueusegem F (2002) Signal transduction during oxidative stress. J Exp Bot 53:1227–1236

Walter MH, Grima-Pettenati J. Grand C, Lamb CJ (1988) Cinnamyl-alcohol dehydrogenase, a molecular marker specific for lignin synthesis: cDNA cloning and mRNA induction by fungal elicitor. Proc Natl Acad Sci USA 85:5546–5550

Walters D, Walsh D, Newton A, Lyon G (2005) Induced resistance for plant disease control: Maximizing the efficacy of resistance elicitors. Phytopathology 95:1368–1373

Wan J, Zhang X-C, Stacey G (2008) Chitin signaling and plant disease resistance. Plant Signal Behav 3:831–833

Wang JY, Cai Y, Gou JY, Mao YB, Xu YH, Jiang WH, Chen XY (2004) VdNEP, an elicitor from *Verticillium dahliae*, induces cotton plant wilting. Appl Environ Microbiol 70:4989–4995

Wang W, Li S, Zhao X, Du Y, Lin B (2008) Oligochitosan induces cell death and hydrogen peroxide accumulation in tobacco suspension cells. Pesticide Biochem Physiol 90:106–113

Wang D, Wang Y, Fu M, Mu S, Han B, Ji H, Cai H, Dong H, Zhang C (2014) Transgenic expression of the functional fragment $Hpa1_{10-42}$ of the harpin protein Hpa1 imparts enhanced resistance to powdery mildew in wheat. Plant Dis 98:448–455

Ward JM, Pei Z-M, Schroeder JI (1995) Roles of ion channels in initiation of signal transduction in higher plants. Plant Cell 7:833–844

Wasternack C (2007) Jasmonates: an update on biosynthesis, signal transduction and action in plant stress response, growth and development. Ann Bot (Lond) 100:681–697

Wasternack C, Hause B (2013) Jasmonates: biosynthesis, perception, signal transduction and action in plant stress response, growth and development. An update to the 2007 review in Annals of Botany. Ann Bot 111:1021–1058

Wathugala DL, Hemsley PA, Moffat CS, Cremelie P, Knight MR, Knight H (2012) The Mediator subunit SFR6/MED16 controls defence gene expression mediated by salicylic acid and jasmonate responsive pathways. New Phytol 195:217–230

Watt SA, Tellstrom V, Patschkowski T, Niehaus K (2006) Identification of the bacterial superoxide dismutase (SodM) as plant-inducible elicitor of the oxidative burst reaction in tobacco cell suspension cultures. J Biotechnol 126:78–86

Wei Z-M, Laby R-J, Zumoff CH, Bauer DW, He SY, Collmer A, Beer SV (1992) Harpin, elicitor of the hypersensitive response produced by the plant pathogen *Erwinia amylovora*. Science 257:85–88

Wei Z-M, Qiu D, Kropp MJ, Schading RI (1998) Harpin, an HR elicitor, activates both defense and growth systems in many commercially important crops. Phytopathology (Suppl) 88:S96

Wen W, Wang J (2001) Cloning and expressing a harpin gene from *Xanthomonas oryzae* pv *oryzae*. Acta Phytopathol Sin 31:296–300

Wendehenne D, Binet M-N, Bleim J-P, Ricci P, Pugin A (1995) Evidence for specific, high-affinity binding sites for a proteinaceous elicitor in tobacco plasma membrane. FEBS Lett 374:203–207

Wendehenne D, Lamotte O, Frachisse J-M, Barbier-Brygoo H, Pugin A (2002) Nitrate efflux is an essential component of the cryptogein signaling pathway leading to defense responses and hypersensitive cell death in tobacco. Plant Cell 14:1937–1951

White PJ (2000) Calcium channels in higher plants. Biochim Biophys Acta 1465:171–189

White PJ (2004) Calcium signals in root cells: the roles of plasma membrane calcium channels. Biologia 59(S13):77–83

White PJ, Broadley MR (2003) Calcium in plants. Ann Bot (Lond) 92:487–511

Wiermer M, Feys BJ, Parker JE (2005) Plant immunity: the EDS1 regulatory node. Curr Opin Plant Biol 8:383–389

Wildermuth MC, Dewdney J, Wu G, Ausubel FM (2001) Isochorismate synthesis is required to synthesize salicylic acid for plant defense. Nature 414:562–565

Willmann R, Lajunen HM, Erbs G, Newman MA, Kolb D, Tsuda K, Katagiri F, Fliegmann J, Bono JJ, Cullimore JV, Jehle AK, Göltz F, Kulik A, Molinaro A, Lipka V, Gust AA, Nürnberger T (2011) Arabidopsis lysin-motif proteins LYM1 LYM3 CERK1 mediate bacterial peptidoglycan sensing and immunity to bacterial infection. Proc Natl Acad Sci U S A 108:19824–19829

Wong HL, Pinontoan R, Hayashi K, Tabata R, Yaeno T, Hasegawa K, Kojima C, Yoshioka H, Iba K, Kawasaki T, Shimamoto K (2007) Regulation of rice NADPH oxidase by binding of Rac GTPase to its N-terminal extension. Plant Cell 19:4022–4034

Wu X, Wu T, Long J, Yin Q, Zhang Y, Chen L, Liu R, Gao T, Dong H (2007) Productivity and biochemical properties of green tea in response to full-length and functional fragments of $HpaG_{Xooc}$, a harpin protein from the bacterial rice leaf streak pathogen *Xanthomonas oryzae* pv. *oryzicola*. J Biosci 32:1119–1131

Wu Y, Zhang D, Chu JY, Boyle P, Wang Y, Brindle ID, De Luca V, Desprès C (2012) The Arabidopsis NPR1 protein is a receptor to the plant defense hormone salicylic acid. Cell Rep 1:639–647

Wypijewski K, Hornyik C, Shaw JA, Stephens J, Goraczniak R, Gunderson SI, Lacomme C (2009) Ectopic 5′ splice sites inhibit gene expression by engaging RNA surveillance and silencing pathways in plants. Plant Physiol 151:955–965

Xing DH, Lai ZB, Zheng ZY, Vinod KM, Fan BF, Chen ZX (2008) Stress- and pathogen-induced *Arabidopsis* WRKY48 is a transcriptional activator that represses plant basal defense. Mol Plant 1:459–470

Xu F, Yang Y, Xie FJ, Liu Z, Qiu DW, Yang XF (2006) Effect of the activator protein from *Magnaporthe grisea* on plant growth and physiological activities. Acta Agriculturae Boreali-Sinica 21:1–5

Xu G, Chen S, Chen F (2010) Transgenic chrysanthemum plants expressing a $harpin_{Xoo}$ gene demonstrate induced resistance to alternaria leafspot and accelerated development. Russian J Plant Physiol 57:548–553

Yakushiji S, Ishiga Y, Inagaki Y, Toyoda K, Shiraishi T, Ichinose Y (2009) Bacterial DNA activates immunity in *Arabidopsis thaliana*. J Gen Plant Pathol 75:227–234

Yamaguchi T, Ito Y, Shibuya N (2000a) Oligosaccharide elicitors and their receptors for plant defense responses. Trends Glyosci Glyotechnol 12:113–120

Yamaguchi T, Yamada A, Hong N, Ogawa T, Ishii T, Shibuya N (2000b) Differences in the recognition of glucan elicitor signals between rice and soybean: β-glucan fragments from the rice blast disease fungus *Pyricularia oryzae* that elicit phytoalexin biosynthesis in suspension-cultured rice cells. Plant Cell 12:817–826

Yamaguchi T, Maehara Y, Kodama O, Matsumura M, Shibuya N (2002) Two purified oligosaccharide elicitors, N-acetylchitoheptaose and tetraglucosyl glucitol, derived from *Magnaporthe grisea* cell walls, synergistically activate biosynthesis of phytoalexin in suspension-cultured rice cells. J Plant Physiol 159:1147–1149

Yamamoto-Katou A, Katou S, Yoshioka H, Doke N, Kawakita K (2006) Nitrate reductase is responsible for elicitin-induced nitric oxide production in *Nicotiana benthamiana*. Plant Cell Physiol 47:728–735

Yang KY, Liu YD, Zhang SQ (2001) Activation of a mitogen-activated protein kinase pathway is involved in disease resistance in tobacco. Proc Natl Acad Sci U S A 98:741–746

Yang B, Shiping T, Jie Z, Yonghong G (2005) Harpin induces local and systemic resistance against *Trichothecium roseum* in harvested Hami melons. Postharvest Biol & Technol 38:183–187

Yokosuka A, Mimaki Y (2009) Steroidal saponins from the whole plants of *Agave utahensis* and their cytotoxic activity. Phytochemistry 70:807–815

Yoo S-D, Cho Y-H, Tena G, Xiong Y, Sheen J (2008) Dual control of nuclear EIN3 by bifurcate MAPK cascades in C_2H_4 signalling. Nature 451:789–795

Yoshioka K, Moeder W, Kang H-G, Kachroo P, Masmoudi K, Berkowitz G, Klessig D-F (2006) The chimeric *Arabidopsis* CYCLIC NUCLEOTIDE-GATED ION CHANNEL11/12 activates multiple pathogen responses. Plant Cell 18:747–763

Yu C, Chen H, Tian F, Leach JF, He C (2014) Differentially-expressed genes in rice infected by *Xanthomonas oryzae* pv. *oryzae* relative to a flagellin-deficient mutant reveal potential functions of flagellin in host-pathogen interactions. Rice 7:20

Zago E, Morsa S, Dat JF, Alard P, Ferrarini A, Inzé D, Delledonne M, Van Breusegem F (2006) Nitric oxide- and hydrogen peroxide-responsive gene regulation during cell death induction in tobacco. Plant Physiol 404–411

Zamioudis C, Peterse CMJ (2012) Modulation of host immunity by beneficial microbes. Mol Plant-Microbe Interact 25:139–150

Zander M, La Camera S, Lamotte O, Metraux JP, Gatz C (2010) *Arabidopsis thaliana* class-II TGA transcription factors are essential activators of jasmonic acid/ethylene-induced defense responses. Plant J 61:200–210

Zaninotto F, Camera SL, Polverari A, Delledonne M (2006) Cross talk between reactive nitrogen and oxygen species during the hypersensitive disease resistance response. Plant Physiol 141:379–383

Zeidler D, Dubery IA, Schmitt-Kopplin P, Von Rad U, Durner J (2010) Lipopolysaccharide mobility in leaf tissue of *Arabidopsis thaliana*. Mol Plant Pathol 11:747–755

Zhang J, Zhou J-M (2010) Plant immunity triggered by microbial molecular signatures. Mol Plant 3:783–793

Zhang S, Liu Y, Klessig DF (2000) Multiple levels of tobacco WIPK activation during the induction of cell death by fungal elicitins. Plant J 23:339–347

Zhang H, Fang Q, Zhang Z, Wang Y, Zheng X (2009) The role of respiratory burst oxidase homologues in elicitor-induced stomatal closure and hypersensitive response in *Nicotiana benthamiana*. J Expt Bot 60:3109–3122

Zhang L, Xiao S, Li W, Feng W, Li J, Wu Z, Gao X, Liu F, Shao M (2011) Overexpression of a harpin-encoding gene *hrf1* in rice enhances drought tolerance. J Exp Bot 62:4229–4238

Zhang Z, Wu Y, Gao M, Zhang J, Kong Q, Liu Y, Ba H, Zhou J, Zhang Y (2012) Disruption of PAMP-induced MAP kinase cascade by a *Pseudomonas syringae* effector activates plant immunity mediated by the NB-LRR protein SUMM2. Cell Host Microbe 11:253–263

Zhang W, Fraiture M, Kolb D, Löffelhardt B, Desaki Y, Boutrot FFG, Tör M, Zipfel C, Gust AA, Brunner F (2013) Arabidopsis RECEPTOR-LIKE PROTEIN30 and receptor-like kinase SUPPRESSOR OF BIR1-1/EVERSHED mediate innate immunity to necrotrophic fungi. Plant Cell 25:4227–4241

Zhang L, Kars I, Essenstam B, Liebrand TW, Wagemakers L, Elberse J, Tagkalaki P, Tjoitang D, van den Ackerveken G, van Kan JA (2014) Fungal endopolygalacturonases are recognized as microbe-associated molecular patterns by the arabidopsis receptor-like protein RESPONSIVENESS TO BOTRYTIS POLYGALACTURONASES1. Plant Physiol 164:352–364

Zhao XM, She XP, Yu W, Liang XM, Du YG (2007) Effects of oligochitosans on tobacco cells and role of endogenous nitric oxide burst in resistance of tobacco to *Tobacco mosaic virus*. J Plant Pathol 89:55–65

Zheng Z, Qamar SA, Chen Z, Mengiste T (2006) Arabidopsis WRKY33 transcription factor is required for resistance to necrotrophic fungal pathogens. Plant J 48:596–605

Zhou J-M, Trifa Y, Silva H, Pontier D, Lam E, Shah J, Klessig DF (2000) NPR1 differentially interacts with members of the TGA/OBF family of transcription factors that bind an element of the *PR-1* gene required for induction by salicylic acid. Mol Plant-Microbe Interact 13:191–202

Zipfel C (2008) Pattern-recognition receptors in plant innate immunity. Curr Opin Immunol 20:10–16

Zipfel C (2009) Early molecular events in PAMP-triggered immunity. Curr Opin Plant Biol 12:414–420

Zipfel C (2014) Plant pattern-recognition receptors. Trends Immunol 35:345–351

Zipfel C, Robatzek S, Navarro L, Oakeley EJ, Jones JDG, Felix G, Boller T (2004) Bacterial disease resistance in *Arabidopsis* through flagellin perception. Nature 428:764–767

Zipfel C, Kunze G, Chinchilla D, Caniard A, Jones JD, Boller T, Felix G (2006) Perception of the bacterial PAMP EF-Tu by the receptor EFR restricts *Agrobacterium*-mediated transformation. Cell 125:749–760

Zitter TA, Beer SV (1998) Harpin for insect control. Phytopathology 88:S104–S105

Zuppini A, Baldan B, Millioni R, Favaron F, Navazio L, Mariani P (2004) Chitosan induces Ca^{2+}-mediated programmed cell death in soybean cells. New Phytol 161:557–568

Zuppini A, Navazio L, Sella L, Castiglioni C, Favaron F, Mariani P (2005) An endopolygalacturonase from *Sclerotinia sclerotiorum* induces calcium-mediated signaling and programmed cell death in soybean cells. Mol Plant-Microbe Interact 18:849–855

Chapter 4
Switching on Plant Immune Signaling Systems Using Pathogen-Induced Molecular Patterns/ Host-Associated Molecular Patterns

Abstract Plant immune system is a sleeping giant and on activation, it can trigger defense responses against a wide range of bacterial, fungal, oomycete, and viral pathogens. Specific signals are required to switch on the innate immune system. Pathogen-associated molecular patterns (PAMPs) are the alarm signal molecules triggering the plant immune system. The microbe-derived elicitors and invading pathogens release pathogen-induced molecular patterns (PIMPs)/host-associated molecular patterns (HAMPs or endogenous elicitors), which are involved in amplifying the PAMP signals to activate host defense responses. The well-characterized host-derived elicitors include oligogalacturonides (OGAs) and Pep proteins (Plant Elicitor Peptides). The OGAs activate cyclic nucleotide gated channels (CNGCs) and glutamate receptors (GLRs) and elicit a rapid elevation in cytosolic Ca^{2+}. OGAs trigger a robust oxidative burst mediated by NADPH oxidase. OGAs trigger NO production by activating both nitrate reductase and NO synthase pathways. OGAs also induce the enzymes involved in MAPK signaling cascade. OGAs trigger the expression of genes involved in SA, JA, ET, and ABA signaling systems. Both the degrees of substitution (methylesterification and/or acetylation) and polymerization determine the efficacy of OGAs in triggering defense responses. Degree of methyl esterification modulates the elicitor activity of OGAs and OGAs with different degrees of polymerization differ in triggering defense responses. Ability of OGAs to trigger defense responses also depends on their level of acetylation. Methyl esterification, degree of polymerization, and level of acetylation are modulated by pectin methylesterases (PMEs), polygalacturonases (PGs), and pectin acetylesterases (PAEs), respectively. PME activity is tightly regulated by an inhibitor protein called pectin methylesterase inhibitor protein (PMEI). Transgenic plants overexpressing genes encoding PME inhibitor proteins show enhanced disease resistance. Transgenic plants expressing an attenuated version of a fungal PG show enhanced resistance against diseases. Polygalacturonase inhibitor proteins (PGIPs) belong to the super family of leucine-rich repeat (LRR) proteins and they play important role in switching on plant defense signaling pathways. Several studies conducted both under greenhouse and field conditions clearly show that *PGIP* gene is a potential tool for engineering disease resistance against various fungal, oomycete and bacterial pathogens. The expression of *PGIP* genes did not affect the agronomic characters of the transformed plants. The expression of PG inhibitor protein neither alters

P. Vidhyasekaran, *Switching on Plant Innate Immunity Signaling Systems*,
Signaling and Communication in Plants, DOI 10.1007/978-3-319-26118-8_4

the physiological performances nor exhibits detrimental effects on the growth of transgenic plants. Pep proteins accumulate at the site of infection and switch on the immune signaling systems upon binding to plasma membrane-localized Pep Receptors (PEPRs). Peps activate CNGC2-dependent plasma membrane inwardly conducting Ca^{2+} permeable channels in mesophyll cells, resulting in elevation of cytosolic Ca^{2+} ($[Ca^{2+}]_{cyt}$) level. This activity is dependent on the Pep Receptors. BAK1 may function as a coreceptor with PEPRs and contribute to Pep immune signaling leading to defense gene expression. PEPR1 has been shown to interact with receptor-like cytoplasmic kinases botrytis-induced kinase 1 (BIK1) and PBS1-like 1 (PBL1) to mediate Pep1-induced defense responses. Downstream from the early Ca^{2+} signal, Ca^{2+}-dependent protein kinases (CDPKs) are involved in decoding the Ca^{2+} signal. CDPK-dependent phosphorylation has been shown to be involved in the Pep signaling cascade, Pep proteins trigger both ROS and NO generation. Pep proteins mediate JA, ET, and SA signaling systems and induce the expression of defense genes. Pep proteins can be developed as potential tools for disease management. Application of Pep proteins induce resistance against fungal and bacterial pathogens. Transgenic plants overexpressing *PROPEP1* and *PROPEP2* genes show enhanced resistance against pathogens. PAMP signals also induce secretion of several peptides called secreted peptides and the secreted peptide precursors are called "prePIPs" (precursors of PAMP-induced Peptides). The prePIPs are cleaved close to the C terminus to form PIPs. PIP1 induces the precursor of the PIMP/HAMP PEP1, PROPEP1, while PEP1 induces the precursor of PIP1, prePIP1. Transgenic plants overexpressing *prePIP1* or *prePIP2* show enhanced resistance against fungal and bacterial diseases. Systemin is a damage/wound induced bioactive peptide and it triggers defense responses against various pathogens and insect pests. Systemin plays a key role in the JA biosynthesis pathway. The transgenic plants overexpressing prosystemin show enhanced resistance against necrotrophic fungal pathogens. PIMPs/HAMPs appear to be powerful tools to engineer disease resistance in field crops.

4.1 Pathogen-Induced Molecular Patterns (PIMPs)/Host-Associated Molecular Patterns (HAMPs)

Plant immune system is a potential basal defense system, which is activated on perception of pathogen-associated molecular patterns (PAMPs) (Kwon and Yun 2014; Lloyd et al. 2014; Macho and Zipfel 2014), The PAMPs are microbe-associated (derived) molecular patterns (MAMPs) or microbe-derived elicitors (Pearce and Ride 1982; Rakwal et al. 2002; Masunaka et al. 2010; Vatsa et al. 2010; Böhm et al. 2014; Noirot et al. 2014; Zhang et al. 2014). Besides the microbe-derived elicitors, some elicitors derived from host plants have been shown to activate the plant innate immune system (Yamaguchi and Huffaker 2011; Vallarino and Osorio 2012; Bellincampi et al. 2014; Hou et al. 2014). The host-derived elicitors are called by different names by different authors. They are called host-associated molecular patterns (HAMPs) as they are of host origin (Galletti et al. 2009; Brutus et al. 2010; Vidhyasekaran 2014), host-derived patterns or host-derived elicitors (Böhm et al. 2014), host damage-associated or danger-associated molecular patterns (DAMPs) as they are also induced by cellular damage or by danger signals (Boller and Felix 2009; Zipfel 2009; Hann et al. 2014; Mott et al. 2014), or endogenous/internal elicitors (Ryan et al. 2007; Huffaker et al. 2011). The endogenous elicitors are induced by pathogen infection (Vidal et al. 1998; Furman-Matarasso et al. 1999; Boudart et al. 2003; Poinssot et al. 2003; Huffaker et al. 2006, 2011; Logemann et al. 2013) and hence they are also called pathogen induced molecular patterns (PIMPs) (Mackey and McFall 2006; Aziz et al. 2007). Most of the endogenous elicitors have the property of inducing expression of their own genes to initiate a feedback mechanism to the original PAMP signals and therefore they can also be called "PAMP amplifiers" (Huffaker and Ryan 2007).

Polygalacturonases and cellulases are produced by a wide range of pathogens and they act as effectors and also function as general elicitors/PAMPs (Rotblat et al. 2002; Boudart et al. 2003; Poinssot et al. 2003). During host-pathogen interaction, many pathogens secrete these cell-wall-degrading enzymes (Vidal et al. 1998; Furman-Matarasso et al. 1999; Boudart et al. 2003; Poinssot et al. 2003; Ferrari et al. 2008). These enzymes can themselves function as elicitors/PAMPs (Rotblat et al. 2002; Poinssot et al. 2003; Zhang et al. 2014), but their enzymatic products are also known to be general elicitors of plant defense responses (Shibuya and Minami 2001). These enzymes degrade the plant cell wall structure and some of the degradation products such as pectin-derived oligogalacturonides (OGAs) (Randoux et al. 2010; Ferrari et al. 2013; Aranega-Bou et al. 2014; Malinovsky et al. 2014) and cellodextrins (Aziz et al. 2007) act as potent elicitors of innate immunity. *P. syringae* pv. *tomato* DC 3000 produces a pectin lyase (PL) and a polygalacturonase (PG), which could generate OGAs during the infection process (Buell et al. 2003). Fungal pectate lyases (Hahn et al. 1981) and bacterial pectate lyases (Vorhőlter et al. 2012) have been shown to release OGAs during infection. Pectate lyases catalyze the cleavage of α-1,4-glycosidic bonds between galacturonic acid residues of homogalacturonans. Likewise, polygalacturonases catalyze the cleavage of the glycosidic bonds between adjacent galacturonic acid residues, but the hydrolysis of the glycosidic linkage results in the addition of a water molecule from the environment (Vorhőlter et al. 2012). Inhibitors of PG activity have been implicated in contributing to the formation of pectin fragments that function as PIMPs (Nuhse 2012; Ferrari et al. 2013). OGAs induce the expression of *PGIP* gene, which encodes a polygalacturonase inhibitor protein (PGIP) involved in formation of active pectin fragments (Ferrari et al. 2008).

Several PAMPs including Flg22, NPP1, and HrpZ and pathogens including *Botrytis cinerea*, *Phytophthora infestans*, and *Pseudomonas syringae* trigger expression of *PROPEP2* gene encoding PROPEP2, the precursor for the PIMP AtPep2 (Huffaker et al. 2006). Similar induced expression of genes encoding PROPEP3 and PROPEP1 due to PAMP and pathogens treatment has also been reported (Huffaker et al. 2006). Treatment with the bacterial PAMP flg22 upregulates the transcription of genes encoding PROPEP family precursors for AtPeps and PEPR receptors

(Zipfel et al. 2004; Ryan et al. 2007). The PIMPs Peps are also induced by fungal pathogen infection (Huffaker et al. 2011).

The host-derived elicitors (PIMPs/HAMPs) function almost in the same fashion as the PAMPs function in plant innate immunity (Denoux et al. 2008). OGAs and the PAMP flg22 activate similar early transcriptional programs that diverge over time (Denoux et al. 2008; Ferrari et al. 2013). Pep family PIMPs and the classical PAMPs activate similar downstream responses using many of the same molecular components (Ryan et al. 2007; Krol et al. 2010; Postel et al. 2010; Qi et al. 2010; Yamaguchi et al. 2010; Huffaker et al. 2011). Both the PAMP flg22 and the PIMP AtPeps bind specific LRR receptors and both activate the same downstream signaling events (Yamaguchi et al. 2006; Huffaker and Ryan 2007; Krol et al. 2010). The endogenous elicitor AtPep1 treatment induces the transcription of FLS2, the PRR for the PAMP flg22 (Ryan et al. 2007). The receptors for both flg22 and AtPep1 associate with the interacting receptor partner, BAK1 (Ma et al. 2009; Postel et al. 2010). Collectively these studies suggest that the PIMPs are functionally similar to classical PAMPs and may act as amplifiers of PAMP-induced signals.

4.2 Oligogalacturonides Switch on Plant Innate Immunity

Oligogalacturonides (OGAs) are the best-characterized plant cell wall-derived elicitors/PIMPs/HAMPs, which switch on plant immune system (Vallarino and Osorio 2012). Pectin is composed mainly of homogalacturonan (HG) that consists of a homopolymer of 1,4-linked α-D-galacturonic acid (GalA) with a degree of polymerization of approximately 100 residues. Pectin is synthesized in the Golgi apparatus and secreted into the cell wall in a highly methyl-esterified form (Mohnen 2008). Homogalacturonan, the most abundant pectic polymer that can be methylesterified or acetylated, can be modified by various enzymes. These modifications of pectin can alter defense responses (Osorio et al. 2008). Both the degrees of substitution (methylesterification and/or acetylation) and polymerization can be controlled by specific enzymes such as pectin methylesterases (PMEs), pectin acetylesterases (PAEs), polygalacturonases (PGs), or pectate lyases-like (PLLs) (Sénéchal et al. 2014). Several oomycete, fungal, and bacterial pathogens produce polygalacturonases (PGs) which hydrolyze the homogalacturonans of the plant cell wall (Vidhyasekaran 2007). The PGs release oligogalacturonides with a degree of polymerization between 10 and 15 by partial degradation of homogalacturonans (Vorwerk et al. 2004). However, not all OGAs are capable of switching on defense responses. Their ability to elicit defense responses depends on length (degree of polymerization), degree of methyl esterification and the level of acetylation (Côté and Hahn 1994; Vidhyasekaran 1997, 2007; Wiethölter et al. 2003; Aziz et al. 2004; Ferrari et al. 2007; Osorio et al. 2008; Vallarino and Osorio 2012).

Oligogalacturonides, the oligomers of α-1,4-linked galacturonosyl residues formed by partial hydrolysis of polygalacturonic acid, have been identified as PIMPs/HAMPs (Ferrari et al. 2013). Degradation of the major component of pectin, homogalacturonan (HGA), which occurs during microbial infections may cause the accumulation of OGAs. OGAs are generated from enzymatic activity encoded by either the pathogen or the host (Ferrari et al. 2013). *P, syringae* pv. *tomato* DC 3000 produces a pectin lyase (PL) and a polygalacturonase (PG), which could generate OGAs during the infection process (Buell et al. 2003). Inhibitors of PG activity have been implicated in contributing to the formation of pectin fragments that function as PIMPs/HAMPs (Nuhse 2012; Ferrari et al. 2013). The PGs release OGAs with a degree of polymerization (DP) between 10 and 15 by partial degradation of homogalacturonans and the OGAs act as sensors for incoming infections (Vorwerk et al. 2004). OGAs trigger expression of several genes involved in the early and late stages of the multiple signaling defense pathways in *Arabidopsis* (Denoux et al. 2008).

OGAs activate several early signaling events in the plant immune system. OGAs are involved in activation of Ca^{2+} signaling system (Fig. 4.1). The OGAs elicited a rapid, but transient, elevation in cytosolic Ca^{2+} in *Arabidopsis* cells (Moscatiello et al. 2006). The genes encoding cyclic nucleotide gated channels (CNGCs) have been found to be activated by the OGA elicitor (Denoux et al. 2008). CNGCs have been shown to play important role in defense signaling and they trigger hypersensitive reaction (HR) in *Arabidopsis* (Balagué et al. 2003; Jurkowski et al. 2004; Baxter et al. 2008).

Glutamate receptors (GLRs) comprise another class of ion channel that provides a calcium-permeable pathway across the plasma membrane (Sanders et al. 2002; Tikhonov and Magazanik 2009; Vatsa et al. 2011; Price et al. 2012; Vincill et al. 2012; Manzoor et al. 2013). OGAs-triggered immune responses have been shown to be dependent on the glutamate receptors (GLRs) in *Arabidopsis*. Twenty glutamate receptor-like channels have been detected in *A. thaliana* genome (Chiu et al. 1999, 2002; Wheeler and Brownlee 2008). GLRs are intimately associated with Ca^{2+} influx. They are partly involved in free cytosolic Ca^{2+} variations ($[Ca^{2+}]$cyt) (Manzoor et al. 2013). The downstream event in the glutamate receptor-mediated signaling pathway included NO production (Vatsa et al. 2011). A calmodulin gene, *CML41*, has been reported to be induced by OGAs (Denoux et al. 2008). The OGA elicitor activated the expression of calcium-dependent phospholipid binding protein (Denoux et al. 2008).

OGAs trigger a robust oxidative burst mediated by NADPH oxidase AtRbohD (Zhang et al. 2007; Galletti et al. 2008). The OGA elicitor induced expression of the gene encoding NADPH oxidase (Denoux et al. 2008; Galletti et al. 2008) and the enzyme is involved in ROS signaling system (Sagi and Fluhr 2006). OGAs induced a rapid and transient generation of H_2O_2 in grapevine (Aziz et al. 2004). The transgenic tobacco plants expressing PG gene, which release endogenous OGA elicitor, showed accumulation of H_2O_2 (Ferrari et al. 2008). *AtPcb* gene encoding peroxidase, and *RbohC* and *RbohD* genes encoding NADPH oxidase are involved in ROS accumulation and OGAs activated expression of these genes in *Arabidopsis* (Denoux et al. 2008).

OGAs triggered fast and long-lasting NO production which was calcium ion – dependent in *Arabidopsis thaliana* challenged with *Botrytis cinerea* (Rasul et al.

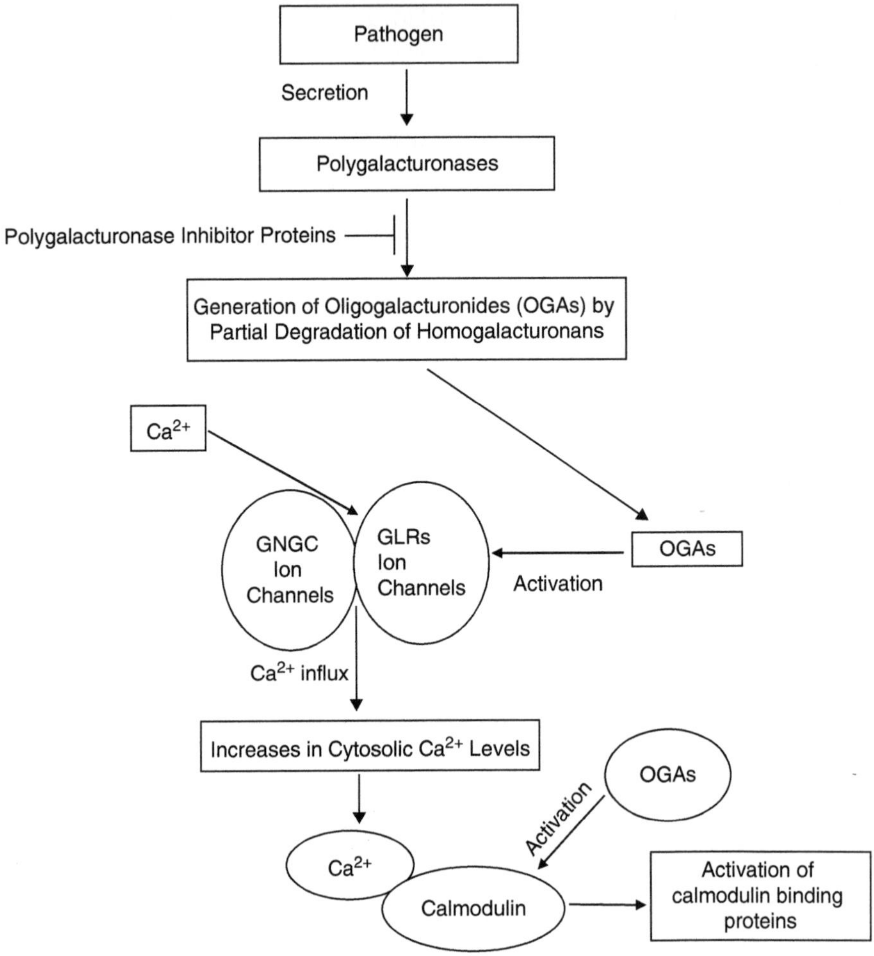

Fig. 4.1 OGAs-triggered Ca^{2+} signaling system in plant immune system (Adapted from Moscatiello et al. 2006; Denoux et al. 2008; Manzoor et al. 2013)

2012a). OGAs triggered an increase of both nitrate reductase (NR) activity and transcript accumulation (Rasul et al. 2012a, b). NO production was also affected by the NO synthase (NOS) inhibitor L-NAME (Rasul et al. 2012b). These results suggest that OGAs may trigger NO production by activating both NR and NOS pathways (Fig. 4.2). It has been further demonstrated that NO modulates RBOHD-mediated ROS production and participates in the regulation of anionic peroxidase (Rasul et al. 2012a, b). *Arabidopsis* plants treated with the NO scavenger cPTIO, were more susceptible to *B. cinerea* (Rasul et al. 2012a). The results suggest NO production mediates OGAs-triggered immunity and resistance to *B. cinerea* in *Arabidopsis*

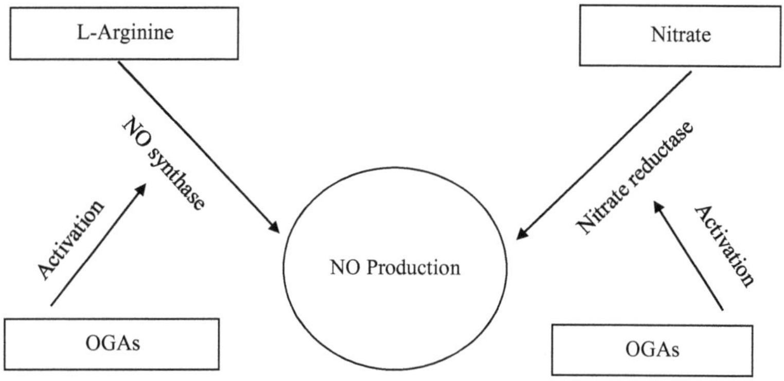

Fig. 4.2 OGAs-trigger NO production through NO synthase and Nitrate reductase pathways (Adapted from Rasul et al. 2012a, b)

OGAs also induced the enzymes involved in MAPK signaling cascade in *Arabidopsis* (Denoux et al. 2008; Galletti et al. 2011). OGAs induce the phosphorylation of two MAPKs, namely AtMPK3 and AtMPK6 (Galletti et al. 2011). AtMPK6 has been shown to be necessary for the early expression of defense genes. It activates CTR1, a MAPKK Kinase (Denoux et al. 2008) and the CTR1 negatively regulates ET signaling (Guo and Ecker 2004). The OGA elicitor induced the expression of MPK4, a MAPK, which acts as a negative regulator of defense responses through a salicylic acid-dependent signaling system (Petersen et al. 2000; Brodersen et al. 2006). MPK4 may also be involved in induction of a subset of ET-regulated genes (Brodersen et al. 2006). The elicitor also activated MKS1 (for MAP Kinase 4 Substrate 1), a substrate for MPK4 (Denoux et al. 2008). MKS1 mediates some effects of MPK4 on SA-mediated resistance responses (Andreasson et al. 2005).

The PIMP OGA triggers the expression of genes involved in SA, JA, ET, and ABA signaling systems (Denoux et al. 2008). Isochorismate synthase (SID2/ICS1) and phenylalanine ammonia-lyase (PAL1) are the key enzymes involved in SA biosynthesis (Garcion et al. 2008) and the PIMP induced the expression of these enzymes (Denoux et al. 2008). EDS1 and PAD4 are the regulatory proteins involved in synthesis of SA (Feys et al. 2001). EDS1 is required for SA production and it controls SA production to amplify defense signals (Eulgem et al. 2004; Song et al. 2004). EDS1 and PAD4 transduce ROS-derived signals leading to SA production (Wiermer et al. 2005). SAG101 is an in vivo signaling partner of the EDS1 and PAD4 complex (Wiermer et al. 2005; Xing and Chen 2006). EDS5 (SID1) is involved in transport of precursors for SA biosynthesis (Nawrath et al. 2002). The *NPR1* gene is a master regulator of the SA-mediated induction of defense genes (Tada et al. 2008). Expression of all these proteins is induced by OGAs (Denoux et al. 2008). The bZIP transcription factor TGA3 is a NPR1 interacting protein (Fitzgerald et al. 2005) and it is activated by the OGA elicitor (Denoux et al. 2008).

The OGA elicitor triggered the key enzymes involved in biosynthesis of JA in Arabidopsis. These enzymes include lipoxygenases (LOX3, LOX4),

12-oxophytodienoate reductase (OPR3) and acyl-CoA oxidase (ACX1) (Denoux et al. 2008). The OGA elicitor induced several ET-signaling components. These included the ET biosynthesis enzyme ACC synthase (ACS7) and several plant cell membrane-bound ethylene receptors such as CTR1, ETR1, ERF1, ERF4, and ERF11 (Denoux et al. 2008). Ethylene is perceived by the receptors and the signal is transmitted to downstream effectors (Qu et al. 2007; Grefen et al. 2008). The ERF1 and ERF4 transcription factors are involved in defense signaling system (Grennan 2008). The OGA elicitor induced the enzyme molybdenum cofactor sulfurase (ABA3), which is involved in the biosynthesis of ABA (Denoux et al. 2008).

Several transcription factors are activated by the OGA elicitor. The transcription factors activated by OGAs include WRKY7, WRKY18, WRKY22, WRKY33, WRKY40, and WRKY53 (Denoux et al. 2008). WRKY18 positively modulates defense-related gene expression (Chen and Chen 2002) whereas SA-inducible WRKY53 transcription factor interacts with a JA-inducible protein and triggers defense gene expression (Miao and Zentgraf 2007). WRKY7 and WRKY33 act as negative regulators of defense-related genes, including *PR1* gene (Kim et al. 2006; Zheng et al. 2006). While WRKY18 positively regulates expression of defense genes, its coexpression with WRKY40 results in suppression of the expression of defense responses (Xu et al. 2006). WRKY40 and WRKY53 transcription factors are both modulators of SA and JA signalling pathways and they function as activators of JA-dependent defense pathways and repressors of SA signalling system (Xu et al. 2006; Zheng et al. 2006). Both the transcription factors are strongly induced by by OGAs (Denoux et al. 2008).

OGAs have been reported to induce the expression of several defense-related genes in *Arabidopsis*. These include *PAL1* encoding phenylalanine ammonia-lyase, the key enzyme involved in biosynthesis of phenolics and phytoalexins, *Thi2.2* encoding thionin, and *HEL* encoding the PR-4 protein (Denoux et al. 2008). The OGAs induced accumulation of *PR-1* transcripts in *Arabidopsis* (Ferrari et al. 2008). The OGAs strongly up-regulated *PAD3* gene, which is involved in biosynthesis of phytoalexin in *Arabidopsis* (Ferrari et al. 2003). The OGA elicitor stimulated chitinase and β-1,3-glucanase activities in grapevine cells (Aziz et al. 2004).

4.3 OGAs with Different Degrees of Polymerization Differ in Triggering Defense Responses

The effectiveness of OGAs as eliciting molecules depends on specific structural requirements, such as their size (Aziz et al. 2004; Vorhőlter et al. 2012; Ferrari et al. 2013) and degree of methyl esterification (Wiethölteret al. 2003; Osorio et al. 2008; Randoux et al. 2010). The pectin fragments with varying degree of polymerization (DP) were found to differ in eliciting defense responses (Vorhőlter et al. 2012; Ferrari et al. 2013). Elicitor active pectin fragments differing in their DPs were obtained by co-incubation of *Capsicum annuum* cell walls with *Xanthomonas*

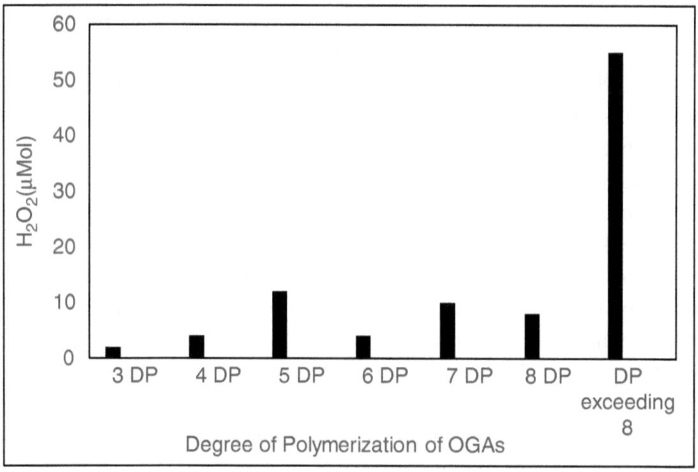

Fig. 4.3 Oxidative burst reactions in tobacco cell cultures after elicitation with OGAs differing in their DPs (Adapted from Vorhőlter et al. 2012)

campestris pv. *campestris* (Vorhőlter et al. 2012). The ability of OGAs with varying DPs to induce oxidative burst reactions in *Nicotiana tabacum* cell suspension cultures differed. The OGAs with DPs of 2, 3, 4 and 6 had only minimal effects on the *N. tabacum* cell cultures. The response to OGAs with DPs of 5, 7, and 8 was slightly stronger but still small. By contrast, OGAs with DP exceeding 8 triggered strong induction of H_2O_2 production (Vorhőlter et al. 2012; Fig. 4.3). OGAs released by fungal enzymes with DPs ranging from 9 to 12 were able to elicit oxidative burst reactions in cucumber hypocotyl segments (Svalheim and Robertsen 1993). OGAs with a DP below 9 were unable to trigger defense responses in soybean cotyledons (Nothnagel et al. 1983). OGAs with DP of 10–15 size have been found to be optimal for the formation of Ca^{2+}-mediated intermolecular cross-links resulting in structures called "egg boxes" (Cabrera et al. 2008), which are necessary for OGA activity (Ferrari et al. 2013).

4.4 Degree of Methyl Esterification of OGAs Modulates the Elicitor Activity of OGAs

Homogalacturonan (HGA) is synthesized in an esterified form in the Golgi apparatus and, subsequently, is secreted into the cell wall where it undergoes a partial de-esterification by the action of pectin methylesterases (PMEs) (Pelloux et al. 2007; Raiola et al. 2011; Ferrari et al. 2013). The degree of esterification of HGA varies widely and OGAs with different degrees of esterification are released under different conditions. Esterification appears to affect the biolological activity of OGAs

(Randoux et al. 2010). Reduced degree of esterification of OGAs has been shown to be necessary for triggering defense responses against *Botrytis cinerea* in strawberry (Osorio et al. 2008). The presence of OGAs with a low degree of methylation in strawberry fruits overexpressing a PME was correlated with the expression of defense responses and a partial resistance against *B. cinerea* (Osorio et al. 2008). Acetylated OGAs, but not de-esterified OGAs, reduce the haustoria formation of *Blumeria graminis* growing on wheat leaves, suggesting that esterification is necessary for some specific responses (Pelloux et al. 2007; Randoux et al. 2010). *Arabidopsis thaliana* plants overexpressing an inhibitor of PME have a high degree of pectin methyl esterification (Lionetti et al. 2007, 2010; Raiola et al. 2011). These plants exhibit enhanced resistance to *Botrytis cinerea* and *Pectobacterium carotovorum* (Lionetti et al. 2012). These studies suggest that OGA structure type plays an important role in triggering plant defense responses. Pectin methyl esterification is critically important in the function of OGAs in manipulation of plant immune responses for crop disease management.

The degree and pattern of pectin methyl esterification plays an important role in host defense responses against soft rot erwinias in potato. Pectin esterification was higher in the resistant genotype than that in susceptible lines (McMillan et al. 1993; Marty et al. 1997). Tomato isolines resistant to *Colletotrichum lindemuthianum* had higher degree of methylesterification than that in susceptible tomato isolines (Boudart et al. 1998). A higher degree of pectin methylation was found to be correlated with resistance to *Ralstonia solanacearum* (Wydra and Beri 2006) in tomato genotypes. Differences in methyl ester distribution in homogalacturons have been found in near- isogenic wheat lines resistant and susceptible to the *Puccinia graminis* f. sp. *tritici*. In particular, a more random distribution of methyl ester was found in isogenic resistant lines compared with a more blockwise distribution found in the susceptible lines (Wiethölteret al. 2003).

The methylesterification of pectin is mainly controlled by pectin methyl esterase (PME) (Willats et al. 2001; Volpi et al. 2011). A pectin methyl esterase inhibitor protein (PMEI) was isolated from kiwi (*Actinidia chinensis*). Transgenic wheat lines expressing *Acpmei* gene encoding AcPMEI protein, were developed (Volpi et al. 2011). The transgenic lines showed a significant increase of degree of pectin methyl esterification of approximately 32 % over wild-type plants (Fig. 4.4; Volpi et al. 2011). The *Acpmei* lines homozygous for the transgene showing a high reduction of endogenous PME activity showed enhanced resistance against the spot blotch disease caused by *Bipolaris sorokiniana* (Fig. 4.5; Volpi et al. 2011). These transgenic plants also showed resistance against the Fusarium head blight pathogen *Fusarium graminearum* (Volpi et al. 2011). These results suggest that *PMEI* gene may be a potential tool to develop disease-resistant plants.

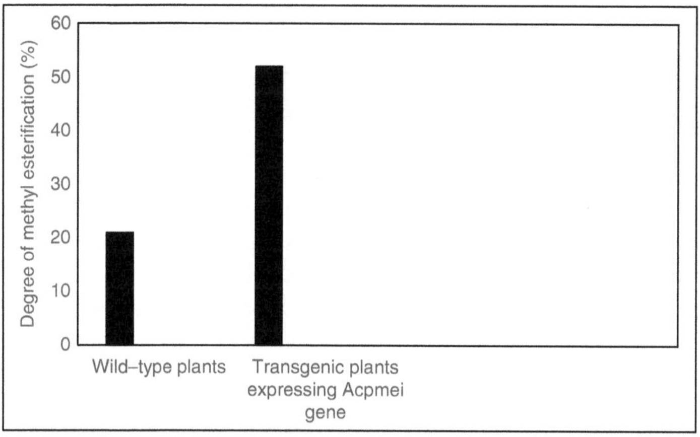

Fig. 4.4 Increases in degree of methyl esterification of cell wall pectin in transgenic wheat plants expressing the kiwi pectin methyl esterase inhibitor *Acpmei* gene (Adapted from Volpi et al. 2011)

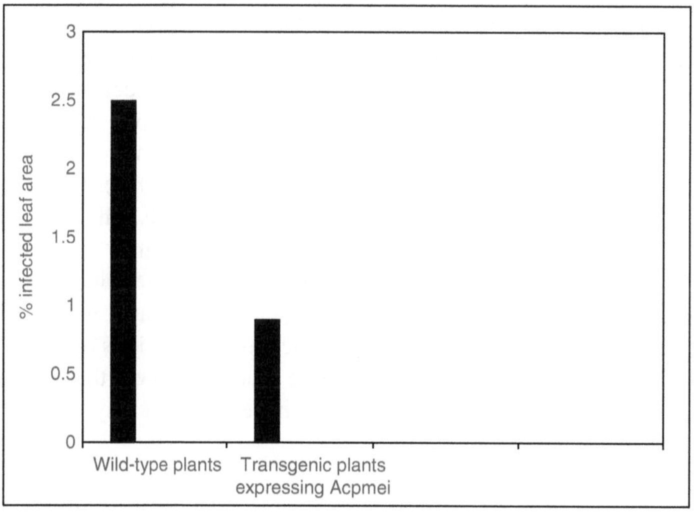

Fig. 4.5 Increased resistance against *Bipolaris sorokiniana* in transgenic wheat plants expressing pectin methyl esterase inhibitor gene (*Acpmei*) from kiwi (*Actinidia chinensis*) (Adapted from Volpi et al. 2011)

4.5 Ability of OGAs to Trigger Defense Responses May Depend on Their Level of Acetylation

Many cell wall polysaccharides are esterified with *O*-acetyl substituents to varying extents (Manabe et al. 2011). Pectic polysaccharides have complex structures that are characterized by a high content of 1,4-linked α-D-galacturonic acid (GalUA) residues. Acetylation of the GalUA residues of pectin is found in homogalacturonans (Mohnen 2008). Acetylation of OGAs takes place in the Golgi apparatus, and acetyl-coA has been identified as a donor substrate for pectin acetylation in plants (Pauly and Scheller 2000). *RWA2* (*REDUCED WALL ACETYLATION2*) gene is involved in polysaccharide *O*-acetylation. Loss-of-function *rwa2* mutants had decreased levels of acetylated cell wall polymers. Acetylation of pectic polysaccharides has been shown to be affected by the *rwa2* mutation. The *Arabidopsis rwa2* mutants showed increased resistance to *Botrytis cinerea* (Manabe et al. 2011). These results suggest that reduced acetylation of OGAs may trigger defense responses and confer disease resistance.

Differences in ability to trigger defense responses were observed between acetylated and non-acetylated fractions of OGAs in wheat against the powdery mildew pathogen *Blumeria graminis* f. sp. *tritici* (Randoux et al. 2009, 2010). Microscopic assessment of the number of haustoria occurring 40 h post-inoculation of *B. graminis* f. sp. *tritici* showed that it was reduced only when acetylated fraction was used (Randoux et al. 2009, 2010). Leaves infiltrated with non-acetylated OGA fragments provided the same distribution of penetration events as control leaves. In contrast, when acetylated OGAs were used, the percentage of strongly fluorescent papillae without haustorium rose from 14.5 to 56 %. The percentage of penetration events showing a haustorium was very much reduced in acetylated OGA fragment (Randoux et al. 2010). Despite different modes of action of the acetylated and non-acetylated pectin fractions, a similar 45 % protective effect occurred in both cases when the OGA fractions were sprayed on wheat plans (Randoux et al. 2009). Foliar spray with both acetylated and non-acetylated OGs significantly reduced *B. graminis* f. sp. *tritici* infection almost similarly (Randoux et al. 2010). Both the form of OGAs triggered almost similar early defense signaling events. The accumulation of H_2O_2 at the penetration site of the fungus increased 21 h after inoculation to the same extent in areas of plantlets infiltrated by both acetylated and non-acetylated OGA fractions (Randoux et al. 2009, 2010). Plantlets infiltrated with acetylated or non-acetylated OGAs showed a very similar increase in activities of peroxidase and oxalate oxidase (Randoux et al. 2010). These studies suggest that acetylation of OGAs may only partially modulate early signaling events, but may affect the expression of defense responses.

4.6 Engineering Pectin Methyl Esterase Genes to Develop Disease Resistant Plants

Pectin, one of the main components of the plant cell wall, is continuously modified and remodeled during plant growth (Osorio et al. 2008; Taurino et al. 2014). Pectin is known to be synthesized with a high degree of methyl esterification of the carboxyl groups and secreted to the cell wall, where it is de-esterified *in muro* by pectin methyl esterases (PME) (Osorio et al. 2008; Lionetti et al. 2012; Bethke et al. 2014). Pectin undergoes depolymerization by the activities of pectin-modifying enzymes such as polygalacturonases and pectate lyases. Pectin-derived oligogalacturonides (OGAs) are essential components for elicitation of defense responses. Modifications of pectin by various pectin-modifying enzymes alter the ability of pectin to elicit defense responses (Osorio et al. 2008).

Transgenic wild strawberry *Fragaria vesca* plants expressing a fruit specific PME gene, *FaPE1*, were developed (Osorio et al. 2008). Expression of the PME gene in *Fragaria* was found to modify pectin structure in the fruits. Pectin from transgenic lines differed from the wild-type with regard to the degree and pattern of methyl esterification, as well as the average size of pectin polymers. Purified oligogalacturonides from the transgenic fruits showed a reduced degree of esterification compared to oligogalacturonides from wild-type fruits (Osorio et al. 2008). Methyl de-esterification is a major factor regulating the extent of pectin depolymerization (Wakabayashi et al. 2003). Purified OGA from the *Fragaria* fruits of transgenic plants caused a significant increase in *PR5* gene expression compared to expression of *PR5* gene in fruits injected with OGA from the wild-type plants (Osorio et al. 2008). The OGA from the transgenic fruit triggered high accumulation of SA (Fig. 4.6; Osorio et al. 2008) and activation of several defense-related genes (Osorio et al. 2011). These results suggest that PME can modify the structure of OGAs and the modified OGAs will be highly active in triggering SA signaling system and defense responses.

The transgenic strawberry (*F. vesca*) lines expressing *PME* gene showed high resistance against the fruit rot pathogen *Botrytis cinerea* (Osorio et al. 2008, 2011). These studies reveal that engineering pectin methyl esterase gene is a powerful tool to develop disease-resistant plants.

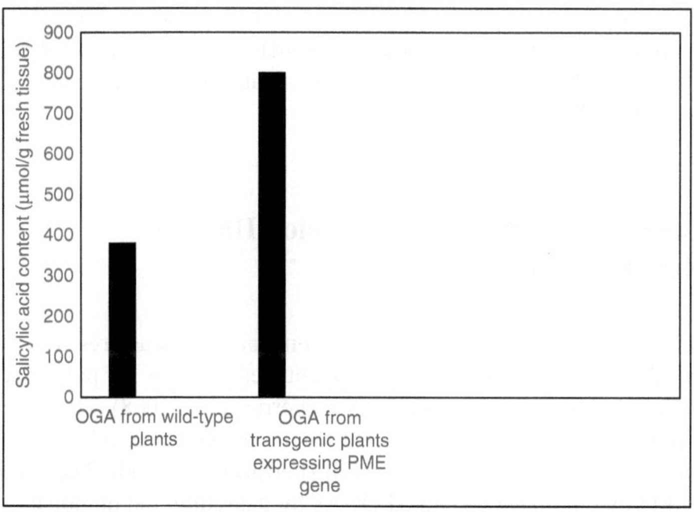

Fig. 4.6 Purified oligogalacturonides from transgenic strawberry fruits expressing PME gene trigger enhanced salicylic acid accumulation in strawberry fruits (Adapted from Osorio et al. 2008)

4.7 Bioengineering Pectin Methyl Esterase Inhibitor Protein for Plant Disease Management

PME activity is tightly regulated by an inhibitor protein called pectin methylesterase inhibitor protein (PMEI) (Giovane et al. 2004). The inhibitor protein PMEI forms a stoichiometric 1:1 complex with PME. It binds with the putative active site of the enzyme PME (Di Matteo et al. 2005). Transgenic *Arabidopsis* plants overexpressing genes encoding two PME inhibitor proteins, AtPMEI-1 and AtPMEI-2, were developed. Both the inhibitor proteins interacted with endogenous PME isoforms, and the transgenic tissues showed a significant reduction of PME activity (Lionetti et al. 2007). These plants showed an increased level of cell wall pectin methyl esterification. These transgenic *Arabidopsis* plants showed enhanced resistance to *Botrytis cinerea* (Lionetti et al. 2007). The gene *CaPMEI1*encoding the pectin methyl esterase inhibitor protein CaPMEI1 was isolated from pepper (*Capsicum annuum*) and transgenic *Arabidopsis* plants expressing the pepper gene were developed (An et al. 2008). Transgenic *Arabidopsis CaPMEI1*-overexpression lines exhibited enhanced resistance to the bacterial pathogen *Pseudomonas syringae* pv. *tomato* (An et al 2008).

Tobacco plants constitutively expressing a *PMEI* gene from *Actinidia chinensis*, showed resistance against *Tobacco mosaic virus* (TMV). In the transgenic plants, systemic movement of TMV was limited and disease symptoms were reduced (Lionetti et al. 2014). Transgenic *Arabidopsis* plants expressing a PME inhibitor gene from Arabidopsis, *AtPMEI2*, showed reduced susceptibility to *Turnip vein clearing virus* (TVCV). A delayed movement of the virus was also observed in the transgenic Arabidopsis plants expressing the PME inhibitor gene (Lionetti et al. 2014). However, these transgenic lines did not show enhanced resistance to the biotrophic oomycete pathogen *Hyaloperonospora parasitica* (An et al. 2008). Collectively these studies suggest that pectin methyl esterase inhibitor gene can be an important tool for management of diseases caused by a wide range of fungal, bacterial, and viral pathogens.

4.8 Engineering *PG* Gene to Develop Disease Resistant Plants

To exploit the potential of OGA elicitor in enhancing disease resistance, tobacco and *Arabidopsis* plants expressing an attenuated version of polygalacturonase (PGII) of *Aspergillus niger* were developed (Ferrari et al. 2008). The PG expressed in these plants had a point mutation that has an estimated activity approximately 20-fold lower than the native *A. niger* PGII (Ferrari et al. 2008). The expression of attenuated PG would have generated OGAs by degrading the homogalacturonans. Partial degradation of homogalacturonans by fungal PGs releases OGAs with a degree of polymerization between 10 and 15 that show elicitor activity (Cervone

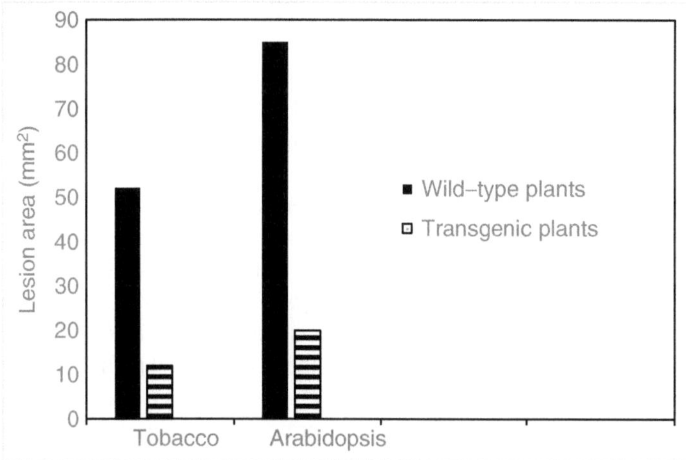

Fig. 4.7 Enhanced resistance to *Botrytis cinerea* in transgenic tobacco and *Arabidopsis* plants expressing fungal PG gene (Adapted from Ferrari et al. 2008)

et al. 1987, 1989). The transgenic *Arabidopsis* plants expressing PG gene showed enhanced expression of *AtPGIP1* when inoculated with the fungal pathogen *Botrytis cinerea* (Ferrari et al. 2008). Polygalacturonase inhibitor protein (PGIP) is known to trigger defense responses in plants (Ferrari et al. 2012; Wang et al. 2013). Both tobacco and *Arabidopsis* transgenic plants expressing PG showed enhanced resistance against the gray mold pathogen *Botrytis cinerea* (Fig. 4.7; Ferrari et al. 2008). The tobacco transgenic plants expressing the PG gene did not show any visible symptoms when inoculated with the bacterial pathogen *Pseudomonas syringae* pv. *tabaci* while the wild-type plants showed collapse of the inoculated area (Ferrari et al. 2008). These studies suggest that OGAs can be exploited to develop plants showing resistance against broad-spectrum of pathogens.

4.9 Engineering *PGIP* Gene to Develop Disease-Resistant Plants

Polygalacturonase-inhibiting proteins (PGIPs) play important role in switching on plant defense signaling pathways (Manfredini et al. 2005; Federici et al. 2006; Alexandersson et al. 2011). Antisense expression of the *Arabidopsis thaliana AtPGIP1* gene has been shown to reduce polygalacturonase-inhibiting protein accumulation and enhance susceptibility to *Botrytis cinerea* in *Arabidopsis* (Ferrari et al. 2006). PGIPs are glycoproteins located in plant cell wall which reduce the hydrolytic activity of polygalacturonases (PGs) and elicit defense responses in plants (Di Matteo et al. 2003, 2006; Di et al. 2006). PGIPs belong to the super family of leucine-rich repeat (LRR) proteins (Di et al. 2006; Kalunke et al. 2014). PGIPs are typically encoded by gene families with a small number of gene copies. Soybean genome contains two *pgip* loci and each locus is composed of three clustered genes that are induced following infection with *Sclerotinia sclerotiorum* (Kalunke et al. 2014). PGIPs selectively inhibit PGs secreted by invading plant pathogenic fungi (Wang et al. 2013). They display differential inhibition towards PGs from different fungi, also towards different isoforms of PGs originating from a specific pathogen (Joubert et al. 2006). However, three members of the pepper *PGIP* gene family acted similarly in inhibiting PGs produced by *Alternaria alternata* and *Colletotrichum nicotianae*. These three PGIP genes acted similarly in conferring resistance to infection by *Phytophthora capsici* in pepper (*Capsicum annuum*) (Wang et al. 2013). Transgenic tobacco plants expressing *CaPGIP1* gene isolated from pepper plants showed enhanced resistance to the fungal pathogens *A. alternata* and *C. nicotianae* (Wang et al. 2013). A *PGIP* gene from *Vitis vinifera* (*Vvpgip*) was isolated and transgenic tobacco plants overexpressing the grapevine gene was developed. Over-expression of the grapevine *PGIP* gene in tobacco showed enhanced resistance against the necrotrophic fungal pathogen *Botrytis cinerea* (Joubert et al. 2006, 2007; Nguema-Ona et al. 2013). Increased lignin levels were observed in the transgenic tobacco plants expressing *Vvpgip1 gene* (Alexandersson et al. 2011).

Transgenic tobacco plants expressing *pgip2* encoding bean polygalacturonase inhibiting protein PGIP2 show enhanced resistance against *Botrytis cinerea* (Fig. 4.8, Ferrari et al. 2008). Transgenic wheat plants expressing a bean PGIP gene (*Pvpgip2*) showed enhanced resistance against the foliar blotch spot disease pathogen *Bipolaris sorokiniana* (Fig. 4.9; Janni et al. 2008). Transgenic wheat plants expressing the bean *Pvpgip2* in their flowers also showed resistance against *Fusarium* head blight and had a significant reduction of symptoms when infected with *Fusarium graminearum* (Ferrari et al. 2012). Transgenic tobacco and

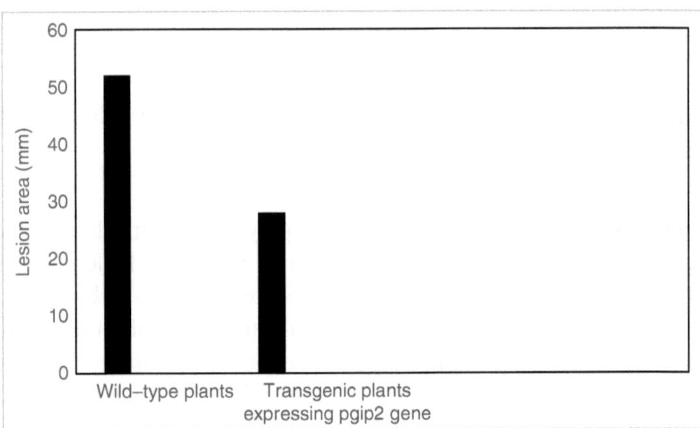

Fig. 4.8 Transgenic tobacco plants expressing *pgip2* encoding bean polygalacturonase inhibiting protein PGIP2 showing resistance to *Botrytis cinerea* (Adapted from Ferrari et al. 2008)

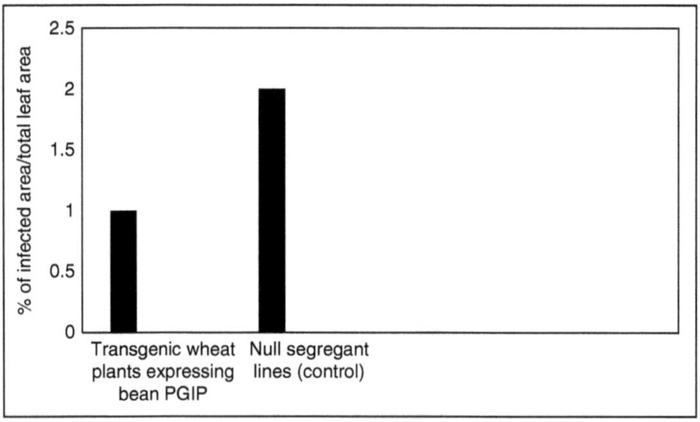

Fig. 4.9 Transgenic wheat plants expressing bean PGIP gene show enhanced resistance against *Bipolaris sorokiniana* (Adapted from Janni et al. 2008)

Arabidopsis plants overexpressing *PvPGIP2* show enhanced resistance against *B. cinerea* (Manfredini et al. 2005).

Transgenic tobacco plants expressing bean PGIP gene (*Pvpgip2*) showed resistance against *Rhizoctonia solani*. The disease symptoms (both seedling death and stem rot) were severe on wild-type plants. Disease symptoms were very limited and less visible in transgenic lines (Borras-Hidalgo et al. 2012). These transgenic tobacco plants also showed resistance to the oomycete pathogens *Phytophthora parasitica* var. *nicotianae* and *Peronospora hyoscyami* f. sp. *tabacina* (Borras-Hidalgo et al. 2012). The efficacy of the bean PGIP gene for engineering disease resistance in tobacco against the blue mold pathogen *P. hyoscyami* was assessed under field conditions. Trials were conducted in the tobacco area in Cuba with a high inoculum pressure where *P. hyoscyami* f. sp. *tabacina* is a significant problem for tobacco production each year. The transgenic plants displayed a high level of resistance against the blue mold pathogen in the field trials (Fig. 4.10; Borras-Hidalgo et al. 2012). Transgenic tomato plants expressing the pear fruit polygalacturonase inhibitor protein (pPGIP) showed enhanced resistance against *Botrytis cinerea* infection. Hetrologous expression of *pPGIP* gene in tomato slows the expansion of disease lesions (Powell et al. 2000).

Pierce's disease caused by the xylem-limiting bacterium *Xylella fastidiosa* is a serious disease in grapevine. The bacterial pathogen causes yellowing and gradual necrosis of grapevine leaves resulting in vine death. Transgenic grapevine cultivars expressing pear fruit PGIP-encoding gene (*pPGIP*) were developed (Agüero et al. 2005). The development of Pierce's disease was delayed in the transgenic grapevine plants expressing increased pPGIP activity. The transgenic lines had reduced leaf scorching, lower *Xylella* titres and better re-growth after pruning than the untransformed controls (Agüero et al. 2005). These transgenic grapevine plants also showed resistance to the fungal pathogen *Botrytis cinerea.*

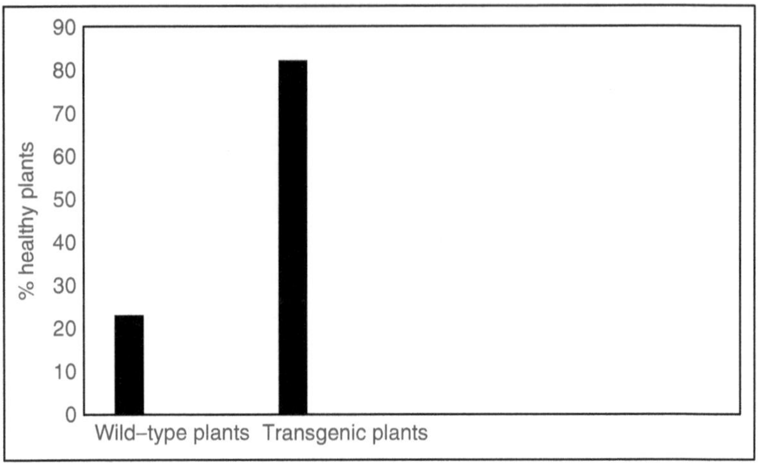

Fig. 4.10 Transgenic tobacco plants expressing a bean gene (*Pvpgip2*) show high resistance against the blue mold pathogen *Peronospora hyoscyami* f. sp. *tabacina* under field conditions (Adapted from Borras-Hidalgo et al. 2012)

Collectively, the studies conducted both under greenhouse and field conditions (Table 4.1) clearly show that PGIP gene is a potential tool for engineering disease resistance against various fungal, oomycete and bacterial pathogens. The expression of PGIP genes did not affect the agronomic characters of the transformed plants. The expression of PG inhibitor protein neither alters the physiological performances nor exhibits detrimental effects on the growth of transgenic plants (Powell et al. 2000; Capodicasa et al. 2004; Agüero et al. 2005; Borras-Hidalgo et al. 2012; Nguema-Ona et al. 2013).

4.10 Manipulation of Oligogalacturonides by Salicylic Acid (SA) Analog to Induce Resistance against Pathogens

SA has been reported to regulate the responsiveness of plants to PAMPs and PRRs (Tateda et al. 2014). Manipulation of PGIP gene expression by a salicylic acid analog has been shown to activate defense responses (Jiang et al. 2008). The pear scab pathogen *Venturia nashicola* grows in pectin-enriched regions of cell walls in pear leaves (Park et al. 2000). It produces polygalacturonase (Isshiki et al. 2000) and the enzyme may be involved in degradation of pectin in plant cell walls (Park et al. 2000). The pectin degradation may be an important infection process for utilizing nutrients by the pathogen (Jiang et al. 2007). Application of the synthetic analog of salicylic acid (SA), acibenzolar-S-methyl (ASM), resulted in interference in pectin degradation process during pathogenesis. Ultrastructural studies revealed that the

Table 4.1 Bioengineering polygalacturonase inhibiting proteins for crop disease management

Gene used for engineering disease resistance	Developed transgenic plants	Pathogen against which resistance is induced by the transgene	References
Bean *Pvpgip2* gene	Wheat	*Fusarium graminearum*	Ferrari et al. (2012)
Bean *Pvpgip2* gene	Wheat	*Bipolaris sorokiniana*	Janni et al. (2008)
Bean *Pvpgip2* gene	Tobacco	*Botrytis cinerea*	Ferrari et al. (2008)
Pepper *CaPGIP1* gene	Tobacco	*Alternaria alternata*	Wang et al. (2013)
Grapevine *Vvpgip* gene	Tobacco	*Botrytis cinerea*	Joubert et al. (2007) and Nguema-Ona et al. (2013)
Bean *Pvpgip2* gene	Tobacco	*Rhizoctonia solani*	Borras-Hidalgo et al. (2012)
Bean *Pvpgip2* gene	Tobacco	*Peronospora hyoscyami* f. sp. *nicotianae*	Borras-Hidalgo et al. (2012)
Pear *pPGIP* gene	Tomato	*Botrytis cinerea*	Powell et al. (2000)
Pear *pPGIP* gene	Grapevine	*Xylella fastidiosa*	Agűero et al. (2005)
Attenuated version of *Aspergillus niger* PGII gene	Tobacco	*Botrytis cinerea*	Ferrari et al. (2008)
Attenuated version of *Aspergillus niger* PGII gene	Tobacco	*Pseudomonas syringae pv. tabaci*	Ferrari et al. (2008)
AtPME1 gene from *Actinidia chinensis*	Tobacco	*Tobacco mosaic virus*	Lionetti et al. (2014)
FAPE1 PME gene from strawberry	Strawberry	*Botrytis cinerea*	Osorio et al. (2008, 2011)
CaPMEI gene from pepper	Arabidopsis	*Pseudomonas syringae* pv. *tomato*	An et al. (2008)
AtPMEI2 gene from *Actinidia chinensis*	Arabidopsis	*Turnip vein clearing virus*	Lionetti et al. (2014)

subcuticular hyphae of the pathogen *V. nashicola* digested the pectin layers in water-treated control leaves much more than in ASM-treated leaves (Jiang et al. 2008). It suggests that activity or production of pectin-degrading enzymes may be inhibited in ASM-treated leaves. The inhibition activity may be due to the expression of pectin-degrading enzyme-inhibitor such as PGIP, which is able to inactivate PG produced by the hyphae (Jiang et al. 2008). The expression of *PGIP* gene encoding PGIP was enhanced in ASM-pretreated pear leaves postinoculated with the pear scab pathogen *Venturia nashicola* (Faize et al. 2004). The expression of PGIP may be the most important defense strategy against the scab pathogen and ASM may induce resistance by enhancing the expression of *pgip* gene in pear (Jiang et al. 2008).

4.11 Switching on Plant Immune Signaling Systems Using Plant Elicitor Peptides (Peps) for Disease Management

4.11.1 Plant Elicitor Peptides

Plant elicitor peptides (Peps) are the peptide signals that can trigger defense responses in plants (Huffaker et al. 2006; Yamaguchi and Huffaker 2011; Ma et al. 2013). Several Peps have been identified in *Arabidopsis* (Huffaker and Ryan 2007). A 23-aa peptide, AtPep1, has been identified as a PIMP/HAMP in *Arabidopsis* (Huffaker et al. 2006). The peptide was found to be derived from a 92-aa precursor encoded within a small gene (Huffaker et al. 2006). The peptide is processed from the precursor protein, PROPEP1. The gene encoding PROPEP1 is a member of a small family composed of six annotated genes and one unannotated gene (Huffaker et al. 2006; Huffaker and Ryan 2007). The six genes encode six peptides, PROPEP1, PROPEP2, PROPEP3, PROPEP4, PROPEP5, and PROPEP6 and synthesize six different AtPep peptides, Pep1-Pep6 (Huffaker and Ryan 2007; Logemann et al. 2013). Peps present initially at the site of infection act as signals that induce immune responses upon binding to plasma membrane-localized PRRs (Pep receptors or PEPRs) (Yamaguchi and Huffaker et al. 2011). The receptor for AtPep1, PEPR1, has been isolated and identified as an LRR receptor kinase (Yamaguchi et al. 2006).

A pathogen-inducible gene orthologous to the Arabidopsis *AtPROPEP1* gene has been identified in maize (*Zea mays*) and called *ZmPROPEP1* (Huffaker et al. 2011). *ZmPROPEP1* encodes a peptide, *Zm*Pep1, which has been identified as a PIMP/HAMP. The native length of *Zm*Pep1 is predicted to be 23 amino acids similar to the *At*Pep peptides of *Arabidopsis thaliana* (Huffaker et al. 2011). Orthologs of AtPROPEP1 have been identified in rice, wheat, barley, maize, potato, soybean, grape, canola, poplar, and alfalfa, suggesting that Pep proteins may be present in wide range of plants (Huffaker et al. 2006).

4.11.2 Peps – Triggered Immune Signaling Systems

Plant elicitor peptides accumulate at the site of infection and switch on the immune signaling systems upon binding to plasma membrane-localized Pep Receptors (PEPRs) (Yamaguchi and Huffaker 2011). Pep binding to the PEPR1 or PEPR2 receptor activates a cyclic GMP-dependent Ca^{2+}-conducting ion channel (cyclic nucleotide gated channel2; CNGC2) (Qi et al. 2010; Ma et al. 2012; Fig. 4.11). CNGCs are Ca^{2+}-permeable cation transport channels (Nawaz et al. 2014) and involved in the uptake of various cations, such as Ca^{2+} (Chen et al. 2015). Peps detected in *Arabidopsis*, AtPeps, activate CNGC2-dependent plasma membrane inwardly conducting Ca^{2+} permeable channels in mesophyll cells, resulting in elevation of cytosolic Ca^{2+} ($[Ca^{2+}]_{cyt}$) level (Qi et al. 2010). This activity was dependent on the Pep Receptors. The leucine-rich repeat receptor-like kinase receptor AtPEPR1

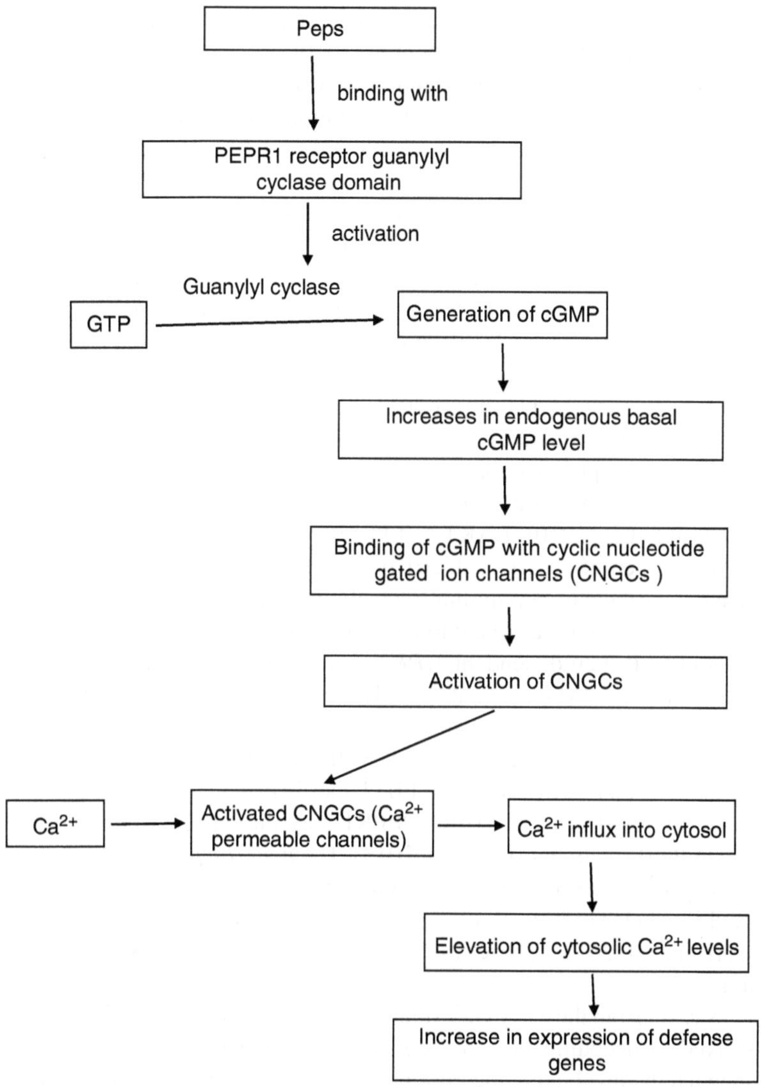

Fig. 4.11 Peps-induced elevation of cytosolic Ca^{2+} levels in plants (Adapted from Qi et al. 2010; Ma et al. 2012, 2013)

has guanylyl cyclase (GC) activity (Qi et al. 2010). GC activity is enhanced over twofold by Ca^{2+} in a concentration-dependent manner (Muleya et al. 2014). Overexpression of a leucine-rich repeat receptor-like kinase in *Arabidopsis* leaf protoplasts raised the endogenous basal cGMP levels over 20-fold, indicating that the receptor has GC activity in vivo (Kwezi et al. 2011). Guanylyl cyclase is involved in generation of cGMP (cyclic guanosine-3′,5′-cyclic monophosphate) from GTP (Qi et al. 2010). PEPRs have a cytosolic guanylyl cyclase (GC) domain that is responsible for cGMP production (Kwezi et al. 2011). Affinity purified PEPR1-GC domain generated cGMP in vitro (Qi et al. 2010). Exogenous application of Pep1 led to in vivo cytosolic cGMP production in *Arabidopsis* plants (Ma et al. 2012). Pep application caused in vivo cGMP generation and downstream signaling that was lost when the predicted PEPR receptor guanylyl cyclase active domain was mutated (Ma et al. 2012). AtCNGCs are gated primarily by binding of cGMP (Yoshioka et al. 2006). The generated cGMP may be involved in activation of CNGC2-dependent cytosolic Ca^{2+} elevation (Qi et al. 2010). The results suggest that PEPR signaling requires an extracellular Ca^{2+} and a functional version of the CGMP-activated Ca^{2+} channel CNGC2 (Ma et al. 2012). The signaling cascade initiated by AtPeps/AtPEPR1 leads to expression of defense genes in a Ca^{2+}-dependent manner (Fig. 4.11; Krol et al. 2010; Qi et al. 2010; Yamaguchi et al. 2010; Kwezi et al. 2011; Yamaguchi and Huffaker 2011; Ma et al. 2012, 2013; Hung et al. 2014).

Brassinosteroid-insensitive1 associated kinase1 (BAK1) is a co-activator of the PAMP-receptor complex to enhance various signaling pathways (Wang et al. 2008). BAK1 has been found to physically interact with PEPRs (PEPR1 and PEPR2) in vitro (Postel et al. 2010). BAK1 and PEPRs can phosphorylate each other after after Pep1 treatment (Schulze et al. 2010). BAK1 may function as a coreceptor with PEPRs and contribute to Pep immune signaling leading to defense gene expression (Ma et al. 2013). BAK1 was found to contribute to the generation of a Ca^{2+} signal in response to PEPR. Pep-dependent *MPK3* expression was substantially impaired in a *bak1* null mutant (Ma et al. 2013). These results suggest the importance of BAK1 in downstream PEPR signaling.

PEPR1 has been shown to interact with receptor-like cytoplasmic kinases botrytis-induced kinase 1 (BIK1) and PBS1-like 1 (PBL1) to mediate Pep1-induced defense responses (Liu et al. 2013). PEPR1 may directly phosphorylate BIK1 in response to Pep1 treatment (Liu et al. 2013). The *pepr1/pepr2* double-mutant seedlings displayed reduced in sensitivity to ethylene. ET-induced expression of defense genes and resistance to *Botrytis cinerea* were compromised in *pepr1/pepr2* and *bik1* mutants (Liu et al. 2013). These results show the importance of BIK1 in Pep-PEPR signaling pathway.

Downstream from the early Ca^{2+} signal, Ca^{2+}-dependent protein kinases (CDPKs) are involved in decoding the Ca^{2+} signal. CDPK-dependent phosphorylation has been shown to be involved in the Pep signaling cascade (Ma et al. 2013). Binding of Pep to its receptor leads to ROS generation in Arabidopsis (Ma et al. 2012, 2013). Pep-evoked ROS generation occurred downstream from cytosolic Ca^{2+} elevation in this signaling cascade (Ma et al. 2013). NO generation was also observed in *Arabidopsis* downstream of Pep signaling. Application of the NO scavenger cPTIO inhibited Pep-dependent defense gene expression, suggesting the importance of NO generation in Pep signaling (Ma et al. 2013). NO generation appears to require the critical and early Ca^{2+} elevation in the Pep signaling pathway (Ma et al. 2013).

All *At*Pep peptides, except AtPep4, strongly induced SA signaling pathway inducing the gene encoding PR-1 protein. H_2O_2 production was found essential to *At*Pep-mediated expression of JA/ET and SA defense signaling pathways (Huffaker et al. 2006; Huffaker and Ryan 2007). These results suggest that ROS signaling system induced by *At*Pep peptides mediate JA, ET, and SA signaling systems and induce the expression of defense genes. Localized Pep application in *Arabidopsis* has been shown to activate genetically separable JA and SA branches in systemic leaves. Local PEPR activation provided a critical step in connecting local to systemic immunity by reinforcing separate defense signaling pathways (Ross et al. 2014).

ZmPep1, the ortholog of AtPep1 detected in maize activates de novo synthesis of JA and ET. It triggered the expression of *AOS* gene encoding allene oxide synthase and *AOC* gene encoding allene oxide cyclase, which are involved in JA biosynthesis (Huffaker et al. 2011). ZmPep1 treated leaves exhibited 30-fold increased induction of *AOC* gene expression, while 3.8-fold greater induction of *AOS* transcripts was observed in ZmPep1-treated leaves (Huffaker et al. 2011). ZmPep1 treatment also triggered ethylene production and expression of the gene encoding 1-amino-cyclopropane-1-carboxylic acid oxidase (ACC oxidase) also responded to ZmPep1 treatment (Huffaker et al. 2011).

*At*Pep1 induced the expression of the defensin gene *PDF1.2* in wild-type *Arabidopsis* leaves but not in ethylene-insensitive or jasmonate-deficient mutants, suggesting that both JA and ET signaling systems are involved in *At*Pep peptides–activated defense signaling system (Huffaker and Ryan 2007). All the six *At*Pep peptides induced the expression of JA and ET-mediated signaling systems (Huffaker and Ryan 2007). The *At*Pep2 induced *PDF1.2* more strongly than *At*Pep1 and the other four peptides only weakly induced the expression of *PDF1.2* (Huffaker and Ryan 2007). ZmPep1 treatment induced several defense genes. Expression of genes encoding endochitinase A, PR-4 and PRm pathogenesis-related proteins and peroxidase was elevated in excised leaves that had been supplied with ZmPep1 (Huffaker et al. 2011).

4.11.3 Management of Crop Diseases Using Pep Proteins

Pep proteins can be developed as potential tools for disease management. Application of Pep proteins induced resistance against *Pseudomonas syringae* pv. *tomato* in *Arabidopsis* (Yamaguchi et al. 2010). A 23-amino acid peptide corresponding to the predicted ZmPep1 active peptide sequence was synthesized. The ZmPep1 treatment improved the plant disease resistance. When the maize plants were treated with ZmPep1 at 18 h prior to inoculation with the southern leaf blight pathogen *Cochliobolus heterostrophus*, the plants showed resistance against the pathogen. Chlorotic lesions spread from the wound sites of infected leaves that had been pretreated only with water, while the lesions were contained at the edge of the wound site and had not spread in the leaves pretreated with ZmPep1 (Fig. 4.12; Huffaker et al. 2011). ZmPep1 treatment effectively reduced the incidence of maize anthracnose stalk rot caused by *Colletotrichum graminicola* (Huffaker et al. 2011). These results reveal the potential of ZmPep1 formulations in management of crop diseases.

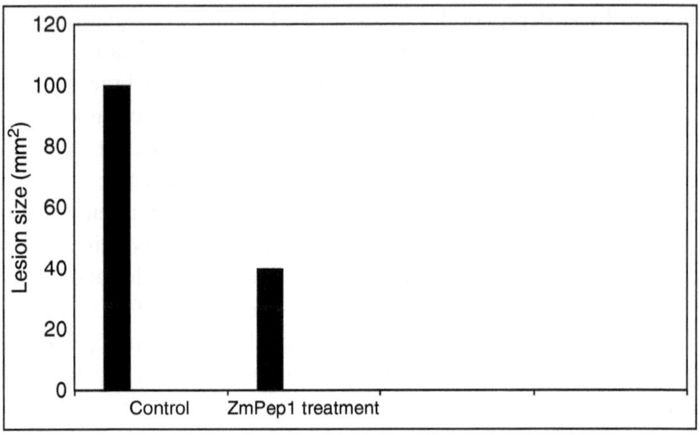

Fig. 4.12 ZmPep1 pretreatment induces resistance against southern leaf blight disease caused by *Cochliobolus heterostrophus* in maize plants (Adapted from Huffaker et al. 2011)

4.11.4 Engineering PROPEP Genes for Disease Management

PROPEP1 and *PROPEP2* genes encode the precursor proteins PROPEP1 and PROPEP2 in Arabidopsis. The plant elicitor peptides AtPep1 and AtPep2 are processed from the precursor proteins, PROPEP1 and PROPEP2 (Huffaker et al. 2006; Huffaker and Ryan 2007). Transgenic *Arabidopsis* plants overexpressing *PROPEP1* and *PROPEP2* genes were developed. These transgenic plants showed enhanced resistance to the oomycete pathogen *Pythium irregulare* (Huffaker et al. 2006). Since orthologs of PROPEP1 have been identified in rice, wheat, barley, maize, potato, soybean, grape, canola, poplar, and alfalfa (Huffaker et al. 2006), *PROPEP* genes may be useful in developing disease-resistant agriculturally important crops.

4.11.5 Engineering prePIP Genes to Amplify Immunity Induced by the PEP1 for Disease Management

PAMP signals are known to trigger immune responses and the induced immune responses are amplified by the PIMP plant elicitor peptides (Peps) (Yamaguchi et al. 2010; Logemann et al. 2013; Hou et al. 2014; Vidhyasekaran 2014). PAMP signals also induce secretion of several peptides called secreted peptides. Some of these secreted peptides are potentially involved in regulation of immunity and they are called secreted peptide elicitors (Hou et al. 2014). The PAMPs flg22- and elf18-induced genes include a gene family of secreted peptide precursors, which are called "prePIPs" (precursors of PAMP-induced Peptides) (Hou et al. 2014). In Arabidopsis, the genes encoding prePIPs, *prePIP1* and *prePIP2,* are induced by a variety of pathogens and elicitors. When *Arabidopsis thaliana* seedlings were exposed to flg22 or chitin, prePIP1 transcription was markedly up-regulated (Hou et al. 2014).

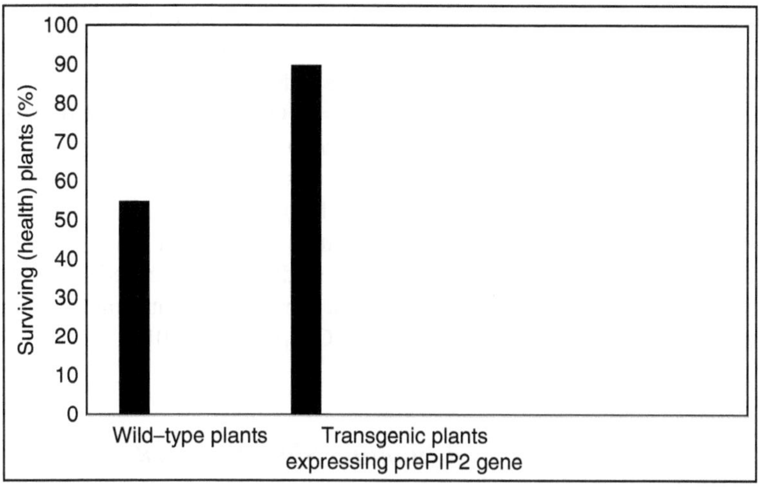

Fig. 4.13 Engineering *prePIP* genes to develop disease resistant plants against *Fusarium oxysporum* f. sp. *conglutinans* (Adapted from Hou et al. 2014)

The prePIPs are cleaved close to the C terminus to form PIPs. PIP1 is the biologically active form derived from prePIP1. PIP1 induces the precursor of PIMP/HAMP PEP1, PROPEP1, while PEP1 induces the precursor of PIP1, prePIP1 (Hou et al. 2014). PIP1 and PEP1 induce the expression of each other's precursor and the two endogenous peptide signaling pathways are interdependent and cooperate to amplify the immune responses (Hou et al. 2014). These results suggest that PIP1 may enhance the function of PEP1 in triggering disease resistance.

Transgenic Arabidopsis plants overexpressing *prePIP1* or *prePIP2* showed enhanced resistance against *Fusarium oxysporum* f. sp. *conglutinans* (Fig. 4.13; Hou et al. 2014), These transgenic plants also showed resistance to the bacterial pathogen *Pseudomonas syringae* pv. *tomato* (Hou et al. 2014). These results suggest that disease resistant plants can be developed by engineering *PIP* gene, which would enhance the expression of *PEP* genes and trigger defense responses against the fungal and bacterial pathogens.

4.12 Switching on Plant Immune Signaling Systems Using Systemin for Disease Management

4.12.1 Systemin

Systemin is an 18 amino acid polypeptide detected in leaves of tomato plants (Pearce et al. 1991; Vidhyasekaran 2007). The 18 amino acids peptide systemin is released from the C-terminal region of a larger precursor of 200 amino acids, called prosystemin (McGurl et al. 1992). Of the 200 amino acid prosystemin, amino acids 179–196 encode systemin. Prosystemin is encoded by a single gene (McGurl and Ryan 1992;

McGurl et al. 1992). Prosystemin gene homologs have been found in potato and the prosystemin gene is wound-inducible (McGurl et al. 1992). Systemin is a damage/ wound induced bioactive peptide and it triggers defense responses against various pathogens and insect pests (Díaz et al. 2002; El Oirdi et al. 2011; Coppola et al. 2014). The necrotic fungal pathogen *Botrytis cinerea* induces the expression of prosystemin in tomato (El Oirdi et al. 2011) and prosystemin is the precursor of systemin biosynthesis. Systemin is released by stress cells to dictate innate immune responses (Howe and Jander 2008; Yamaguchi and Huffaker 2011; Albert 2013; Coppola et al. 2014) and can be considered as a HAMP or host damage-associated molecular pattern. Systemin peptide has been recognized as a damage-associated molecular pattern (DAMP) triggering defense signaling system (Coppola et al. 2014).

4.12.2 Systemin-Triggered Immune Signaling Systems

Systemin triggers early signaling systems involved in plant innate immunity. A systemin receptor has been identified on the surface of *Lycopersicon peruvianum* suspension cultured cells (Scheer and Ryan 1999, 2002). The systemin receptor is a 160 kDa protein and the Systemin Receptor SR160 contains 25 leucine-rich repeats (LRRs) on the extracellular surface, one transmembrane domain, a Thr/Ser kinase domain on the intracellular portion of the receptor, and a 68 amino acid non-LRR region separating the twenty-first and twenty-second LRRs (Scheer and Ryan 2002). Systemin binds to its receptor within minutes and activates several immune response signaling systems. It activates Ca^{2+} influx resulting in an increase of intracellular Ca^{2+} concentration (Ryan 2000). It induces a depolarization of the plasma membrane (Ryan 2000). It triggers production of ROS and activates mitogen-activated protein kinases (MAPKs) (Ryan 2000; Orozco-Cárdenas et al. 2001; Sun et al. 2011; Coppola et al. 2014),

Systemin plays a key role in JA biosynthesis pathway (Fig. 4.14). Release of linoleic acid from cell membranes is an important step in the JA biosynthesis pathway (Wang et al. 2000). Systemin triggers an increase in phospholipase A2 activity (Ryan 2000) and the phospholipase A is known to mediate the release of linolenic acid from the membrane (Wasternack 2007). Linoleic acid is the starting point in the lipoxygenase pathway leading to JA biosynthesis (Balbi and Devoto 2008). Systemin induces ROS accumulation and ROS is involved in the induction of JA biosynthesis (Vandenabeele et al. 2003). The gene encoding lipoxygenase (*Lox1*) is induced by ROS. The genes encoding lipase, lipoxygenase, and 12-oxo-phytodienoate reductase (OPDA reductase), which are involved in JA biosynthesis, are activated in tobacco leaves accumulating ROS (Vandenabeele et al. 2003). Systemin also activates allene oxide cyclase (AOC), which is involved in OPDA biosynthesis (Fig. 4.14). Collectively these studies suggest that systemin-induced ROS accumulation may trigger the JA synthesis pathway. Systemin also activates MAPK signaling system in tomato (Kandoth et al. 2007). Systemin-induced MPK1and MPK2 have been shown to be involved in JA biosynthesis. Co-silencing of *MPK1* and *MPK2* genes reduced JA biosynthesis, suggesting the role of the MAPKs in JA biosynthesis (Kandoth

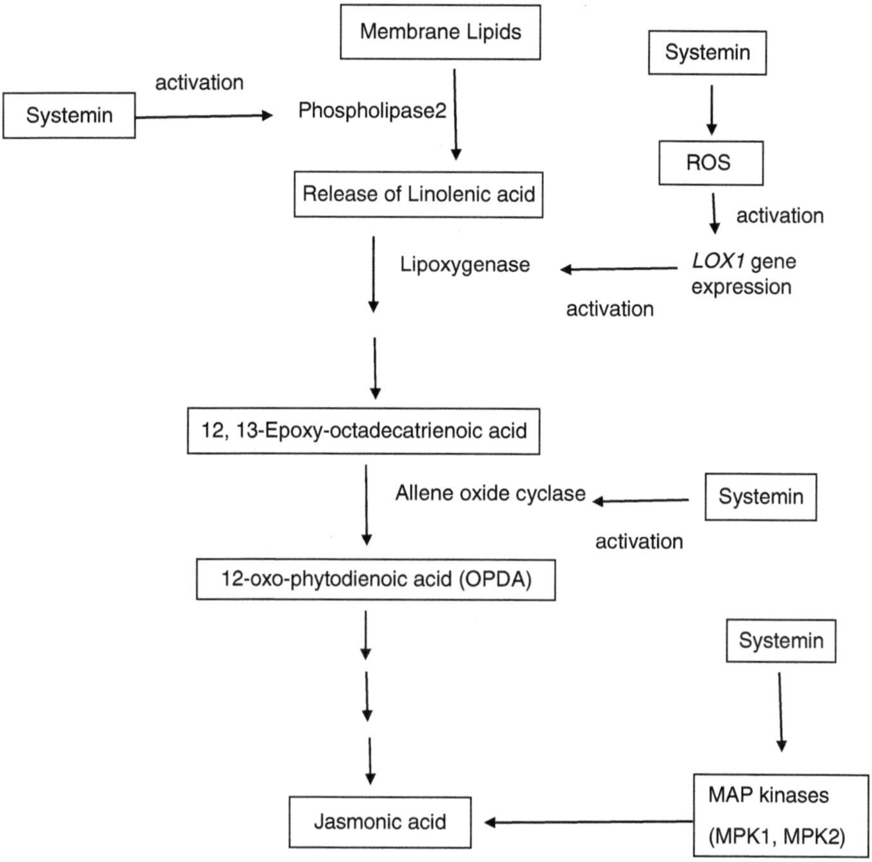

Fig. 4.14 Systemin-induced ROS, MAP kinases and phospholipases in induction of JA biosynthesis (Adapted from Orozco-Cárdenas et al. 2001; Zhao and Sakai 2003; Kandoth et al. 2007; Wasternack 2007)

et al. 2007). Systemin generates and amplifies a systemic signal, activating the JA pathway (McGurl et al. 1994; Ryan 2000; Li et al. 2002; Lee and Howe 2003; Stratmann 2003; Schilmiller and Howe 2005; Vidhyasekaran 2007).

4.12.3 Engineering Prosystemin Gene to Develop Disease-Resistant Plants

Systemin peptide has been shown to be a plant elicitor peptide triggering JA signaling system (Li et al. 2003). JA signaling is involved in triggering defense responses against necrotrophic pathogens (Zheng et al. 2006; Flors et al; 2008; Zander et al. 2010; Méndez-Bravo et al. 2011; Vidhyasekaran 2015). Systemin has been detected only in solanaceous plants (Pearce et al. 1991; Díaz et al. 2002; El Oirdi et al. 2011).

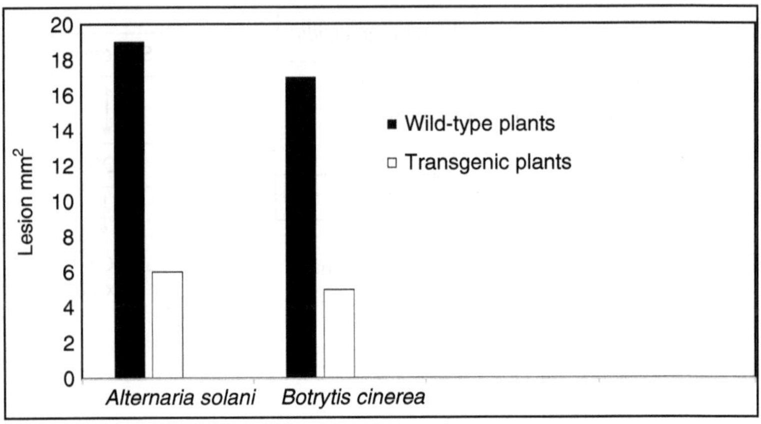

Fig. 4.15 Transgenic tomato plants overexpressing prosystemin show enhanced resistance to *Alternaria solani* and *Botrytis cinerea* (Adapted from Coppola et al. 2014)

Prosystemin gene is involved in biosynthesis of prosystemin, which is a precursor of systemin. Hence attempts were made to develop transgenic tomato plants over-expressing prosystemin gene for effective management of necrotrophic fungal pathogens (Díaz et al. 2002; Coppola et al. 2014). Prosystemin-overexpressing tomato plants showed high resistance against *Botrytis cinerea* (Díaz et al. 2002). Tomato plants were stably transformed via *Agrobacterium* with a construct containing prosystemin cDNA under the control of the *Cauliflower mosaic virus* 35S RNA promoter (Coppola et al. 2014). The transgenic plants showed enhanced resistance to the fungal pathogens *Botrytis cinerea* and *Alternaria solani* (Fig. 4.15; Coppola et al. 2014). These transgenic plants also showed resistance to insect pests, suggesting that the bioengineering technology would be a powerful tool for management of tomato diseases and pests.

References

Agüero CB, Uratsu SL, Greve C, Powell ALT, Labavitch JM, Meredith CP, Dandekar AM (2005) Evaluation of tolerance to Pierce's disease and *Botrytis* in transgenic plants of *Vitis vinifera* L. expressing the pear PGIP gene. Mol Plant Pathol 6:43–51

Albert M (2013) Peptides as triggers of plant defence. J Exp Bot 64:5269–5279

Alexandersson E, Becker JVW, Jacobson D, Nguema-Ona E, Steyn C, Denby KJ, Vivier MA (2011) Constitutive expression of a grapevine polygalacturonase-inhibiting protein affects gene expression and cell wall properties in uninfected tobacco. BMC Res Notes 4:493

An SH, Sohn KH, Choi HW, Hwang IS, Lee SC, Hwang BK (2008) Pepper pectin methylesterase inhibitor protein CaPMEI1 is required for antifungal activity, basal resistance and abiotic stress tolerance. Planta 228:61–78

Andreasson E, Jenkins T, Brodersen P, Thorgrimsen S, Petersen NHT, Zhu S, Qiu JL, Micheelsen P, Rocher A, Petersen M, Newman A, Nielsen HB, Hirt H, Somssich I, Mattsson O, Mundy J (2005) The MAP kinase substrate MKS1 is a regulator of plant defense responses. EMBO J 24:2579–2589

Aranega-Bou P, de la O Leyva M, Finiti I, Garcia-Agustin P, Gonzălez-Bosch C et al (2014) Priming of plant resistance by natural compounds. Hexanoic acid as a model. Front Plant Sci 5:488. doi:10.3389/tpls.2014.00488

Aziz A, Heyraud A, Lambert B (2004) Oligogalacturonide signal transduction, induction of defence-related responses and protection of grapevine against *Botrytis cinerea.* Planta 218:767–774

Aziz A, Gauthier A, Bezier A, Poinssot B, Joubert JM, Pugin A, Heyraud A, Baillieu F (2007) Elicitor and resistance-inducing activities of β-1,4 cellodextrins in grapevine, comparison with β-1,3 glucans and α-1,4 oligogalacturonides. J Exp Bot 58:1463–1472

Balagué C, Lin B, Alcon C, Flottes G, Malmstrom S, Kohler C, Neuhaus G, Pelletier G, Gaymard F, Roby D (2003) HML1, an essential signaling component in the hypersensitive response, is a member of the cyclic nucleotide-gated ion channel family. Plant Cell 15:365–379

Balbi V, Devoto A (2008) Jasmonate signalling network in *Arabidopsis thaliana*: crucial regulatory nodes and new physiological scenarios. New Phytol 177:301–318

Baxter J, Moeder W, Urquhart W, Shahinas D, Chin K, Christendat D, Kang H-G, Angelova M, Kato N, Yoshioka K (2008) Identification of a functionally Bardoel BWessential amino acid for Arabidopsis cyclic nucleotide gated ion channels using the chimeric *AtCNGC11/12* gene. Plant J 56:457–469

Bellincampi D, Cervone F, Lionetti V (2014) Plant cell wall dynamics and wall-related susceptibility in plant-pathogen interactions. Front Plant Sci 5:228. doi:10.3389/fpls.2014.00228

Bethke G, Grundman RE, Sreekanta S, Truman W, Katagiri E, Glazebrook J (2014) Arabidopsis PECTIN METHYLESTERASEs contribute to immunity against *Pseudomonas syringae.* Plant Physiol 164:1093–1107

Böhm H, Albert I, Fan L, Reinhard A, Nürnberger T (2014) Immune receptor complexes at the plant surface. Curr Opin Plant Biol 20:47–54

Boller T, Felix G (2009) A renaissance of elicitors: perception of miceobe-associated molecular patterns and danger signals by pattern-recognition receptors. Annu Rev Plant Biol 60:379–406

Borras-Hidalgo O, Caprari C, Hernandez-Estevez I, De Lorenzo G, Cervone F (2012) A gene for plant protection: expression of a bean polygalacturonase inhibitor in tobacco confers a strong resistance against *Rhizoctonia solani* and two oomycetes. Front Plant Sci 3:1–6

Boudart G, Lafitte C, Barthe JP, Frasez D, Esquerre-Tugayė MT (1998) Differential elicitation of defence responses by pectin fragments in bean seedlings. Planta 206:86–94

Boudart G, Charpentier M, Lafitte C, Martinez Y, Jauneau A, Gaulin E, Esquerré-Tugayé M-T, Dumas B (2003) Elicitor activity of a fungal endopolygalacturonase in tobacco requires a functional catalytic site and cell wall localization. Plant Physiol 131:93–101

Brodersen P, Petersen M, Bjorn Nielsen H, Zhu S, Newman MA, Shokat KM, Rietz S, Parker J, Mundy J (2006) *Arabidopsis* MAP kinase 4 regulates salicylic acid- and jasmonic acid/ethylene-dependent responses via EDS1 and PAD4. Plant J 47:532–546

Brutus A, Sicilia F, Macone A, Cervone F, De Lorenzo G (2010) A domain swap approach reveals a role of the plant wall-associated kinase 1 (WAK1) as a receptor of oligogalacturonides. Proc Natl Acad Sci U S A 107:9452–9457

Buell CR, Joardar V, Lindeberg M, Selengut J, Paulsen IT, Gwinn ML, Dodson RJ, Deboy RT, Durkin AS, Kolonay JF, Madupu R, Daugherty S, Brinkac L, Beanan MJ, Haft DH, Nelson WC, Davidsen T, Zafar N, Zhou L, Liu J, Yuan Q, Khouri H, Fedorova N, Tran B, Russell D, Berry K, Utterback T, Van Aken SE, Feldblyum TV, D'Ascenzo M, Deng W-L, Ramos AR, Alfano JR, Cartinhour S, Chatterjee AK, Delaney TP, Lazarowitz SG, Martin GB, Schneider DJ, Tang X, Bender CL, White O, Fraser CM, Collmer A (2003) The complete genome sequence of the *Arabidopsis* and tomato pathogen *Pseudomonas syringae* pv. *tomato* DC3000. Proc Natl Acad Sci U S A 100:10181–10186

Cabrera JC, Boland A, Messiaen J, Cambier P, Cutsem PV (2008) Egg box conformation of oligogalacturonides: the time-dependent stabilization of the elicitor-active conformation increases its biological activity. Glycobiology 18:473–482

Capodicasa C, Vairo D, Zabotina O, McCartney L, Caprari C, Mattei B, Manfredini C, Aracri B, Benen J, Knox JF, De Lorenzo G, Cervone F (2004) Targeted modification of homogalacturonan by transgenic expression of a fungal polygalacturonase alters plant growth. Plant Physiol 135:1294–1304

Cervone F, De Lorenzo G, Degra L, Salvi G, Bergami M (1987) Purification and characterization of a polygalacturonase-inhibiting protein from *Phaseolus vulgaris* L. Plant Physiol 85:631–637

Cervone F, Hahn MG, De Lorenzo G, Darvill A, Albersheim P (1989) Host-pathogen interactions. XXXIII. A plant protein converts a fungal pathogenesis factor into an elicitor of plant defense responses. Plant Physiol 90:545–548

Chen C, Chen Z (2002) Potentiation of developmentally regulated defense response by AtWRKY18, a pathogen-induced *Arabidopsis* transcription factor. Plant Physiol 129:706–716

Chen J, Yin H, Gu J, Li L, Liu Z, Jiang X, Zhou H, Wei S, Zhang S, Wu J (2015) Genomic characterization, phylogenetic comparison and differential expression of the cyclic nucleotide-gated channels gene family in pear (*Pyrus bretchneideri* Rehd,). Genomics 105:39–52

Chiu J, DeSalle R, Lam HM, Meisel L, Coruzzi G (1999) Molecular evolution of glutamate receptors: a primitive signaling mechanism that existed before plants and animals diverged. Mol Biol Evol 16:826–838

Chiu JC, Brenner ED, DeSalle R, Nitabach MN, Holmes TC, Coruzzi GM (2002) Phylogenetic and expression analysis of the glutamate-receptor-like gene family in *Arabidopsis thaliana.* Mol Biol Evol 19:1066–1082

Coppola M, Corrado G, Coppola V, Cascone P, Martinelli R, Digilio MC, Pennacchio F, Rao R (2014) Prosystemin overexpression in tomato enhances resistance to different biotic stresses by activating genes of multiple signaling pathways. Plant Mol Biol Rep 33:1270–1285. doi:10.1007/s11105-014-0834-x

Côté F, Hahn MG (1994) Oligosaccharins: structures and signal transduction. Plant Mol Biol 26:1379–1411

Denoux C, Galletti R, Mammarella N, Gopalan S, Werck D, De Lorenzo G, Ferrari S, Ausubel FM, Dewdney J (2008) Activation of defense response pathways by OGs and Flg22 elicitors in Arabidopsis seedlings. Mol Plant 1:423–445

Di Matteo A, Federici L, Mattei B, Salvi G, Johnson KA, Savino C, De Lorenzo G, Tsernoglou D, Cervone F (2003) The crystal structure of polygalacturonase-inhibiting protein (PGIP), a leucine-rich repeat protein involved in plant defense. Proc Natl Acad Sci U S A 100:10124–10128

Di Matteo A, Giovane A, Raiola A, Camardella L, Bonivento D, De Lorenzo G, Cervone F, Bellincampi D, Tsernoglou D (2005) Structural basis for the interaction between pectin methylesterase and a specific inhibitor protein. Plant Cell 17:849–858

Di Matteo A, Bonivento D, Tsernoglou D, Federici L, Cervone F (2006) Polygalacturonase-inhibiting protein (PGIP) in plant defence: a structural view. Phytochemistry 67:526–533

Di C, Zhang M, Xu S, Cheng T, An L (2006) Role of poly-galacturonase inhibiting protein in plant defense. Crit Rev Microbiol 32:91–100

Díaz J, ten Have A, van Kan JAL (2002) The role of ethylene and wound signaling in resistance of tomato to *Botrytis cinerea.* Plant Physiol 129:1341–1351

El Oirdi M, El Rahman TA, Rigano L, El Hadrami A, Rodriguez MC, Daay F, Vojnov A, Bouarab K (2011) *Botrytis cinerea* manipulates the antagonistic effects between immune pathways to promote disease development in tomato. Plant Cell 23:2405–2421

Eulgem T, Weighman VJ, Chang H-S, McDowell JM, Holub EB, Glazebrook J, Zhu T, Dangl JL (2004) Gene expression signatures from three genetically separable resistance gene signaling pathways for downy mildew resistance. Plant Physiol 135:1129–1144

Faize M, Faize L, Koike N, Ishizaka M, Ishii H (2004) Acibenzolar-S-methyl-induced resistance to Japanese pear scab is associated with potentiation of multiple defense responses. Phytopathology 94:604–612

Federici L, Di Matteo A, Fernandez-Recio J, Tsernoglou D, Cervone F (2006) Polygalacturonase-inhibiting proteins: players in plant immunity? Trends Plant Sci 11(2):65–70

Ferrari S, Plotnokova JM, De Lorenzo G, Ausubel FM (2003) *Arabidopsis* local resistance to *Botrytis cinerea* involves salicylic acid and camalexin and requires *EDS4* and *PAD2*, but not *SID2*, *EDS5* or *PAD4*. Plant J 35:193–205

Ferrari S, Galletti R, Vairo D, Cervone F, De Lorenzo G (2006) Antisense expression of the *Arabidopsis thaliana AtPGIP1* gene reduces polygalacturonase-inhibiting protein accumulation and enhances susceptibility to *Botrytis cinerea*. Mol Plant Microbe Interact 19:931–936

Ferrari S, Galletti R, Denoux C, De Lorenzo G, Ausubel FM, Dewdney J (2007) Resistance to *Botrytis cinerea* induced in Arabidopsis by elicitors is independent of salicylic acid, ethylene, or jasmonate signaling but requires PHYTOALEXIN DEFICIENT3. Plant Physiol 144:367–379

Ferrari S, Galletti R, Pontiggia D, Manfredni D, Lionetti V, Bellincampi D, Cervone F, De Lorenzo G (2008) Transgenic expression of a fungal *endo*-polygalacturonase increases plant resistance to pathogens and reduces auxin sensitivity. Plant Physiol 146:669–681

Ferrari S, Sella L, Janni M, De Lorenzo G, Favaron F, D'Ovidio R (2012) Transgenic expression of polygalacturonase-inhibitimg proteins in *Arabidopsis* and wheat increases resistance to the flower pathogen *Fusarium graminearum*. Plant Biol 14:31–38

Ferrari S, Savatin DV, Sicilia F, Gramegna G, Cervone F, Lorenzo GD (2013) Oligogalacturonides: plant damage-associated molecular patterns and regulators of geowth and development. Front Plant Sci 4:49. doi:10.3389/fpls.2013.00049

Feys BJ, Moisan LJ, Newman MA, Parker JE (2001) Direct interaction between the Arabidopsis disease resistance signaling proteins, EDS1 and PAD4. EMBO J 20:5400–5411

Fitzgerald HA, Canlas PE, Chern M-S, Ronald PC (2005) Alteration of TGA factor activity in rice results in enhanced tolerance to *Xanthomonas oryzae* pv. *oryzae*. Plant J 43:335–347

Flors V, Ton J, van Doorn R, Garcia-Agustin P, Mauch-Mani B (2008) Interplay between JA, SA, and ABA signaling during basal and induced resistance against *Pseudomonas syringae* and *Alternaria brassicicola*. Plant J 54:81–92

Furman-Matarasso N, Cohen E, Du Q, Chejanovsky N, Hanania U, Avni A (1999) A point mutation in the ethylene-inducing xylanase elicitor inhibits the beta-1-4-endoxylanase activity but not the elicitation activity. Plant Physiol 121:345–352

Galletti R, Denoux C, Gambetta S, Dewdney J, Ausubel FM, De Lorenzo G, Ferrari S (2008) The AtrbohD-mediated oxidative burst elicited by oligogalacturonides in Arabidopsis is dispensable for the activation of defense responses effective against *Botrytis cinerea*. Plant Physiol 148:1695–1706

Galletti R, De Lorenzo G, Ferrari S (2009) Host-derived signals activate plant innate immunity. Plant Signal Behav 4:33–34

Galletti R, Ferrari S, De Lorenzo G (2011) *Arabidopsis* MPK3 and MPK6 play different roles in basal and oligogalacturonide- or flagellin-induced resistance against *Botrytis cinerea*. Plant Physiol 157:804–814

Garcion C, Lohmann A, Lamodiere E, Catinot J, Buchala A, Doermann P, Métraux J-P (2008) Characterization and biological function of the *ISOCHORISMATE SYNTHASE2* gene of the Arabidopsis. Plant Physiol 147:1279–1287

Giovane A, Servillo L, Balestrieri C, Raiola A, D'Arvino R, Tamburrini M, Ciardiello MA, Camardella L (2004) Pectin methylesterase inhibitor. Biochim Biophys Acta 1696:245–252

Grefen C, Städele K, Ruzicka K, Obrdlik P, Harter K, Horák J (2008) Subcellular localization and in vivo interactions of the *Arabidopsis thaliana* ethylene receptor family members. Mol Plant 1:308–320

Grennan AK (2008) Ethylene response factors in jasmonate signaling and defense response. Plant Physiol 146:1457–1458

Guo H, Ecker JR (2004) The ethylene signaling pathway; new insights. Curr Opin Plant Biol 7:40–49

Hahn MG, Darvill AG, Albersheim P (1981) Host-pathogen interactions: XIX. The endogenous elicitor, a fragment of a plant cell wall polysaccharide that elicits phytoalexin accumulation in soybeans. Plant Physiol 68:1161–1169

Hann CT, Bequette CJ, Dombrowski JE, Stratmann JVV (2014) Methanol and ethanol modulate responses to danger- and microbe-associated molecular patterns. Front Plant Sci 5:550. doi:10.3389/fpls.2014.00550

Hou S, Wang X, Chen D, Yang X, Wang M, Turrà D, Di Pietro A (2014) The secreted peptide PIP1 amplifies immunity through receptor-like kinase 7. PLoS Pathog 10(9):e1004331. doi:10.1371/journal.ppat.1004331

Howe GA, Jander G (2008) Plant immunity to insect herbivores. Annu Rev Plant Biol 59:41–46

Huffaker A, Ryan CA (2007) Endogenous peptide defense signals in *Arabidopsis* differentially amplify signaling for the innate immune response. Proc Natl Acad Sci U S A 104:10732–10736

Huffaker A, Pearce G, Ryan CA (2006) An endogenous peptide signal in *Arabidopsis* activates components of the innate immune response. Proc Natl Acad Sci U S A 103:10098–10103

Huffaker A, Dafoe NJ, Schmelz EA (2011) ZmPep1, an ortholog of Arabidopsis elicitor peptide 1, regulates maize innate immunity and enhances disease resistance. Plant Physiol 155:1325–1338

Hung CY, Aspesi P Jr, Hunter MR, Lomax AVV, Perera IY (2014) Phosphoinositide-signaling is one component of a robust plant defense response. Front Plant Sci 5:267. doi:10.3389/fpls.2014.00267

Isshiki A, Akimitsu K, Ishii A, Yamamoto H (2000) Purification of polygalacturonases produced by pear scab pathogens, *Venturia nashicola* and *Venturia pirina*. Physiol Mol Plant Pathol 56:263–271

Janni M, Sella L, Favaron F, Blechl AE, De Lorenzo G, D'Ovidio R (2008) The expression of a bean PGIP in transgenic wheat confers increased resistance to the fumgal pathogen *Bipolaris sorokiniana*. Mol Plant Microbe Interact 21:171–177

Jiang S, Park P, Ishii H (2007) Ultrastructural study on scab resistance expressed in epidermal pectin layers of pear leaves. J Gen Plant Pathol 73:314–323

Jiang S, Park P, Ishii H (2008) Ultrastructural study on acibenzolar-S-methyl-induced scab resistance in epidermal pectin layers of Japanese pear leaves. Phytopathology 98:585–591

Joubert DA, Slaughter AR, Kemp G, Becker JV, Krooshof GH, Bergmann C, Benen J, Pretorius IS, Vivier MA (2006) The grapevine polygalacturonase-inhibiting protein (VvPGIP1) reduces *Botrytis cinerea* susceptibility in transgenic tobacco and differentially inhibits fungal polygalacturonases. Transgenic Res 15:702–887

Joubert DA, Kars I, Wagemakers L, Bergmann C, Kemp G, Vivier MA, van Kan JA (2007) A polygalacturonase-inhibiting protein from grapevine reduces the symptoms of the endopolygalacturonase BcPG2 from *Botrytis cinerea* in *Nicotiana benthamiana* leaves without any evidence for in vitro interaction. Mol Plant Microbe Interact 20:392–402

Jurkowski GI, Smith RK, Yu I-C, Ham JH, Sharma SB, Klessig DF, Fengler KA, Bent AF (2004) Arabidopsis *DND2*, a second cyclic nucleotide-gated ion channel gene for which mutation causes the "Defense, No Death" phenotype. Mol Plant Microbe Interact 17:511–520

Kalunke RM, Cenci A, Volpi C, Sullivan DM, Sella L, Favaron F, Cervone F, De Lorenzo G, D'Ovidio R (2014) The pgip family in soybean and three other legume species: evidence for a birth-and-death model of evolution. BMC Plant Biol 14(1):189

Kandoth PK, Ranf S, Pancholi SS, Jayanty S, Walla MD, Miller W, Howe GA, Lincoln DE, Stratmann JW (2007) Tomato MAPKs LeMPK1, LeMPK2, and LeMPK3 function in the systemin-mediated defense response against herbivorous insects. Proc Natl Acad Sci U S A 104:12205–12210

Kim KC, Fan B, Chen Z (2006) Pathogen-induced Arabidopsis WRKY7 is a transcriptional repressor and enhances susceptibility to *Pseudomonas syringae*. Plant Physiol 142:1180–1192

Krol E, Mentzel T, Chinchilla D, Boller T, Felix G, Kemmerling B, Postel S, Arents M, Jeworutzki E, Al-Rasheid KAS, Becker D, Hedrich R (2010) Perception of the Arabidopsis danger signal peptide 1 involves the pattern recognition receptor AtPEPR1 and its close homologue AtPEPR2. J Biol Chem 285:13471–13479

Kwezi L, Ruzvidzo O, Wheeler JI, Govender K, Iacuone S, Thompson PE, Gehring C, Irving HR (2011) The phytosulfokine (PSK) receptor is capable of guanylate cyclase activity and enabling cyclic GMP-dependent signaling in plants. J Biol Chem 286:222580–222588

Kwon C, Yun HS (2014) Plant exocytic secretion of toxic compounds for defense. Toxicol Res 30:77–81

Lee GI, Howe GA (2003) The tomato mutant *spr1* is defective in systemin perception and the production of a systemic wound signal for defense gene expression. Plant J 33:567–576

Li L, Li CY, Lee GI, Howe GA (2002) Distinct roles for jasmonate synthesis and action in the systemic wound response of tomato. Proc Natl Acad Sci U S A 99:6416–6421

Li C, Liu G, Xu C, Lee GI, Bauer P, Ling HQ, Ganal MW, Howe GA (2003) The tomato *Suppressor of prosystemin-mediated responses2* gene encodes a fatty acid desaturase required for the biosynthesis of jasmonic acid and the production of a systemic wound signal for defense gene expression. Plant Cell 15:1646–1661

Lionetti V, Raiola A, Camardella L, Giovane A, Obel N, Pauly M, Favaron F, Cervone F, Bellincampi D (2007) Overexpression of pectin methylesterase inhibitors in Arabidopsis restricts fungal infection by *Botrytis cinerea*. Plant Physiol 143:1871–1880

Lionetti V, Francocci F, Ferrari S, Volpi C, Bellincampi D, Galletti R, D'Ovidio R, de Lorenzo G, Cervone F (2010) Engineering the cell wall by reducing de-methyl-esterified homogalacturonan improves saccharification of plant tissues for bioconversion. Proc Natl Acad Sci U S A 107:616–621

Lionetti V, Cervone F, Bellincampi D (2012) Methyl esterification of pectin plays a role during plant-pathogen interactions and affects plant resistance to diseases. Plant Physiol 169:1623–1630

Lionetti V, Raiola A, Cervone F, Bellincampi D (2014) Transgenic expression of pectin methylesterase inhibitors limits tobamovirus spread in tobacco and Arabidopsis. Mol Plant Pathol 15:265–274

Liu Z, Wu Y, Yang F, Zhang Y, Chen S, Xie Q, Tian X, Zhou JM (2013) BIK1 interacts with PEPRs to mediate ethylene-induced immunity. Proc Natl Acad Sci U S A 110:6205–6210

Lloyd SR, Schoonbeek H, Trick M, Zipfel C, Ridout CJ (2014) Methods to study PAMP-triggered immunity in *Brassica* species. Mol Plant Microbe Interact 27:286–295

Logemann E, Birkenbihl RP, Rawat V, Schneeberger K, Schmelzer E, Somssich IE (2013) Functional dissection of the *PROPEP2* and *PROPEP3* promoters reveals the importance of WRKY factors in mediating microbe-associated molecular pattern-induced expression. New Phytol 198:1165–1177

Ma W, Smigel A, Verma R, Berkowitz GA (2009) Cyclic nucleotide gated channels and related signaling components in plant innate immunity. Plant Signal Behav 4:277–282

Ma Y, Walker RK, Zhao Y, Berkowitz GA (2012) Linking ligand perception by PEPR pattern recognition receptors to cytosolic Ca^{2+} elevation and downstream immune signaling in plants. Proc Natl Acad Sci U S A 109:195852–195857

Ma Y, Zhao Y, Walker RK, Berkowitz GA (2013) Molecular steps in the immune signaling pathway evoked by plant elicitor peptides: Ca^{2+}-dependent protein kinases, nitric oxide, and reactive oxygen species are downstream from the early Ca^{2+} signal. Plant Physiol 163:1459–1471

Macho AP, Zipfel C (2014) Plant PRRs and the activation of innate immune signalling. Mol Cell 54:263–272

Mackey D, McFall AJ (2006) MAMPs and MIMPs: proposed classification for inducers of innate immunity. Mol Microbiol 61:1365–1371

Malinovsky FG, Fangel JU, Willats GT (2014) The role of the cell wall in plant immunity. Front Plant Sci 5:178. doi:10.3369/fpls.2014.00178

Manabe Y, Nafisi M, Verhertbruggen Y, Orfila C, Gille S, Rautengarten C, Cherk C, Marcus SE, Somerville S, Pauly M, Knox JP, Sakuragi Y, Scheller HV (2011) Loss-of-function mutation of REDUCED WALL ACETYLATION2 in Arabidopsis leads to reduced cell wall acetylation and increased resistance to *Botrytis cinerea*. Plant Physiol 155:1068–1078

Manfredini C, Sicilla F, Ferrari S, Pontiggia D, Salvi G, Caprari C, Lorito M, De Lorenzo G (2005) Polygalacturonase-inhibiting protein 2 of *Phaseolus vulgaris* inhibits BcPG1, a polygalacturonase of *Botrytis cinerea* important for pathogenicity, and protects transgenic plants from infection. Physiol Mol Plant Pathol 67:108–115

Manzoor H, Kelloniemi J, Chiltz A, Wendehenne D, Pugin A, Poinssot B, Garcia-Brugger A (2013) Involvement of the glutamate receptor AtGLR3.3 in plant defense signaling and resistance to *Hyaloperonospora arabidopsidis*. Plant J 76:466–480

Marty P, Jouan B, BertheauY VB, Goldberg R (1997) Charge density in stem cell walls of *Solanum tuberosum* genotypes and susceptibility to blackleg. Phytochemistry 44:1435–1441

Masunaka A, Sekiguchi H, Takahashi H, Takenaka S (2010) Distribution and expression of elicitin-like protein genes of the biocontrol agent *Pythium oligandrum*. J Phytopathol 158:417–426

McGurl B, Ryan CA (1992) The organization of the prosystemin gene. Plant Mol Biol 20:405–409

McGurl N, Pearce G, Orozco-Cardenas M, Ryan CA (1992) Structure, expression and antisense inhibition of the systemin precursor gene. Science 255:1570–1573

McGurl B, Orozco-Cardenas M, Pearce G, Ryan CA (1994) Overexpression of the prosystemin gene in transgenic tomato plants generates a systemic signal that constitutively induces proteinase inhibitor synthesis. Proc Natl Acad Sci U S A 91:9799–9802

McMillan GP, Hedley D, Fyffe L, Pĕrombelon MC (1993) Potato resistance to softrot erwinias is related to cell wall pectin esterification. Physiol Mol Plant Pathol 42:279–289

Méndez-Bravo A, Calderón-Vázquez C, Ibarra-Laclette E, Raya-González J, Ramirez-Chávez E, Molina-Torres J, Guevara-Garcia AA, López-Bucio J, Herrera-Estrella L (2011) Alkamides activate jasmonic acid biosynthesis and signaling pathways and confer resistance to *Botrytis cinerea* in *Arabidopsis thaliana*. PLoS One 6(11):e27251

Miao Y, Zentgraf U (2007) The antagonist function of *Arabidopsis* WRKY53 and ESR/ESP in leaf senescense is modulated by the jasmonic and salicylic acid equilibrium. Plant Cell 19:819–830

Mohnen D (2008) Pectin structure and biosynthesis. Curr Opin Plant Biol 11:266–277

Moscatiello R, Mariani P, Sanders D, Maathuis FJ (2006) Transcriptional analysis of calcium-dependent and calium-independent signaling pathways induced by ologogalacturonides. J Exp Bot 57:2847–2865

Mott GA, Middleton MA, Desveux D, Guttmann DS (2014) Peptides and small molecules of the plant-pathogen apoplastic arena. Front Plant Sci 5:677

Muleya V, Wheeler JI, Ruzvidzo O, Freihat L, Manallack DT, Gehring C, Irving HR (2014) Calcium is the switch in the moonlighting dual function of the ligand-activated receptor kinase phytosulfokine receptor 1. Cell Commun Signal 12:60

Nawaz Z, Kakar KU, Saand MA, Shu QY (2014) Cyclic nucleotide-gated ion channel gene family in rice, identification, characterization and experimental analysis of expression response to plant hormones, biotic and abiotic stresses. BMC Genomics 15:853. doi:10.1186/1471-2164-15-853

Nawrath C, Heck S, Parinthawong N, Metraux JP (2002) EDS5, an essential component of salicylic acid-dependent signaling for disease resistance in Arabidopsis, is a member of MATE transporter family. Plant Cell 14:275–286

Nguema-Ona E, Moore JP, Fagerström AD, Fangel JU, Willats WG, Hugo A, Vivier MA (2013) Overexpression of the grapevine PGIP1 in tobacco results in compositional changes in the leaf arabinoxyloglucan network in the absence of fungal infection. BMC Plant Biol 13:46. doi:10.1186/1471-2229-13-46

Noirot E, Der C, Lherminier J, Robert F, Moricova P, Kiêu K, Leborgne-Castel N, Simon-Plas F, Bouhidel K (2014) Dynamic changes in the subcellular distribution of the tobacco ROS-producing enzyme RBOHD in response to the oomycete elicitor cryptogein. J Exp Bot 65:5011–5022

Nothnagel EA, McNeil M, Albersheim P, Dell A (1983) Host-pathogen interactions. XXII. A galacturonic acid oligosaccharide from plant cell walls elicits phytoalexins. Plant Physiol 71:916–926

Nuhse TS (2012) Cell wall integrity signaling and innate immunity in plants. Front Plant Sci 3:280

Orozco-Cárdenas ML, Narvaez-Vasquez J, Ryan CA (2001) Hydrogen peroxide acts as a second messenger for the induction of defense genes in tomato plants in response to wounding, systemin, and methyl jasmonate. Plant Cell 13:1047–1061

Osorio S, Castillejo C, Quesada MA, Medina-Escobar N, Brownsey GJ, Suau R, Heredia A, Botella MA, Valpuesta V (2008) Partial demethylation of oligogalacturonides by pectin methylesterase 1 is required for eliciting defence responses in wild strawberry (*Fragaria vesca*). Plant J 54:43–55

Osorio S, Bombarely A, Giavalisco P, Usadel B, Stephens C, Aragüez I, Medina-Escobar N, Botella MA, Fernie AR, Valpuesta V (2011) Demethylation of oligogalacturonides by *FaPE1* in the fruits of the wild strawberry *Fragaria vesca* triggers metabolic and transcriptional changes associated with defence and development of the fruit. J Exp Bot 62:2855–2873

Park P, Ishii H, Adachi Y, Kanematsu S, Ieki H, Umemoto S (2000) Infection behavior of *Venturia nashicola*, the cause of scab on Asian pears. Phytopatholoy 90:1209–1216

Pauly M, Scheller HV (2000) O-acetylation of plant cell wall polysaccharides: identification and partial characterization of a rhamnogalacturonan *O*-acetyl-transferase from potato suspension cultured cells. Planta 210:659–667

Pearce RB, Ride JP (1982) Chitin and related compounds as elicitors of lignification response in wounded wheat leaves. Physiol Plant Pathol 20:119–123

Pearce G, Strydom D, Johnson S, Ryan CA (1991) A polypeptide from tomato leaves induced wound-inducible inhibitor proteins. Science 253:895–898

Pelloux J, Rustérucci C, Mellerowicz EJ (2007) New insights into pectin methylestrase structure and function. Trends Plant Sci 12:267–277

Petersen M, Brodersen P, Naested H, Andreasson E, Lindhart U, Johansen B, Nielsen HB, Lacy M, Austin MJ, Parker JE, Sharma SB, Klessig DF, Martienssen R, Mattsson O, Jensen AB, Mundy J (2000) *Arabidopsis* MAP kinase 4 negatively regulates systemic acquired resistance. Cell 103:1111–1120

Poinssot B, Vandelle E, Bentejac M, Adrian M, Levis C, Brygoo Y, Garin J, Sicilia F, Coutos-Thevenot P, Pigin A (2003) The endopolygalacturonase I from *Botrytis cinerea* activates grapevine defense reactions unrelated to its enzymatic activity. Mol Plant Microbe Interact 16:553–564

Postel S, Kufner I, Beueter C, Mazzotta S, Schwedt A, Borlotti A, Halter T, Kemmerling B, Nürnberger T (2010) The multifunctional leucine-rich repeat receptor kinase BAK1 is implicated in *Arabidopsis* development and immunity. Eur J Cell Biol 89:169–174

Powell AL, van Kan J, ten Have A, Visser J, Greve LC, Bennett AB, Labavitch JM (2000) Transfenic expression of pear PGIP in tomato limits fungal colonization. Mol Plant Microbe Interact 13:942–950

Price MB, Jelesko J, Okumoto S (2012) Glutamine receptor homologs in plants: functions and evolutionaryorigins. Front Plant Sci 3:235. doi:10.3389/fpls.2012.00235

Qi Z, Verma R, Gehring C, Yamaguchi Y, Zhao Y, Ryan CA, Berkowitz GA (2010) Ca^{2+} signaling by plant *Arabidopsis thaliana* Pep peptides depends on AtPepR1, a receptor with guanyl cyclase activity, and CGMP-activated Ca^{2+} channels. Proc Natl Acad Sci U S A 107:21193–21198

Qu X, Hall BP, Gao Z, Schaller GE (2007) A strong constitutive ethylene-response phenotype conferred on Arabidopsis plants containing null mutations in the ethylene receptors *ETR1* and *ERS1*. BMC Plant Biol 7:3. doi:10.1186/1471-2229-7-3

Raiola A, Lionetti V, Elmaghraby I, Immerzeel P, Mellerowicz EJ, Salvi G, Cervone F, Bellincampi D (2011) Pectin methylesterase is induced in *Arabidopsis* upon infection and is necessary for a successful colonization by necrotrophic pathogens. Mol Plant Microbe Interact 24:432–440

Rakwal R, Tamogani S, Agrawal GK, Iwahashi H (2002) Octadecanoid signaling component 'burst' in rice (*Oryza sativa* L.) seedlings leaves upon wounding by cut and treatment with fungal elicitor chitosan. Biochem Biophys Res Commun 295:1041–1045

Randoux B, Renard-Merlier D, Duyme F, Sanssené J, Courtois J, Durand R, Reignault P (2009) Oligogalacturonides induce resistance in wheat against powdery mildew. Commun Agric Appl Biol Sci 74:681–685

Randoux B, Renard-Merlier D, Mulard G, Rossard S, Duyme F, Sanssené J, Courtois J, Durand R, Reignault P (2010) Distinct defenses induced in wheat against powdery mildew by acetylated and nonacetylated oligogalacturonides. Phytopathology 100:1352–1363

Rasul S, Dubreuil-Maurizi C, Lamotte O, Koen E, Poinssot B, Alcaraz G, Wendehenne D, Jeandroz S (2012a) Nitric oxide production mediates oligogalacturonide-triggered immunity and resistance to *Botrytis cinerea* in *Arabidopsis thaliana*. Plant Cell Environ 35:483–499

Rasul S, Wendehenne D, Jeandroz S (2012b) Study of oligogalacturonides-triggered nitric oxide (NO) production provokes new questioning about the origin of NO biosynthesis in plants. Plant Signal Behav 7:1031–1033

Ross A, Yamada K, Hiruma K, Yamashita-Yamada M, Lu X, Takano Y, Tsuda K, Saito Y (2014) The Arabidopsis PEPR pathway couples local and systemic plant immunity. EMBO J 33:62–75

Rotblat B, Enshel-Seijffers D, Gershoni MJ, Schuster S, Avni A (2002) Identification of an essential component of the elicitation active site of the EIX protein elicitor. Plant J 32:1–7

Ryan CA (2000) The systemin signaling pathway: differential activation of plant defensive genes. Biochim Biophys Acta 1447:112–121

Ryan CA, Huffaker A, Yamaguchi Y (2007) New insights into innate immunity in Arabidopsis. Cell Microbiol 9:1902–1908

Sagi M, Fluhr R (2006) Production of reactive oxygen species by plant NADPH oxidases. Plant Physiol 141:336–340

Sanders D, Pelloux J, Brownlee C, Harper JF (2002) Calcium at the cross-roads of signaling. Plant Cell 14:S401–S417

Scheer JM, Ryan CA (1999) A 160-kDa systemin receptor on the surface of Lycopersicon peruvianum suspension-cultured cells. Plant Cell 11:1525–1536

Scheer JM, Ryan CA (2002) The systemin receptor 160 from *Lycopersicon peruvianum* is a member of the LRR receptor kinase family. Proc Nat Acad Sci U S A 99:9585–9590

Schilmiller AL, Howe GA (2005) Systemic signaling in the wound response. Curr Opin Plant Biol 8:369–377

Schulze B, Mentzel T, Jehle A, Mueller K, Beeler S, Boller T, Felix G, Chinchilla D (2010) Rapid heteromerization and phosphorylation of ligand-activated plant transmembrane receptors and their associated kinase BAK1. J Biol Chem 285:9444–9451

Sénéchal F, Wattier C, Rustérucci C, Pelloux J (2014) Homogalacturonan-modifying enzymes: structure, expression, and roles in plants. J Exp Bot 65:5125–5160

Shibuya N, Minami E (2001) Oligosaccharide signalling for defence responses in plant. Physiol Mol Plant Pathol 59:223–233

Song JT, Lu H, McDowell JM, Greenberg JT (2004) A key role for *ALD1* in activation of local and systemic defenses in *Arabidopsis*. Plant J 40:200–212

Stratmann JW (2003) Long distance run in the wound response – Jasmonic acid is pulling ahead. Trends Plant Sci 8:247–250

Sun JQ, Jiang HL, Li CY (2011) Systemin/Jasmonate-mediated systemic defense signaling in tomato. Mol Plant 4:607–615

Svalheim O, Robertsen B (1993) Elicitation of H_2O_2 production in cucumber hypocotyl segments by oligo-1,4-alpha-D-galacturonides and an oligo-beta-glucan preparation from cell walls of *Phytophthora megasperma* f.sp. *glycinea*. Physiol Plant 88:675–681

Tada Y, Spoel SH, Pajerowska-Mukhtar K, Mou Z, Song J, Dong X (2008) Plant immunity requires conformational changes of NPR1 via S-nitrosylation and thioredoxin. Science 321:952–956

Tateda C, Zhang Z, Shrestha J, Jelenska J, Chinchilla D, Greenberg JT (2014) Salicylic acid regulates Arabidopsis microbial pattern receptor kinase levels and signaling. Plant Cell 26:4171–4187

Taurino M, Abelenda JA, Rio-Alvarez I, Navarro C, Vicedo B, Farmaki T, Jiménez P, Garcia-Agustin P, López-Solanilla E, Prat S, Rojo E, Sánchez-Serrano JJ, Sanmartin M (2014) Jasmonate-dependent modifications of the pectin matrix during potato development function as a defense mechanism targeted by *Dickeya dadantii* virulence factors. Plant J 77:418–429
Tikhonov DB, Magazanik LG (2009) Origin and molecular evolution of ionotropic glutamate receptors. Neurosci Behav Physiol 39:763–773
Vallarino JG, Osorio S (2012) Signaling role of oligogalacturonides derived during cell wall degradation. Plant Signal Behav 7:1447–1449
Vandenabeele S, Van Der Kelen K, Dat J, Gadjev I, Boonefaes T, Morsa S, Rottiers P, Slooten L, Van Montagu M, Zabeau M, Inzé D, Van Breusegem F (2003) A comprehensive analysis of hydrogen peroxide-induced gene expression in tobacco. Proc Natl Acad Sci U S A 100:16113–16118
Vatsa P, Sanchez L, Clement C, Baillieul F, Dorey S (2010) Rhamnolipid biosurfactants as new players in animal and plant defense against microbes. Int J Mol Sci 11:5095–5108
Vatsa P, Chiltz A, Bourgue S, Wendehenne D, Garcia-Brugger A, Pugin A (2011) Involvement of putative glutamate receptors in plant defence signaling and NO production. Biochimie 93:2095–2101
Vidal S, Erikkson ARB, Montesano M, Dencke J, Palva T (1998) Cell wall-degrading enzymes from *Erwinia carotovora* cooperate in the salicylic acid-independent induction of a plant defence response. Mol Plant Microbe Interact 11:23–32
Vidhyasekaran P (1997) Fungal pathogenesis in plants and crops: molecular biology and host defense mechanisms. Marcel Dekker, New York, p 553
Vidhyasekaran P (2007) Fungal pathogenesis in plants and crops: molecular biology and host defense mechanisms, 2nd edn. CRC Press, Taylor Francis Group, Boca Raton, p 510
Vidhyasekaran P (2014) PAMP signals in plant innate immunity: signal perception and transduction. Springer, Dordrecht, p 442
Vidhyasekaran P (2015) Plant hormone signaling systems in plant innate immunity. Springer, Dordrecht, p 458
Vincill ED, Bieck AM, Spalding EP (2012) Ca^{2+} conduction by an amino acid-gated ion channel related to glutamate receptors. Plant Physiol 159:40–46
Volpi C, Janni M, Lionetti V, Bellincampi D, Favaron F, D'Ovidio R (2011) The ectopic expression of a pectin methyl esterase inhibitor increases pectin methyl esterification and limits fungal diseases in wheat. Mol Plant Microbe Interact 24:1012–1019
Vorhőlter F-J, Wiggerich H-G, Scheidle H, Sidhu VK, Mrozek K, Kűster H, Pűhler A, Niehaus K (2012) Involvement of bacterial TonB-dependent signaling in the generation of an oligogalacturonide damage-associated molecular pattern from plant cell walls exposed to *Xanthomonas campestris* pv. *campestris* pectate lyases. BMC Microbiol 12:239
Vorwerk S, Somerville S, Somerville C (2004) The role of plant cell wall polysaccharide composition in disease resistance. Trends Plant Sci 9:203–209
Wakabayashi K, Oson T, Huber DJ (2003) Methyl de-esterification as a major factor regulating the extent of pectin depolymerization during fruit ripening: a comparison of the action of avocado (*Persea americana*) and tomato (*Lycopersicon esculentum*) polygalacturonases. J Plant Physiol 160:667–673
Wang C, Zien C, Afitlhile M, Welti R, Hildebrand DF, Wang X (2000) Involvement of phospholipase D in wound-induced accumulation of jasmonic acid in Arabidopsis. Plant Cell 12:2237–2246
Wang X, Kota U, He K, Blackburn K, Li J, Goshe MB, Clouse SD (2008) Sequential transphorylation of the BRI1/BAK1 receptor kinase complex impacts early events in brassinosteroid signaling. Dev Cell 15:220–235
Wang X, Zhu X, Tooley P, Zhang X (2013) Cloning and functional analysis of three genes encoding polygalacturonase-inhibiting proteins from *Capsicum annuum* and transgenic CaPGIP1 in tobacco in relation to increased resistance to fungal pathogens. Plant Mol Biol 81:379–400
Wasternack C (2007) Jasmonates: an update on biosynthesis, signal transduction and action in plant stress response, growth and development. Ann Bot (Lond) 100:681–697

Wheeler GL, Brownlee C (2008) Ca^{2+} signaling in plants and green algae: changing channels. Trends Plant Sci 13:506–514

Wiermer M, Feys BJ, Parker JE (2005) Plant immunity: the EDS1 regulatory node. Curr Opin Plant Biol 8:383–389

Wiethölter N, Graessner B, Mierau M, Mort AJ, Moerschbacher BM (2003) Differences in the methyl ester distribution of homogalacturonans from near-isogenic wheat lines resistant and susceptible to the wheat stem rust fungus. Mol Plant Microbe Interact 16:945–952

Willats WG, Orfila C, Limberg G, Buchholt HC, Van Alebeek GJ, Voragen AG, Marcus SE, Christensen TM, Mikkelsen JD, Murray BS, Knox JP (2001) Modulation of the degree and pattern of methyl-esterification of pectic homogalacturonans in plant cell walls. Implications for pectin methyl esterase action, matrix properties, and cell adhesion. J Biol Chem 276:19404–19413

Wydra K, Beri H (2006) Structural changes of homogalacturonan, rhamnogalacturonan I and arabinogalactan protein in xylem cell walls of tomato genotypes in reaction to *Ralstonia solanacearum*. Physiol Mol Plant Pathol 68:41–50

Xing D, Chen Z (2006) Effects of mutations and constitutive overexpression of *EDS1* and *PAD4* on plant resistance to different types of microbial pathogens. Plant Sci 171:251–262

Xu X, Chen C, Fan B, Chen Z (2006) Physical and functional interactions between pathogen-induced *Arabidopsis* WRKY18, WRKY40, and WRKY60 transcription factors. Plant Cell 18:1310–1326

Yamaguchi Y, Huffaker A (2011) Endogenous peptide elicitors in higher plants. Curr Opin Plant Biol 14:351–357

Yamaguchi Y, Pearce G, Ryan CA (2006) The cell surface leucine-rich repeat receptor for *At*Pep1, an endogenous peptide elicitor in *Arabidopsis*, is functional in transgenic tobacco cells. Proc Natl Acad Sci U S A 103:10104–10109

Yamaguchi Y, Huffaker A, Bryan AC, Tax FE, Ryan CA (2010) PEPR2 is a second receptor for the Pep1 and Pep2 peptides and contributes to defense responses in *Arabidopsis*. Plant Cell 22:508–522

Yoshioka K, Moeder W, Kang H-G, Kachroo P, Masmoudi K, Berkowitz G, Klessig D-F (2006) The chimeric *Arabidopsis* CYCLIC NUCLEOTIDE-GATED ION CHANNEL11/12 activates multiple pathogen responses. Plant Cell 18:747–763

Zander M, La Camera S, Lamotte O, Metraux JP, Gatz C (2010) *Arabidopsis thaliana* class-II TGA transcription factors are essential activators of jasmonic acid/ethylene-induced defense responses. Plant J 61:200–210

Zhang J, Shao F, Li Y, Cui HT, Chen LJ, Li HT, Zou Y, Long CZ, Lan LF, Chai JJ, Chen S, Tang X, Zhou J-M (2007) A *Pseudomonas syringae* effector inactivates MAPKs to suppress PAMP-induced immunity in plants. Cell Host Microbe 1:175–185

Zhang L, Kars I, Essenstam B, Liebrand TW, Wagemakers L, Elberse J, Tagkalaki P, Tjoitang D, van den Ackerveken G, van Kan JA (2014) Fungal endopolygalacturonases are recognized as microbe-associated molecular patterns by the arabidopsis receptor-like protein RESPONSIVENESS TO BOTRYTIS POLYGALACTURONASES1. Plant Physiol 164:352–364

Zhao J, Sakai K (2003) Multiple signalling pathways mediate fungal elicitor-induced B-thujalplicin biosynthesis in *Cupressus lusitanica* cell cultures. J Exp Bot 54:647–656

Zheng Z, Qamar SA, Chen Z, Mengiste T (2006) Arabidopsis WRKY33 transcription factor is required for resistance to necrotrophic fungal pathogens. Plant J 48:596–605

Zipfel C (2009) Early molecular events in PAMP-triggered immunity. Curr Opin Plant Biol 12:414–420

Zipfel C, Robatzek S, Navarro L, Oakeley EJ, Jones JDG, Felix G, Boller T (2004) Bacterial disease resistance in *Arabidopsis* through flagellin perception. Nature 428:764–767

Chapter 5
Switching on Plant Immune Signaling Systems Using Pattern Recognition Receptor Complex

Abstract Plants are endowed with innate immune system to resist pathogen attack. Pathogen-associated molecular patterns (PAMPs) and host-associated molecular patterns (HAMPs) are danger/alarm signals to switch on the plant immune systems. Plants possess receptor systems called plant pattern recognition receptors (PRRs) for sensing pathogen- or host-derived patterns to trigger inducible immune defenses. Most of the PRRs identified are receptor-like kinases (RLKs) and receptor-like proteins (RLPs). Plants use the PRRs to defend themselves from microbial pathogens. PRRs do not act alone but rather function as part of multi-protein complexes at the plasma membrane. The PRRs interact with several transmembrane LRR receptor-like kinase proteins that act as signaling adapters or amplifiers to achieve full functionality. The PRRs are localized at the plasma membrane and the PAMPs activate expression of the genes encoding various PRRs and PRR interacting regulators to trigger defense responses against pathogens. PRRs play important role in triggering defense responses, PAMP perception by PRR is required for full immunity and plants deficient in the specific PRR become more susceptible to pathogens. However, pathogens secrete effectors, which directly target the PRRs and suppress the PRR-activated downstream immune responses. Early and robust activation of PRRs before the pathogens invade and secrete virulence effectors seems to be necessary for triggering strong defense responses and for effective management of crop diseases. Transgenic plants constitutively expressing various PRRs including FLS2, EF-Tu Receptor, XA21, and WAK1 show enhanced resistance against various fungal, oomycete, and bacterial diseases. Pathogens that are adapted to a particular host plant may be adept at suppressing the PRRs of that host by their effectors. The effectors of the pathogens might not recognize PRRs from other host plants and development of transgenic plants expressing PRRs from other plant species may provide good resistance against various pathogens. Transfer of Arabidopsis and rice PRRs to various economically important crop plants has revealed high potential of the PRRs for crop disease management. Engineering PRR-interacting protein complexes has also shown to be useful technology for crop disease management.

P. Vidhyasekaran, *Switching on Plant Innate Immunity Signaling Systems*,
Signaling and Communication in Plants, DOI 10.1007/978-3-319-26118-8_5

5.1 Pattern Recognition Receptors (PRRs)

Pathogen-associated molecular patterns (PAMPs) and Pathogen-induced molecular patterns (PIMPs)/host-associated molecular patterns (HAMPs) are perceived by plants as alarm signals by specific receptors called pattern recognition receptors (PRRs) (Nicaise et al. 2009; Brutus et al. 2010; Petutschnig et al. 2010; Shinya et al. 2010; Schulze et al. 2010; Segonzac and Zipfel 2011; Hann et al. 2014). Plants employ plasma membrane and cytoplasmic receptor systems for sensing microbe- or host-derived patterns to trigger inducible immune defenses (Böhm et al. 2014). Several receptors for the PAMPs have been recognized in plasma membrane of plant cells (Fig. 5.1; Mithöfer et al. 2000; Montesano et al. 2003; Fliegmann et al. 2004; Ron and Avni 2004; Fritz-Laylin et al. 2005; Altenbach and Robatzek 2007; He et al. 2007; Lloyd et al. 2014). The repertoire of the PRRs varies not only between plant species but also between different accessions of *Arabidopsis thaliana* (Zipfel et al. 2004; Vetter et al. 2012). Plants utilize the PRRs to recognize PAMPs/PIMPs (Macho and Zipfel 2014; Zhang et al. 2014). These receptors recognize PAMPs and activate signaling pathways that lead to immunity (Mentlak et al. 2012). Most of the PRRs identified are receptor-like kinases (RLKs) and receptor-like proteins (RLPs) (Yang et al. 2012; Zhang et al. 2013; Zipfel 2014). The RLK PRRs consist of an extracellular ligand-binding domain, a single transmembrane domain, and a cytosolic protein kinase domain (Seifert and Blaukopf 2010). The RLPs and RLKs have similar structures, except that the RLPs lack the cytoplasmic kinase domain (Wang et al. 2010). The RLK and RLP PRRs harbor an extracellular domain consisting of leucine-rich repeat (LRR) or lysine motifs (LysM) (Saijo 2010; Segonzac and Zipfel 2011).

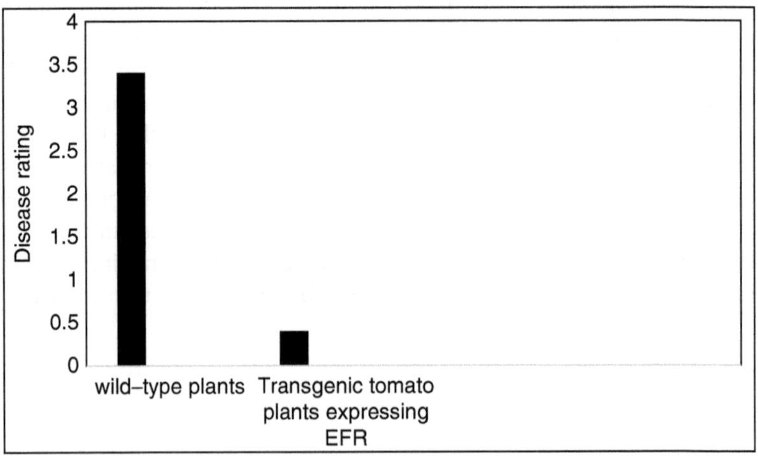

Fig. 5.1 Transgenic tomato plants expressing EFR from Arabidopsis show enhanced resistance against *Ralstonia solanacearum* (Adapted from Lacombe et al. 2010)

Table 5.1 PAMP/PIMP-PRR pairs involved in PAMP-triggered immunity

PAMP/PIMP	PRR	Structure of PRRs
Flg22	FLS2	LRR-RLK
EF-Tu	EFR	LRR-RLK
Chitin	CERK1	LysM-RLK
	CEBiP	LysM-RLP
Chitosan	CEBiP	LysM-RLP
β-glucans	GBP	Glycoside hydrolases
Peptidoglycan	LYM1 LYM3	LysM-receptor kinase
Mannose	MBL receptor	Lectin receptor kinase
Ax21	XA21	LRR-RLP
INF1 elicitin	NbLRK1	Lectin-like receptor kinase
Capsicein elicitin	NgRLK1	Receptor-like kinase
EIX1	LeEIX1	LRR-RLP
	LeEIX2	LRR-RLP
Lipopolysaccharides	Receptor	RLK
SCFE1	RLP30	RLP
EndoPG	RBPG1	LRR-RLP
eMax	ReMAX	RLP
Oligogalacturonides	WAK1	RLK
Pep1	PEPR1	LRR-RLK
Pep2	PEPR2	LRR-RLK

FLS2 (FLAGELLIN-SENSING 2) is the PRR for the PAMP flg22 in *Arabidopsis*. FLS2 is a leucine-rich repeat receptor-like kinase (LRR-RLK) (Chinchilla et al. 2006; Sun et al. 2012). It consists of an extracellular domain with 28 LRR motifs, a transmembrane domain, and a cytoplasmic Ser/Thr kinase domain (Gómez-Gómez and Boller 2000). Flg22 directly binds to FLS2 and physically interacts with the flg22 epitope (Chinchilla et al. 2006; Dunning et al. 2007). The PRR FLS2 homologous to *Arabidopsis* FLS2 has been detected in tomato, tobacco, rice, barley, and *Brassica* species (Chinchilla et al. 2006, 2007a; Dunning et al. 2007; Hann and Rathjen 2007; Robatzek et al. 2007; Shen et al. 2007; Takai et al. 2008; Shinya et al. 2010; Hao et al. 2014). The PRR FLS2 is found localized at the plasma membrane. When activated by the PAMP flg22, FLS2 is translocated to endocytic compartments (Robatzek et al. 2006). Endocytosis of the PRR FLS2 has been shown to be important for the PAMP flg22-induced defense signaling system (Chinchilla et al. 2007b).

The PRR FLS2 binds with the PAMP flg22. Binding of flg22 peptide with FLS2 was observed in all tested 45 genotypes of *A. thaliana*. However, a significant variation for flg22 binding was observed among the *A. thaliana* genotypes, and also among the tested 21 species of plants belonging to Brassicaceae family (Vetter et al. 2012). FLS2 protein abundance was also assessed in different *A. thaliana* genotypes. The genotypes C24, Yo-O, and Lov-5 displayed higher FLS2 protein abundance than the genotypes Col-O, Sei-O, and Eri. The differences in FLS2 abundance matched differences in flg22 binding. Variation in flg22-binding activity reflected variation in abundance of the FLS2 receptor. Therefore, changes in FLS2 function appeared to be primarily quantitative (Vetter et al. 2012). Higher levels of flg22 binding in most of the *A. thaliana* genotypes was found to be associated with reduced proliferation of the bacterial pathogen *Pseudomonas syringae* pv. *tomato*. The observed variation correlated with both the severity of elicited defense responses and bacterial proliferation (Vetter et al. 2012). The extensive variation in flg22 perception in different *A. thaliana* genotypes would have resulted from changes in FLS2 protein abundance. (Vetter et al. 2012). Sensitivity of different ecotypes and mutants of *Arabidopsis thaliana* toward the bacterial elicitor flagellin correlated with the presence of receptor-binding sites (Bauer et al. 2001).

Two PAMPs have been detected from flagellin, flg22 and flgII-28 (Veluchamy et al. 2014). FlgII-28 appears to be specifically recognized by Solanaceae species (Cai et al. 2011). Perception of flg22 occurs via the PRR FLS2 (Gómez-Gómez and Boller 2002), while FLS3 (Flagellin sensing 3) is involved in the perception of flgII-28 in the Solanaceae species (Clarke et al. 2013) and activation of defenses (Felix et al. 1999; Bauer et al. 2001; Vetter et al. 2012).

EF-Tu is the abundant bacterial protein and is involved in translation of bacterial mRNAs. It acts as a bacterial PAMP in *Arabidopsis* and other members of the family Brassicaceae (Kunze et al. 2004; Zipfel 2008; Serrano et al. 2012; Furukawa et al. 2014). The N-acetylated peptide comprising the N-terminal 18 aminoacids of EF-Tu has been recognized as the PAMP epitope. Responsiveness to elf18 has been observed only in various Brassicaceae species but not in members of other plant families (Kunze et al. 2004; Zipfel et al. 2006).

EFR (EF-Tu Receptor) has been identified as the PRR in Arabidopsis for binding the PAMP EF-Tu. EFR is a receptor-like kinase with a 21-LRR extracellular domain, a transmembrane domain, and a cytoplasmic Ser/Thr kinase domain (Albert and Felix 2010; Albert et al. 2010). EFR physically interacts with the first 18 amino acids of the N-terminus of EF-Tu, the elf 18 (Zipfel et al. 2006). EF-Tu is recognized by the EFR, a Brassicaceae-specific PRR (Zipfel et al. 2006). It has been suggested that the restricted perception of EF-Tu may be due to the fact that no orthologs of EFR exist in other plant families (Boller and Felix 2009; Furukawa et al. 2014). Transient heterologous expression of AtEFR in *Nicotiana benthamiana*, a plant that normally lacks elf18 responsiveness, restores elf18 binding and responses (Zipfel et al. 2006).

XA21 has been identified as the PRR for the bacterial PAMP Ax21. XA21 is a receptor kinase, which consists of LRR, transmembrane, juxtamembrane (JM) and intracellular kinase domains (Song et al. 1995). XA21 belongs to subfamily XII of the LRR-RKs and is highly similar to EFR. Similar to FLS2 and EFR, XA21 possesses a non-RD kinase, whose presence has been correlated with a role in innate immunity across kingdoms (Darick and Ronald 2006). The activity of the PRR XA21is negatively regulated by various PRR binding proteins (Chen et al. 2010). Upon perception of the PAMP Ax21, the PRR is dissociated from the negative regulators (Chen et al. 2010). Thus the PAMP Ax21 activates the PRR XA21 by removing the action of negative regulators of PRR (Chen et al. 2010).

A receptor-like kinase, designated CERK1 (for Chitin Elicitor Receptor Kinase 1) in *Arabidopsis* has been identified as a PRR for the fungal PAMP chitooligosaccharides (Miya et al. 2007; Iizasa et al. 2010; Le et al. 2014). CERK1 possesses three extracellular Lysine Motif (LysM) domains instead of LRRs (Miya et al. 2007; Wan et al. 2008). The CERK1 ectodomain binds chitin and partially de-acetylated chitosan directly without any requirement for interacting proteins (Eckardt 2008; Lohmann et al. 2010; Petutschnig et al. 2010). A LysM receptor-like kinase similar to *Arabidopsis* CERK1 has been detected in rice. It was designated OsCERK1 and it showed high homology with *Arabidopsis* CERK1 (Shimizu et al. 2010). OsCERK1 contained only one LysM motif in the extracellular domain, while CERK1 contained three LysM motifs in its extracellular domain (Shimizu et al. 2010).

The *cerk1* mutants, however, were not impaired in their responsiveness to flg22, elf18, LPS, or PGN (Gimenez-Ibanez et al. 2009a), suggesting that CERK1 is involved in the recognition of yet unknown bacterial PAMP. Gimenez-Ibanez et al. (2009b) reported reduced activation of a PAMP-induced defense response on plants lacking the CERK1 gene after treatment with crude extracts of the bacterial pathogen *P. syringae* pv. *tomato* DC3000. This strengthens the earlier findings that CERK1 mediates perception of an unknown bacterial PAMP in *Arabidopsis*. OsCERK1 mediates the signal of a fungal cell wall component, chitin, by coordinating with a lysin motif (LysM)-containing protein CEBiP. The *OsCERK1*-disrupted rice lines also showed lowered responsiveness to a bacterial cell wall component, peptidoglycan. It was also found that OsCERK1 interacts with the LysM-containing proteins LYP4 and LYP6 (Kouzai et al. 2014). The results suggest that the chitin receptor OsCERK1 in rice mediates the signaling pathways of both fungal and bacterial molecular patterns by integrating with different LysM-containing receptor-like proteins (Kouzai et al. 2014).

Another PRR protein for the perception of chitin has been recognized in rice and designated chitin elicitor-binding protein (CEBiP) (Kaku et al. 2006; Shinya et al. 2010). It is a receptor-like protein (RLP), unlike the PRR OsCERK1, which is a receptor-like kinase (RLK). CEBiP contains extracellular LysM motifs for chitin-binding but lacks an intracellular kinase domain, which is characteristically present in OsCERK1 (Kaku et al. 2006). Rice requires both CEBiP and OsCERK1 for chitin signaling (Shimizu et al. 2010).

Many β-glucan components of fungal and oomycete cell walls act as PAMPs (Shibuya and Minami 2001; Silipo et al. 2010). A β-glucan-binding protein (GBP) has been identified as a PRR in soybean. It lacks a transmembrane domain and predominantly localizes to the cytoplasmic face of the plant cell wall (Fliegmann et al. 2004). GBP is a glycoside hydrolase (Fliegmann et al. 2005). GBP initially acts on oomycete/fungus-derived heptaglucosides as a glucan hydrolase, releasing β-glucans that subsequently are perceived by a different domain of GBP (Fliegmann et al. 2004).

The bacterial PAMP lipopolysaccharide (LPS) was found to bind to tobacco cells and become internalized into endocytic vesicles, suggesting a receptor-mediated process (Gross et al. 2005). It has been suggested that the receptor of the LPS may be a receptor-like kinase (RLK) (Sanabria and Dubery 2006). The LysM-receptor kinase LYM1 LYM3 has been identified as the PRR which recognizes bacterial peptidoglycan (Willmann et al. 2011). The LysM motif present in several receptor kinases and transmembrane proteins in plants can bind peptidoglycan (Zhang et al. 2007; Buist et al. 2008; Nicaise et al. 2009).

LeEIX1 and LeEIX2 are the two different PRRs identified in tomato for the perception of the fungal PAMP EIX (Ethylene-Inducing Xylanase elicitor). These PRRs are cell-surface receptors without kinase domain in plants (Ron and Avni 2004; Kaku et al. 2006). The structure of these EIX receptors is similar to a family of receptor-like proteins (RLPs) (He et al. 2007). Both the PRRs are able to bind the PAMP EIX, but only LeEIX2 is involved in triggering defense responses (Bar et al. 2011).

An *Arabidopsis thaliana* leucine-rich repeat receptor-like protein, AtRLP42 that recognizes fungal endopolygalacturonases (PGs) has been shown to act as a MAMP receptor. The receptor-like protein was called RESPONSIVENESS TO BOTRYTIS POLYGALACTURONASES1 (RBPG1). RBPG1 recognizes several PGs from the necrotrophic fungal pathogen *Botrytis cinerea* as well as one from the saprotrophic fungus *Aspergillus niger* (Bellincampi et al. 2014; Zhang et al. 2014). The Arabidopsis gene *RBPG1* gene has been shown to be essential for the responsiveness of Arabidopsis to PGs. Pretreating leaves of the transgenic *Arabidopsis* plants expressing the *RBPG1* gene with a PG resulted in increased resistance to *Hyaloperonospora arabidopsidis* (Zhang et al. 2014). The results suggest that RBPG1 is a PRR, which takes part in PAMP-PRR signaling complex triggering PAMP-triggered immunity.

RLP30 (RECEPTOR –LIKE PROTEIN30) has been identified as a receptor for SCFE1 (SCLEROTINIA CULTURE FILTRATE ELICITOR1) elicitor from *Sclerotinia sclerotiorum* in *Arabidopsis* (Zhang et al. 2013). The SCFE1-triggered immune responses engage a signaling pathway involving the receptor-like protein RLP30 and the regulatory receptor-like kinases BAK1 (BRASSINOSTEROID INSENSITIVE1-ASSOCIATED RECEPTOR KINASE1) and SOBIR1/EVR (SUPPRESSOR OF BIR1-1/EVERSHED) (Zhang et al. 2013).

NbLRK1 is the PRR identified for the PAMP elicitin INF1 of *Phytophthora infestans in Nicotiana benthamiana* (Kanzaki et al. 2008; Singh and Zimmerti 2013). NbLRK1 is a lectin-like receptor kinase and it is a typical RD kinase (Kanzaki et al. 2008). A PRR for the elicitin capsicein has been identified in tobacco (*Nicotiana glutinosa*) and it was designated NgRLK1. NgRLK1 has a domain structure similar to that of all plant RLKs (Kim et al. 2010).

Some mannose-binding lectin (MBL) receptor kinases have been identified as PRRs for perception of PAMPs. Function of MBL appears to be pattern recognition. MBL recognizes carbohydrate patterns, found on the surface of a large number of pathogenic microorganisms. MBL binds to carbohydrates (specifically mannose residues). The lectin receptor kinases (LecRKs) contain an extracellular domain of mannose specific binding lectin and an intracellular Ser/Thr kinase catalytic domain (Barre et al. 2002). A mannose-binding lectin (MBL) gene, *CaMBL1*, has been isolated from pepper (*Capsicum annuum*) leaves (Hwang and Hwang 2011). The *CaMBL1* gene contains a predicted *Galanthus nivalis* agglutinin-related lectin domain responsible for the recognition of high-mannose *N*-glycans. The CaMBL1 protein exhibits binding specificity for mannose and is mainly localized to the plasma membrane. Mannose has been recognized as a PAMP found in fungal pathogens (Kogel et al. 1998).

A PRR of *Arabidopsis thaliana* with specificity for the bacterial PAMP eMax (for enigmatic MAMP of *Xanthomonas*) from xanthomonads including *X. axonopodis* pv. *citri*, *X. campestris* pv. *juglandis*, *X. campestris* pv. *malvacearum* has been identified and termed ReMax (for Receptor for eMax) (Jehle et al. 2013a, b). ReMAX is also called RECEPTOR-LIKE PROTEIN1 (RLP1). ReMAX/RLP1 is a typical RLP with structural similarity to the tomato RLP EIX2, which detects fungal xylanase as a PAMP/MAMP (Jehle et al. 2013b). Response to eMAX is restricted to the Brassicaceae family (Jehle et al. 2013b). ReMAX has no ortholog in *Nicotiana benthamiana* and this species does not respond to eMAX even when transformed with ReMAX. However, interfamily transfer of eMAX perception was successful with a chimeric form of ReMAX where the C-terminal part of the protein was replaced by the corresponding part of the tomato RLP EIX2. Functionality of ReMAX also depended on the presence of the receptor kinase SOBIR (SUPPRESSOR OF BIR1-1) (Jehle et al. 2013a).

5.2 Importance of PRRs in Triggering Defense Responses against Pathogens

PRRs play important role in triggering defense responses, Flagellin-deficient mutant of *Xanthomonas oryzae* pv. *oryzae* causes increased disease (longer lesion lengths) in rice leaves relative to wild-type *X. oryzae* pv. *oryzae* (Tian et al. 2015). The results suggest that flagellin is involved in triggering defense responses in rice against the bacterial blight pathogen. Flagellin perception is required for full immunity against bacteria, because plants deficient in FLS2 are more susceptible to adapted and nonadapted bacterial pathogens (Zipfel 2009). In *Arabidopsis*, pretreatment with flg22 restricts growth of the pathogenic bacterium *Pseudomonas syringae* pv. *tomato* DC3000 and *fls2* mutants are more susceptible to this bacterium (Zipfel et al. 2004). Lack of flagellin recognition by the PRR FLS2 allows more growth of the nonadapted bacteria *P. syringae* pv. *tabaci* and *P. syringae* pv. *phaseolicola* (Li et al. 2005; de Torres et al. 2006). Successful bacterial pathogens need to avoid or suppress PAMP-triggered immunity induced by flagellin (Zipfel 2009). Some virulence effectors from phytopathogenic bacteria do so by directly targeting the PRR FLS2 (Göhre et al. 2008).

Arabidopsis efr mutants are more susceptible to *Agrobacterium tumefaciens* (Zipfel et al. 2006). The results suggest the importance of the PRR EFR in activation of plant innate immunity. The knock-out mutants for *CERK1* completely lost the ability to respond to the chitin elicitor, including MAPK activation, ROS generation, and gene expression (Miya et al. 2007). The complete loss of the gene responses induced by the chitin elicitor indicates that CERK1 serves as the 'master switch' of the signaling cascade. The mutation in *CERK1* gene blocked the induction of almost all chitooligosaccharide-responsive genes and led to more susceptibility to fungal pathogens (Wan et al. 2008). *Arabidopsis cerk1* mutants are more susceptible to fungal pathogens (Miya et al. 2007; Wan et al. 2008). Exogenously applied chitooligosaccharides enhanced resistance against both fungal and bacterial pathogens in the wild-type plants but not in the mutant. These results suggest that CERK1 (LysM RLK1) is essential for chitin signaling in plants as part of the receptor complex and is involved in chitin-mediated plant innate immunity (Wan et al. 2008). CERK1 was also involved in bacterial recognition, as *cerk1* mutants are more susceptible to *P. syringae* pv. *tomato* DC3000 (Gimenez-Ibanez et al. 2009a). CERK1, a LysM receptor kinase, is essential for chitin elicitor signaling in Arabidopsis (Miya et al. 2007). OsCERK1 is a rice receptor-like kinase that mediates the signal of chitin by coordinating with a lysin motif (LysM)-containing protein CEBiP (Kouzai et al. 2014). OsCERK1 was localized around the primary infection hyphae of *Magnaporthe oryzae* in rice. Disruption of *OsCERK1* led to increased hyphal growth of the rice blast pathogen *M. oryzae* (Kouzai et al. 2014). The results suggest that OsCERK1 is indispensable for chitin perception and participates in innate immunity in rice (Kouzai et al. 2014). RLP30 is a receptor for SCFE1 elicitor from *Sclerotinia sclerotiorum* in *Arabidopsis.* Mutants of *RLP30* were found to be more susceptible to *S. sclerotiorum* and *Botrytis cinerea*, suggesting the importance of the receptor in triggering immune responses against pathogens (Zhang et al. 2013).

Besides the PAMPs, PIMPs/HAMPs also play important role in triggering plant immunity. The endogenous molecular patterns are present only when the tissue is infected or damaged (De Lorenzo et al. 2011). Oligogalacturonides (OGAs) and Plant elicitor peptides (Peps) are the well-characterized PIMPs/HAMPs (Krol et al. 2010; Vallarino and Osorio 2012; Ma et al. 2013; Ross et al. 2014). The Pep1 receptor, PEPR1 is a typical LRR receptor kinase, having an extracellular LRR domain and an intracellular protein kinase domain (Shiu et al. 2004; Qi et al. 2010; Tintor et al. 2013). The second Pep1 receptor, PEPR2, is a plasma membrane LRR receptor kinase and has 76 % amino acid similarity to PEPR1 (Yamaguchi et al. 2010). PEPR1 has been identified as a receptor for the HAMPs *At*Pep1, *At*Pep2, *At*Pep3, *At*Pep4, *At*Pep5, and *At*Pep6 and PEPR2 is a receptor for *At*Pep1 and *At*Pep2 (Yamaguchi et al. 2010).

The wall-associated kinase 1 (WAK1) has been identified as a receptor of OGAs in *Arabidopsis thaliana* (Brutus et al. 2010). WAK1 belongs to the huge family of 610 receptor-like kinases identified in the *Arabidopsis thaliana* genome (Shiu and Bleecker 2001). WAKs display a typical plant Ser/Thr kinase signature and an extracytoplasmic domain (ectodomain) containing several EGF (Epidermal Growth Factor)-like repeats. WAK1 binds *in vitro* to OGAs through the N-terminal non-EGF portion of the ectodomains (Decreux et al. 2006; Cabrera et al. 2008). The binding of WAK1 in vitro to homogalacturonan (HGA), elicitor-active OGAs occurs under conditions that are compatible with the formation of calcium-induced "egg box" structures (Decreux and Messiaen 2005; Cabrera et al. 2008). WAK1 is induced by wounding, bacterial infection, and salicylic acid treatment (He et al.

1998; Wagner and Kohorn 2001). WAK1 is up-regulated in response to OGAs (Denoux et al. 2008). WAK1 is capable to sense OGs in vivo and trigger a defense response (Brutus et al. 2010).

Another WAK gene, *SlWAK*1, has been reported to be the receptor of OGAs in tomato (Rosli et al. 2013). Tomato recognition of PAMPs/MAMPs leads to increased expression of *SlWAK1* and an increase in SlWAK1 receptors at the plasma membrane (Rosli et al. 2013). A cDNA encoding a novel WAK has been isolated from rice and was named OsWAK1 (Li et al. 2009). The deduced amino acid sequence of OsWAK1 showed 27.6 % identity to WAK2 from *Arabidopsis* (Li et al. 2009). Infection of the rice blast fungus, *Magnaporthe oryzae* induced the *OsWAK1* transcripts. The accumulation of *OsWAK1* mRNA occurred earlier and was abundant in rice leaves infected with an incompatible race than with a compatible race of *M. oryzae*.

5.3 Engineering PRRs for Disease Management

5.3.1 Engineering the PRR EFR for Crop Disease Management

EFR (EF-Tu Receptor) is a PRR detected in *Arabidoposis*, which binds the bacterial PAMP Elongation factor Thermo unstable (EF-Tu). The PAMP EF-Tu has been detected in several bacterial species causing diseases in a wide range of plant species (Zipfel 2008). However, EFR is a Brassicaceae-specific PRR (Zipfel et al. 2006). The perception of the PAMP EF-Tu by the PRR was found to be restricted to Brassicaceae and it has been suggested that the restricted perception of EF-Tu may be due to the fact that no orthologs of EFR exist in other plant families (Boller and Felix 2009; Furukawa et al. 2014). Transient heterologous expression of AtEFR in *Nicotiana benthamiana*, a plant that normally lacks EF-Tu responsiveness, restores EF-Tu binding and responses (Zipfel et al. 2006). Transfer of EFR from *Arabidopsis* to various crop plants may be highly useful for crop disease management. Pathogens that are adapted to a particular host plant may be adept at suppressing the PRRs of that host by their effectors. The effectors of the pathogens might not recognize PRRs from other host plants and development of transgenic plants expressing PRRs from other plant species may provide good resistance against various bacterial pathogens possessing the PAMP EF-Tu (Lacombe et al. 2010). Transgenic tomato plants expressing *EFR* gene from *Arabidopsis thaliana* show enhanced resistance against the bacterial wilt pathogen *Ralstonia solanacearum* (Fig. 5.1; Lacombe et al. 2010).

Transgenic tomato plants expressing *EFR* were more resistant to *Xanthomonas axonopodis* pv. *vesicatoria* (revised name *X. perforans*), the causal agent of the bacterial leaf spot (Lacombe et al. 2010). Transgenic *Nicotiana benthamiana* expressing *EFR* gene from *Arabidopsis* showed resistance against the crown gall

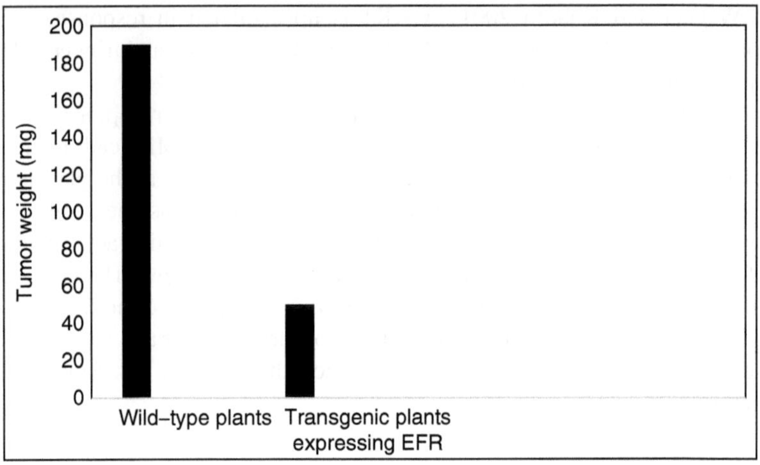

Fig. 5.2 Transgenic *Nicotiana benthamiana* plants expressing *EFR* gene from *Arabidopsis* show enhanced resistance against *Agrobacterium tumefaciens* (Adapted from Lacombe et al. 2010)

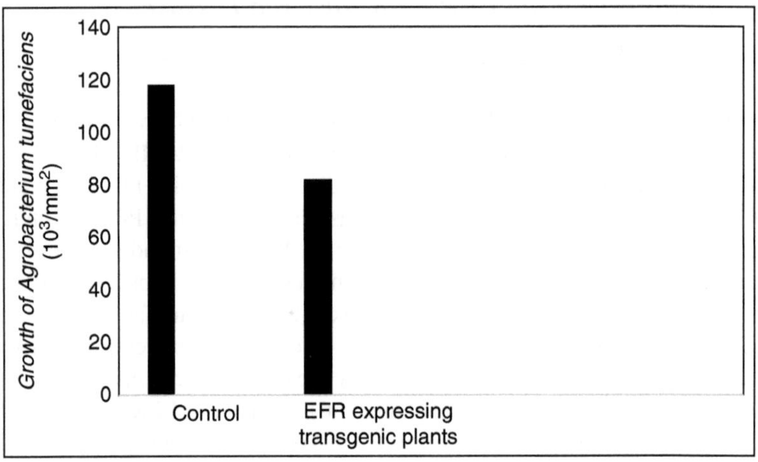

Fig. 5.3 Growth of *Agrobacterium tumefaciens* in tobacco tissues transiently expressing EFR protein (Adapted from Brutus et al. 2010)

bacterial pathogen *Agrobacterium tumefaciens* (Fig. 5.2; Lacombe et al. 2010). EF-Tu perception by EFR in *N. benthamiana* restricted gall formation caused by a virulent tumorigenic *A. tumefaciens* strain. Stab-inoculated stems of transgenic *N. benthamiana* plants expressing EFR developed tumors 75 % smaller than the ones present on wild-type stems (Lacombe et al. 2010).

Transgenic tobacco (*Nicotiana tabacum*) expressing EFR also showed resistance against *A. tumefaciens* (Fig. 5.3; Brutus et al. 2010). Transgenic *N. benthamiana* plants expressing the *Arabidopsis EFR* gene showed resistance against *Pseudomonas syringae* pv. *syringae*. However, the transgenic *N. benthamiana* plants expressing

EFR gene did not show resistance against the fungal wilt pathogen *Verticillium dahliae*, which does not carry the PAMP EF-Tu (Lacombe et al. 2010). Collectively, these studies suggest that engineering *EFR* gene may be a potential tool to develop disease resistant plants against a wide spectrum of bacterial pathogens that carry the recognized EF-Tu PAMP.

5.3.2 Engineering the PRR FLS2 for Plant Disease Management

FLS2 is the PRR involved in perception of the bacterial PAMP flg22 to trigger defense responses. *FLS2* genes have been detected in wide range of plants including Arabidopsis, tomato, tobacco, rice, and barley and the PAMP flg22 has been detected in large number of bacterial species. Early and robust activation of PRRs is necessary to trigger strong defense responses (Saijo 2010; Vidhyasekaran 2014). Transgenic *Arabidopsis* plants expressing *FLS2* gene have been developed and these transgenic plants showed increased resistance against *Pseudomonas syringae* pv. *tomato* DC3000, which carries flagellin recognized by FLS2. Infected transgenic plants showed a significantly slower and less severe development of disease symptoms than a FLS2 natural mutant, *Arabidopsis* ecotype Wassilewskija (Ws-O) plants. The weaker symptoms correlated with a lower number of bacteria in the transgenic plants (De Lorenzo et al. 2011). Treatment with flg22 induced H_2O_2 production and callose deposition in the *FLS2* transgenic plants but not in untransformed Ws-O plants (De Lorenzo et al. 2011). These results suggest that the PRR FLS2 can be exploited for crop disease management by developing transgenic plants.

5.3.3 Engineering the PRR XA21 for Crop Disease Management

XA21 is the PRR detected in rice and it recognizes a sulfated peptide PAMP, called Ax21, which is present in all *Xanthomonas* and *Xylella* species (Lee et al. 2009; Park et al. 2010b). *XA21* gene has been used for developing transgenic plants with increased disease resistance against various *Xanthomonas* spp. possessing the PAMP Ax21. Tramsgenic rice plants overexpressing *Xa21* gene showed enhanced resistance against the rice bacterial blight pathogen *Xanthomonas oryzae* pv. *oryzae* (Fig. 5.4; Park et al. 2008, 2010a; Chen et al. 2014b).

Transgenic *Citrus sinensis* plants expressing the rice PRR *XA21* gene were developed and the transgenic plants showed enhanced resistance against the citrus canker pathogen *Xanthomonas axonopodis* pv. *citri* (Mendes et al. 2010). Banana Xanthomonas wilt caused by the bacterium *Xanthomonas campestris* pv.

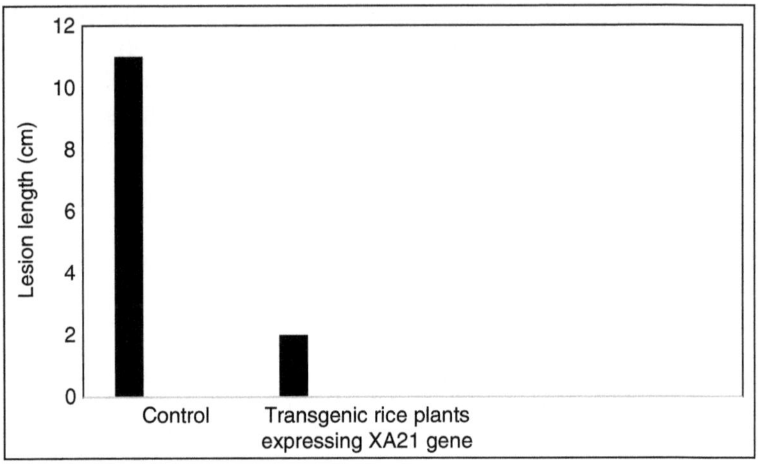

Fig. 5.4 Transgenic rice plants expressing the PRR *XA21* gene show enhanced resistance to *Xanthomonas oryzae* pv. *oryzae* (Adapted from Park et al. 2008)

musacearum is the most devastating disease of banana in Africa, Transgenic banana plants expressing the rice PRR *XA21* gene were developed. About 50 % of the transgenic lines showed complete resistance to *X. campestris* pv. *musacearum*. In contrast, all of the nontransgenic control plants showed severe symptoms that progressed to complete wilting (Tripathi et al. 2014). These results suggest that PRRs are potential tools for effective management of devastating diseases of crop plants.

5.3.4 Engineering WAK1 Receptors for Crop Disease Management

OsWAK1 is a HAMP receptor detected in rice involved in perception of OGAs (Li et al. 2009). Infection of the rice blast fungus, *Magnaporthe oryzae* induces the *OsWAK1* transcripts. The accumulation of *OsWAK1* mRNA occurred earlier and was abundant in rice leaves infected with an incompatible race than with a compatible race of *M. oryzae*. Transgenic rice plants were developed to express the *OsWAK1* gene constitutively. Six transgenic rice lines with constitutive expression of *OsWAK1* showed resistance to the compatible race (Li et al. 2009). SA treatment enhances transcription of *OsWAK1* gene (Li et al. 2009) and SA signaling system is known to induce resistance against *M. oryzae* (Wang et al. 2008b). Transgenic *Arabidopsis thaliana* plants overexpressing WAK1 showed increased resistance against the fungal pathogen *Botrytis cinerea* (Fig. 5.5; Brutus et al. 2010). These transgenic plants overexpressing *WAK1* show levels of *WAK1* transcripts about 10 times higher than those present in untransformed plants and exhibit enhanced responsiveness to OGAs (Brutus et al. 2010). These transgenic plants showed

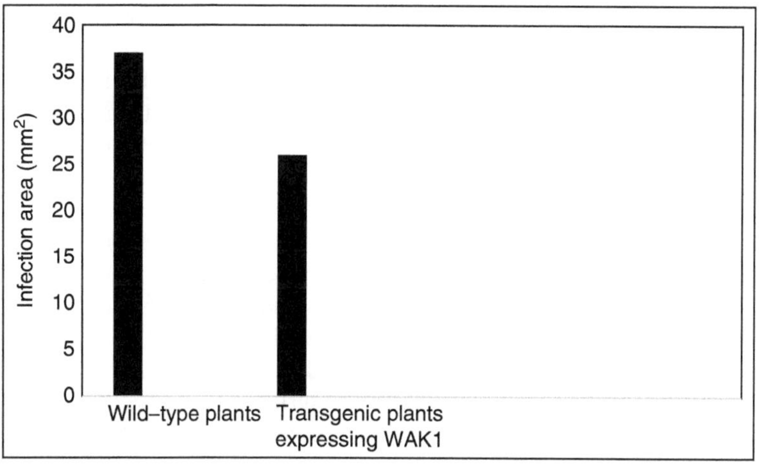

Fig. 5.5 Disease development in transgenic *Arabidopsis* plants expressing *WAK1* gene inoculated with *Botrytis cinerea* (Adapted from Brutus et al. 2010)

resistance against *Pectobacterium carotovorum* (formerly *Erwinia carotovora*), the causal agent of black leg and soft rot (De Lorenzo et al. 2011). These results suggest that the HAMP receptor is a potential tool for management of crop diseases.

5.4 PRR-Interacting Protein Complexes

PRRs do not act alone but rather function as part of multi-protein complexes at the plasma membrane (Monaghan and Zipfel 2012). The PRRs interact with several transmembrane LRR receptor-like kinase proteins that act as signaling adapters or amplifiers to achieve full functionality (Zipfel 2009). These transmembrane proteins include BAK1 (for BRI1 (BRASSINOSTEROID INSENSITIVE1)-ASSOCIATED KINASE1), BIK1 (BOTRYTIS-INDUCED KINASE1), and BIR1 (BRANCHING INHIBITING RECEPTOR1) (Postel et al. 2010; Zhang and Zhou 2010; Zhang et al. 2010; Kørner 2013; Chen et al. 2014a; Lin et al. 2014a). BAK1 is a multifunctional leucine-rich repeat receptor-like kinase (LRR-RLK) that exerts its function by interacting with multiple ligand immune receptors (PRRs) (Chinchilla et al. 2007b; Heese et al. 2007; Ryan et al. 2007; Zipfel 2008; Gao et al. 2009; Postel et al. 2010; Schulze et al. 2010; Halter et al. 2014a). BAK1 is a positive regulator of PAMP-triggered plant immunity and it acts as an adapter of multiple LRR-RKs that act in defense signaling (Ma et al. 2013). BAK1 is required for the immune responses triggered by multiple PAMPs including flagellin, the bacterial elongation factor EF-Tu, peptidoglycans, lipopolysaccharides, cold-shock protein, HrpZ (harpin), and the oomycete elicitor INF1 in *Arabidopsis* and tobacco (Chinchilla et al. 2007b; Heese et al. 2007; Shan et al. 2008). A BAK1 ortholog has been detected in rice (Li et al. 2009) and it may be involved in the PAMP Ax21-mediated immunity (Segonzac and Zipfel 2011). BAK1 is also required for the activation of another PRR CERK1 in chitin signaling (Zhang and Zhou 2010). BAK1 acts as a co-receptor of the PRRs, FLS2 and EFR (Kim et al. 2013). BAK1 is a co-activator of the PAMP-receptor complex to enhance various signaling pathways (Wang et al. 2008a). It has been found to physically interact with PEPRs (PEPR1 and PEPR2) in vitro (Postel et al. 2010). BAK1 and PEPRs can phosphorylate each other after Pep1 treatment (Schulze et al. 2010).

The PAMP flg22 perception by the PRR FLS2 triggers an interaction between FLS2 and BAK1 (Chinchilla et al. 2007b; Heese et al. 2007). The flg22 induced FLS2-BAK1 association occurs within seconds and is accompanied by increased phosphorylation on both FLS2 and BAK1 (Schulze et al. 2010). Flg22 binding induces FLS2 heteromerization with BAK1 and their reciprocal activation. A conserved and a nonconserved site from the inner surface of the FLS2 solenoid recognize the C- and N-terminal segment of flg22, respectively, without oligomerization or conformational changes in the FLS2 ectodomain (Sun et al. 2013). Besides directly interacting with FLS2, BAK1 acts as a co-receptor by recognizing the C terminus of the FLS2-bound flg22 (Sun et al. 2013). The BAK1 kinase activity is required for FLS2-mediated signaling but not flg22- induced association of FLS2 and BAK1 (Schulze et al. 2010). It suggests that BAK1 acts downstream of FLS2 perception to regulate signaling (Chinchilla et al. 2007b).

BIK1 is a receptor-like cytoplasmic kinase (RLCK). The PAMP flg22 interacts with BAK1 to phosphorylate BIK1, which seems to act as positive regulator of the PAMP signaling pathway (Lu et al. 2010; Zhang et al. 2010). BIK1 is an essential component in PAMP signal transduction, which links the PAMP receptor complex to downstream intracellular signaling (Lu et al. 2010). BIK1 forms a complex with unstimulated FLS2 in plants, and flg22 induces a rapid phosphorylation of BIK1 in both FLS2- and BAK-dependent manner (Zhang et al. 2010). BIK1 is phosphorylated within 1 min upon flagellin perception (Lu et al. 2010).

BIK1 associates with both FLS2 and BAK1. BIK1 is a substrate of BAK1, whereas BAK1 and FLS2 are also substrates of BIK1, suggesting transphosphorylation events between BIK1 and the FLS2/BAK1 complex. BIK1 is phosphorylated by BAK1, and BIK1 also directly phosphorylates BAK1 and FLS2 *in vitro*. The flagellin phosphorylation site Thr^{287} of BIK1 is required for its phosphorylation on BAK1 and FLS2, suggesting that BIK1 is likely first phosphorylated upon flagellin perception and subsequently BIK1 transphosphorylates FLS2/BAK1 to propagate flagellin signaling (Lu et al. 2010). BIK1 appears to function downstream of FLS2/BAK1 complex formation and phosphorylation because BIK1 phosphorylation requires not only the presence of both FLS2 and BAK1, but also their kinase activity (Lu et al. 2010). In addition to phosphorylation, flg22 induces a dissociation of BIK1 from FLS2 (Zhang et al. 2010; Lu et al. 2010). The dissociation of the phosphorylated BIK1 proteins from FLS2 allows the activation of other components downstream of BIK1 (Zhang et al. 2010). The ligand-induced EFR-BAK1 interaction has been reported in the PAMP EF-Tu signaling (Schulze et al. 2010) and BIK1 has been shown to be required for interaction with the EFR (Zhang et al. 2010). BIK1 plays a critical role in the integration of signals from multiple PRRs (Lu et al.

2010). The BIK1 is an essential component in PAMP signal transduction, which links the PAMP receptor complex to downstream intracellular signaling (Lu et al. 2010). BIK1 also interacts with the HAMPs PEPRs to mediate ethylene-induced immunity (Liu et al. 2013). PEPR1 has been shown to interact with BIK1 and PBS1-like 1 (PBL1) to mediate Pep1-induced defense responses (Liu et al. 2013).

BIR1 is a BAK-1 interacting receptor-like kinase. It negatively regulates defense responses (Gao et al. 2009). BAK1 forms a complex with BIR1 (for BRANCHING INHIBITING RECEPTOR 1) to negatively regulate defense responses (Gao et al. 2009). BIR1 is a BAK-1 interacting receptor-like kinase. Knocking out *BIR1* leads to activation of constitutive defense responses. A mutant, which suppresses the activity of *BIR1* has been obtained. The gene *SOBIR1* (suppressor of *bir1*) encodes another receptor-like kinase whose overexpression activates defense responses (Gao et al. 2009). SOBIR1 functions as a specific regulator of resistance activated by *bir1* mutation. SOBIR1 is not required for flg22-mediated defense responses. SOBIR1 and BIR1 did not interact with each other. SOBIR1 has been shown to be a positive regulator of innate immunity and it activates plant immune responses (Gao et al. 2009).

Two BAK1-interacting RLKs, BIR2 and BIR3, have also been identified in *Arabidopsis* (Halter et al. 2014a, b). BIR2 is a BAK1 interacting protein, which is also a LRR-RLK, but in contrast to BAK1, negatively regulates BAK1-dependent PAMP responses (Halter et al. 2014a). BAK1 exists in preformed complexes with its ligand binding receptor partners (Halter et al. 2014a). BIR2 is released from BAK1 after ligand perception, increasing the pool of free BAK1 that is available to form complexes with activated ligand binding receptors. Overexpression of BIR2 results in reduced complex formation of BAK1 with FLS2, showing that BIR2 negatively regulates BAK1 complex formation with ligand (Halter et al. 2014a). BIR2 is unidirectionally phosphorylated by BAK1. BIR2 interacts constitutively with BAK1, thereby preventing interaction with the ligand-binding LRR-RLK FLS2. PAMP perception leads to BIR2 release from the BAK1 complex and enables the recruitment of BAK1 into the FLS2 complex (Halter et al. 2014b).

RLP30 (RECEPTOR –LIKE PROTEIN30 (RLP30) has been identified as a receptor for SCFE1 elicitor from *Sclerotinia sclerotiorum* (Zhang et al. 2013). The SCFE1-triggered immune responses engage a signaling pathway involving the receptor-like protein RLP30 and the regulatory receptor-like kinases BAK1 and SOBIR1/EVR (SUPPRESSOR OF BIRI1-1/EVERSHED) (Liebrand et al. 2013, 2014; Zhang et al. 2013). Mutants of *RLP30*, *BAK1*, and *SOBIR1* were found to be more susceptible to *Sclerotinia sclerotiorum* and *Botrytis cinerea*, suggesting the importance of them in PAMP-triggered immunity (Zhang et al. 2013).

The leucine-rich repeat receptor-like protein SOBIR1 (for SUPPRESSOR OF BIR1) has been shown to form a complex with PG (PAMP) and RBPG1 (PRR) in *Nicotiana benthamiana*. The *sobir1* mutant plants did not induce necrosis in response to PGs and were compromised in PG-induced resistance to the biotrophic oomycete pathogen *Hyaloperonospora arabidopsidis* (Zhang et al. 2014). The results suggest the importance of SOBIR1 in the PAMP-PRR signaling system in conferring disease resistance.

IK1 and PUB13 have been identified as BAK1-interacting proteins which interact with the PAMP-PRR complexes to transduce the signaling downstream (Kim et al. 2013). Zhang et al. (2010) identified a number of the *Arabidopsis* resistance gene PBS1-like (PBL) cytoplasmic receptor-like kinases (RLCKs) as components in PAMP- signaling pathways. AtPHOS32, AtPHOS34, and AtPHOS43 have also been reported to interact with PRRs (Peck et al. 2001; Merkouropoulos et al. 2008). Another LRR-RLK, ERECTA also has been shown to interact with PRRs (Godiard et al. 2003; Llorente et al. 2005). It is suggested that ERECTA may function in signal perception and/or in transduction (Godiard et al. 2003). It is involved in triggering resistance against the necrotrophic fungus *Plectosphaerella cucumerina* (Llorente et al. 2005) and the bacterium *Ralstonia solanacearum* in *Arabidopsis* (Godiard et al. 2003).

Some lectin receptor kinases (LecRKs) also have been shown to act as cell surface receptors involved in plant immunity (Bouwmeester and Govers 2009; Bouwmeester et al. 2011, 2014; Huang et al. 2013; Singh and Zimmerli 2013). LecRKs are receptor-like kinases (RLKs) characterized by an extracellular lectin motif (Bouwmeester and Govers 2009; Vaid et al. 2012). The Arabidopsis LecRK-VI.2 has been shown to be associated with the PRR FLS2 and it primes *Nicotiana benthamiana* pattern-triggered immunity (PTI) (Huang et al. 2014). LecRK genes appear to be induced upon treatment with PAMPs and pathogen infection (Bouwmeester and Govers 2009). The cell surface receptor LecRK-1.9, which belongs to the family of L-type lectin receptor kinases was found to be involved in triggering defense responses against *Phytophthora brassicae* in *Arabidopsis* (Bouwmeester et al. 2011). Expression of the *Arabidopsis LecRK-1,9* gene in potato and *Nicotiana benthamiana* resulted in enhanced resistance against the late blight pathogen *Phytophthora infestans* (Bouwmeester et al. 2014).

5.5 Engineering PRR-Interacting Protein Complexes for Crop Disease Management

5.5.1 Engineering SOBIR1 Gene Encoding Receptor-Like Kinase Interacting with PRRs for Inducing Disease Resistance

SOBIRI encodes an RLK with four extracellular LRRs and a cytoplasmic kinase domain, suggesting that it could be involved in signal perception and transduction (Gao et al. 2009). SOBIR1 encodes an RLK with similar structure as BAK1. BIR interacts with BAK1 and BIR is a negative regulator of PAMP-triggered immunity. BIR1 negatively regulates PAMP-PRR signaling system in activating immune responses. The *bir1* mutation activates defense responses. SOBIR1 acts as a specific regulator of resistance activated by *bir1* mutation (Gao et al. 2009). Transgenic plants overexpressing *SOBRI1* showed resistance against the bacterial wilt pathogen

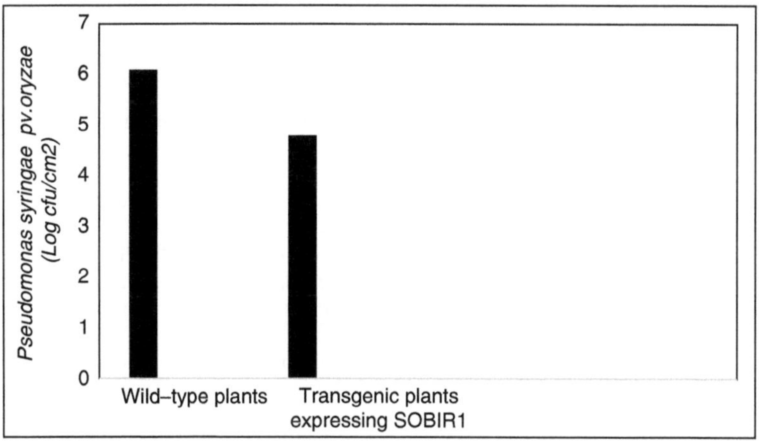

Fig. 5.6 Transgenic *Arabidopsis* plants expressing *SOBIR1* gene show resistance against the bacterial pathogen *Pseudomonas syringae* pv. *tomato* (Adapted from Gao et al. 2009)

Pseudomonas syringae pv. *tomato* DC3000 strain (Fig. 5.6; Gao et al. 2009). The transgenic *Arabidopsis* plants showed enhanced expression of defense-related *PR-1* and *PR-2* genes (Gao et al. 2009). The results suggest that the PRR-interacting protein SOBIRI can be engineered to develop disease-resistant plants.

5.5.2 Engineering the PRR-Interacting ERECTA Gene for Disease Management

ERECTA is a LRR Receptor-like kinase, which has been shown to interact with PRRs (Godiard et al. 2003; Llorente et al. 2005). ERECTA is involved in triggering resistance against the necrotrophic fungus *Plectosphaerella cucumerina* (Llorente et al. 2005). Transgenic Arabidopsis plants expressing *ERECTA* gene have been developed. These transgenic plants showed enhanced resistance against *Ralstonia solanacearum* (Godiard et al. 2003). The studies reveal the potential use of *ERECTA* gene for developing disease resistant plants.

References

Albert M, Felix G (2010) Chimeric receptors of the *Arabidopsis thaliana* pattern recognition receptors EFR and FLS2. Plant Signal Behav 5:1430–1432

Albert M, Jehle AK, Mueller K, Eisele C, Lipschis M, Felix G (2010) *Arabidopsis thaliana* pattern recognition receptors for bacterial elongation factor Tu and flagellin can be combined to form functional chimeric receptors. J Biol Chem 285:19035–19042

Altenbach DD, Robatzek S (2007) Pattern recognition receptors: from the cell surface to intracellular dynamics. Mol Plant-Microbe Interact 20:1031–1039

Bar M, Sharfman M, Avni A (2011) LeEix1 functions as a decoy receptor to attenuate LeEix2 signaling. Plant Signal Behav 6:455–457

Barre A, Herve C, Lescure B, Rougé P (2002) Lectin receptor kinases in plants. Crit Rev Plant Sci 21:379–399

Bauer Z, Gómez-Gómez L, Boller T, Felix G (2001) Sensitivity of different ecotypes and mutants of *Arabidopsis thaliana* toward the bacterial elicitor flagellin correlates with the presence of receptor-binding sites. J Biol Chem 276:45669–45676

Bellincampi D, Cervone F, Lionetti V (2014) Plant cell wall dynamics and wall-related susceptibility in plant-pathogen interactions. Front Plant Sci 5:228. doi:10.3389/fpls.2014.00228

Böhm H, Albert I, Fan L, NürnbergerT RA (2014) Immune receptor complexes at the plant surface. Curr Opin Plant Biol 20:47–54

Boller T, Felix G (2009) A renaissance of elicitors: perception of miceobe-associated molecular patterns and danger signals by pattern-recognition receptors. Annu Rev Plant Biol 60:379–406

Bouwmeester K, Govers F (2009) Arabidopsis L-type lectin receptor kinases: phylogeny, classification, and expression profiles. J Exp Bot 60:4383–4396

Bouwmeester K, de Sain M, Weide R, Gouget A, Klamer S, Canut H, Govers F (2011) The lectin receptor kinase LecRK-1.9 is a novel *Phytophthora* resistance component and a potential host target for a RXLR effector. PLoS Pathog 7(3):e1001327. doi:10.1371/journal.ppat.1001327

Bouwmeester K, Han M, Blanco-Portales R, Song W, Weide R, Guo LY, van der Vossen EA, Govers F (2014) The *Arabidopsis* lectin receptor kinase LecRK-1.9 enhances resistance to *Phytophthora infestans* in Solanaceous plants. Plant Biotechnol J 12:10–16

Brutus A, Sicilia F, Macone A, Cervone F, De Lorenzo G (2010) A domain swap approach reveals a role of the plant wall-associated kinase 1 (WAK1) as a receptor of oligogalacturonides. Proc Natl Acad Sci U S A 107:9452–9457

Buist G, Steen A, Kok J, Kuipers OP (2008) LysM, a widely distributed protein motif for binding to (peptido)glycans. Mol Microbiol 68:838–847

Cabrera JC, Boland A, Messiaen J, Cambier P, Cutsem PV (2008) Egg box conformation of oligogalacturonides: the time-dependent stabilization of the elicitor-active conformation increases its biological activity. Glycobiology 18:473–482

Cai R, Lewis J, Yan S, Liu H, Clarke CR, Campanile F, Almeida NF, Studholme DJ, Lindeberg M, Schneider D, Zaccardelli M, Setubal JC, Morales-Lizcano NP, Bernal A, Coaker G, Baker C, Bender CL, Leman S, Vinatzer BA (2011) The plant pathogen *Pseudomonas syringae* pv. *tomato* is genetically monomorphic and under strong selection to evade tomato immunity. PLoS Pathog 7:e1002130

Chen X, Chern M, Canlas PE, Ruan D, Jiang C, Ronald PC (2010) An ATPase promotes autophosphorylation of the pattern recognition receptor XA21 and inhibits XA21-mediated immunity. Proc Natl Acad Sci U S A 107:8029–8034

Chen C-W, Panzeri D, Yeh Y-H, Kadota Y, Huang P-Y, Tao C-N, Roux M, Chien SC, Chin T-C, Chu P-W, Zipfel C, Zimmerli L (2014a) The *Arabidopsis* malectin-like leucine-rich repeat receptor-like kinase IOS1 associates with the pattern recognition receptors FLS2 and EFR and is critical for priming of pattern-triggered immunity. Plant Cell 26:3201–3219

Chen X, Zuo S, Schwessinger B, Chern M, Canlas PE, Ruan D, Zhou X, Wang J, Daudi A, Petzold CJ (2014b) An XA21-associated kinase (OsSERK2) regulates immunity mediated by the XA21 and XA3 immune receptors. Mol Plant 7:874–892

Chinchilla D, Bauer Z, Regenass M, Boller T, Felix G (2006) The *Arabidopsis* receptor kinase FLS2 binds flg22 and determines the specificity of flagellin perception. Plant Cell 18:465–476

Chinchilla D, Boller T, Robatzek S (2007a) Flagellin signaling in plant immunity. Adv Exp Med Biol 598:358–371

Chinchilla D, Zipfel C, Robatzek S, Kemmerling B, Nürnberger T, Jones JDG, Felix G, Boller T (2007b) A flagellin-induced complex of the receptor FLS2 and BAK1 initiates plant defence. Nature 448:497–500

Clarke CR, Chinchilla D, Hind SR, Taguchi F, Miki R, Ichinose Y, Martin GB, Leman S, Felix G, Vinatzer BA (2013) Allelic variation in two distinct *Pseudomonas syringae* flagellin epitopes modulates the strength of plant immune responses but not bacterial mobility. New Phytol 200:847–860

Dardick C, Ronald P (2006) Plant and animal pathogen recognition receptors signal through non-RD kinases. PLoS Pathog 2:e2

De Lorenzo G, Brutus A, Savatin DY, Sicilia F, Cervone F (2011) Engineering plant resistance by constructing chimeric receptors that recognize damage-associated molecular patterns (DAMPs). FEBS Lett 585:1521–1526

de Torres M, Mansfield JW, Grabov N, Brown IR, Ammouneh H, Tsiamis G, Forsyth A, Robatzek S, Grant M, Boch J (2006) *Pseudomonas syringae* effector AvrPtroB suppresses basal defence in Arabidopsis. Plant J 47:368–382

Decreux A, Messiaen J (2005) Wall-associated kinase WAK1 interacts with cell wall pectins in a calcium-induced conformation. Plant Cell Physiol 46:268–278

Decreux A, Thomas A, Spies B, Brasseur R, Van Cutsem P, Messiaen J (2006) In vitro characterization of the homogalacturonan-binding domain of the wall-associated kinase WAK1 using site-directed mutagenesis. Phytochemistry 67:1068–1079

Denoux C, Galletti R, Mammarella N, Gopalan S, Werck D, De Lorenzo G, Ferrari S, Ausubel FM, Dewdney J (2008) Activation of defense response pathways by OGs and Flg22 elicitors in Arabidopsis seedlings. Mol Plant 1:423–445

Dunning FM, Sun W, Jansen KL, Helft L, Bent AF (2007) Identification and mutational analysis of *Arabidopsis* FLS2 leucine-rich repeat domain residues that contribute to flagellin perception. Plant Cell 19:3297–3317

Eckardt NA (2008) Chitin signaling in plants: insights into the perception of fungal pathogens and rhizobacterial strains. Plant Cell 20:241–243

Felix G, Duran JD, Volko S, Boller T (1999) Plants have a sensitive perception system for the most conserved domain of bacterial flagellin. Plant J 18:265–276

Fliegmann J, Mithöfer A, Wanner G, Ebel J (2004) An ancient enzyme domain hidden in the putative β-glucan elicitor of soybean may play an active part in the perception of pathogen-associated molecular patterns during broad host resistance. J Biol Chem 279:1132–1140

Fliegmann J, Montel E, Djulić A, Cottaz S, Driguez H, Ebel J (2005) Catalytic properties of the bifunctional soybean beta-glucan-binding protein, a member of family 81 glycoside hydrolases. FEBS lett 5:579(29):6647–6652

Fritz-Laylin LK, Krishnamurthy N, Tor M, Sjolander KV, Jones JDG (2005) Phylogenomic analysis of the receptor-like proteins of rice and Arabidopsis. Plant Physiol 138:511–623

Furukawa T, Inagaki H, Takai R, Hirai H, Che F-S (2014) Two distinct EF-TU epitopes induce immune responses in rice and *Arabidopsis*. Mol Plant Microbe Interact 27:113–124

Gao M, Wang X, Wang D, Xu F, Ding X, Zhang Z, Bi D, Cheng YT, Chen S, Li X, Zhang Y (2009) Regulation of cell death and innate immunity by two receptor-like kinases in *Arabidopsis*. Cell Host Microbe 23:34–44

Gimenez-Ibanez S, Hann DR, Ntoukakis V, Petutschnig E, Lipka V, Rathjen JP (2009a) AvrPtoB targets the LysM receptor kinase CERK1 to promote bacterial virulence on plants. Curr Biol 19:423–429

Gimenez-Ibanez S, Ntoukakis V, Rathjen JP (2009b) The LysM receptor kinase CERK1 mediates bacterial perception in Arabidopsis. Plant Signal Behav 4:539–541

Godiard L, Sauviac L, Torii KU, Grenon O, Mangin B, Grimsley NH, Marco Y (2003) ERECTA, an LRR receptor-like kinase protein controlling development pleiotropically affects resistance to bacterial wilt. Plant J 36:353–365

Göhre V, Spallek T, Häweker H, Mersmann S, Mentzel T, Boller T, de Torres M, Mansfield JW, Robatzek S (2008) Plant pattern-recognition receptor FLS2 is directed for degradation by the bacterial ubiquitin ligase AvrPtoB. Curr Biol 18:1824–1832
Gómez-Gómez L, Boller T (2000) FLS2: an LRR- receptor-like kinase involved in the perception of the bacterial elicitor flagellin in *Arabidopsis*. Mol Cell 5:1003–1011
Gómez-Gómez L, Boller T (2002) Flagellin perception: a paradigm for innate immunity. Trends Plant Sci 7:251–256
Gross A, Kapp D, Nielsen T, Niehaus K (2005) Endocytosis of *Xanthomonas campestris* pathovar *campestris* lipopolysaccharides in non-host plant cells of *Nicotiana tabacum*. New Phytol 165:215–226
Halter T, Imkampe J, Blaum BS, Stehle T, Kemmerling B (2014a) BIR2 affects complex formation of BAK1 with ligand binding receptors in plant defense. Plant Signal Behav 29:9, pii:e28944
Halter T, Imkampe J, Mazotta S, Wierzba M, Postel S, Bücherl C, Kiefer C, Stahl M, Chinchilla D, Wang X, Nürnberger T, Zipfel C, Clouse S, Borst JW, Boeren S, de Vries SC, Tax F, Kemmerling B (2014b) The leucine-rich repeat receptor kinase BIR2 is a negative regulator of BAK1 in plant immunity. Curr Biol 24:134–143
Hann DR, Rathjen JP (2007) Early events in the pathogenecity of *Pseudomonas syringae* on *Nicotiana benthamiana*. Plant J 49:607–618
Hann CT, Bequette CJ, Dombrowski JE, Stratmann JVV (2014) Methanol and ethanol modulate responses to danger- and microbe-associated molecular patterns. Front Plant Sci 5:550. doi:10.3389/fpls.2014.00550
Hao G, Pitino M, Ding F, Lin H, Stover E, Duan Y (2014) Induction of innate immune responses by flagellin from the intracellular bacterium from the intracellular bacterium, '*Candidatus* Liberibacter solanacearum'. BMC Plant Biol 14:14211
He DY, Yazaki Y, Nishizawa Y, Takai R, Yamada K, Sakano K, Shibuya N, Minami E (1998) Gene activation by cytoplasmic acidification in suspension-cultured rice cells in response to the potent elicitor, *N*-acetylchitoheptaose. Mol Plant Microbe Interact 11:1167–1174
He P, Shan L, Sheen J (2007) Elicitation and suppression of microbe-associated molecular patttern-triggered immunity in plant-microbe interactions. Cel Microbiol 9:1385–1396
Heese A, Hann DR, Gimenez-Ibanez S, Jones AME, He K, Li J, Schroeder JI, Peck SC, Rathjen JP (2007) The receptor-like kinase SERK3/BAK1 is a central regulator of innate immunity in plants. Proc Natl Acad Sci U S A 104:12217–12222
Huang P, Ju H, Min J-H, Zhang X, Kim SH, Yang K-Y, Kim CS (2013) Overexpression of L-type lectin-like protein kinase 1 confers pathogen resistance and regulates salinity response in *Arabidopsis thaliana*. Plant Sci 203–204:98–106
Huang PY, Yeh YH, Liu AC, Cheng CP, Zimmerli L (2014) The *Arabidopsis* LecRK-VI.2 associates with the pattern-recognition receptor FLS2 and primes *Nicotiana benthamiana* pattern-triggered immunity. Plant J 79:243–255
Hwang IS, Hwang BK (2011) The pepper mannose-binding lectin gene *CaMBL1* is required to regulate cell death and defense responses to microbial pathogens. Plant Physiol 155:447–463
Iizasa E, Mitsutomi M, Nagano Y (2010) Direct binding of a plant LysM receptor-like kinase, LysM RLK1/CERK1, to chitin in vitro. J Biol Chem 285:2996–3004
Jehle AK, Fürst U, Lipschis M, Albert M, Felix G (2013a) Perception of the novel MAMP eMax from different *Xanthomonas* species requires the *Arabidopsis* receptor-like protein ReMAX and the receptor-kinase SOBIR. Plant Signal Behav 8:e27408, PMID 24384530
Jehle AK, Lipschis M, Albert M, Fallahzadeh-Mannaghani V, Fürst U, Mueller K, Felix G (2013b) The receptor-like protein ReMAX of *Arabidopsis* detects the microbe-associated molecular pattern eMax from *Xanthomonas*. Plant Cell 25:2330–2340
Kaku H, Nishizawa Y, Ishii-Minami N, Akimoto-Tomiyama C, Dohmae N, Takio K, Minami E, Shibuya N (2006) Plant cells recognize chitin fragments for defense signaling through a plasma membrane receptor. Proc Natl Acad Sci U S A 103:11086–11091

Kanzaki H, Saitoh H, Takahashi Y, Berberich T, Ito A, Kamoun S, Terauchi R (2008) NbLRK1, a lectin-like receptor kinase protein of *Nicotiana benthamiana*, interacts with *Phytophthora infestans* INF1 elicitin and mediates INF1-induced cell death. Planta 228:977–987

Kim Y-T, Oh J, Kim K-H, Uhm J-Y, Lee B-M (2010) Isolation and characterization of *NgRLK1*, a receptor-like kinase of *Nicotiana glutinosa* that interacts with the elicitin of *Phytophthora capsici*. Mol Biol Rep 37:717–727

Kim BH, Kim SY, Nam KH (2013) Assessing the diverse functions of BAK1 and its homologs in arabidopsis, beyond BR signaling and PTI responses. Mol Cells 35:7–16

Kogel G, Beissmann B, Reisener HJ (1998) A single glycoprotein from *Puccinia graminis* f. Sp. *tritici* cell walls elicits the hypersensitive lignification response in wheat. Physiol Mol Plant Pathol 33:173–185

Kørner CJ, Klauser D, Niehl A, Dominguez-Ferreras A, Chinchilla D, Boller T, Heinlein M, Hann DR (2013) The immunity regulator *BAK1* contributes to resistance against diverse RNA viruses. Mol Plant Microbe Interact 26:1271–1280

Kouzai Y, Mochizuki S, Nakajima K, Desaki Y, Hayafune M, Miyazaki H, Yokotani N, Ozawa K, Minami E, Kaku H, Shibuya N, Nishizawa Y (2014) Targeted gene disruption of OsCERK1 reveals its indispensable role in chitin perception and involvement in the peptidoglycan response and immunity in rice. Mol Plant Microbe Interact 27:975–982

Krol E, Mentzel T, Chinchilla D, Boller T, Felix G, Kemmerling B, Postel S, Arents M, Jeworutzki E, Al-Rasheid KAS, Becker D, Hedrich R (2010) Perception of the Arabidopsis danger signal peptide 1 involves the pattern recognition receptor AtPEPR1 and its close homologue AtPEPR2. J Biol Chem 285:13471–13479

Kunze G, Zipfel C, Robatzek S, Niehaus K, Boller T, Felix G (2004) The N terminus of bacterial elongation factor Tu elicits innate immunity in *Arabidopsis* plants. Plant Cell 16:3496–3507

Lacombe S, Rougon-Cardoso A, Sherwood E, Peeters N, Dahlbeck D, van Esse HP, Smoker M, Rallapalli G, Thomma BPHJ, Stakawicz B, Jones JDG, Zipfel C (2010) Interfamily transfer of a plant pattern-recognition receptor confers broad-spectrum bacterial resistance. Nat Biotechnol 28:365–369

Le MH, Cao Y, Zhang X-C, Stacey G (2014) LIK1, a CERK1-interacting kinase, regulates plant immune responses in Arabidopsis. PLoS One 9(7):e102245. doi:10.1371/journal.pone.0102245

Lee SW, Han S, Sririyanum M, Park C, Seo Y, Ronald PC (2009) A type 1-secreted, sulfated peptide triggers XA21-mediated innate immunity. Science 326:850–853

Li XY, Lin H, Zhang W, Zou Y, Zhang J, Tang X, Zhou JM (2005) Flagellin induces innate immunity in nonhost interactions that is suppressed by *Pseudomonas syringae* effectors. Proc Natl Acad Sci U S A 102:12990–12995

Li H, Zhou SY, Zhao WS, Su SC, Peng YL (2009) A novel wall-associated receptor-like protein kinase gene, *OsWAK1*, plays important roles in rice blast disease resistance. Plant Mol Biol 69:337–346

Liebrand TWH, van den Berg GCM, Zhang Z, Smit P, Cordewener JHG, America AHP, Sklenar J, Jones AME, Tameling WIL, Robatzek S, Thomma BPHJ, Joosten MHAJ (2013) Recptor-like kinase SOBIR1/EVR interacts with receptor-like proteins in plant immunity against fungal infection. Proc Natl Acad Sci U S A 110:10010–10015

Liebrand TWH, van den Berg GCM, Joosten MHAJ (2014) Two for all: receptor-associated kinases SOBIR1 and BAK1. Trends Plant Sci 19:123–132

Lin W, Li B, Lu D, Chen S, Zhu N, He P, Shan L (2014) Tyrosine phosphorylation of protein kinase complex BAK1/BIK1 mediates *Arabidopsis* innate immunity. Proc Natl Acad Sci U S A 111:3632–3637

Liu Z, Wu Y, Yang F, Zhang Y, Chen S, Xie Q, Tian X, Zhou JM (2013) BIK1 interacts with PEPRs to mediate ethylene-induced immunity. Proc Natl Acad Sci U S A 110:6205–6210

Llorente F, Alonso-Blanco C, Sanchez-Rodriquez C, Jorda L, Molina A (2005) ERECTA receptor-like kinase and heterotrimeric G protein from Arabidopsis are required for resistance to the necrotrophic fungus *Plectosphaerella cucumerina*. Plant J 43:165–180

Lloyd SR, Schoonbeek H, Trick M, Zipfel C, Ridout CJ (2014) Methods to study PAMP-triggered immunity in *Brassica* species. Mol Plant Microbe Interact 27:286–295

Lohmann GV, Shimoda Y, Nielsen W, Jørgensen FG, Grossmann C, Sandal N, Sørensen K, Thirup S, Madsen LH, Tabata S, Sato S, Stougaard J, Radutoiu S (2010) Evolution and regulation of the *Lotus japonica* LysM receptor gene family. Mol Plant-Microbe Interact 23:510–521

Lu D, Wu S, Gao X, Zhang Y, Shan L, He P (2010) A receptor-like cytoplasmic kinase, BIK1, associates with a flagellin receptor complex to initiate plant innate immunity. Proc Natl Acad Sci U S A 107:495–501

Ma Y, Zhao Y, Walker RK, Berkowitz GA (2013) Molecular steps in the immune signaling pathway evoked by plant elicitor peptides: Ca^{2+}-dependent protein kinases, nitric oxide, and reactive oxygen species are downstream from the early Ca^{2+} signal. Plant Physiol 163:1459–1471

Macho AP, Zipfel C (2014) Plant PRRs and the activation of innate immune signalling. Mol Cell 54:263–272

Mendes BMJ, Cardoso SC, Boscariol-Camargo RL, Cruz RB, Mourao Filho FAA, Bergamin Filho A (2010) Reduction in susceptibility to *Xanthomonas axonopodis* pv. *citri* in transgenic *Citrus sinensis* expressing the rice *Xa21* gene. Plant Pathol 59:68–75

Mentlak TA, Kombrink A, Shinya T, Ryder LS, Otomo I, Saitoh H, Terauchi R, Nishizawa Y, Shibuya N, Thomma BPHJ, Talbot NJ (2012) Effector-mediated suppression of chitin-triggered immunity by *Magnaporthe oryzae* is necessary for rice blast disease. Plant Cell 24:322–335

Merkouropoulos G, Andreasson F, Hess D, Boller T, Peck SC (2008) An *Arabidopsis* protein phosphorylated in response to microbial elicitation, *AtPHOS32*, is a substrate of MAP kinases 3 and 6. J Biol Chem 283:10493–10499

Mithöfer A, Fliegmann J, Neuhaus-Url G, Schwarz H, Ebel J (2000) The hepta-β-glucoside elicitor-binding proteins from legumes represent a putative receptor family. Biol Chem 381:705–713

Miya A, Albert P, Shinya T, Desaki Y, Ichimura K, Shirasu K, Narusaka Y, Kawakami N, Kaku H, Shibuya N (2007) CERK1 a LysM receptor kinase is essential for chitin elicitor signaling in *Arabidopsis*. Proc Natl Acad Sci U S A 104:19613–19618

Monaghan J, Zipfel C (2012) Plant pattern recognition receptor complexes at the plasma membrane. Curr Opin Plant Biol 15:349–357

Montesano M, Brader G, Palva ET (2003) Pathogen-derived elicitors: searching for receptors in plants. Mol Plant Pathol 4:73–78

Nicaise V, Roux M, Zipfel C (2009) Recent advances in PAMP-triggered immunity against bacteria: Pattern recognition receptors watch over and raise the alarm. Plant Physiol 150:1638–1647

Park C-J, Peng Y, Chen X, Dardick C, Ruan D, Bart R, Canlas PE, Ronald PC (2008) Rice XB15 a protein phosphatase 2C negatively regulates cell death and XA21-mediated innate immunity. PLoS Biol 6:e231

Park C-J, Bart R, Chern M, Canlas PE, Bai W, Ronald PC (2010a) Overexpression of the endoplasmic reticulum chaperone BiP3 regulates XA21-mediated innate immunity in rice. PLoS One 5(2):e9262

Park C-J, Han S-W, Chen X, Ronald PC (2010b) Elucidation of XA21-mediated innate immunity. Cell Microbiol 12:1017–1025

Peck SC, Nühse TS, Hess D, Iglesias A, Meins F, Boller T (2001) Directed proteomics identifies a plant-specific protein rapidly phosphorylated in response to bacterial and fungal elicitors. Plant Cell 13:1467–1475

Petutschnig EK, Jones AME, Serazetdinova L, Lipka U, Lipka V (2010) The Lysin motif receptor-like kinase (LysM-RLK) is a major chitin binding protein in *Arabidopsis thaliana* and subject to chitin-induced phosphorylation. J Biol Chem 285:28902–28911

Postel S, Kufner I, Beueter C, Mazzotta S, Schwedt A, Borlotti A, Halter T, Kemmerling B, Nürnberger T (2010) The multifunctional leucine-rich repeat receptor kinase BAK1 is implicated in *Arabidopsis* development and immunity. Eur J Cell Biol 89:169–174

Qi Z, Verma R, Gehring C, Yamaguchi Y, Zhao Y, Ryan CA, Berkowitz GA (2010) Ca^{2+} signaling by plant *Arabidopsis thaliana* Pep peptides depends on AtPepR1, a receptor with guanyl cyclase activity, and CGMP-activated Ca^{2+} channels. Proc Natl Acad Sci U S A 107:21193–21198

Robatzek S, Chinchilla D, Boller T (2006) Ligand-induced endocytosis of the pattern recognition receptor FLS2 in *Arabidopsis*. Genes Dev 20:537–542

Robatzek S, Bittel P, Chinchilla D, Köchner P, Felix G, Shiu SH, Boller T (2007) Molecular identification and characterization of the tomato flagellin receptor LeFLS2, an orthologue of Arabidopsis FLS2 exhibiting characteristically different perception specificities. Plant Mol Biol 64:539–547

Ron M, Avni A (2004) The receptor for the fungal elicitor ethylene-inducing xylanase is a member of a resistance-like gene family in tomato. Plant Cell 16:1604–1615

Rosli HG, Zheng Y, Pombo MA, Zhong S, Bombarely A, Fei Z, Collmer A, Martin GB (2013) Transcriptomics-based screen for genes induced by flagellin and repressed by pathogen effectors identifies a cell wall-associated kinase involved in plant immunity. Genome Biol 14:R139

Ross A, Yamada K, Hiruma K, Yamashita-Yamada M, Lu X, Takano Y, Tsuda K, Saito Y (2014) The Arabidopsis PEPR pathway couples local and systemic plant immunity. EMBO J 33:62–75

Ryan CA, Huffaker A, Yamaguchi Y (2007) New insights into innate immunity in Arabidopsis. Cell Microbiol 9:1902–1908

Saijo Y (2010) ER quality control of immune receptors and regulators in plants. Cell Microbiol 12:716–724

Sanabria NM, Dubery IA (2006) Differential display profiling of the *Nicotiana* response to LPS reveals elements of plant basal resistance. Biochem Biophys Res Commun 344:1001–1007

Schulze B, Mentzel T, Jehle A, Mueller K, Beeler S, Boller T, Felix G, Chinchilla D (2010) Rapid heteromerization and phosphorylation of ligand-activated plant transmembrane receptors and their associated kinase BAK1. J Biol Chem 285:9444–9451

Segonzac C, Zipfel C (2011) Activation of plant pattern-recognition receptors by bacteria. Curr Opin Microbiol 14:54–61

Seifert GJ, Blaukopf C (2010) Irritable walls: the plant extracellular matrix and signaling. Plant Physiol 153:467–478

Serrano M, Kanehara K, Torres M, Yamada K, Tintor N, Kombrink E, Schulze-Lefert P, Saijo Y (2012) Repression of sucrose/ultraviolet B light-induced flavonoid accumulation in microbe-associated molecular pattern-triggered immunity in Arabidopsis. Plant Physiol 158:408–422

Shan L, He P, Li J, Heese A, Peck SC, Nümberger T, Martin GB, Sheen J (2008) Bacterial effectors target the common signaling partner BAK1 to disrupt multiple MAMP receptor-signaling complexes and impede plant immunity. Cell Host Microbe 4:17–27

Shen QH, Saijo Y, Mauch S, Biskup C, Bieri S, Keller B, Seki H, Ulker B, Somssich IE, Schulze-Lefert P (2007) Nuclear activity of MLA immune receptors links isolate-specific and basal disease resistance responses. Science 315:1098–1103

Shibuya N, Minami E (2001) Oligosaccharide signalling for defence responses in plant. Physiol Mol Plant Pathol 59:223–233

Shimizu T, Nakano T, Takamizawa D, Desaki Y, Ishi-Minami N, Nishizawa Y, Minami E, Okada K, Yamane H, Kaku H, Shibuya N (2010) Two LysM receptor molecules, CEBiP and OsCERK1, cooperatively regulate chitin elicitor signaling in rice. Plant J 64:204–214

Shinya T, Osada T, Desaki Y, Hatamoto M, Yamanaka Y, Hirano H, Takai R, Che F-S, Kaku H, Shibuya N (2010) Characterization of receptor proteins using affinity cross-linking with biotinylated ligands. Plant Cell Physiol 51:262–270

Shiu S-H, Bleecker AB (2001) Plant receptor-like kinase gene family: diversity, function, and signaling. Sci STKE 113:RE22

Shiu SH, Karlowski WM, Pan R, Tzeng YH, Mayer KE, Li WH (2004) Comparative analysis of the receptor-like kinase family in Arabidopsis and rice. Plant Cell 16:1220–1234

Silipo A, Erbs G, Shinya T, Dow JM, Parrilli MT, Lanzetta Shibuya N, Newman M-A, Molinaro A (2010) Glyco-conjugates as elicitors or suppressors of plant innate immunity. Glycobiology 20:406–419

Singh P, Zimmerli L (2013) Lectin receptor kinases in plant innate immunity. Front Plant Sci 4:124

Song W-Y, Wang G-L, Chen L-L, Kim H-S, Pi L-Y, Holsten T, Gardner J, Wang B, Zhai W-X, Zhu L-H, Fauquet C, Ronald P (1995) A receptor kinase-like protein encoded by the rice disease resistance gene *Xa21*. Science 270:1804–1806

Sun W, Cao Y, Labby KJ, Bittel P, Boller T, Bent AF (2012) Probing the *Arabidopsis* flagellin receptor: FLS2-FLS2 association and contributions of specific domains to signaling function. Plant Cell 24:1096–1113

Sun Y, Li L, Macho AP, Han Z, Hu Z, Zipfel C, Zhou JM, Chai J (2013) Structural basis for flg22-induced activation of the Arabidopsis FLS2-BAK1 immune complex. Science 342:624–628

Takai R, Isogai A, Takayama S, Che F-S (2008) Analysis of flagellin perception mediated by flg22 receptor OsFLS2 in rice. Mol Plant Microbe Interact 21:1635–1642

Tian F, Yu C, Li HY, Wu XL, Li B, Chen HM, Wu M, He C (2015) Alternative sigma factor RpoN2 is required for flagellar motility and full virulence of *Xanthomonas oryzae* pv. *oryzae*. Microbiol Res 170:177–183

Tintor N, Ross A, Kanehara K, Yamada K, Fan L, Kemmerling B, Nürnberger T, Tsuda K, Saijo Y (2013) Layered pattern receptor signaling via ethylene and endogenous elicitor peptides during Arabidopsis immunity to bacterial infection. Proc Natl Acad Sci U S A 110:6211–6216

Tripathi JN, Lorenzen J, Bahar O, Ronald P, Tripathi L (2014) Transgenic expression of the rice *Xa21* pattern-recognition receptor in banana (*Musa* sp.) confers resistance to *Xanthomonas campestris* pv. *musacearum*. Plant Biotech J 12:663–673

Vaid N, Pandey PK, Tuteja N (2012) Genome-wide analysis of lectin receptor-like kinase family from Arabidopsis and rice. Plant Mol Biol 80:365–388

Vallarino JG, Osorio S (2012) Signaling role of oligogalacturonides derived during cell wall degradation. Plant Signal Behav 7:1447–1449

Veluchamy S, Hind SR, Dunham DM, Martin GB, Panthee DR (2014) Natural variation for responsiveness to flg22, flgII-28, and CSP22 and *Pseudomonas syringae* pv. *tomato* in heirloom tomatoes. PLoS One 9(9):e105119

Vetter MM, Kronholm I, He F, Häweker H, Reymond M, Bergelson J, Robatzek S, de Meaux J (2012) Flagellin perception varies quantitatively in *Arabidopsis thaliana* and its relatives. Mol Biol Evol 29:1655–1667

Vidhyasekaran P (2014) PAMP signals in plant innate immunity: signal perception and transduction. Springer, Dordrecht, p 442

Wagner TA, Kohorn BD (2001) Wall-associated kinases are expressed throughout plant development and are required for cell expansion. Plant Cell 13:303–318

Wan J, Zhang X-C, Stacey G (2008) Chitin signaling and plant disease resistance. Plant Signal Behav 3:831–833

Wang X, Kota U, He K, Blackburn K, Li J, Goshe MB, Clouse SD (2008a) Sequential transphorylation of the BRI1/BAK1 receptor kinase complex impacts early events in brassinosteroid signaling. Dev Cell 15:220–235

Wang Y, Gao M, Li Q, Wang L, Jeon J-S, Qu N, Zhang Y, He Z (2008b) *OsRAR1* and *OsSGT1* physically interact and function in rice basal disease resistance. Mol Plant-Microbe Interact 21:294–303

Wang G, Fiers M, Ellendorff U, Wang Z, deWit PJGM, Angenent GC, Thomma BPHJ (2010) The diverse roles of extracellular leucine-rich repeat-containing receptor-like proteins in plants. Crit Rev Plant Sci 29:285–299

Willmann R, Lajunen HM, Erbs G, Newman MA, Kolb D, Tsuda K, Katagiri F, Fliegmann J, Bono JJ, Cullimore JV, Jehle AK, Göltz F, Kulik A, Molinaro A, Lipka V, Gust AA, Nürnberger T (2011) Arabidopsis lysin-motif proteins LYM1 LYM3 CERK1 mediate bacterial peptidoglycan sensing and immunity to bacterial infection. Proc Natl Acad Sci U S A 108:19824–19829

Yamaguchi Y, Huffaker A, Bryan AC, Tax FE, Ryan CA (2010) PEPR2 is a second receptor for the Pep1 and Pep2 peptides and contributes to defense responses in *Arabidopsis*. Plant Cell 22:508–522

Yang X, Deng F, Ramonell KM (2012) Receptor-like kinases and receptor –like proteins; keys to pathogen recognition and defense signaling in plant innate immunity. Front Biol 7(2):155–166. doi:10.1007/s11515-011-1185-8

Zhang J, Zhou J-M (2010) Plant immunity triggered by microbial molecular signatures. Mol Plant 3:783–793

Zhang X-C, Wu X, Findley S, Wan J, Libault M, Nguyen HT, Cannon SB, Stacey G (2007) Molecular evolution of LysM type receptor-like kinases in plants. Plant Physiol 144:623–636

Zhang J, Li W, Xiang T, Liu Z, Laluk K, Ding X, Zou Y, Gao M, Zhang X, Chen S, Mengiste T, Zhang Y, Zhou JM (2010) Receptor-like cytoplasmic kinases integrate signaling from multiple plant immune receptors and are targeted by a *Pseudomonas syringae* effector. Cell Host Microbe 7:290–301

Zhang W, Fraiture M, Kolb D, Löffelhardt B, Desaki Y, Boutrot FFG, Tör M, Zipfel C, Gust AA, Brunner F (2013) Arabidopsis RECEPTOR-LIKE PROTEIN30 and receptor-like kinase SUPPRESSOR OF BIR1-1/EVERSHED mediate innate immunity to necrotrophic fungi. Plant Cell 25:4227–4241

Zhang L, Kars I, Essenstam B, Liebrand TW, Wagemakers L, Elberse J, Tagkalaki P, Tjoitang D, van den Ackerveken G, van Kan JA (2014) Fungal endopolygalacturonases are recognized as microbe-associated molecular patterns by the arabidopsis receptor-like protein RESPONSIVENESS TO BOTRYTIS POLYGALACTURONASES1. Plant Physiol 164:352–364

Zipfel C (2008) Pattern-recognition receptors in plant innate immunity. Curr Opin Immunol 20:10–16

Zipfel C (2009) Early molecular events in PAMP-triggered immunity. Curr Opin Plant Biol 12:414–420

Zipfel C (2014) Plant pattern-recognition receptors. Trends Immunol 35:345–351

Zipfel C, Robatzek S, Navarro L, Oakeley EJ, Jones JDG, Felix G, Boller T (2004) Bacterial disease resistance in *Arabidopsis* through flagellin perception. Nature 428:764–767

Zipfel C, Kunze G, Chinchilla D, Caniard A, Jones JD, Boller T, Felix G (2006) Perception of the bacterial PAMP EF-Tu by the receptor EFR restricts *Agrobacterium*-mediated transformation. Cell 125:749–760

Yamaguchi Y, Huffaker A, Bryan AC, Tax FE, Ryan CA (2010) PEPR2 is a second receptor for the Pep1 and Pep2 peptides and contributes to defense responses in *Arabidopsis*. Plant Cell 22:508–522

[illegible]

Chapter 6
Molecular Manipulation of Transcription Factors, the Master Regulators of PAMP-Triggered Signaling Systems

Abstract Transcription factors (TFs) are the 'master switches', which switch on the gene transcription processes. They play important role in initiating transcription of defense genes regulating expression of defense genes in plant innate immune systems. Many TFs have been shown to be involved in the defense signaling system and they confer resistance against several bacterial, oomycete, fungal, and viral pathogens. Induction of expression of genes encoding TFs appears to be critical in inducing defense-related genes and conferring disease resistance. Pathogen-associated molecular patterns (PAMPs) switch on the expression of various transcription factor genes. Messages generated by PAMP-PRR signaling complex are sensed by calcium ion, which acts as a signal carrier. Calcium signatures are recognized by different calcium sensors to transduce calcium-mediated signals into downstream events. Calmodulins (CaMs) are the important Ca^{2+} sensors identified in plants. Several transcription factors are involved in activation of calmodulin and they bind with calmodulins and decode the message from calmodulin proteins to activate defense responses in plants. PAMP triggers rapid and transient production of H_2O_2, which has been shown to regulate expression of several transcription factors in plants. Activation of these transcription factors may result in activation of transcription of several defense genes. Several mitogen-activated protein kinases (MAPKs) modulate phosphorylation of transcription factors to trigger transcription of defense genes. The MAPKs transduce extracellular stimuli into intracellular transcription factors, which enhance expression of defense-related genes. Some transcription factors trigger biosynthesis of salicylic acid. SA induces enhanced expression of several WRKY, ERF, and TGA transcription factors to activate transcription of defense genes. Transcription factors play an important role in triggering SA-dependent systemic acquired resistance (SAR). Some transcription factors are involved in activation of JA biosynthesis pathway. JA may trigger expression of several transcription factors. JA may also downregulate expression of some transcription factors. Transcription factors also regulate ethylene signaling system in plant innate immunity. Technologies have been developed to bioengineer the transcription factor genes (*OsWRKY13, OsWRKY22, OsWRKY30, OsWRKY31, OsWRKY42, OsWRKY45, OsWRKY47, OsWRKY53, OsWRKY71, OsWRKY89, TaWRKY45, VvWRKY1, VvWRKY2, VpWRKY3, MdWRKY1, GhWRKY15, AtMYB44, RF2a and RF2b, CabZIP2, Pti5, Pti4, GbERF2, NtERF5, Tsi1, OsBIERF3, CaPF1, OPBP1, HvRAF, ERF1, OsEREBP1*, OsNAC111, OsNAC6, *TaNAC1* and *CaATL1*) for management of viral, bacterial, fungal, and oomycete

P. Vidhyasekaran, *Switching on Plant Innate Immunity Signaling Systems*,
Signaling and Communication in Plants, DOI 10.1007/978-3-319-26118-8_6

diseases in rice, wheat, barley, tomato, grapevine, pepper, amd tobacco. Several transcription factors trigger "priming" of defense responses. The priming results in a faster and stronger induction of defense mechanisms after pathogen attack. Some plant defense activators, including benzothiadiazole (BTH) and β-aminobutyric acid (BABA), and some rhizobacteria such as *Pseudomonas fluorescens* WCS417r trigger priming of the transcription factors and defense responses. The primed genes may be poised for enhanced activation of gene expression by histone modification. The histone modifications might create a memory of the primary infection that is associated with an amplified reaction to a second stress conditions. The priming can be inherited epigenetically from disease-exposed plants and descendants of primed plants exhibit next-generation systemic acquired resistance. DNA methylation may also play important role in transgenerational SAR. These plant defense activators and rhizobacteria, which modulate the priming of transcription factors, are potential tools for management of crop diseases.

6.1 Transcription Factors as 'Master Switches' Regulating Expression of Defense Genes in Plant Immune Signaling Systems

Transcription factors (TFs) are the master regulators of expression of genes involved in plant innate immune systems. Transcription factors regulate the gene transcription processes by modulating the rate of transcription initiation of target genes (Du et al. 2009). The TFs are composed of four discrete domains, DNA binding domain, nuclear localization signal (NLS), transcription activation domain, and oligomerization site, which operate together to regulate the gene transcription processes by modulating the rate of transcription initiation of target genes (Du et al. 2009). TFs play important role in regulating defense gene expression and plant defense responses either positively (Liu et al. 2007a; Marchive et al. 2007; Qiu et al. 2007; Century et al. 2008; Moreau et al. 2012) or negatively (Journot-Catalino et al. 2006; Li et al. 2006; Nurmberg et al. 2007; Zheng et al. 2006, 2007; Kim et al. 2008; Oh et al. 2008; Qiu et al. 2008a; Peng et al. 2010; Sun et al. 2010; Moreau et al. 2012). Some of the TFs may positively regulate a particular signaling pathway while negatively regulating another signaling pathway (Nurmberg et al. 2007). The TFs are classified into different families or super-families based on the type of their DNA binding domain. The important families of transcription factors which participate in the regulation of plant defense responses include WRKY family that contains the amino-acid sequence WRKYGQK (Dong et al. 2003; Knoth et al. 2007; Ross et al. 2007; Higashi et al. 2008; Lai et al. 2008; Peng et al. 2010; Ren et al. 2010; Hwang et al. 2011; Alves et al. 2014; Takatsuji 2014; Tsuda and Somssich 2015), basic leucine zipper domain (bZIP) family (Després et al. 2000; Zhang et al. 2003; Ndamukong et al. 2007; Lim et al. 2015), APETALA2/ETHYLENE-RESPONSIVE ELEMENT BINDING FACTORS family (AP2/ERF) (Zhang et al. 2003; McGrath et al. 2005; Oñate-Sánchez et al. 2007; Grennan 2008; Tsuda and somssich 2015), myeloblastosis related proteins (MYB) (Deluc et al. 2006; Raffaele et al. 2008;

Van der Ent et al. 2008), myelocytomatosis related family (MYC) (Lorenzo et al. 2004; Dombrecht et al. 2007), NAM-ATAF-CUC (NAC) transcription factors (for No apical meristem [NAM], Arabidopsis transcription activation factor [ATAF], and Cup-shaped cotyledon [CUC] transcription factors) (Yoshii et al. 2010; Chen et al. 2013; Nuruzzaman et al. 2013; Su et al. 2013; Donze et al. 2014; Alves et al. 2014; Sun et al. 2015; Wang et al. 2015), and homeodomain transcription factors (Coego et al. 2005; Luo et al. 2005a, b; Park et al. 2007).

Specific TFs have been shown to take part in various defense signaling events in plants. Some TFs were shown to bind calmodulin (Onions et al. 2000; Bouché et al. 2002; Yang and Poovaiah, 2000, 2002; Choi et al. 2005; Park et al. 2005) and activate certain Ca^{2+} transduction pathways in plants (Park et al. 2005; Yoo et al. 2005). Some TFs such as WRKY25 and WRKY33 function as downstream components of the MAP kinase-mediated signaling pathway (Andreasson et al. 2005). The tobacco transcription factor WRKY1 has been shown to act downstream of the MAPK signaling system (Menke et al. 2005). MAPKs phosphorylate WRKY proteins, which enhance their DNA-binding activities (Yang et al. 1999; Menke et al. 2005).

Several signaling pathways have been shown to mediate expression of genes encoding TFs. MAPK cascade has been shown to activate TFs involved in defense signaling system (Asai et al. 2002). MAPKs appear to function upstream of NtWRKY1 and NtWRKY3 in a defense-signaling cascade in *Nicotiana* (Kim and Zhang 2004; Liu et al. 2004). Some TFs such as WRKY70 may function downstream of ROS-signaling system (Knoth et al. 2007).

Several TFs factors are known to take part in the regulation of SA (Kim et al. 2007; Xing et al. 2008), JA (Lee et al. 2001; Zheng et al. 2006; Mao et al. 2007; van Verk et al. 2008; Yoshii et al. 2010; Murmu et al. 2013), ET (Berrocal-Lobo and Molina 2004; Zhang et al. 2004; Cao et al. 2006; Zhou et al. 2008; Liu et al. 2011) and ABA (Abe et al. 2003; Finkler et al. 2007) signaling systems. SA induces expression of several transcription factors (Zhang et al. 2007; Grennan 2008). Some of these TFs may act both upstream and downstream of SA signaling system. For example, a rice WRKY gene, OsWRKY13, activates both SA synthesis-related genes and SA-responsive genes (Qiu et al. 2007). A transcription factor in *Arabidopsis*, AtMYC2, plays a predominant role in JA pathway (Boter et al. 2004) and it functions as a downstream factor of the MAPK cascade in JA signaling (Takahashi et al. 2007). JA may also act upstream of some TFs (Zheng et al. 2006; Wasternack 2007; Kazan and Manners 2008). Ethylene response factors (ERF) belonging to the EREBP transcription factor family function downstream in ethylene signaling system (Grennan 2008). Some TFs have been shown to regulate cross-talk between SA- and JA-regulated disease response pathways (Zheng et al. 2006).

Many TFs have been shown to be involved in the defense signaling system and they confer resistance against several bacterial, oomycete, fungal, and viral pathogens (Yang et al. 1999; Dellagi et al. 2000; Beyer et al. 2001; Asai et al. 2002; Berrocal-Lobo et al. 2002; Chen et al. 2002; Deslandes et al. 2002; Singh et al. 2002; Yoda et al. 2002; Dong et al. 2003; Kalde et al. 2003; Berrocal-Lobo and Molina, 2004; Zheng et al. 2006; Kim et al. 2007; Knoth et al. 2007; Mzid et al. 2007; Mabrouk et al. 2008; Bhattarai et al. 2010; Li et al. 2010; Jingyuan et al. 2011; Murmu et al. 2013; Alves et al. 2014; Li et al. 2014; Lim et al. 2015).

Induction of expression of genes encoding TFs appears to be critical in inducing disease resistance genes and conferring disease resistance (Cao et al. 2006; Waller et al. 2006; Zhang et al. 2007; van Verk et al. 2008).

6.2 PAMPs and PIMPs/HAMPs Trigger Expression of Transcription Factors

PAMPs and PIMPs/HAMPs switch on the expression of various transcription factor genes (Laquitaine et al. 2006; Chujo et al. 2007; Denoux et al. 2008; Higashi et al. 2008; Chujo et al. 2013; McLellan et al. 2013). Mixture of *Phytophthora* PAMPs triggers the accumulation of transcripts of two NAC transcription factors in *Nicotiana benthamiana* (McLellan et al. 2013). The fungal PAMP ergosterol induced WRKY gene expression in grapevine plantlets (Laquitaine et al. 2006). The bactereial PAMP flg22 treatment consistently induced the expression of WRKY70 around 20-fold in *Arabidopsis thaliana.* WRKY70 has been shown to function downstream of ROS and SA signaling (Knoth et al. 2007). It induces expression of SA-responsive PR genes (Li et al. 2004). The WRKY70 transcription factor acts as a node of convergence for jasmonate-mediated and salicylate-mediated signals in plant defense (Li et al. 2004). It modulates the selection of signaling pathways in plant defense (Li et al. 2006).

Flg22 induces activation of several other transcription factors in *Arabidopsis* (Table 6.1; Denoux et al. 2008). It activates WRKY33 and WRKY40 transcription factors (Denoux et al. 2008). These transcription factors function as activators of JA signaling system and repressors of SA signaling system (Zheng et al. 2006). WRKY33 positively regulates JA-mediated responses, while it retards the SA-mediated PR genes expression (Zheng et al. 2006). Flg22 induced the transcription factor gene *WRKY40* in *Arabidopsis* and the induction of gene expression increased at increasing concentrations of the PAMP (Denoux et al. 2008). Flg22 triggered the expression of several other WRKY transcription factors, including WRKY6, WRKY7, WRKY18, WRKY22, WRKY29, WRKY53, and WRKY60 in *Arabidopsis* (Denoux et al. 2008). Overexpression of the AtWRKY29 constitutively activates the plant defense response against bacterial invasion (Asai et al. 2002). Flg22 also triggered the expression of MYB51, ERF1, ERF4 and TGA1, and TGA4 transcription factors in Arabidopsis (Table 6.1)

Flg22 treatment up-regulated the expression of the ET-responsive transcription factor ERF1 in *Arabidopsis* (Clay et al. 2009). ERF1 is a downstream component of ethylene signaling system (Berrocal-Lobo and Molina 2004). ET signaling is required for the full induction of ERF1 in response to flg22 (Clay et al. 2009).

The fungal PAMP chitin induced the expression of ERF5, an ethylene-responsive element-binding factor. It induces defense against *Pseudomonas syringae* pv. *tomato* in *Arabidopsis* (Son et al. 2012). *WRKY22*, *WRKY29*, *WRKY33*, and *WRKY53*, which encode four WRKY transcription factors that recognize TTGAC(C/T) W-box elements in promoters of several defense-related genes, were

Table 6.1 The PAMP flg22 - triggered expression of transcription factors in *Arabidopsis*

Transcription factors	Fold increase in expression of transcription factor
MYB51	47.1
TGA3	2.3
TGA4	2.2
WRKY6	28.1
WRKY7	3.1
WRKY18	4.7
WRKY22	19.3
WRKY33	23.5
WRKY40	34.1
WRKY53	23.5
WRKY60	5.0
ERF1	5.0
ERF4	3.2

Adapted from Denoux et al. (2008)

Table 6.2 Chitin-induced enhanced expression of transcription factor genes in *Arabidopsis*

Transcription factors	Fold increase in expression of transcription factor genes induced by chitin
AP2/ERE transcription factor	39.2
Basic Helix-Loop-Helix (bHLH) transcription factor	2.6
bZIP transcription factor	7.9
C2H2 zinc finger transcription factor	17.2
Homeobox transcription factor	3.7
MYB domain transcription factor	5.6
NAC domain transcription factor	11.1
WRKY transcription factor	21.9
GRAS transcription factor	4.5

Adapted from Libault et al. 2007

up-regulated by chitin treatment (Wan et al. 2004). Chitin (chitooctaose) treatment induced 118 TF genes representing 29 TF families in Arabidopsis. Among the chitooctaose-responsive TF genes, AP2/ERE, C2H2 zinc finger, MYB, and WRKY gene families were overrepresented (Libault et al. 2007). Chitin also triggered expression of genes encoding GRAS, bHLH, bZIP, Homeobox transcription factor, and NAC domain transcription factors by several-fold (Libault et al. 2007; Table 2.2). Nine WRKY genes were induced by the PAMP Nep1 treatment, while WRKY65 was repressed in *A. thaliana* (Bae et al. 2006). Keates et al. (2003) detected the induction of WRKY18 within 15 min after treatment with the PAMP Nep1.

Table 6.3 OGs-induced enhanced expression of transcription factors in *Arabidopsis*

Transcription factors	Fold increase in expression of transcription factor
MYB51	5.3
TGA3	1.5
TGA4	1.3
WRKY6	1.3
ERF1	1.1
WRKY18	2.0
WRKY33	1.8
WRKY40	1.2
WRKY53	2.1
WRKY60	1.3
ERF11	1.4
ERF4	1.3

Adapted from Denoux et al. (2008)

The PIMP/HAMPs also have been shown to trigger the expression of several transcription factors involved in plant defense responses. The HAMP oligogalacturonides triggered the expression of WRKY6, WRKY7, WRKY 18, WRKY22, WRKY33, WRKY53, WRKY60, WRKY70, ERF1, ERF4, TGA3, TGA4, and MYB51 transcription factors (Table 6.3; Denoux et al. 2008). WRKY29, WRKY33, and WRKY53 are induced by the HAMP Pep1 and they are positive regulators of defense responses (Asai et al. 2002; Zheng et al. 2006; Murray et al. 2007). Pep1 induced transcriptional induction of *WRKY29*, *WRKY33*, and *WRKY53* in *Arabidopsis* (Yamaguchi et al. 2010).

6.3 Role of Transcription Factors in Regulation of Ca^{2+} Signaling System

Messages generated by PAMP-PRR signaling complex trigger Ca^{2+} signaling system (Blume et al. 2000; Kwaaitaal et al. 2011; Ranf et al. 2011). Calcium ion acts as a signal carrier (Allen et al. 2000). Calcium signaling is modulated by specific calcium signatures (Luan et al. 2002; Lecourieux et al. 2006). Specific calcium signatures are recognized by different calcium sensors to transduce calcium-mediated signals into downstream events (Harmon et al. 2000; Rudd and Franklin-Tong 2001; Sanders et al. 2002; Reddy and Reddy 2004; Reddy et al. 2003, 2011a, b; Wang et al. 2012; Hashimoto et al. 2012). Calmodulins (CaMs) are the important Ca^{2+} sensors identified in plants (Snedden and Fromm 2001; Reddy and Reddy 2004; Kang et al. 2006; Takabatake et al. 2007). CaM has no catalytic activity of its own, but upon binding Ca^{2+}, it activates numerous target proteins involved in a variety of cellular processes (Snedden and Fromm 2001; Reddy et al. 2011a). CaMs contain an autoinhibitory domain that occludes the active site in the resting state.

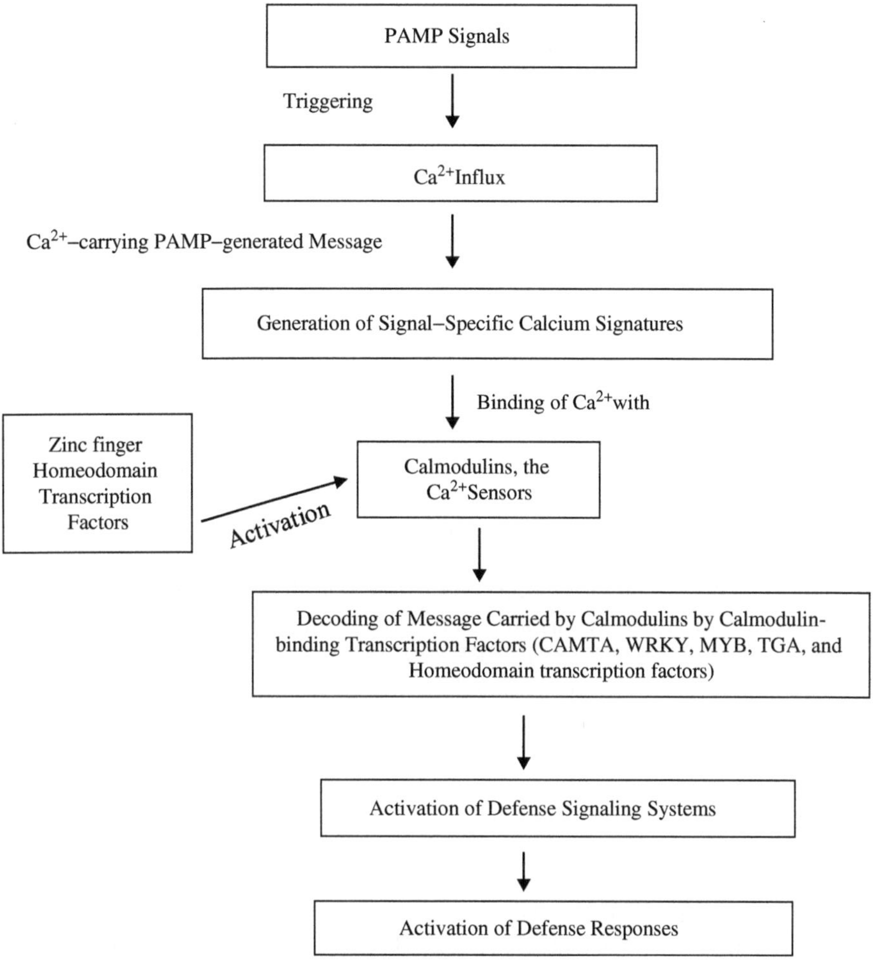

Fig. 6.1 Transcription factors activating calmodulin genes and downstream defense responses as calmodulin-binding proteins (Adapted from Bouché et al. 2002; Jakoby et al. 2002; Choi et al. 2005; Park et al. 2005, 2007; Popescu et al. 2007)

Ca^{2+} binds to a site near or overlapping the autoinhibitory domain, thereby releasing it from the active site and activating the protein (Luan et al. 2002).

Calmodulin genes are activated during pathogenesis and transcription of these genes occurs within a few minutes of pathogen invasion. Transcription of calmodulin isoform 4 (*GmCaM4*) is rapidly induced within 30 min after pathogen (*Pseudomonas syringae* pv. *glycinea*) stimulation in soybean (Park et al. 2007). Two zinc finger homeodomain transcription factors, GmZF-HD1 and GmZF-HD2 proteins have been shown to activate the *GmCaM4* gene expression in response to the bacterial pathogen (Fig. 6.1; Park et al. 2007). The pathogen induced binding of GmZF-HD1 and GmZF-HD2 to repeats of ATTA homeodomain binding site in the *GMCaM4* promoter (Park et al. 2007).

Calmodulin binds with several transcription factors to transmit the message generated by the PAMP signals and carried by the Ca^{2+} signal carrier for activation of downstream host defense responses (Zhang and Yuan 1998). CaM is an acidic EF-hand protein and is composed of 148 amino acids arranged in two globular domains connected with a long flexible helix. Each globular domain contains a pair of intimately linked EF hand (Snedden and Fromm 2001; Rainaldi et al. 2007). Ca^{2+}-free CaM exhibits a flat, hydrophilic molecular surface, while the Ca^{2+}-saturated form of the protein contains a Met-rich cavity containing hydrophobic surface in each domain. These hydrophobic surfaces are largely responsible for the binding of CaM to its target proteins such as transcription factors. The unique flexibility and high polarizability of the Met residues located at the entrance of each hydrophobic pocket together with other hydrophobic amino acid residues create adjustable, sticky interaction surface areas that can accommodate CaM-binding proteins (Zhang and Yuan 1998). The binding of calcium to calmodulin induces a conformational change that exposes hydrophobic binding sites that interact with CAM binding proteins (Harmon 2003).

CaM binds to the Ca^{2+}- dependent CaM-binding domain of several WRKY transcription factors. CaM binds specifically to the Ca^{2+}-dependent CaM-binding domain of AtWRKY7 transcription factor (Park et al. 2005). WRKY7 is a member of the WRKYIId subfamily, and all members of this subfamily including WRKY11, WRKY15, WRKY17, WRKY21, WRKY39, and WRKY74 were found to interact with Ca^{2+}/CaM (Park et al. 2005). WRKY43, WRKY45, WRKY50, and WRKY 53 have been shown to interact with different isoforms of CaM in a Ca^{2+}-dependent manner (Popescu et al. 2007). WRKY transcription factors have been shown to positively or negatively regulate the plant's defense responses (Journot-Catalino et al. 2006; Kim et al. 2006; Shimono et al. 2007; Hu et al. 2012; Matsushita et al. 2012; Shim et al. 2013).

One family of CaM-binding proteins, designated as the calmodulin-binding transcription activator (CAMTA) family resembles a group of putative transcription activators identified in the human genome (Bouché et al. 2002; Galon et al. 2010). This family of proteins contains a transcription activation domain and two types of DNA-binding domains designated the CG1 domain and the transcription factor immunoglobulin domain, ankyrin repeats, and a varying number of IQ CaM-binding motifs (Bouché et al. 2002). *Arabidopsis thaliana* contains six *CAMTA* genes (*AtCAMTA1* – *AtCAMTA6*). CAMTAs comprise a conserved family of transcription factors (Bouché et al. 2002). The C-terminal CaM binding domain of CAMTAs mediates interactions with calmodulin (Kudla et al. 2010). CAMTA3 directly interacts with the promoter of the *EDS1* gene, a regulator of salicylic acid levels, and represses its expression (Du et al. 2009).

A CaM-binding transcription factor, OsCBT (*Oryza sativa* CaM-binding transcription factor), has been isolated from rice (Choi et al. 2005). It contains a CG-1 homology DNA binding domain, three ankyrin repeats, a putative transcriptional activation domain, and five putative CaM-binding motifs (Choi et al. 2005).

OsCBT has two different types of functional CaM-binding domains, an IQ motif, and a Ca^{2+}-dependent motif (Choi et al. 2005). Some ethylene-responsive genes-encoded proteins in tobacco, tomato, rapeseed, *Arabidopsis* have structural similarity to the rice CaM-binding transcription factor, OsCBT (Choi et al. 2005). The tobacco early ethylene-responsive gene *NtER1* encodes a CaM-binding protein (Yang and Poovaiah 2000). One *NtER1* homolog (*AtSR1*) and five related genes (*AtSR2-6*) have been identified in *Arabidopsis* and they are designated as *AtSR* (for *Arabidopsis thaliana Signal Responsive*) genes (Yang and Poovaiah 2002). Ca^{2+}/CaM binds to a 23-mer peptide in all AtSRs that corresponds to the CaM-binding region of NtER1. Each AtSR has a conserved structural feature with a DNA-binding domain in the N terminus and a CaM-binding domain in the C terminus. AtSR1 targets the nucleus and specifically recognizes a novel 6-bp CGCG box (A/C/G) CGCG (G/T/C). The multiple CGCG *cis*-elements are found in promoters of various genes involved in ET- and ABA signaling systems (Yang and Poovaiah 2002).

CaM binds to several other transcription factors (Fig. 6.1) involved in plant immune systems. Some transcription factors of the basic helix-loop-helix family were shown to bind calmodulin (Onions et al. 2000). Several members of the MYB class of transcription factors have been shown to bind Ca^{2+}/CaM (Popescu et al. 2007). Direct interaction of a CaM isoform with the transcription factor MYB2 has been reported in *Arabidopsis* (Yoo et al. 2005). Some R2R3-MYB subfamily of transcription factors has been found to be CaM-binding proteins (Stracke et al. 2001). A soybean CaM, Gm-Cam4, activates an R2R3-MYB2 transcription factor (Yoo et al. 2005). A grapevine R2R3-MYB transcription factor gene, *VvMYB5a*, induces the expression of genes controlling the biosynthesis of defense-related phenylpropanoids (Deluc et al. 2006). A R2R3-MYB-like transcription factor, MYB72, has been shown to be involved in triggering defense responses against broad-spectrum of pathogens in *Arabidopsis thaliana* (Van der Ent et al. 2008). *Arabidopsis* MYB2 has been reported to function as transcriptional activator in abscisic acid signaling (Abe et al. 2003). The Arabidopsis *BOS1* (*BOTRYTIS SUSCEPTIBLE 1*) gene, which encodes a R2R3-MYB transcription factor, induces disease resistance against *P. syringae* and disruption of the gene enhances disease symptom development after infection by *P. syringae* (Mengiste et al. 2003).

Eighteen bZIP family members have been identified as CaM binding proteins in *Arabidopsis* (Popescu et al. 2007). Several TGA proteins have been shown as CaM-binding proteins. TGA3, a member of a family of basic leucine zipper (bZIP) transcription factors, has been identified as a CaM binding protein that binds the promoter of CaM3 (Jakoby et al. 2002). An abscisic acid (ABA) - responsive bZIP transcription factor, ABF2, has been shown to bind CaM (Popescu et al. 2007). These studies suggest that several transcription factors are involved in activation of calmodulin and they are also involved in activation of defense responses by binding with calmodulins and decoding the message from the Ca^{2+} sensor calmodulin proteins to activate defense responses in plants (Fig. 6.1).

6.4 ROS-Regulated Expression of Transcription Factors

PAMP treatment triggers rapid and transient production of reactive oxygen species (ROS) in plants (Faize et al. 2004; Asada 2006; Sagi and Fluhr 2006; Lehtonen et al. 2012; Vidhyasekaran 2014). The reactive oxygen species include hydrogen peroxide (H_2O_2), superoxide (O_2^-), singlet oxygen ($^1O_2$), and hydroxyl radical (OH°) (Grant and Loake 2000). H_2O_2 activates several plant defense signaling systems (León et al. 1995; Murata et al. 2001; Asai et al. 2002; Vranová et al. 2002; Vandenabeele et al. 2003; Nakagami et al. 2006; Wang et al. 2006; Zhang et al. 2011) resulting in enhanced disease resistance.

H_2O_2 has been shown to regulate expression of several transcription factors in plants. The transcription factors induced by H_2O_2 include ethylene-responsive element binding protein (EREBP) in *Arabidopsis* (Desikan et al. 2001). H_2O_2 up-regulates four ethylene-responsive EREBPs, the ethylene-responsive transcription factor (ERF1), and a CEO1-like protein, which is a potential cofactor of EREBP transcription factors in tobacco (Vandenabeele et al. 2003). H_2O_2 also induces MYB-related transcription factor, and zinc finger transcription factor in *Arabidopsis* (Desikan et al. 2001) and WRKY 11 in tobacco (Vranová et al. 2002).

Vandenabeele et al. (2003) identified four main classes of transcription factors which are regulating downstream gene expression in ROS signaling system in tobacco: MYB family, WRKY, AP2, and SCARECROW. The identified MYB transcription factor was homologous with the N-terminal MYB domain of a JA-dependent transcription factor. The induced WRKY transcription factors contained a redox-sensitive zinc-finger DNA-binding domain (Vandenabeele et al. 2003). Activation of these transcription factors may result in activation of transcription of several defense genes. These transcription factors may be direct targets for redox modification by H_2O_2 (Hancock et al. 2006).

The ROS signal functions are manifested as a consequence of their ability to act as mobile carriers of an unpaired electron (Forman et al. 2004). In redox signaling, the reaction of the ROS with the target protein is reminiscent of on-off signaling associated with phosphorylation/dephosphorylation events. Redox signaling occurs when at least one step in a signaling event involves one of its components being specifically modified by a reactive oxygen species through a reaction that is chemically reversible under physiological conditions and/or enzymatically catalyzed (Forman et al. 2004). Signaling through the redox active molecule H_2O_2 is important in inducing plant defense responses in SA-mediated signaling pathway (Desikan et al. 2005).

Redox signaling plays an important role in activation of TGA transcription factors to trigger defense responses (Fig. 6.2). NPR1 (for Nonexpressor of Pathogenesis Related genes 1), the transcriptional regulatory cofactor interacts with TGA transcriptional factors to activate defense responses (Fig. 6.2). In unstressed cells, NPR1 is maintained in the cytoplasm in a large complex comprising disulfide-bonded intermolecular oligomers (Mou et al. 2003). NPR1 is activated when SA accumulates in cells in response to PAMP/stress signals. Upon activation, the intermolecular disulfide bonds are reduced by redox signaling system, releasing monomeric NPR1.

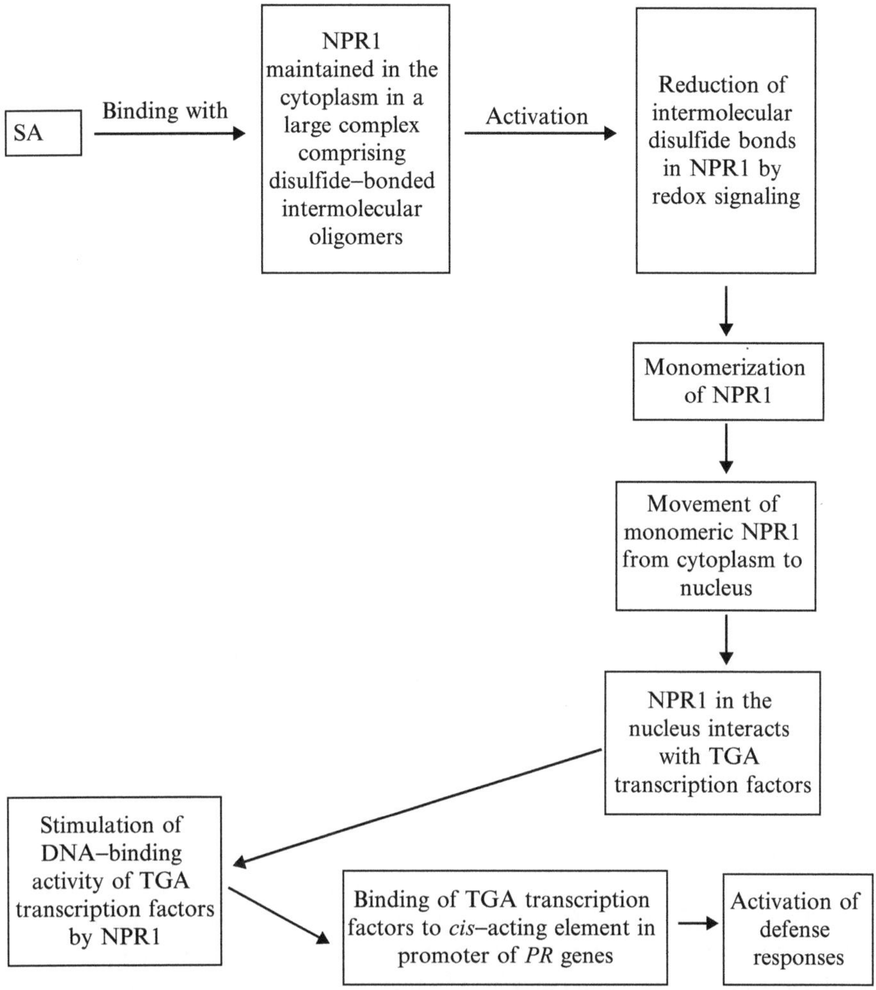

Fig. 6.2 Stimulation of DNA-binding activity of TGA transcription factors by NPR1 to trigger defense responses (Adapted from Fobert and Després 2005; Fedoroff 2006)

The released NPR1 then moves into the nucleus to interact with TGA transcription factors and activate defense gene expression (Fedoroff 2006). The intramolecular disulfide bond between Cys260 and Cys266 in TGA1 prevents interaction with NPR1 (Després et al. 2003). Reduction of the disulfide bond permits TGA1 to interact with NPR1, which in turn stimulates its DNA-binding activity (Fedoroff 2006). SA promotes the reduction of NPR1 and TGA1 (Fobert and Després 2005), by producing H_20_2 (Torres et al. 2006). Along with increase in H_20_2, the transcript levels of genes encoding antioxidant proteins, such as peroxidase and glutathione-S-transferases have increased (Joo et al. 2005). The build-up of antioxidants that enhance the general cellular reducing capacity would have reduced the NPR1 complex

(Cumming et al. 2004). Reduction of conserved cysteines in TGA1 and TGA4 enables their interaction with NPR1, which acts as a redox-sensitive cofactor in stimulating TGA1 DNA-binding activity (Fobert and Després 2005; Fig. 6.2).

6.5 MAPKs-Modulated Phosphorylation of Transcription Factors in Activation of Plant Immune Responses

Several mitogen-activated protein kinases (MAPKs) modulate phosphorylation of transcription factors to trigger transcription of defense genes (Andreasson et al. 2005; Djamei et al. 2007; Liu et al. 2010; Ishihama et al. 2011; Mao et al. 2011). The MAPKs transduce extracellular stimuli into intracellular transcription factors, which enhance expression of defense-related genes in plant innate immune system (Djamei et al. 2007; Vidhyasekaran 2014). WRKY33 is a pathogen-inducible transcription factor, whose expression is regulated by the MPK3/MPK6 cascade. WRKY33 is a substrate of MPK3/MPK6. It has been demonstrated that WRKY33 is phosphorylated by MPK3/MPK6 *in vivo* in response to *Botrytis cinerea* infection in *Arabidopsis*. The phosphorylated WRKY33 transcription factor has been shown to induce the phytoalexin camalexin biosynthesis in Arabidopsis (Mao et al. 2011). The MPK3/MPK6 pathway is also involved in activation of WRKY22 and WRKY29 transcription factors (Asai et al. 2002; Takahashi et al. 2007).

Nicotiana benthamiana WRKY8 transcription factor has been shown to be phosphorylated by the MAPKs, SIPK, WIPK, and NTF4 (Ishihama et al. 2011). Clustered Pro-directed Ser residues, which are conserved in group 1 WRKY proteins, in the N-terminal region of WRKY8 were phosphorylated by these MAPKs in vitro. The interaction of WRKY8 with the MAPKs depended on its D domain, which is a MAPK-interacting motif, and this interaction was required for effective phosphorylation of WRKY8 in plants. Phosphorylation of WRKY8 increased its DNA binding activity to the cognate W-box sequence (Fig. 6.3; Ishihama et al. 2011). Transgenic expression of WRKY8 induced defense-related genes. By contrast, silencing of WRKY8 decreased the expression of defense-related genes and increased susceptibility to the oomycete pathogen *Phytophthora infestans* and the fungal pathogen *Colletotrichum orbiculare* (Ishihama et al. 2011). These results suggest that MAPK-mediated phosphorylation of WRKY8 has an important role in triggering downstream immune responses.

The rice TF OsWRKY53 has been identified as a substrate of OsMPK3/MPK6. The OsMPK3/MPK6 activated by OsMKK4 phosphorylated OsWRKY53 recombinant protein at its multiple clustered serine-proline residues (SP cluster). When *OsWRKY53* was coexpressed with a constitutively active mutant of *OsMKK4*, the enhanced transactivation activity of *OsWRKY53* was found to be dependent on phosphorylation of the SP cluster (Chujo et al. 2014). These results suggest that the OsMKK4-OsMPK3/OsMPK6 cascade regulates transactivation activity of OsWRKY53. Overexpression of OsWRKY53 resulted in a major change to the rice

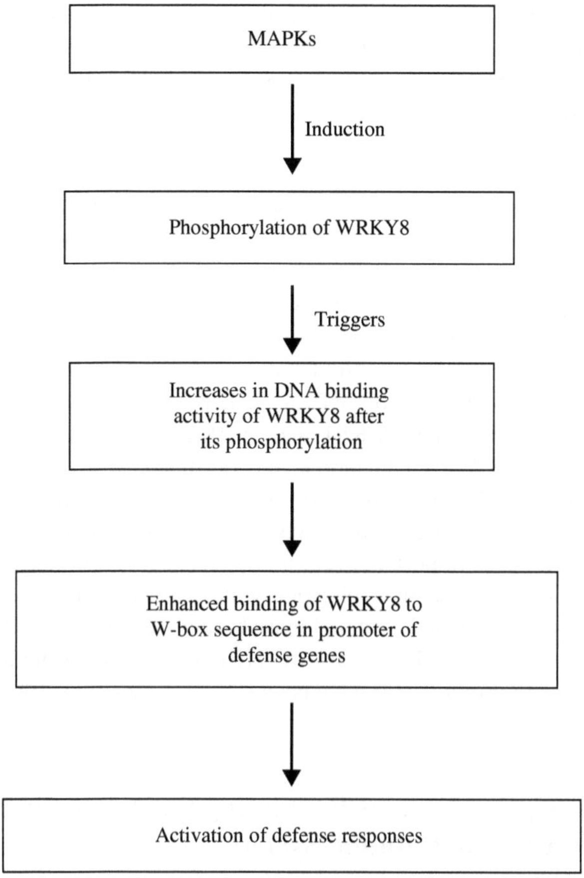

Fig. 6.3 MAPK-mediated phosphorylation of WRKY8 transcription factor activates expression of defense-related genes (Adapted from Ishihama et al. 2011)

transcriptome at steady state that led to activation of a defense response against the blast pathogen *Magnaporthe oryzae* in rice plants (Chujo et al. 2014).

The tobacco MAP kinase WIPK phosphorylates and activates NtWIF, a transcription factor. The transgenic tobacco plants overexpressing *NtWIF* exhibited constitutive accumulation of transcripts for *PR* genes, *PR-1a* and *PR-2* (Waller et al. 2006). The MAPK MPK3 phosphorylates VirE2-interacting protein 1 (VIP1), a bZIP transcription factor (Liu et al. 2010). Upon phosphorylation by MPK3, VIP1 relocalizes from the cytoplasm to the nucleus (Djamei et al. 2007). The bZIP transcription factor VIP1 is phosphorylated by MPK3 and the phosphorylated transcription factor regulates the expression of pathogenesis-related genes (Djamei et al. 2007; Liu et al. 2010).

BWMK1 (Blast- and wounding- activated MAPK 1), a rice mitogen-activated protein kinase is targeted to the nucleus. This protein phosphorylates the rice transcription factor OsEREBP1 (Oryza sativa ethylene-responsive element-binding protein 1). EREBPs are known to bind to the GCC box DNA motif (AGCCGCC) that is located in the promoter of several *PR* genes. In vitro phosphorylation of OsEREBP1 by BWMK1 enhanced its ability to bind to the GCC box. Ectopic expression of the BWMK1 in tobacco plant induced the expression of a broad spectrum of *PR* genes (Cheong et al. 2003).

MKS1 (for MAP kinase 4 substrate 1) is a substrate for the *Arabidopsis* MAPK MPK4. MKS1 interacts with the transcription factor WRKY33 (Andreasson et al. 2005). The interaction of MKS1 with WRKY33 has been shown to be dependent on the phosphorylation status of MKS1 induced by MPK4 (Qiu et al. 2008b). In the absence of pathogens, inactivated MPK4 forms a ternary complex with MKS1 and WRKY33 in the nucleus, which prevents WRKY33 from functioning as a transcription factor. MPK4 is activated by the PAMP treatment or pathogen inoculation in *Arabidopsis*. Upon activation of MPK4, MKS1 is phosphorylated by MPK4. Subsequently, phosphorylated MKS1 and WRKY33 proteins are released from MPK4. The unbound WRKY33 targets the promoter of a defense-related gene *PAD3* (*PHYTOALEXIN DEFICIENT3*) for transcriptional activation (Qiu et al. 2008b, c). Collectively, these results suggest that phosphorylation of transcription factors by MAPKs is an important event in triggering expression of defense-related genes. Protein kinases also may phosphorylate transcription factors involved in transcription of defense genes. Protein kinase C-mediated phosphorylation activates a basic leucine zipper transcription factor G/HBF-1, enabling its binding to the chalcone synthase *Ch15* promoter from soybean (Dröge-Laser et al. 1997).

6.6 Transcription Factors Regulating Salicylate Signaling in Plant Innate Immune System

6.6.1 Transcription Factors Triggering SA Biosynthesis

Several MYB (MYELOBLASTOSIS) transcription factors have been reported in plants. Some MYB transcription factors may be involved in salicylic acid (SA) signaling system. The MYB gene *MYB96* has been shown to be involved in biosynthesis of salicylic acid (Seo and Park 2010). Two different SA biosynthesis pathways have been reported in plants. SA may be synthesized via isochorismate pathway (Wildermuth et al. 2001; Garcion et al. 2008) and/or phenylalanine pathway (Ferrari et al. 2003; Ogawa et al. 2006; Sawada et al. 2006). The bacterial PAMP flg22 treatment rapidly induced the expression of the *MYB96* gene and the transcript abundance reached the peak within 1 h. The PAMP treatment induced the expression of *SID2* gene encoding the key SA biosynthesis enzyme isochrorismate synthase (ICS) (Fig. 6.4; Seo and Park 2010). The *SID2* gene was significantly up-regulated in the activation-tagging *myb96-1d* line but slightly suppressed in the *myb96-1* mutant (Seo and Park 2010).

SA is also synthesized from phenylalanine by phenylalanine ammonia lyase (PAL) activity. The *PAL1* gene was induced moderately in the activation tagging

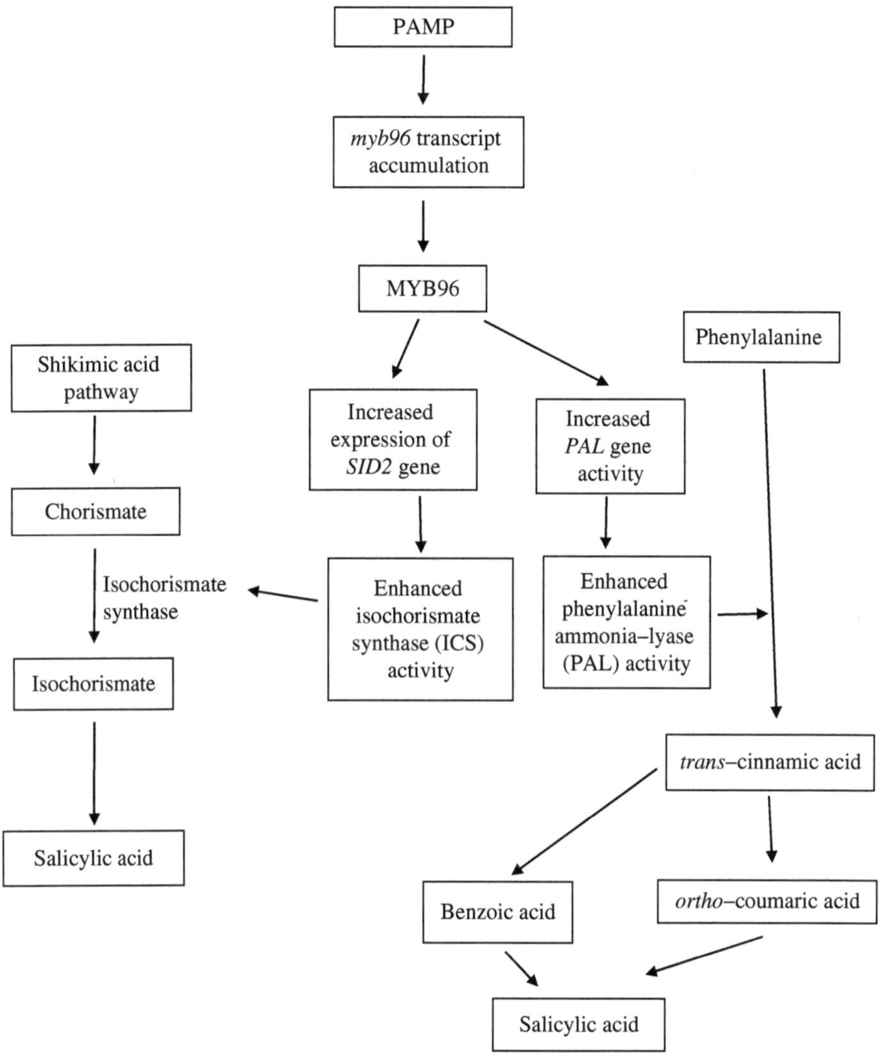

Fig. 6.4 Role of MYB96 transcription factor in induction of salicylic acid biosynthesis through isochorismate and phenylalanine pathways in *Arabidopsis* (Adapted from Seo and Park 2010)

myb96-1d line (Seo and Park 2010; Figure 6.4). SA biosynthesis was elevated in the activation-tagging *myb96-1d* line and the endogenous concentration of SA was seven-fold higher in the activation-tagging *myb96-1d* line (Seo and Park 2010). These results suggest that PAMP triggers *MYB96* induction and the induced *MYB96* gene positively regulates the *SID2* gene and also *PAL1* gene to certain extent, which would result in SA accumulation.

Another transcription factor AtMYB30 has been shown to be involved in SA biosynthesis. Alterations of *AtMYB30* expression (overexpression, depletion by

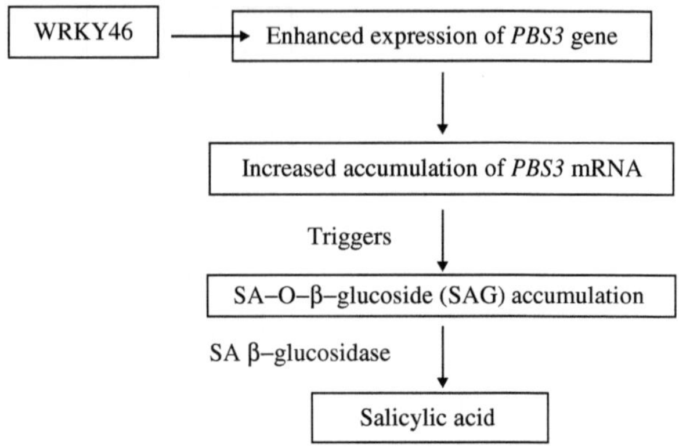

Fig. 6.5 Role of WRKY46 transcription factor in induction of *PBS3* gene involved in biosynthesis of SAG and SA (Adapted from van Verk et al. 2011)

antisense strategy, T-DNA insertion mutant) have been shown to modulate SA levels in *Arabidopsis* (Raffaele et al. 2006). *SID1*/*EDS5* and *SID2* which are involved in SA biosynthesis may be activated by AtMYB30 (Raffaele et al. 2006). WRKY28 positively regulates the expression of *ICS1* involved in SA biosynthesis pathway. WRKY28 binding sites positioned -445 and -460 base pairs upstream of the transcription start site have been identified in the *ICS1* promoter (van Verk et al. 2011).

WRKY46 induces enhanced expression of *PBS3* gene. Overexpression of *WRKY46* resulted in increased accumulation of *PBS3* mRNA, suggesting that WRKY46 is a transcriptional activator of *PBS3* (van Verk et al. 2011; Fig. 6.5). PBS3 (GH3.12) is involved in pathogen-induced accumulation of SA-O-β-glucoside (SAG) and it regulates SA-dependent defense responses (Nobuta et al. 2007).

Some transcription factors may also act upstream of SA signaling and induce SA biosynthesis pathway. NtWIF is a transcription factor activated upon phosphorylation by wound-induced protein kinase (WIPK) in tobacco plants. The transgenic tobacco plants overexpressing *NtWIF* exhibited 50-fold higher levels of SA (Waller et al. 2006), suggesting that the transcription factor acts upstream of SA biosynthesis.

6.6.2 SA Induces Enhanced Expression of Transcription Factors to Activate Transcription of Defense Genes

SA induces enhanced expression of several WRKY, ERF, and TGA transcription factors to activate transcription of defense genes (Després et al. 2003; Knoth et al. 2007; Mao et al. 2007; Zhang et al. 2007; Johnson et al. 2008; van Verk et al. 2008; Fig. 6.6). SA induces increased expression of several WRKY transcription factors including WRKY6, WRKY7, WRKY11, WRKY17, WRKY18, WRKY25, WRKY26, WRKY 29, WRKY33, WRKY38, WRKY40, WRKY46, WRKY48, WRKY53, WRKY54, WRKY60, WRKY62, WRKY63, WRKY64, WRKY67, WRKY70, and WRKY75 in *Arabidopsis* (Chen and Chen 2002; Dong et al. 2003; Kalde et al. 2003; Li et al. 2004; Miao and Zentgraf 2007; Zheng et al. 2006, 2007),

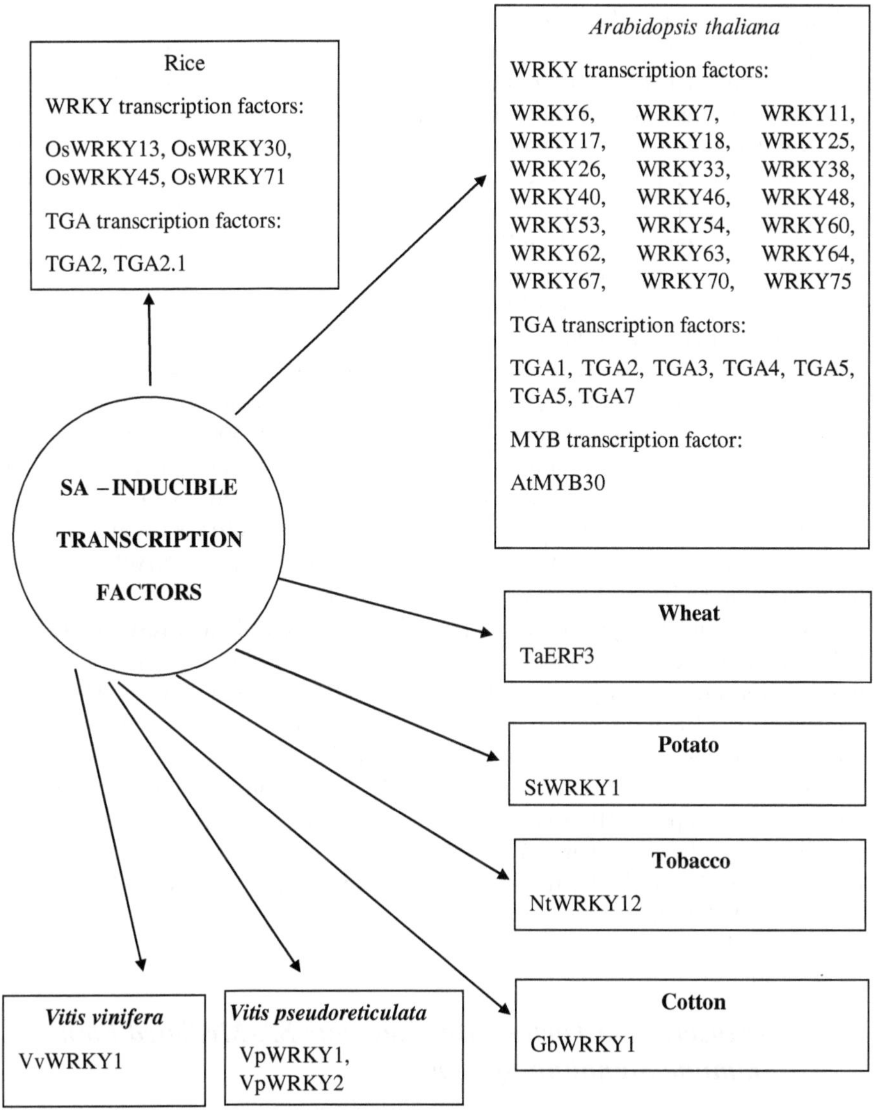

Fig. 6.6 SA-inducible transcription factors in plants

OsWRKY13, OsWRKY71, WRKY45 in rice (Qiu et al. 2007; Shimono et al. 2007), NtWRKY12 in tobacco (van Verk et al. 2008), VvWRKY1 in *Vitis vinifera* (Marchive et al. 2007), and StWRKY in potato (Dellagi et al. 2000).

WRKY18 is a SA-inducible transcription factor. It positively modulates defense-related gene expression (Chen and Chen 2002). WRKY18 physically interacts with structurally related *Arabidopsis* WRKY40 and WRKY60, resulting in altered DNA-binding activities (Xu et al. 2006). While constitutive expression of WRKY18 enhanced resistance to *Pseudomonas syringae*, its coexpression with WRKY40 or WRKY60 made plants more susceptible to the pathogen (Xu et al. 2006). SA-inducible WRKY53 transcription factor interacts with a JA-inducible protein, which triggers resistance against bacterial and fungal pathogens in *Arabidopsis* (Miao and Zentgraf 2007). SA may induce the activation of WRKY1 through its action on SA-inducible protein kinase (SIPK) in tobacco. SIPK phosphorylates and activates WRKY1.and coexpression of WRKY1 and SIPK results in more rapid induction of defense-related cell death than overexpression of either alone in tobacco (Menke et al. 2005). Transcripts of WRKY30 are rapidly inducible by SA (Peng et al. 2012). Expression of WRKY62 is induced by SA (Mao et al. 2007). A *Vitis vinifera* transcription factor VvWRKY1 is induced by SA. It encodes a polypeptide of 151 amino acids and it specifically interacts with the W-box in various nucleotide contexts (Marchive et al. 2007). SA treatment induces expression of two genes encoding transcription factors, VpWRKY1 and VpWRKY2, in *Vitis pseudoreticulata* (Li et al. 2010). Ectopic expression of *VpWRKY1* or *VpWRKY2* in *Arabidopsis* enhanced resistance to the powdery mildew pathogen *Erysiphe cichoracearum* (Li et al. 2010). The potato transcription factor StWRKY1 is also induced by SA (Dellagi et al. 2000). A novel WRKY transcription factor, NtWRKY12 has been shown to be induced by SA (Van Verk et al. 2008).

SA induces another type of transcription factors, called ethylene response factors (ERF). ERF transcriptional factors belong to a subfamily of the APETELA2 (AP2) transcription factor family and contain a single DNA-binding domain. The target sequence for ERF transcription factors is the GCC box that is found in several promoters of PR genes (Grennan 2008). SA induces expression of *TaERF3* in wheat (*Triticum aestivum*) (Zhang et al. 2007). TGA family members of basic-region leucine zipper (bZIP) transcription factors are the other group of transcription factors induced by SA (Zhang et al. 1999; Després et al. 2000; Chern et al. 2001; Fitzgerald et al. 2005). There are 10 TGA transcription factors in *Arabidopsis* (Jakoby et al. 2002).

6.6.3 Transcription Factors May Regulate SA-Mediated Plant Immune Signaling Systems

WRKY transcription factors are important regulators of SA-dependent defense responses (Maleck et al. 2000; Wang et al. 2006). WRKY proteins have been shown to be strongly involved in alteration of gene expression in response to SA treatment (Dellagi et al. 2000; Asai et al. 2002; Chen and Chen 2002; Yoda et al. 2002; Kalde et al. 2003). The promoters of a large number of defense-related genes including *PR* genes contain W-box sequences that are recognized by WRKY proteins (Yu et al. 2001). SA-inducible NtWRKY12 belongs to the class of transcription factors in which the WRKY sequence is followed by a GKK rather than a GQK sequence. The binding sequence of NtWRKY12 (WK box TTTTCCAC) deviated significantly from the consensus sequence (W box TTGAC[C/T]) shown to be recognized by WRKY factors with the GQK sequence (van Verk et al. 2008). NtWRKY12 binds with specific binding sites in the *PR-1a* promoter in positions -564 (box WK1) and -859 (box WK2). NtWRKY12 induces PR-1a expression (van Verk et al. 2008).

In rice, SA triggers *WRKY45* gene expression (Takatsuji 2014), Two defense-related genes, encoding a glutathione *S*-transferase and a cytochrome P450, were found to be regulated downstream of the SA-inducible *WRKY45* in rice (Shimono et al. 2007). The transcription factor WRKY45 plays important role in SA-dependent induced systemic disease resistance (ISR) in rice. Rice plants overexpressing *WRKY45* show extremely high resistance against the blast pathogen *Magnaporthe oryzae* and the bacterial blight pathogen *Xanthomonas oryzae* pv. *oryzae* (Takatsuji 2014).

Expression of WRKY70 is activated by SA and overexpression of WRKY70 induced enhanced expression of *SA*-responsive PR genes (Li et al. 2004). WRKY46 is specifically induced by SA in *Arabidopsis* (Hu et al. 2012). It coordinates with WRKY70 and WRKY53 in activating immune responses against the bacterial pathogen *Pseudomonas syringae* (Hu et al. 2012).

Some of the WRKY transcription factors may act as negative regulators of SA-mediated defense gene expression and they have been shown to negatively regulate pathogen-induced *PR* gene expression (Journot-Catalino et al. 2006; Wang et al. 2006; Xu et al. 2006; Zheng et al. 2007). WRKY7 is a negative regulator of SA-mediated defense responses in Arabidopsis. SA induces *WRKY7* gene expression in *Arabidopsis thaliana.* Transgenic plants overexpressing *WRKY7* showed reduced expression of defense-related genes, including *PR1*, suggesting that WRKY 7 is a transcriptional repressor in *Arabidopsis* (Kim et al. 2006). WRKY48 is another negative regulator of SA-induced *PR1* gene expression in *A. thaliana* (Xing et al. 2008). Transcription of *WRKY33*, which is induced by SA, negatively regulates SA-induced defense gene expression (Zheng et al. 2006). Overexpression of *WRKY25* in *Arabidopsis* reduced expression of the SA-regulated *PR1* gene expression (Zheng et al. 2007). *OsWRKY 45-1* and *OsWRKY45-2*, which encode proteins with a 10-amino acid difference have been shown to play opposite roles in resistance against the two different *Xanthomonas oryzae* pathovars, *X. oryzae* pv. *oryzae* and *X. oryzae* pv. *oryzicola* in rice plants (Tao et al. 2009). *OsWRKY45-1* overexpressing plants showed increased susceptibility and *OsWRKY45-1* knockout plants showed enhanced resistance to both *X. oryzae* pv. *oryzae* and *X. oryzae* pv. *oryzicola.* In contrast, *OsWRKY45-2* overexpressing plants showed enhanced resistance and *OsWRKY45-2*-suppressing plants showed increased susceptibility to *X. oryzae* pv. *oryzae* and *X. oryzae* pv. *oryzicola* (Tao et al. 2009). These results suggest that OsWRKY45-1 is a negative regulator and OsWRKY45-2 is a positive regulator of resistance against *X. oryzae* pv. *oryzae* and *X. oryzae* pv. *oryzicola.*

The SA-inducible TGA family of transcription factors binds tandem repeats of a *cis*-element within the promoters called activation sequence-1 (*as-1*), which contains a TGACG motif (Lebel et al. 1998; Després et al. 2000). The *as-1* elements are responsible for SA responsiveness of these promoters. SA treatment increases the TGACG/*as-1* binding activity (Zhou et al. 2000). Reduction of two Cys residues that are uniquely present in TGA1 and TGA4 has been suggested to be involved in SA-activated defense responses in Arabidopsis (Després et al. 2003). TGA1, TGA 3, TGA4 and TGA6 have a positive role in *PR* gene expression. However, TGA2 has repressor activity on *PR* gene expression and it acts as a negative regulator of *PR* genes (Kesarwani et al. 2007). The PR1 promoter contains both positive (LS7) and negative (LS4 and LS5) *cis*-elements. Before induction, TGA2 represses the *PR* genes through interaction with the negative elements. TGA1, TGA3, and TGA6 may bind to the positive element in the uninduced state. This does not lead to *PR* gene expression in the presence of TGA2. It is suggested that upon induction, TGA2 repression is removed and transcription is activated by TGA1, TGA3, TGA5, and TGA6 (Kesarwani et al. 2007). In rice, four TGA and TGA-like factors have been identified (Chern et al. 2001). Rice TGA2.1 binds to oligonucleotides containing the as-1-like (for 'activation sequence-1-like') element from the PR-1 gene promoter and to the promoter of the rice chitinase gene, RCH10 (Chern et al. 2001).

6.7 Transcription Factors Regulating Jasmonate Signaling System in Plant Innate Immunity

6.7.1 Transcription Factors Triggering JA Biosynthesis

Some transcription factors have been reported to be involved in activation of JA biosynthesis pathway. A rice *WRKY* gene, *OsWRKY30*, has been shown to activate JA biosynthesis pathway in rice plants. Transgenic rice plants overexpressing *OsWRKY30* gene were developed. Constitutive expression of the *OsWRKY30* gene increased the endogenous JA accumulation and was associated with increased expression of JA-synthesis-related genes *LOX* and *AOS2* (Peng et al. 2012). *LOX* encodes the key JA biosynthesis enzyme lipoxygenase. Linolenic acid is oxygenated by lipoxygenases (LOXs) to hydroperoxide derivatives. Further reaction is catalysed by allene oxide synthase (AOS) that converts 13-hydroperoxy-linolenic acid (13-HPOT) to an unstable allene oxide intermediate, which is in turn modified by an allene oxide cyclase (AOC) to form 12-oxo-phytodienoic acid (OPDA). OPDA is an important intermediate in JA biosynthesis pathway (Balbi and Devoto 2008; Sun et al. 2011; Kombrink 2012; Wasternack and Hause 2013). *AOS2* encodes allene oxide synthase and *OsWRKY30* triggers the expression of both *LOX* and *AOS2* gene expression in rice plants (Fig. 6.7; Peng et al. 2012).

A pair of allelic genes, *OsWRKY45-1* and *OsWRKY45-2* in rice played important role in JA biosynthesis. Transgenic rice plants overexpressing *WRKY45-1* and *WRKY45-2* showed increased accumulation of JA and induced increased expression of defense-related genes (Tao et al. 2009). Some transcription factors may negatively regulate enzymes involved in JA biosynthesis. The NAC transcription factor RIM1 negatively regulates JA biosynthesis in rice plants. The expression of genes

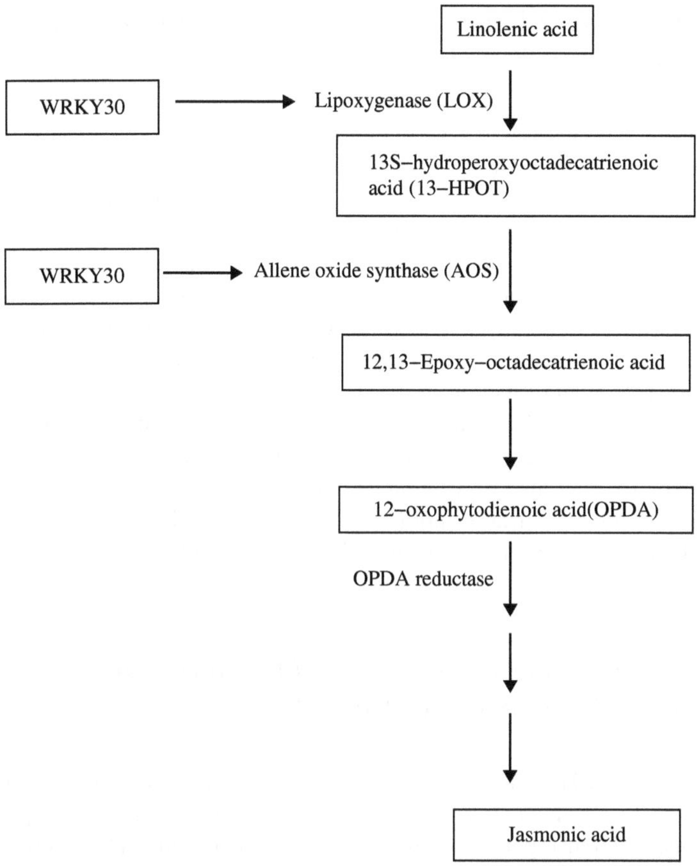

Fig. 6.7 Role of OsWRKY30 in activation of JA biosynthesis in rice plants (Adapted from Peng et al. 2012)

encoding JA biosynthetic enzymes lipoxygenase (*LOX*), allene oxide synthase2 (*AOS2*) and OPDA reductase7 (*OPR7*) was up-regulated in the *rim1* mutants (Yoshii et al. 2010). The results suggest the NAC transcription factor negatively regulates JA biosynthesis.

TCP (TEOSINTE BRANCHED/CYCLOIDEA/PCF) transcription factor genes have also been shown to be involved in JA biosynthesis pathway (Schommer et al. 2008). TCPs positively regulate *LOX2* promoter activity. The TCPs directly regulate expression of *LOX2* (Schommer et al. 2008). Lipoxygenase encoded by *LOX2* is the key enzyme in biosynthesis of JA (Vidhyasekaran 2007, 2014). Some transcription factors may negatively regulate JA biosynthesis. The transcription factor from rice OsWRKY13 suppresses the expression of JA synthesis-related genes (Qiu et al. 2007).

6.7.2 JA Induces Enhanced Expression of Transcription Factors

JA may trigger expression of several transcription factors. Expression and function of the transcription factor WRKY62 was found to be induced by methyl jasmonate (MeJA) (Mao et al. 2007). Transcripts of the rice OsWRKY30 gene accumulate rapidly in response to JA treatment (Peng et al. 2012). Expression of the gene (*GbWRKY1*) encoding the cotton WRKY1 transcription factor is induced rapidly by methyl jasmonate (Li et al. 2014). MeJA treatment induced the expression of the NAC transcription factors ONAC122 and ONAC131 in rice plants (Sun et al. 2013). MeJA treatment has been shown to trigger *PtrWRKY* genes in *Populus* (Jiang et al. 2014). Expression and function of the transcription factor WRKY62 was found to be induced by MeJA (Mao et al. 2007). Grapevine (*Vitis vinifera*) NAC1 transcription factor (*Vv*NAC1) expression is stimulated by MeJA treatment (Le Hénanff et al. 2013). JA may also downregulate expression of some transcription factors. Expression of the NAC transcription factor ATAF1 was downregulated after treatment with jasmonic acid in *Arabidopsis* (Wang et al. 2009a).

6.7.3 Transcription Factors Triggering Expression of JA-Responsive Defense genes

Several ERF, bHLH, WRKY, MYB, NAC, and bZIP transcription factors have been shown to act downstream of JA signaling system triggering expression of JA-responsive defense genes (Memelink et al. 2001; Lorenzo et al. 2004; Lorenzo and Solano 2005; Zheng et al. 2006; Nurmberg et al. 2007; Wasternack 2007; Zander et al. 2010; Cheng et al. 2011; Le Hénanff et al. 2013). Transcription factors belonging to AP2/EREBP transcription factor family which are involved in inducing JA-responsive genes include ERF1, ERF2, ERF4, ERF5, ERF6, ERF14, ERF15, ERF59 (ORA59), ERF98, ORA37, and ORA47 (Berrocal-Lobo and Molina 2004; Gfeller et al. 2010; Zander et al. 2010; Zarei et al. 2011; Cevik et al. 2012; Wager and Browse 2012; Wasternack and Hause 2013). MeJA treatment also activated the expression of *VvNAC1* and induced the expression of JA-responsive *PDF1.2* gene in grapevine plants (Le Hénanff et al. 2013). The Arabidopsis class-II TGA transcription factors TGA2, TGA5, and TGA6 are involved in activation of JA- dependent defense mechanisms (Zander et al. 2010).

Transgenic rice plants overexpressing *OsWRKY30* gene showed increased expression of the pathogenesis-related (*PR*) genes, *PR3* and *PR10*, which are JA-responsive genes (Peng et al. 2012). The gene encoding the PR-12 protein, *PDF1.2*, is induced by JA signaling (Vidhyasekaran 2007; Zander et al. 2010; Kravchuk et al. 2011; Po-Wen et al. 2013). While the transcription factors ERF1 and ORA59 activate expression of the *PDF1.2* gene in *Arabidopsis* (Lorenzo et al. 2003; Pré et al. 2008; Zander et al. 2010; Zarei et al. 2011; Cevik et al. 2012), MYC2 negatively regulate *PDF1.2* gene expression (Zander et al. 2010; Wager and Browse 2012; Wasternack and Hause 2013).

6.8 Transcription Factors Regulating Ethylene Signaling System in Plant Innate Immunity

Ethylene-responsive transcription factors (ERFs) belonging to EREBP transcription factor family are the important group of transcription factors functioning downstream in ethylene signaling system in plant innate immunity (Gutterson and Reuber 2004; Grennan 2008). The important ERF transcription factors involved in defense signaling system include TaERF1 and TaERF3 in wheat (Xu et al. 2007; Zhang et al. 2007), OsBIERF3 in rice (Cao et al. 2006), CaPF1 and CaERFLP 1 in pepper (Lee et al. 2004; Yi et al. 2004), Pti4 and Pti5 in tomato (Zhou et al. 1997), and HvRAF in barley (Jung et al. 2007). ERF2, ERF2 and ERF4 in canola (*Brassica napus*) (Yang et al. 2007), ERF3, and ERF4 in tobacco (Koyama et al. 2003), and AtERF1, AtERF2, AtERF4, and AtERF14 in Arabidopsis (McGrath et al. 2005; Grennan 2008), The transcription factor ERF1 induces transcription of several pathogenesis-related (*PR*) genes including *PR-1*, *PRB-1b*, *PR-2* (β-1,3-glucanase), *ChiB* (PR-3; basic chitinase), *PR-4* (hevein-like protein), *PR-5* (osmotin), *PDF1.2* (PR-12; defensin), *Thi2.1* (PR-13; thionin), and germin-like proteins in *Arabidopsis* (Frye et al. 2001; Lorenzo et al. 2003; Tang et al. 2005; Zhou et al. 2005; Johansson et al. 2006; Jung et al. 2007; Oñate-Sánchez et al. 2007). An ERF transcription factor (*CaPF1*) detected in pepper induced an increase in transcript levels of *PDF1.2* and *GST* (Yi et al. 2004). Expression of a ERF-type transcription factor from barley in *Arabidopsis* plants induced the activation of *PR1*, *PR5*, *PDF1.2*, and *GSH1* (Jung et al. 2007). AtERF14 positively induces defense gene *ChiB* encoding the basic chitinase in *Arabidopsis* (Oñate-Sánchez et al. 2007). AtERF2 positively regulates the defense gene expression (Brown et al. 2003; McGrath et al. 2005). In tobacco, *ERF2* and *ERF4* act as positive regulators of defense responses (Ohme-Takagi and Shinshi 1995; Kitajima et al. 2000; Koyama et al. 2003).

Although most ERFs studied so far are positive regulators/activators, 14 *Arabidopsis* ERF proteins contain an ERF-associated amphiphilic repression (EAR) motif, which has been shown to function as a repression domain (Fujimoto et al. 2000; Ohta et al. 2001; McGrath et al. 2005). In *Arabidopsis*, AtERF4, an EAR-containing ERF, negatively regulates the induction of *PDF1.2* and *CHIB* genes, (Brown et al. 2003; McGrath et al. 2005). In tobacco, the *ERF3* gene acts as a negative regulator of defense responses (Ohme-Takagi and Shinshi 1995; Kitajima et al. 2000; Koyama et al. 2003).

6.9 Transcription Factors May Trigger "Priming" of Defense Responses

6.9.1 What Is Priming?

Several transcription factors trigger "priming" of defense responses and induce "Systemic Acquired resistance (SAR)" in plants (Pozo et al. 2008; Chavan and Kamble 2013; Nakayama et al. 2013). The "priming" of defense responses is defined as the ability to induce cellular defense responses more rapidly and effectively to subsequent infection. When a treatment puts a plant in a state of increased alertness with no or only minimal gene induction it is called priming (Conrath et al. 2002, 2006). Some chemicals may sensitize plant cells for a faster and stronger response to further pathogen attack (Orober et al. 2002) and this phenomenon is called "priming". Infection of plants by necrotizing pathogens, which induce the accumulation of salicylic acid (SA), or treatment of plants with synthetic compounds, which are able to trigger SA signaling, causes "priming" (Conrath 2009, 2011; Camañes et al. 2012; Slaughter et al. 2012). The WRKY45-overexpressing rice plants showed strong resistance to the rice blast pathogen *Magnaporthe oryzae* in spite of the lack of constitutive expression of the defense-related genes before the pathogen infection suggesting that WRKY45 triggers primed state in plants (Shimono et al. 2007).

6.9.2 Histone Modifications in Chromatin Structure May Be Involved in the Priming Process

Chromatin is the combination of DNA and proteins that make up the contents of the nucleus of a cell. Chromatin structure is important for the regulation of gene expression (Quang and Gill 2009; Jaskiewicz et al. 2011). The important function of chromatin is to package DNA into a smaller volume to fit in the cell. The primary protein components of chromatin are histones that compact the DNA (Bender 2004). The histone proteins are subject to various covalent modifications including acetylation, methylation, phosphorylation, and ADP-ribosylation (Berger 2002; Britton et al. 2011; Chen et al. 2011). Jaskiewicz et al. (2011) showed that histone modifications are systemically set during a priming event and these modifications may be involved in the priming process. These histone modifications might create a memory of the primary infection that is associated with an amplified reaction to a second stress conditions. It is known that chromatin states could control cellular memory (Zhang 2008). The primed genes may be poised for enhanced activation of gene expression by histone modification (Jaskiewicz et al. 2011).

6.9.3 Priming in Systemic Acquired Resistance

After localized foliar infections by diverse pathogens or by treatment with synthetic compounds, plants develop whole-plant immunity, called systemic acquired resistance (SAR) (Beckers and Conrath 2007; Conrath 2009; Mukherjee et al. 2010). In this process, distal (systemic) leaves become primed to activate stronger defense response upon secondary infection (Beckers et al. 2009; Slaughter et al. 2012). SAR is associated with priming of defense (Jung et al. 2009; Luna et al. 2012) and the priming results in a faster and stronger induction of defense mechanisms after pathogen attack (Conrath 2011). Priming of pathogen-induced gene allows the plant to react more effectively to a subsequent invader (Conrath et al. 2006; Conrath 2009). In the SAR response, defense genes in the infected and distant tissue show the 'priming' phenomenon; they are able to respond faster and/or to a greater extent to a subsequent challenge (Kohler et al. 2002; Conrath 2009).

6.9.4 Plants May Inherit the Priming Phenomenon to Next-Generation SAR

The priming can be inherited epigenetically from disease-exposed plants and descendants of primed plants exhibit next-generation systemic acquired resistance (Luna et al. 2012; Slaughter et al. 2012). The descendants of *Arabidopsis thaliana* plants that had been primed with the plant defense activator β-amino-butyric acid (BABA) showed a faster and higher accumulation of transcripts of defense-related genes in SA signaling pathway and enhanced disease resistance upon challenge inoculation with the bacterial pathogen *Pseudomonas syringae* (Slaughter et al. 2012). The progeny of the primed plants was also more resistant against the oomycete pathogen *Hyaloperonospora arabidopsidis*. When transgenerationally primed plants were subjected to an additional priming treatment, their descendants displayed an even stronger primed phenotype (Slaughter et al. 2012). The results suggest that plants can inherit a sensitization for the priming phenomenon and the primed state of plants is transferred to their progeny. The transgenerational SAR was found to be sustained over one stress-free generation, indicating an epigenetic basis of the phenomenon (Luna et al. 2012). SA-inducible promoters of the transcription factors *WRKY6*, and *WRKY53* in the progeny from plants showing transgenerational SAR were found to be enriched with acetylated histone H3 at lysine 9, a chromatin mark associated with a permissive state of transcription (Luna et al. 2012). The results suggest the importance of chromatin remodeling in the transgenerational SAR (Luna et al. 2012).

DNA methylation may also play important role in transgenerational SAR (Luna et al. 2012). DNA methylation is initiated by DRM (Domains-Rearranged methyltransferase) and CMT3 (Chromomethyltransferase3) that are methyltransferases (Henderson et al. 2010; Huang et al. 2010; Greenberg et al. 2011). Neither *drm* nor *cmt3* mutants affected the maintenance of pre-established RNA-directed CpG methylation. However, *drm* mutants showed a nearly complete loss of asymmetric methylation and partial loss of CpNpG methylation. The remaining asymmetric and CpNpG methylation was dependent on the activity of CMT3, showing that DRM and CMT3 act redundantly to maintain non-CpG methylation. These DNA methyltransferases appear to act downstream of siRNAs, since *drm1 drm2 cmt3* triple mutants show a lack of non-CpG methylation but elevated levels of siRNAs. Further it has been demonstrated that DRM activity is required for the initial establishment of DNA methylation in all sequence contexts including CpG, CpNpG, and asymmetric sites (Cao et al. 2003). The *drm1drm2cmt3* triple mutant has been shown to be affected in non-CpG DNA methylation. This mutant mimicked the transgenerational SAR (Luna et al. 2012). These results suggest that transgenerational SAR is transmitted by hypomethylated genes that direct priming of SA-dependent defenses in the following generations.

6.9.5 Priming of Transcription Factors in Plant Defense System

SA or its functional analog benzo(1,2,3)thiadiazole-7-carbothioic acid S-methyl ester (BTH), a SA biosynthesis –inducing chemical probenazole and a nonprotein amino acid β-aminobutyric acid (BABA) are known to induce priming and confer resistance against viral (Mandal et al. 2008), bacterial (Flors et al. 2008), oomycete (Hamiduzzaman et al. 2005; Ziadi et al. 2008) and fungal (Nakashita et al. 2002; Hukkanen et al. 2007; Umemura et al. 2009) diseases. BTH treatment induced a typical priming effect on *WRKY 29*, *WRKY6*, and *WRKY53* genes in *Arabidopsis* (Jaskiewicz et al. 2011). The expression of the *WRKY* genes was enhanced in BTH-treated plants when the plants were stressed. Water infiltration, a stress inducing cell death after BTH treatment resulted in strongly enhanced gene activation, compared with plants that were stressed without previous BTH treatment. On the WRKY29 promoter, trimethylation (H3K4me3) and dimethylation (H3K4me2) of H3K4 and all acetylations tested increased after BTH application although this did not induce *WRKY29* transcription. Thus, chromatin marks normally associated with active genes are set by the priming stimulus before gene activation. After previous priming, a stress stimulus enhanced some of the modifications – H3K4me2, H3K9ac and H4K12ac – on WRKY 29 (Jaskiewicz et al. 2011). These results suggest that pre-stress application of BTH induced chromatin modifications on *WRKY* gene promoters that facilitate the activation of gene expression by subsequent stress (Jaskiewicz et al. 2011).

Localized *Pseudomonas syringae* pv. *maculicola* infection in *Arabidopsis* primed the *WRKY* gene promoters in remote leaves to an augmented response to secondary stress and the transcriptional responses in distal leaves were similar to those observed with BTH. On the three *WRKY29*, *WRKY6*, and *WRKY53* gene promoters, clear increases in histone modifications were observed after pathogen infection. The response amplitude after perception of the systemic signals for SAR was similar to that observed after BTH treatment. Thus, pathogen exposure induces one or more systemic signals that are stored in the form of histone modifications. Enhanced trimethylation of H3K4 concomitant with gene priming was a common feature of the assayed *WRKY* promoters (Jaskiewicz et al. 2011). These studies suggest that there may be a tight correlation between histone modification patterns and gene priming and also there may be a histone memory for information storage in the plant stress response (Jaskiewicz et al. 2011).

Overexpression of rice *WRKY45* markedly enhanced blast resistance. BTH induction of blast resistance was markedly compromised in *WRKY45*-knocked down plants (Shimono et al. 2007). The results suggest that rice WRKY45 plays a crucial and predominant role in BTH-inducible defense responses that strongly enhances resistance to blast infection (Shimono et al. 2007). When plants are pretreated by BTH, induction of cellular defense responses upon pathogen infection occurs more rapidly or to a greater degree than in untreated plants, thereby accounting for the enhanced disease resistance in primed plants. Under primed conditions, most defense genes remain silent and become upregulated only after pathogen infection (Conrath et al. 2006).

In BABA-pretreated B*rassica carinata* plants challenged with *Alternaria brassicae*, priming of expression of *BcWRKY70*, *BcWRKY11* and *BcWRKY53* genes was observed (Chavan and Kamble 2013). BABA-primed expression of *BcWRKY70*, *BcWRKY11* and *BcWRKY53* genes was strongly correlated with enhanced expression of the defense genes in *Brassica* (Chavan and Kamble 2013). These results suggest that priming of transcription factors plays an important role in expression of defense responses.

6.10 Bioengineering WRKY Transcription Factors for Rice Disease Management

6.10.1 WRKY Transcription Factors Regulating Plant Immune Responses

WRKY proteins constitute one of the largest transcription factor families in plants (Pandey and Somssich 2009; Rushton et al. 2010; Agarwal et al. 2011; Kavum et al. 2015; Schluttenhofer and Yuan 2015). The WRKY transcription factor family is defined by a domain of 60 amino acids, which contains the amino acid sequence WRKY (tryptophan-arginine-lysine-tyrosine) at its amino-terminal end and a putative zinc finger motif at its carboxy-terminal end (Rushton et al. 2010; Wang et al. 2011; Alves et al. 2014). This sequence is required for proper folding of the zinc finger and for DNA binding (Maeo et al. 2001). Most of the WRKY proteins contain one WRKY domain, while some of the WRKY proteins have two WRKY domains (Eulgem et al. 2000; Maeo et al. 2001; Zheng et al. 2006). WRKYs are subdivided into three subgroups based on the number of WRKY domains in the protein and the structure of their zinc fingers (Eulgem et al. 2000). Members of group I have two WRKY domains, whereas members of groups II and III have one WRKY domain. Group II genes have been subdivided into IIa, IIb, IIc, IId and IIe on the basis of their amino acid sequence (Eulgem et al. 2000; Wang et al. 2011). Group III domains contain a $Cx_7CX_{23}HXC$ pattern of zinc ligands which is distinct from the $Cx_{4\text{-}5}CX_{22\text{-}23}HXH$ zinc finger pattern of group I and group II WRKY domains (Maeo et al. 2001; Knoth et al. 2007). Members of the *Arabidopsis* WRKY group III transcription factors are part of different plant defense signaling pathways (Kalde et al. 2003).

More than 70 *WRKY* genes have been reported in *Arabidopsis* (Eulgem et al. 2000; Robatzek and Somssich 2001; Dong et al. 2003; Eulgem and Somssich 2007; Bhattarai et al. 2010). About 109 *WRKY* genes have been identified in rice (Qu and Zhu, 2006; Liu et al. 2005, 2007a; Qiu et al. 2007; Ross et al. 2007). Sixty-six *WRKY* genes have been identified in papaya. Sixty-eight *WRKY* genes have been detected in sorghum, while 104 *WRKY* genes have been reported in poplar (Pandey and Somssich 2009). Ling et al. (2011) identified 55 *WRKY* genes in the cucumber genome. Eight *BrWRKY* genes showed differential expression after *Pectobacterium carotovorum* subsp. *carotovorum* infection and five genes after *Fusarium oxysporum* f. sp. *conglutinans* infection in *Brassica rapa* (Kavum et al. 2015). A total number of 59 putative VvWRKY transcription factors have been identified in grapevine (Wang et al. 2014). A total of 61 *SmWRKY* genes have been cloned from *Salvia miltiorrhiza* (Li et al. 2015a).

WRKY is localized to the nucleus of plant cells and recognizes DNA molecules containing the *TTGACC* W-box sequence (Zheng et al. 2006). The WRKY domain binds specifically to various W box elements containing a (C/TTGACC/T) core sequence (Eulgem and somssich 2007; Ruston et al. 2012). The promoters of a large number of defense-related genes contain W-box sequences that are recognized by WRKY proteins and the WRKY transcription factors have been shown to be necessary for the inducible expression of these defense genes (Eulgem et al. 1999, 2000; Yu et al. 2001; Ulker and Somssich 2004; Shimono et al. 2007; van Verk et al. 2008). WRKY transcription factors function via binding to a cis-element known as the "W-box" that is present in the promoters of target genes (Rushton et al. 1996; Eulgem et al. 1999). A large number of plant defense or defense-related genes including pathogenesis-related (PR) genes and the regulatory NPR1 gene contain W box sequences in their promoters that are recognized by WRKY proteins (Yu et al. 2001; Eulgem and Somssich 2007).

In *Arabidopsis*, there are 72 expressed *WRKY* genes, and many of them are implicated in the regulation of the plant immune responses positively or negatively via modulation of the JA/SA signaling pathways (Eulgem and Somssich 2007). Multiple W-box DNA elements were predicted in the promoter region of *PEPR1* and *PROPEP1-5* genes in *Arabidopsis* and therefore, the WRKY transcription factors may play an important role in the amplification of the HAMP Pep peptide signal (Yamaguchi et al. 2010). WRKY transcription factors also regulate the expression of their own genes and/or other WRKY genes (Eulgem and Somssich 2007).

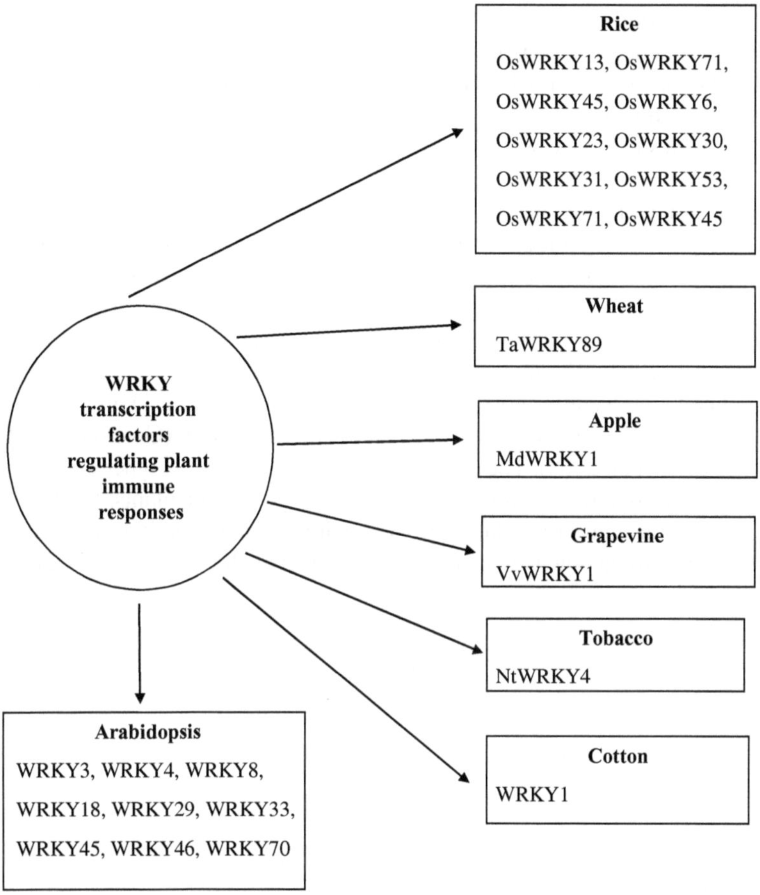

Fig. 6.8 WRKY transcription factors involved in regulation of plant defense responses against pathogens in various crop plants and *Arabidopsis*

WRKY transcription factors are well known for regulating plant immune systems (Schluttenhofer and Yuan 2015). Several WRKY transcription factors have been shown to regulate either positively or negatively plant immune responses against various fungal, oomycete, bacterial and viral diseases in wide range of crop plants and *Arabidopsis* (Fig. 6.8).

6.10.2 Engineering *OsWRKY13* Gene

One of the rice WRKY genes, *OsWRKY13*, showed increased expression after incompatible pathogen infection in several rice lines conferring resistance to the rice bacterial blight pathogen *Xanthomonas oryzae* pv. *oryzae* and the rice blast pathogen *Magnaporthe grisea*. However, there was no change in expression of *OsWRKY13* gene in compatible interactions (Wen et al. 2003). It suggests that increased expression of *OsWRKY13* may induce resistance against diseases caused by various pathogens.

The *OsWRKY13* gene was cloned and transgenic rice plants overexpressing the gene were obtained (Qiu et al. 2007). The transgenic plants showed changes in expression of various genes involved in defense signaling systems (Fig. 6.9; Qiu et al. 2007). The transgenic plants accumulated significantly more *PAD4* (phytoalexin-deficient 4) than did the wild-type plants (Qiu et al. 2007). PAD4 is a key regulator of SA signaling system and it acts upstream of SA (Lippok et al. 2007). It has been suggested that PAD4 transduces ROS-derived signals leading to SA production (Mateo et al. 2004; Wiermer et al. 2005). The *OsWRKY13*-overexpressing plants accumulated significantly more ICS1 (isochorismate synthase 1) and ICS plays an important role in SA biosynthesis via isochorismate pathway (Garcion and Métraux 2006; Garcion et al. 2008). The transgenic plants also showed higher expression levels of *NH1* (Arabidopsis NPR1 homolog 1) than in wild type plants. NPR1 has been reported to act as master regulator of SA signaling system and it acts as a transcriptional co-activator at downstream of SA (Rochon et al. 2006; Chern et al. 2008). The

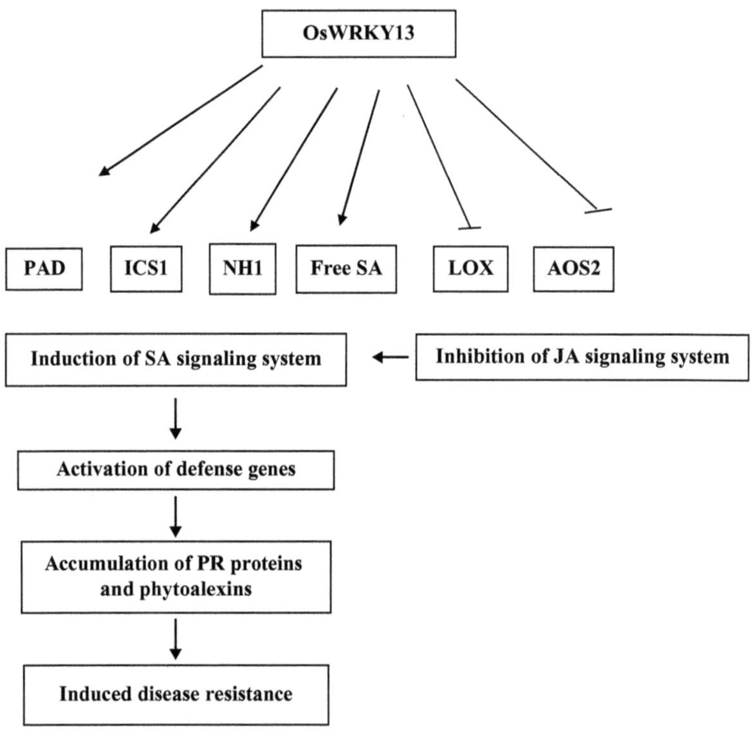

Fig. 6.9 Transgenic rice plants overexpressing *OsWRKY13* modulate SA and JA signaling systems to trigger defense responses (Adapted from Qiu et al. 2007)

transgenic plants also showed accumulation of free salicylic acid (Qiu et al. 2007). These results suggest that OsWRKY13 triggers SA signaling system.

The transgenic plants showed reduced expression of *AOS2* (allene oxide synthase 2) (Qiu et al. 2007) and AOS2 is an important enzyme in JA biosynthesis (Delker et al. 2007; Balbi and Devoto, 2008). When the transgenic plants were infected by pathogens, lipoxygenase (*LOX*) expression was also suppressed (Qiu et al. 2007). LOX is the enzyme involved in oxygenation of linoleic acid to hydroperoxide derivatives and further reaction is catalysed by AOS leading to JA biosynthesis (Balbi and Devoto 2008). Inhibition of both *AOS2* and *LOX* in transgenic rice plants expressing *OsWRKY13* suggests that JA signaling pathway is suppressed in the transgenic plants. The transgenic plants overexpressing *OsWRKY13* showed increased expression of CHS (chalcone synthase), the key enzyme involved in biosynthesis of phytoalexins (Qiu et al. 2007). The transgenic plants also showed enhanced accumulation of *PR-1a* transcripts (Qiu et al. 2007).

WRKY13 is a transcriptional repressor (Xiao et al. 2013). It can bind to the promoters of both *WRKY45-1*/*WRKY45* and *WRKY45-2*, which are alleles encoding proteins with a 10-amino acid difference, suggesting that *WRKY13* may regulate the functions of the *WRKY45* alleles in defense signaling (Tao et al. 2009; Xiao et al. 2013). WRKY42 negatively regulates rice response to *M. oryzae* by suppressing JA signaling related genes (Cheng et al. 2015). WRKY45-2 transcriptionally activates *WRKY13*, whose encoding protein in turn transcriptionally suppresses *WRKY42*, which is known to suppress defense responses against *M. oryzae* (Cheng et al. 2015). These results suggest that WRKY13 may regulate other WRKY transcription factors and confer resistance against the rice blast pathogen.

Transgenic rice plants overexpressing OsWRKY13 showed reduced blast disease incidence (Fig. 6.10; Cheng et al. 2015). The transgenic rice plants overexpressing *OsWRKY13* showed enhanced resistance to the rice blast pathogen *M. oryzae* both

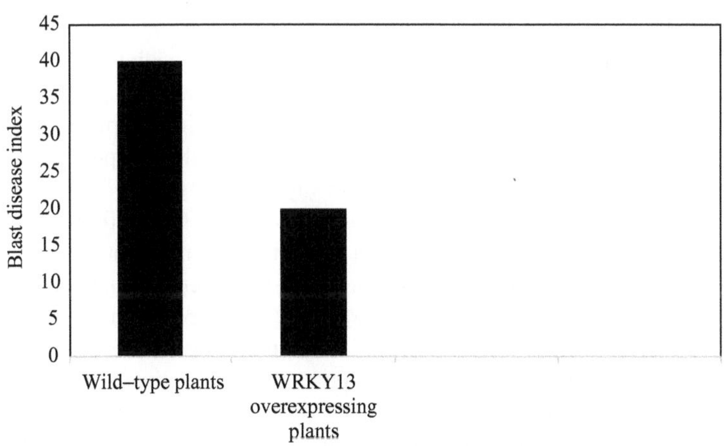

Fig. 6.10 Transgenic rice plants overexpressing WRKY13 show reduced blast disease intensity (Adapted from Cheng et al. 2015)

under greenhouse and field conditions (Qiu et al. 2007). These transgenic plants also showed resistance to the bacterial blight pathogen *Xanthomonas oryzae* pv. *oryzae*. When these transgenic plants were inoculated with the bacterial pathogen, growth rate of the bacteria on the plants was 5.5 to 13.9-fold lower than that on wild-type plants. The lesion area on the bacteria-infected leaves was also reduced in the transgenic plants. The *OsWRKY13* gene conferred resistance against both the bacterial and fungal diseases at both seedling and adult stages in rice without causing any deleterious effects (Qiu et al. 2007). These studies indicate that OsWRKY13 provides a transgenic tool for engineering wide-spectrum disease resistance in rice.

6.10.3 Engineering *OsWRKY22* Gene

OsWRKY22 gene has been isolated from rice plants and characterized (Abbruscato et al. 2012). The mutant *wrky22* knockout plants showed enhanced susceptibility to both virulent and avirulent *Magnaporthe oryzae* strains. These mutant plants also showed altered cellular responses to the nonhost pathogens *Magnaporthe grisea* and *Blumeria graminis*. Several defense-related genes were found to be co-expressed with *OsWRKY 22* (Abbruscato et al. 2012). These results suggest that the transcription factor OsWRKY22 plays an important role in the host defense responses against the rice blast pathogen *M. oryzae*. Transgenic rice plants overexpressing OsWRKY22 were developed and these plants showed enhanced blast disease resistance (Abbruscato et al. 2012), Engineering *OsWRKY22* gene may be a potential tool to develop disease-resistant plants.

6.10.4 Engineering *OsWRKY30* Gene

WRKY30 is another transcription factor involved in SA-mediated signaling pathway. Its transcripts are rapidly inducible by SA (Peng et al. 2012). OsWRKY30 is involved in both SA and jasmonate signaling pathways in rice plants. Its overexpression induces the expression of *WRKY45*, the SA pathway-specific gene (Han et al. 2013) and *LOX* and *AOS2*, the JA pathway specific genes (Peng et al. 2012). Overexpression of OsWRKY30 in rice plants activates OsPR10a promoter (Lee et al. 2013). WRKY30-overexpressing rice plants were resistant to the rice blast pathogen *Magnaporthe oryzae*, the sheath blight pathogen *Rhizoctonia solani* and the bacterial blight pathogen *Xanthomonas oryzae* pv. *oryzae* (Peng et al. 2012; Han et al. 2013; Lee et al. 2013). The C-terminal region of OsWRKY30 was found to be sufficient to confer enhanced resistance to pathogens and activate the expression of defense-related genes (Lee et al. 2013). The endogenous expression of *OsWRKY30* was abolished in SA-nonaccumulating *NahG* transgenic rice plants (Han et al. 2013). The results suggest that SA signaling pathway is important in triggering expression of *OsWRKY30* in rice plants. Overexpression of *OsWRKY30* in rice upregulates *OsWRKY45* and enhances resistance against the rice bacterial blight disease caused by *Xanthomonas oryzae* pv. *oryzae* (Han et al. 2013).

6.10.5 Engineering OsWRKY31 Gene

OsWRKY31 gene was isolated from rice plants. The gene expression was induced in rice plants infected with the rice blast pathogen *Magnaporthe oryzae* (Zhang et al. 2008). This gene encodes a polypeptide of 211 amino-residues. Transgenic rice plants overexpressing *OsWRKY31* were developed and these transgenic plants showed enhanced resistance to *M. oryzae*. The transgenic plants showed constitutive expression of several defense-related genes including *PBZ1* (Zhang et al. 2008). PBZ1 is a probenazole-inducible gene (Nakashita et al. 2002). It is a PR-10 protein, which is an acidic intracellular protein (Midoh and Iwata 1996; Nishizawa et al. 2003; Vidhyasekaran 2007). It plays an important role in conferring resistance against *M. oryzae* (Vidhyasekaran 2007; Fukuoka et al. 2015). The results suggest that *OsWRKY31* may be another tool for developing blast disease-resistant rice variety

6.10.6 Engineering OsWRKY42 Gene

The rice transcription factor OsWRKY42 is a negative regulator of defense responses against the rice blast pathogen *M. oryzae* in rice plants (Cheng et al. 2015). Transgenic rice plants overexpressing *WRKY42* were developed and the *WRKY42*-overexpressing rice plants showed increased susceptibility to *M. oryzae*. The JA contents in different rice plants appeared to be negatively associated with the transcript levels of WRKY42 and with the increased susceptibility to *M. oryzae* infection. The WRKY42 overexpressing plants showed reduced JA levels. Allene oxide synthase is a key enzyme involved in JA biosynthesis (Balbi and Devoto 2008; Wasternack and Hause 2013; Vidhyasekaran 2015). The expression of *Allene Oxide Synthase2* (*AOS2*), encoding an allene oxide synthase was suppressed in the *WRKY42*-overexpressing plants (Fig. 6.11; Cheng et al. 2015). OsJAZ8 is a rice jasmonate ZIM-domain protein, which is involved in JA signaling system. It is highly up-regulated by JA. OsJAZ8 interacts with a putative OsCOI1, which is a component of the SCF(COI1) E3 ubiquitin ligase complex, in a coronatine-dependent manner (Yamada et al. 2012). *JAZ8* functioning in JA-dependent signaling (Yamada et al. 2012) was also suppressed in the *WRKY42*-overexpressing plants (Cheng et al. 2015). It has been suggested that the negative regulation of rice resistance to *M. oryzae* may be related to the decreased JA level and JA-dependent signaling (Cheng et al. 2015). These results suggest that WRKY42 suppresses resistance to *M. oryzae* by suppressing JA signaling system.

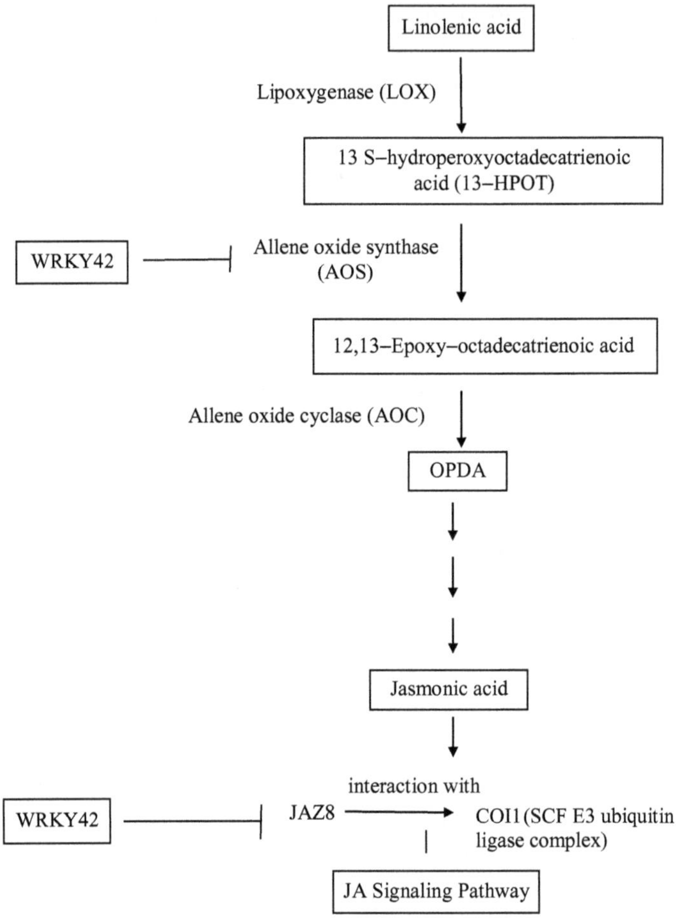

Fig. 6.11 Suppression of JA biosynthesis and JA-dependent signaling by WRKY42 transcription factor (Adapted from Cheng et al. 2015)

WRKY42-suppressing (*WRKY42*-RNA interference [RNAi]) rice plants were developed and these plants showed increased resistance to *M. oryzae* (Fig. 6.12; Cheng et al. 2015). The fungal growth rate in *WRKY42*-RNAi plants was significantly lower. The *WRKY42*-RNAi plants showed increased JA levels and expression of *AOS2* and *JAZ8* was also increased in the *WRKY42*-RNAi plants (Cheng et al. 2015). The results suggest that *WRKY42*-RNAi triggers resistance against *M. oryzae* by triggering JA signaling system. JA signaling system has been shown to play important role in triggering resistance against the rice blast pathogen *M. oryzae* (Mei et al. 2006; Reimann et al. 2013). Collectively, these results suggest that development of *WRKY42*-RNAi transgenic plants is a potential technology for developing disease-resistant rice plants.

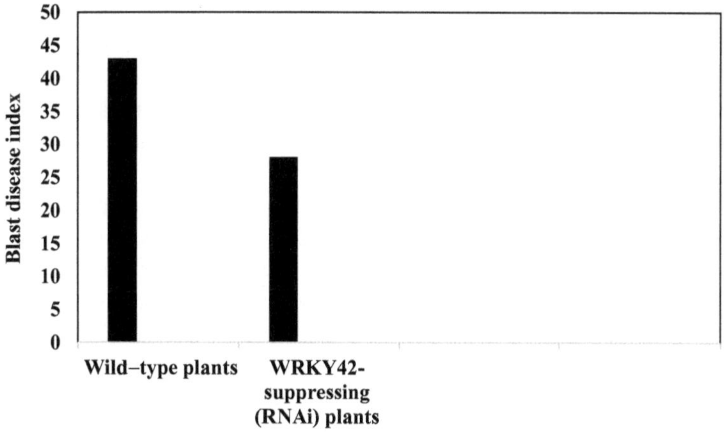

Fig. 6.12 WRKY42-suppressing (*WRKY42*-RNAi) plants show enhanced resistance against the rice blast pathogen *Magnaporthe oryzae* (Adapted from Cheng et al. 2015)

6.10.7 Engineering OsWRKY45 Gene

OsWRKY45 belongs to group III of the WRKY family (Eulgem et al. 2000). WRKY45 is transcriptionally inducible by the plant activator benzothidiazole (BTH) (Shimono et al. 2007). The rice WRKY transcription factor, WRKY45, has been employed to develop disease-resistant plants. Transgenic rice plants overexpressing the transcription factor *WRKY45* showed enhanced resistance to the blast pathogen *M. oryzae* (Shimono et al. 2007). Fungal invasion into rice cells was blocked at most attempted invasion sites (pre-invasive defense) in *WRKY45*-overexpressing (*WRKY45*-ox) rice plants. The leaf blast resistance of *WRKY45*-ox rice plants was much higher than that of other known blast-resistant varieties. *WRKY45*-ox plants also showed strong panicle blast resistance (Shimono et al. 2012). Microscopic analyses demonstrated that the blast resistance in *WRKY45*-overexpressing (*WRKY45*-ox) rice plants is based on a two-layered mechanism: pre-invasive defense, which prevents the invasion of fungal hyphae into rice cells, and post-invasive defense, which accompanies HR cell death (Shimono et al. 2012).

WRKY45 also plays a role in blast resistance mediated by *Panicle blast 1* (*Pb1*), a blast resistance gene with a coiled-coil nucleotide binding leucine-rich repeat structure, through a protein-protein interaction (Inoue et al. 2013). *Pb1* confers durable, broad-spectrum resistance to *Magnaporthe oryzae* races. The Pb1 protein interacted with WRKY45. *Pb1*-mediated panicle blast resistance was largely compromised when *WRKY45* was knocked down in a *Pb1*-containing rice cultivar. Leaf blast resistance by *Pb1* overexpression (*Pb1*-ox) was also compromised in *WRKY45* knockdown/*Pb1*-ox rice plants (Inoue et al. 2013). It was also found that the blast resistance of Pb1 depended on its interaction with WRKY45 in the nucleus (Inoue et al. 2013).

In most cases, enhanced disease resistance conferred by overexpression of transcription factor (TF) genes accompanies growth defects resulting in reduced productivity (Chen and Chen 2002). This is presumably due to constitutive activation of defense responses under the regulation of introduced transcription factors. However, the overexpression of *WRKY45* imposed only relatively small adverse effect on plant growth, presumably due to the primed state established in these plants. The growth defects could be further reduced without major loss of blast resistance by optimizing the level of *WRKY45* expression (Shimono et al. 2007; Goto et al. 2015). The expression of *WRKY45* could be optimized to induce resistance against both the fungal pathogen *Magnaporthe oryzae* and the bacterial pathogen *Xanthomonas oryzae* pv. *oryzae* without affecting agronomic traits by using OsUbi1 promoter to drive the expression of *WRKY45* in rice plants (Goto et al. 2015). These studies suggest that WRKY45 may serve as an effective tool for management of rice diseases.

6.10.8 Engineering *OsWRKY47* Gene

Early response of rice varieties to *M. oryzae* infection included upregulation of WRKY transcription factors. Among these WRKYs, OsWRKY47 was found to play important role in disease resistance. *OsWRKY47* expression was up-regulated 2.4 to 3.3 fold in the two disease-resistant varieties, while the up-regulation was only 1.8 fold in the susceptible variety (Wei et al. 2013). Transgenic rice plants overexpressing *OsWRKY47* were developed. Typical blast lesions with grey center occurred within 7 days after inoculation with *M. oryzae* and the lesions further spread on the leaves of wild-type rice plants. In contrast, no visible lesions were observed on the leaves of transgenic plants (Wei et al. 2013). These results suggest that transgenic plants showed enhanced blast disease resistance. These transgenic plants showed significantly higher expression of the *PR10* gene encoding the pathogenesis-related PR 10 protein (Fig. 6.13; Wei et al. 2013). These studies show that *OsWRKY47* is another potential tool to develop blast disease-resistant rice varieties.

6.10.9 Engineering *OsWRKY53* Gene

OsWRKY53 is a fungal PAMP-induced WRKY gene involved in defense responses in rice (Chujo et al. 2009). Several defense-related genes including pathogenesis-related protein genes such as *PBZ1*, were upregulated in rice cells overexpressing *OsWRKY53* (Chujo et al. 2007). OsWRKY53 has been identified as a substrate of OsMPK3/MPK6. The OsMPK3/MPK6 activated by OsMKK4 phosphorylated OsWRKY53 recombinant protein at its multiple clustered serine-proline residues (SP cluster) (Chujo et al. 2014). OsMKK4-OsMPK3/OsMPK6 cascade has been shown

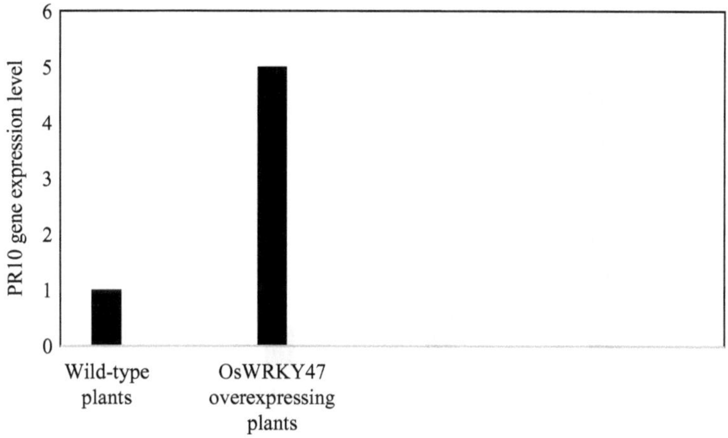

Fig. 6.13 Overexpression of *OsWRKY47* gene enhances expression of the defense-related *PR10* gene in rice plants (Adapted from Wei et al. 2013)

to regulate transactivation activity of OsWRKY53. The enhanced transactivation activity of OsWRKY53 was found to be dependent on phosphorylation in the SP cluster. A phospho-mimic mutant of OsWRKY53, OsWRKY53SD had enhanced transactivation activity (Chujo et al. 2014). The results suggest that phosphorylation of OsWRKY53 within the SP cluster increases its transactivation activity.

Transgenic rice plants overexpressing *OsWRKY53* and *OsWRKY53SD* were developed (Chujo et al. 2007, 2014). Overexpression of *WRKY53* in rice plants resulted in enhanced resistance to the rice blast pathogen *Magnaporthe oryzae* (Chujo et al. 2007). However, overexpression of *OsWRKY53SD*, resulted in further-enhanced disease resistance to *M. oryzae* compared to *WRKY53*-overexpressing rice plants (Fig. 6.14; Chujo et al. 2014). The degree of disease resistance to *M. oryzae* appears to be correlated with the difference in transactivation activity between *OsWRKY53* and *OsWRKY53SD*.

OsWRKY53-overexpressing transgenic rice plants showed increased expression of several defense-related genes, including β-1,3-glucanase, chitinase, and *PR-5*. These defense-related genes were found to be more up-regulated in *OsWRKY53SD*-overexpressing plants compared to *OsWRKY53*-overexpressing rice plants (Chujo et al. 2014). These studies suggest that a modified form of *OsWRKY53* will be more useful in developing disease-resistant plants.

6.10.10 Engineering OsWRKY71 Gene

Another rice transcription factor, OsWRKY71, has been shown to be induced by pathogen infection in rice (Liu et al. 2007a; Chujo et al. 2008). It was also induced by SA, methyl jasmonate, and 1-aminocyclo-propane-1-carboxylic acid (ACC) treatments, suggesting that the transcription factor may act downstream of SA-, JA-,

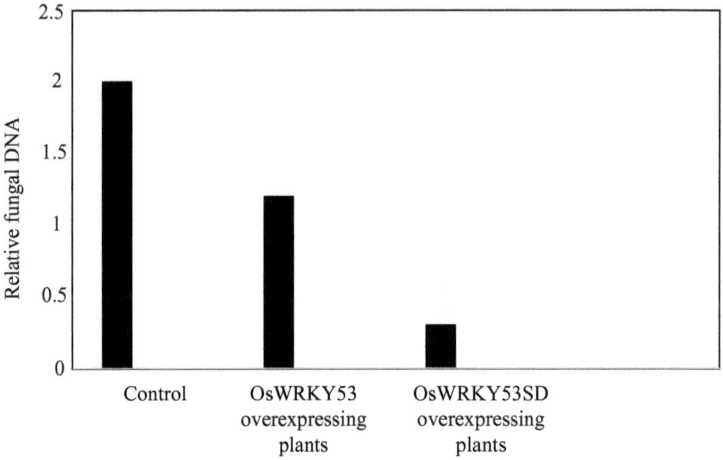

Fig. 6.14 Overexpression of *OsWRKY53SD* confers higher resistance than overexpression of *OsWRKY53* against *Magnaporthe oryzae* in transgenic rice plants (Adapted from Chujo et al. 2014)

and ET-signaling systems. The two marker genes, *OsNPR1* and *OsPR1b*, were constitutively expressed in *OsWRKY71*-overexpressing transgenic rice plants. It suggests that *OsWRKY71* may function upstream of *OsNPR1* and *OsPR1b* in the defense signaling pathway (Liu et al. 2007a). Overexpression of *OsWRKY71* caused up-regulation of several PAMP-induced defense-related genes (Chujo et al. 2008). Transgenic rice plants overexpressing *OsWRKY71* showed enhanced resistance to the bacterial blight pathogen *Xanthomonas oryzae* pv. *oryzae* (Liu et al. 2007a). The OsWRKY71 may be another potential tool in developing disease-resistant plants.

6.10.11 Engineering *OsWRKY89* Gene

Another WRKY gene, *OsWRKY89* was found to play an important role in triggering defense responses against the rice blast pathogen *M. oryzae* (Wang et al. 2007). OSWRKY89 is targeted to nuclei. The 67 C-terminal amino acids of OsWRKY89 are required for the transcriptional activation of the transcription factor and the leucine zipper region at the N-terminus enhances its transcriptional activity. Transgenic rice plants overexpressing *OsWRKY89* were developed and these transgenic plants showed enhanced resistance to the blast pathogen *M. oryzae* (Wang et al. 2007). The transgenic plants showed an increase in SA levels. An increase in lignification was observed in culms of the overexpressor lines (Wang et al. 2007). These results suggest that OsWRKY89 is another potential transcription factor which can be exploited to develop disease resistant plants.

6.11 Bioengineering *WRKY* Transcription Factors for Wheat Disease Management

TaWRKY45 gene encoding a WRKY transcription factor has been isolated from wheat. The wheat TaWRKY45 protein was found to be orthologous to rice OsWRKY45. *TaWRKY45* was constitutively expressed in various organs and throughout the life time of the plant. The *TaWRKY45* gene expression was upregulated in wheat plants inoculated with the head blight pathogen *Fusarium graminearum* (Bahrini et al. 2011b). The gene expression was also upregulated in the SA analog benzothiadiazole (BTH)-treated plants. Transgenic wheat plants overexpressing TaWRKY45 were developed. The transgenic plants overexpressing the *TaWRKY45* transgene showed enhanced resistance against *F. graminearum* (Bahrini et al. 2011b).

The transgenic wheat plants overexpressing *TaWRKY45* also showed enhanced resistance against the powdery mildew pathogen *Blumeria graminis* and the leaf rust pathogen *Puccinia triticina* (Bahrini et al. 2011a). These results suggest that the transcription factor TaWRKY45 gene can be exploited to develop transgenic plants showing resistance against wide range of pathogens.

6.12 Bioengineering *WRKY* Transcription Factors for Tobacco Disease Management

6.12.1 Engineering VvWRKY1 Gene from Vitis vinifera

A transcription factor in grapevine, *VvWRKY*, has been found to be a potential tool to develop disease resistant plants. Both SA and ET treatments induced increases in the level of *VvWRKY* transcripts in grapevine (*Vitis vinifera*) leaves within 2–8 h after treatment (Marchive et al. 2007). The results suggest that the transcription factor is active downstream of the hormone signaling systems. The *VvWRKY* gene was isolated from grapevine berries and transgenic tobacco plants overexpressing *VvWRKY* gene were developed employing the 35S promoter of *Cauliflower mosaic virus* (Marchive et al. 2007).

The transgenic tobacco plants expressing *VvWRKY1* showed resistance to the powdery mildew fungal pathogen *Golovinomyces cichoracearum* (formerly known as *Erysiphe cichoracearum*) (Fig. 6.15) (Marchive et al. 2007). These transgenic tobacco plants were also less affected by the downy mildew pathogen, *Peronospora tabacina* (Fig. 6.16; Marchive et al. 2007). The sporulation of the pathogen was significantly reduced in the transgenic plants. All the transgenic plants also showed enhanced resistance to *Pythium* root rot caused by *Pythium* sp. (Fig. 6.17). The transgenic tobacco plants developed necrosis on 20 % of total root surfaces, whereas the wild-type plants developed necrosis on 60 % of total root surfaces due to infection with *Pythium* sp. (Marchive et al. 2007).

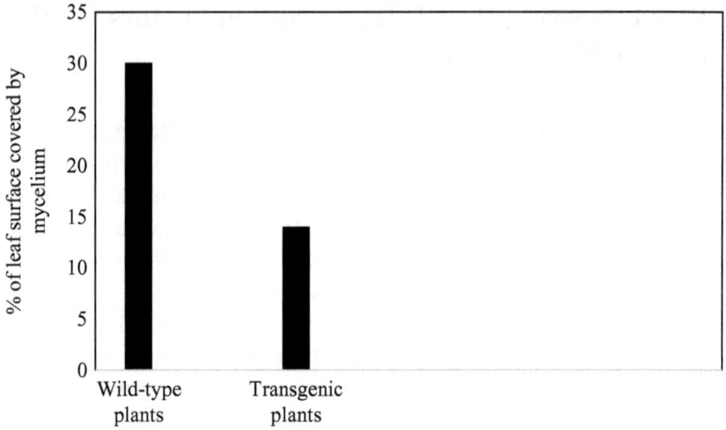

Fig. 6.15 Heterologous expression of grapevine transcription factor *VvWRKY1* in tobacco triggers resistance against the powdery mildew pathogen *Golovinomyces cichoracearum* (Adapted from Marchive et al. 2007)

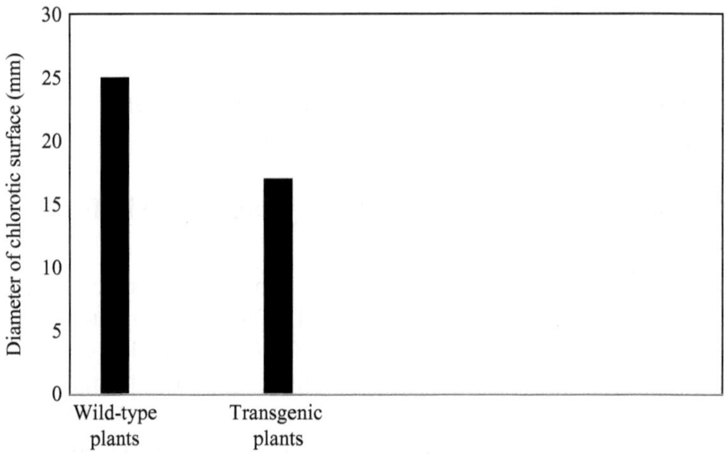

Fig. 6.16 Heterologous expression of grapevine transcription factor *VvWRKY1* in tobacco triggers resistance against the downy mildew pathogen *Peronospora tabacina* (Adapted from Marchive et al. 2007)

It has been widely reported that transgenic plants overexpressing transcription factors show altered leaf morphologies and changes in flowering time (Chen and Chen 2002; Robatzek and Somssich 2002; Li et al. 2004). However, the transgenic tobacco plants overexpressing *VvWRKY1* did not show any phenotypical changes compared with control plants (Marchive et al. 2007). These observations suggest that the transcription factor VvWRKY1 may be useful in developing disease resistant plants.

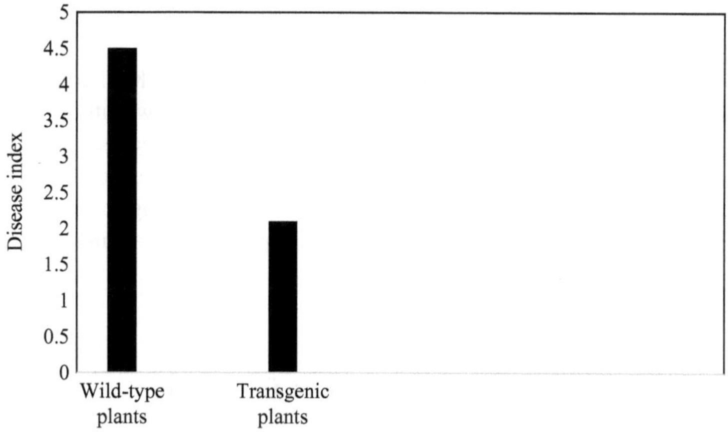

Fig. 6.17 Heterologous expression of grapevine transcription factor *VvWRKY1* in tobacco triggers resistance against *Pythium* root rot (Adapted from Marchive et al. 2007)

6.12.2 Engineering VvWRKY2 Gene from Vitis vinifera

Besides *VvWRKY1*, a full length cDNA was isolated from *Vitis vinifera* and was designated *VvWRKY2*. It encodes a polypeptide of 536 amino acids and shows the structural features of group I of WRKY protein family. The grapevine downy mildew pathogen triggers expression of *VvWRKY2* in infected grapevine leaves (Mzid et al. 2007). Transgenic tobacco plants expressing the grapevine WRKY transcription factor VvWRKY2 were developed. The transgenic tobacco plants expressing the heterologous grapevine *VvWRKY2* gene showed resistance to pathogens infecting different parts of tobacco plants. *Botrytis cinerea* infects leaves, while *Pythium* spp. infect roots and *Alternaria tenuis* infects seeds of tobacco plants. The *VvWRKY2* expressing transgenic tobacco plants showed resistance against all three groups of fungal pathogens (Mzid et al. 2007). The studies suggest that VvWRKY2 is another potential WRKY transcription factor, which can be engineered for developing disease-resistant plants.

6.12.3 Engineering VpWRKY3 Gene from Chinese Wild Grapevine

VpWRKY3 is a transcription factor isolated from Chinese wild grapevine *Vitis pseudoreticulata* (Zhu et al. 2012). The full length cDNA is 1280 bp, encoding a WRKY protein of 320 amino acids. VpWRKY3 specifically accumulated in response to infection by *Erysiphe necator* in *Vitis pseudoreticulata*. SA and ethylene treatments also triggered accumulation of *VpWRKY3* in the wild grapevine leaves. Transgenic tobacco plants expressing VpWRKY3 were developed and these transgenic plants showed resistance against the bacterial pathogen *Ralstonia solanacearum* (Zhu et al. 2012).

Two other WRKY proteins from *V. pseudoreticulata*, VpWRKY1 and VpWRKY2 were found to induce resistance against the powdery mildew pathogen *Erysiphe necator* in *V. pseudoreticulata.* Transgenic *Arabidopsis* plants overexpressing *VpWRKY1* and *VpWRKY2* from *V. pseudoreticulata* showed resistance against *Erysiphe cichoracearum* (Li et al. 2010). These studies suggest that VpWRKY proteins may play an important role in conferring resistance against pathogens in different plants.

6.12.4 Engineering MdWRKY1 Gene from Apple

A WRKY transcription factor gene, *MdWRKY1*, was isolated from the Chinese apple. When the apple plants were infected with the pathogen *Alternaria alternata* f. sp. *mali*, *MdWRKY1* was induced dramatically. Treatment with SA and methyl jasmonate induced transcription of *MdWRKY1* significantly (Fan et al. 2011). Transgenic tobacco plants expressing apple *MdWRKY1* gene were developed. These transgenic tobacco plants showed enhanced resistance to the black shank pathogen *Phytophthora parasitica* var. *nicotianae* (Fan et al. 2011). The results suggest that the serious tobacco black shank disease in tobacco can be managed by developing transgenic plants expressing the heterologous apple transcription factor gene.

6.12.5 Engineering GhWRKY15 Gene from Cotton

GhWRKY15, a member of the WRKY transcription factor family, has been isolated from cotton. The *GhWRKY15* gene was engineered to develop transgenic tobacco plants overexpressing the cotton gene. Transgenic tobacco plants expressing *GhWRKY15* gene showed increased resistance to *Tobacco mosaic virus* (TMV) and *Cucumber mosaic virus* (CMV) infection (Yu et al. 2012). The expression level of the coat protein (CP) gene was much lower in both the TMV- and CMV-inoculated transgenic plants than in the wild-type plants. The TMV and CMV titres were less in the transgenic plants compared to that in the wild-type plants (Figure 6. 18; Yu et al. 2012). Following infection with TMV, the expression of *PR* genes, including *PR1*, *PR2*, *PR4* and *PR5*, was greater in the transgenic tobacco plants overexpressing *GhWRKY15* than the expression in the wild-type plants (Yu et al. 2012). Many WRKY transcription factors have been shown to trigger expression of *PR* genes (Li et al. 2004; van Verk et al. 2008). The promoters of *PR* genes contain W-box sequences that are recognized by WRKY proteins (Yu et al. 2001). Enhanced expression of *NPR1* involved in SA signaling system was observed in TMV-infected transgenic plants. Expression of ACC synthase (ACS) and ACC oxidase (ACO)

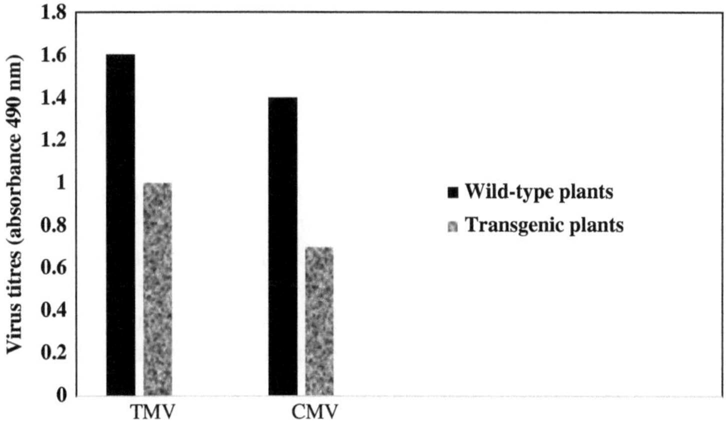

Fig. 6.18 Transgenic tobacco plants overexpressing cotton *GhWRKY15* gene show enhanced resistance to virus diseases (Adapted from Yu et al. 2012)

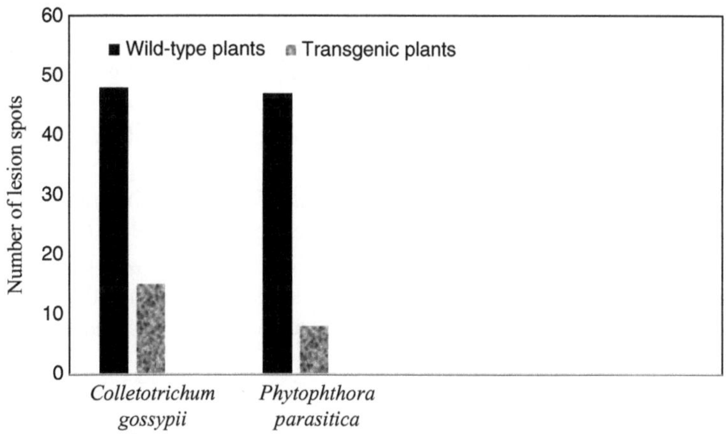

Fig. 6.19 Transgenic tobacco plants overexpressing cotton *GhWRKY15* gene show enhanced resistance to fungal (*Colletotrichum gossypii*) and oomycete (*Phytophthora parasitica*) diseases (Adapted from Yu et al. 2012)

involved in ethylene biosynthesis was also found to be enhanced in the transgenic plants compared to that in wild-type plants (Yu et al. 2012) (Fig. 6.18).

GhWRKY15-overexpressing tobacco plants show enhanced resistance against the fungal pathogen *Colletotrichum gossypii*. Leaves of these transgenic tobacco plants also showed more resistance to the oomycete pathogen *Phytophthora parasitica* (Fig. 6.19; Yu et al. 2012). These results suggest that GhWRKY15 is a potential tool to develop transgenic plants expressing resistance against wide range of pathogens.

6.12.6 Engineering *GhWRKY39-1* and *GhWRKY39* Genes from Cotton

GhWRKY39-1 gene isolated from cotton has been bioengineered to develop transgenic *Nicotiana benthamiana* plants overexpressing *GhWRKY39-1* gene (Shi et al. 2014a). Constitutive overexpression of *GhWRKY39-1* in *N. benthamiana* showed resistance against the bacterial pathogen *Ralstonia solanacearum* and the fungal pathogen *Rhizoctonia solani.* These transgenic plants exhibited elevated mRNA levels of several pathogenesis-related genes, including *PR1c*, *PR2*, and *PR4* (Shi et al. 2014a). Another *GhWRKY39* gene has been isolated from cotton and transgenic *N. benthamiana* plants overexpressing the cotton gene were developed. These transgenic plants also showed enhanced resistance against the bacterial and fungal pathogens (Shi et al. 2014b).

6.12.7 Engineering *GhWRKY44* Gene from Cotton

GhWRKY44, a group I WRKY gene from cotton was isolated and it was found to be localized to the nucleus (Li et al. 2015b). The expression of the gene was found to be induced by pathogen infection. Transgenic *Nicotiana benthamiana* plants expressing the cotton gene were developed. Overexpression of *GhWRKY44* in *N. benthamiana* exhibited enhanced resistance to the bacterial pathogen *Ralstonia solanacearum* and the fungal pathogen *Rhizoctonia solani.* The transgenic plants also showed enhanced expression of defense-related *PR-1*, *PR-2*, *PR-4*, and *PR-5* genes. Expression of *NPR1* gene involved in SA signaling was also induced in transgenic plants (Li et al. 2015b). These results suggest that the cotton gene *GhWRKY44* can be employed to develop disease-resistant *N. benthamiana* plants.

6.12.8 Engineering *CaWRKY27* Gene from *Capsicum annuum*

CaWRKY27 is a subgroup IIe WRKY transcription factor protein in pepper (*Capsicum annuum*) plants. The bacterial wilt pathogen *Ralstonia solanacearum* induced the expression of *CaWRKY27* transcript in a wilt-resistant pepper cultivar. In contrast, *R. solanacearum* infection downregulated the expression of *CaWRKY27* transcript in the susceptible cultivar. *CaWRKY27* transcript levels were also increased by treatments with salicylic acid, methyl jasmonate, ethephon (ethylene) (Dang et al. 2014). The results suggest that the CaWRKY27 transcription factor plays an important role in plant innate immunity. Transgenic tobacco plants overexpressing the pepper *CaWRKY27* gene were developed. These transgenic plants showed increased resistance against *R. solanacearum* (Dang et al. 2014). These studies show that the pepper transcription factor gene is a potential tool for management of diseases of even heterologous plant species.

6.13 Bioengineering WRKY Transcription Factors for Management of Grapevine Diseases

The transcription factor gene, *VvWRKY1*, isolated from *Vitis vinifera* were used to develop transgenic grapevine plants. Genes up-regulated in *VvWRKY1* overexpressors included two lipoxygenase genes (*LOXO* and *LOXA*) involved in JA biosynthesis and two genes encoding proteins similar to JAZ1 and JAZ2, which are transcriptional repressors in JA signaling (Marchive et al. 2013). The transgenic plants showed resistance to the downy mildew pathogen *Plasmopara viticola*. The wild-type grapevine plants showed a disease index of 3.44, while the transgenic plants showed disease index of 2.07, 40 % decrease in the transgenic plants. There was a 36 % decrease of sporulation of the oomycete pathogen on the transgenic line compared to the wild-type plants (Marchive et al. 2013). The results suggest that VvWRKY1 is a potential tool to develop downy mildew-resistant grapevine variety.

6.14 Search for *Arabidopsis* Transcription Factor Genes for Using as Tools for Engineering Disease-Resistant Plants

6.14.1 WRKY33 Transcription Factor

Several transcription factors, which act as positive regulators of defense signaling have been detected in *Arabidopsis thaliana* and these transcription factors have been exploited to develop transgenic plants expressing increased disease resistance. The basic studies have been conducted only with the model plant *Arabidopsis*. However, it has been demonstrated that the gene encoding transcription factors from one plant species can induce defense responses in another plant species when the gene is overexpressed in the transgenic plant (Marchive et al. 2007; Mzid et al. 2007; Li et al. 2010; Fan et al. 2011; Hwang et al. 2011; Yu et al. 2012; Zhu et al. 2012). Hence, the studies with *Arabidopsis* transcription factors may provide powerful tools to develop disease-resistant crop plants.

The *Arabidopsis* transcription factor encoded by *WRKY33* has been shown to play important role in the defense signaling system (Birkenbihl et al. 2012; Liu et al. 2015). *WRKY33* gene encodes a transcription factor with two WRKY domains.

It is expressed at low levels in healthy, uninfected plants, while its transcript levels increased several fold in infected plants. To study the role of *WRKY33* gene in the signaling system, T-DNA insertion alleles of the *WRKY33* gene and transgenic lines constitutively overexpressing *WRKY33* were developed and expression of various defense-related genes was assessed in those plants after infection with the fungal pathogen *Botrytis cinerea* (Zheng et al. 2006). The studies revealed that WRKY33 suppresses expression of various genes involved in SA- signaling system while it enhances expression of genes involved in JA-mediated signaling system (Table 6.4; Zheng et al. 2006). The regulatory gene *PAD4* is a key regulator of SA signaling system (Lippok et al. 2007). The genes encoding PR-1, PR-2, and PR-5 proteins are also associated with SA-regulated defense responses (Yang et al. 1997; van Verk et al. 2008). WRKY33 suppressed the pathogen-induced *PAD4* expression in *Arabidopsis*. It also suppressed the expression of *PR-1*, *PR-2*, and *PR-5* genes as evidenced by their enhanced expression of these genes in *wrky33* mutants after infection by the pathogen (Zheng et al. 2006).

While WRKY33 suppresses SA-mediated signaling system, it induces JA-mediated signaling system. Expression of the gene encoding PR-12 defensin, *PDF1.2*, is specifically induced by JA-mediated signaling system. *PDF1.2* expression was strongly induced in wild-type plants in response to *Botrytis* infection but its induction was significantly reduced in *wrky33* mutant (Zheng et al. 2006). The pathogen-induced expression of the JA-regulated *PR-3* (Lorenzo et al. 2003), chitinase- and osmotin-like genes was reduced or delayed in the *wrky33* mutant (Zheng et al. 2006). These results suggest that the WRKY33 is a positive regulator of these JA-regulated defense genes.

JA-mediated signaling system is known to confer resistance against various necrotrophic fungal pathogens (McGrath et al. 2005; Korolev et al. 2008; Pré et al. 2008). Since WRKY33 is a positive regulator of JA-mediated responses, it may be exploited to develop plants resistant to necrotrophic fungal pathogens. Zheng et al. (2006) developed transgenic *Arabidopsis* plants overexpressing *WRKY33* gene and

Table 6.4 Expression of defense-related genes in *WRKY33* overexpressing and *wrky33* mutant Arabidopsis plants during *Botrytis cinerea* infection

Defense genes	*WRKY33* overexpressing plants	*wrky33* mutant plants
SA-inducible genes:		
PAD4	+	+++
PR-1	–	+++
PR-2	+	+
PR-5	+	++
JA-inducible genes:		
PDF1.2	+++	+
PR-3	+++	+

Adapted from Zheng et al. (2006)

+, ++, +++ indicate increasing intensity of expression of the defense-related genes

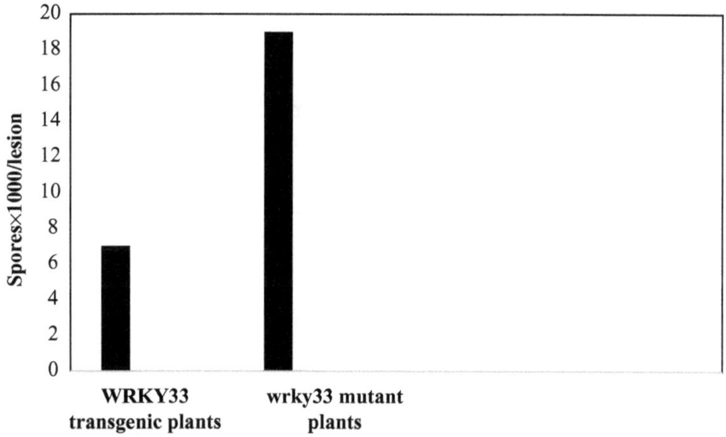

Fig. 6.20 Induced resistance to *Alternaria brassicicola* by overexpression of WRKY33 transcription factor in *Arabidopsis thaliana* (Adapted from Zheng et al. 2006)

showed that these plants expressed enhanced resistance to the necrotrophic fungal pathogen *Alternaria brassicicola* (Fig. 6.20). The transgenic plants also showed reduced fungal growth and symptom development when inoculated with *Botrytis cinerea* (Zheng et al. 2006). These results suggest that *WRKY33* gene can be used to develop disease-resistant crop plants. However, this gene may not be useful to manage heterotrophic bacterial diseases. It has been demonstrated that disease resistance against bacterial diseases can be induced only by activating SA-mediated signaling systems (Zhang et al. 2007; Zheng et al. 2006, 2007; Genger et al. 2008). Since *WRKY33* is a negative regulator of SA signaling system, *WRKY33* gene may be inefficient in the management of bacterial diseases.

6.14.2 WRKY70 Transcription Factor

The *Arabidopsis* transcription factor, WRKY70, is involved in cross-talk between SA and JA signaling pathways. WRKY70 acts as a positive regulator of SA-responsive genes, while it acts as negative regulator of the expression of JA-responsive genes downstream of JA (Li et al 2004, 2006; Journot-Catalino et al. 2006). WRKY70 activates SA-inducible *PR* genes and represses the JA-inducible gene *PDF2.1* (Li et al. 2006). WRKY70-mediated suppression of most of the JA-responsive genes appeared to require functional NPR1 (Li et al. 2006). JA negatively regulates *WRKY70* transcript levels, whereas SA regulates *WRKY70* transcription positively (Li et al. 2004). It has been shown that WRKY70 operates downstream from ROS production and SA accumulation in pathogen-induced defense signaling system (Knoth et al. 2007).

Transgenic *Arabidopsis* plants overexpressing *WRKY70* have been developed. These plants showed high resistance to the powdery mildew pathogen *Golovinomyces cichoracearum* (= *Erysiphe cichoracearum*). However these transgenic plants showed enhanced susceptibility to the necrotrophic fungal pathogen *Alternaria brassicicola* (Li et al. 2006). WRKY 70 was also not effective in inducing resistance against bacterial pathogens (Li et al. 2004, 2006). WRKY70 appears to be a potential tool to manage biotrophic fungal pathogens.

6.14.3 WRKY18 Transcription Factor

A transcription factor, WRKY18, is involved in defense responses to *Pseudomonas syringae* and this transcription factor was found useful in developing disease resistant plants. The transcription factor is rapidly induced in *Arabidopsis* plants upon treatment with SA (Yu et al. 2001). WRKY18 induces expression of *PR* genes. The functional activity of WRKY18 in plant defense response was dependent on NPR1 (Chen and Chen 2002). Transgenic *Arabidopsis* plants constitutively expressing *WRKY18* were developed. The transgenic plants showed constitutive and enhanced expression of *PR-1*, *PR-2*, and *PR-5* genes and enhanced resistance to the bacterial pathogen *P. syringae* at mature/reproductive stages, but not in young seedlings (Fig. 6.21; Chen and Chen 2002). The transgenic lines with high expression of *WRKY18* were severely stunted and the seed set was also markedly reduced. However, the transgenic lines expressing *WRKY18* at low levels grew and developed normally (Chen and Chen 2002). These studies suggest that proper selection of transgenic plants expressing optimum level of the transcription factor gene would help in developing disease-resistant crop plants.

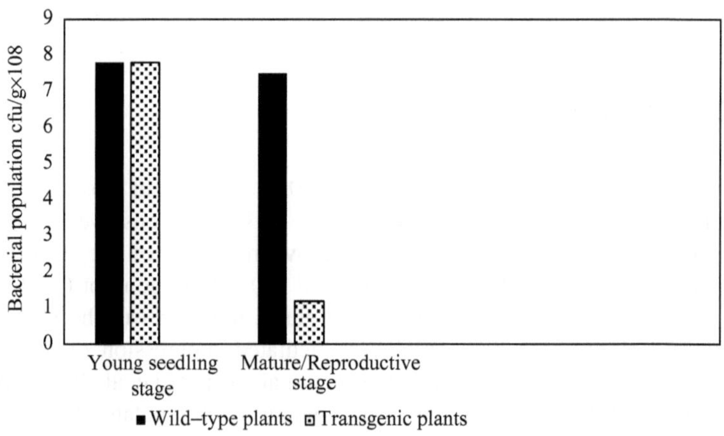

Fig. 6.21 Relative growth of *Pseudomonas syringae* in wild-type and transgenic *Arabidopsis* plants expressing *WRKY18* at seedling and mature/reproductive stage of the plants (Adapted from Chen and Chen 2002)

6.14.4 WRKY29 Transcription Factor

The *Arabidopsis* WRKY29 transcription factor is another positive regulator of plant defense responses. It is up-regulated on recognition of pathogen-associated molecular patterns (PAMPs) by plant pattern-recognition receptors (PRRs). MAPK cascade operates upstream of the transcription factor. Transgenic *Arabidopsis* plants overexpressing *WRKY29* were developed. These plants showed enhanced resistance to both the fungal pathogen *Botrytis cinerea* and the bacterial pathogen *P. syringae* pv. *maculicola* (Asai et al. 2002). It appears that WRKY29 is another tool to develop plants with broad-spectrum disease resistance.

6.14.5 WRKY7 Transcription Factor

Some Arabidopsis WRKY transcription factors negatively regulate defense responses. WRKY7 is a transcriptional repressor and enhances plant susceptibility to *Pseudomonas syringae* (Kim et al. 2006). Transgenic plants that constitutively overexpress *WRKY7* showed reduced expression of defense-related genes, including *PR-1* gene. The WRKY7 might function as a negative regulator of SA signaling system. The loss-of-function T-DNA insertion and RNAi mutants for *WRKY7* were developed. These plants showed enhanced resistance to *P. syringae*. The enhanced resistance in the loss-of function mutants was associated with increased induction of SA-regulated *PR1* gene by the bacterial pathogen (Kim et al. 2006). The results suggest that WRKY7 can be exploited to develop disease resistant plants by developing loss-of-function mutant plants.

6.14.6 WRKY25 Transcription Factor

Another *Arabidopsis* transcription factor, WRKY25, negatively regulates defense responses against the bacterial pathogen *P. syringae. Arabidopsis* constitutively overexpressing *WRKY25* displayed increased disease symptom severity as compared to wild-type plants. T-DNA insertion mutants for WRKY25 developed reduced disease symptoms after infection with the bacterial pathogen (Zheng et al. 2007).

6.14.7 WRKY 48 Transcription Factor

The transcription factor WRKY48 is a negative regulator of resistance to *P. syringae* pv. *tomato* in *Arabidopsis*. Transgenic Arabidopsis plants overexpressing *WRKY48* showed increased susceptibility to the bacterial pathogen. Loss-of-function T-DNA insertion mutant for *WRKY48* showed enhanced disease resistance. The loss-of-function mutants showed increased induction of SA-regulated *PR-1*, *PR-2*, and *PR-5* genes, while the *WRKY48*-overexpressing plants showed highly reduced expression of the *PR* genes (Xing et al. 2008). All these studies reveal a possibility of using the transcription factor genes with negative role in defense response by silencing their function.

6.15 Manipulation of OsWRKY45 Transcription Factor–Dependent Priming Process Using Benzothiadiazole Compounds for Rice disease Management

Several benzothiadiazole compounds have been found useful in manipulation of OsWRKY45 in rice plants for effective fungal and bacterial diseases. These synthetic compounds include acibenzolar-*S*-methyl (ASM; Benzo[1,2,3]-thiadiazole-7-carbothionic acid-*S*-methylester), 3-allyloxy-1,2-benzisothiazole-1,1-dioxide (probenazole) and 3,4-dichloro-*N*-(2-cyanophenyl)-1,2-thiazole-5-carboxamide (tiadinil, TDL) (Shimono et al. 2007; 2012). Acibenzolar-*S*-methyl (ASM) is a benzothiadiazole compound (Lopez and Lucas 2002). Benzothiadiazole (BTH) is a synonym of ASM. BTH was commercialized as a host plant defense inducer in 1989 by Ciba Geigy (Novartis). It was classified and registered as a 'reduced risk compound' in the USA and marketed under the trade name Actigard in USA by Syngenta Crop Protection Inc., Greensboro, NC and BION in Europe by Syngenta Ltd., Basel, Switzerland (Mandal et al. 2008). BTH induces systemic resistance against various pathogens, including fungi, bacteria, and viruses (Lawton et al. 1996; Tally et al. 1999; Oostendorp et al. 2001; Rohilla et al. 2002; Momol et al. 2004; Herman et al. 2008; Mandal et al. 2008; Bengtsson et al. 2009; Hong et al. 2011; Maffi et al. 2011; Shimono et al. 2012; Sillero et al. 2012; Trejo-Saavedra et al. 2013; Barilli et al. 2015). BTH treatment induced resistance against the rice blast pathogen *Magnaporthe oryzae* (Fig. 6.22; Schweizer et al. 1999) by activating *OsWRKY45* gene expression in rice plants (Shimono et al. 2007, 2012). BTH treatment also effectively controlled rice bacterial blight disease caused by *Xanthomonas oryzae* pv. *oryzae* by activating *WRKY45* expression (Shimono et al. 2012).

BTH induces resistance by activating the SA signaling pathway. OsWRKY45 is involved in benzothiadiazole (BTH)-induced disease resistance in rice plants (Shimono et al. 2007, 2012). When BTH is applied to plants at high dosages, it induces constitutive activation of defense responses (van Hulten et al. 2006). By contrast, when BTH is applied at relatively low dosages, the defense responses are not activated immediately but they are activated only after pathogen infection (Conrath et al. 2002). This mode of action is called potentiation or priming (Shimono et al. 2007). The transcription factor OsWRKY45 has been shown to play a crucial role in BTH-induced priming in rice (Shimono et al. 2007).

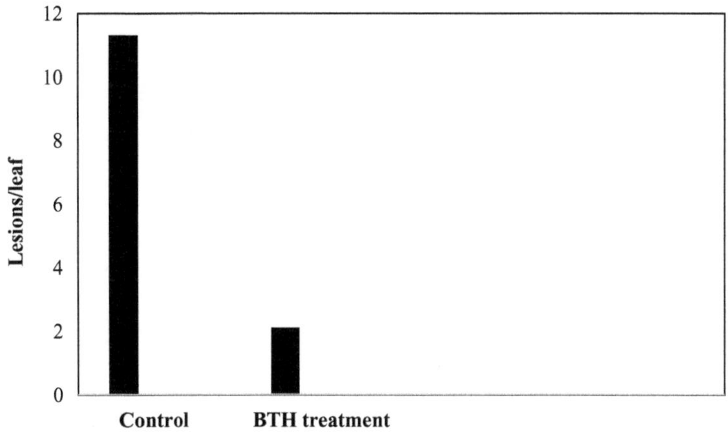

Fig. 6.22 Management of rice blast disease using BTH which enhances Os*WRKY45* gene expression (Adapted from Schweizer et al. 1999)

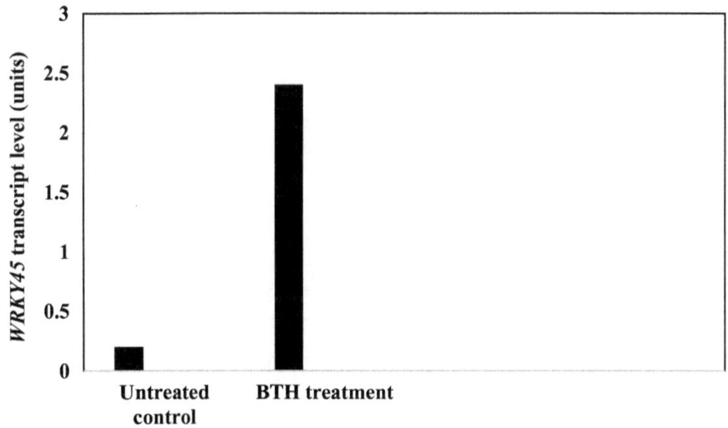

Fig. 6.23 Enhanced expression of *WRKY45* induced by BTH (Adapted from Akagi et al. 2014)

OsWRKY45 is transcriptionally upregulated by BTH (Fig. 6.23; Akagi et al. 2014). Management of rice blast, induced by BTH was markedly reduced in *WRKY45*-knock-down (*WRKY45*-kd) rice plants (Shimono et al. 2012). BTH-induced resistance to *Xanthomonas oryzae* pv. *oryzae* was compromised in *WRKY45*-kd rice plants, whereas *WRKY45*-ox plants were highly resistant to the bacterial pathogen (Shimono et al. 2012). The results indicate that Os*WRKY45* plays an important role in BTH-induced disease resistance.

BTH treatment triggers the expression of Os*WRKY45*, which upregulated several defense-related genes (Fig. 6.24). Nakayama et al. (2013) identified 260 genes as *WRKY45*-dependent BTH-responsive genes in rice plants. Among these genes,

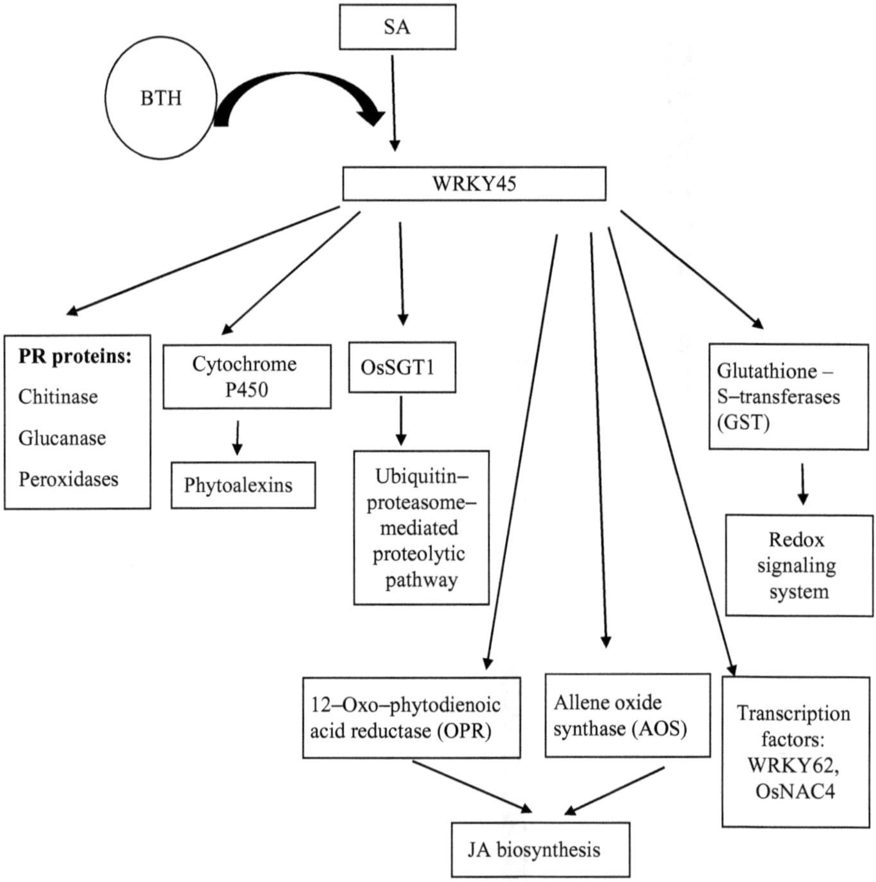

Fig. 6.24 BTH-induced OsWRKY45-dependent defense-related genes (Adapted from Shimono et al. 2007; Nakayama et al. 2013)

220 genes were upregulated in response to BTH and 40 genes were downregulated. Several genes encoding pathogenesis-related (PR) proteins, such as chitinase, glucanase, and peroxidases, and those encoding leucine-rich-repeat-containing putative resistance (R) proteins were found to be OsWRKY45-dependent BTH-responsive genes. Cytochrome P450 genes are involved in biosynthesis of antimicrobial phytoalexins and these genes were also upregulated by OsWRKY45 on rice plants (Nakayama et al. 2013). The OsWRKY45-dependent genes included 11 genes encoding family 1 GTs (UDP-glucuronosyl/UDP-glucosyltransferases) (Nakayama et al. 2013). One of these, salicylic acid glucosyltransferase (OsSGT1) catalyzes the conversion of free SA into SA-O-beta-glucoside (SAG) (Umemura et al. 2009). OsSGT1 is a key mediator of development of chemically induced disease resistance. The *OsSGT1* gene may contribute to the SA signaling mechanism by inducing up-regulation of SAG in rice plants (Umemura et al. 2009).

The rice genome contains 59 genes encoding putative glutathione S-transferases (GST) and 22 of them were found to be upregulated by OsWRKY45 (Nakayama et al. 2013). GSTs are involved in redox signaling involved in SA signaling system (Vidhyasekaran 2015). Genes encoding allene oxide synthase (AOS) and 12-oxophytodienoic acid reductase (OPR) were also upregulated by OsWRKY45 (Nakayama et al. 2013). These genes are involved in JA biosynthesis (Vidhyasekaran 2015). Several genes encoding transcription factors were OsWRKY45-dependently upregulated by BTH. These include *WRKY62*, *OsNAC4*, and *osHSF1* (Nakayama et al. 2013). In rice, WRKY62 negatively regulates basal and *Xa21*-mediated resistance to bacterial blight disease (Peng et al. 2008; Park and Ronald 2012). OsNAC4 is a positive regulator of programmed cell death associated with the hypersensitive reaction (HR) (Kaneda et al. 2009; Taga et al. 2009). OsNAC4 regulates HR cell death in disease resistance induced by the plant activators (Kaneda et al. 2009). OsHSF1 negatively regulates plant cell death by decreasing cellular ROS levels (Yamanouchi et al. 2002; Kojo et al. 2006). OsHSF1 may suppress plant cell death by protecting plant cells from oxidative cellular damage caused by ROS (Nakayama et al. 2013). Necrotic plant cell death may be beneficial for the pathogenesis of the hemibiotrophic rice blast pathogen *Magnaporthe oryzae*. Prevention of the necrotic cell death may be an important defense response against *M. oryzae* (Vidhyasekaran 2007; Nakayama et al. 2013). Collectively these studies suggest the transcription factors regulated by OsWRKY45 induced by BTH may trigger defense responses by modulating hypersensitive cell death and programmed cell death.

HR lesions formed on rice leaves when *M. oryzae* was inoculated onto rice plants pretreated with probenazole (Iwai et al 2007). Post-invasive defense accompanying HR cell death occurred WRKY45-dependently in probenazole–treated rice plants after *M. oryzae* infection (Shimono et al. 2012). HR cell death occurred only after inoculation in probenazole-treated rice (Shimono et al. 2012). OsNAC1 regulates HR cell death through its downstream genes, OsHSP90 and IREN, which are involved in the loss of plasma membrane integrity and nuclear DNA fragmentation, respectively (Kaneda et al. 2009; Taga et al. 2009). It seems that post-translational regulation is required for OsNAC4 to induce HR cell death (Kaneda et al. 2009) and the results are consistent with the lack of HR cell death in *WRKY45*-overexpressing rice plants in the absence of pathogen inoculation (Nakayama et al. 2013).

OsWRKY45 plays a major role in BTH-induced defense program in rice. WRKY45 is constantly degraded by the nuclear ubiquitin proteasome system (UPS) to suppress defense responses in the absence of defense signals. UPS-dependent OsWRKY45 degradation occurs in the nuclei. Transcriptional activity of OsWRKY45 after SA treatment was impaired by proteasome inhibition. The same C-terminal in OsWRKY45 was essential for both transcriptional activity and UPS-dependent degradation (Matsushita et al. 2012). The results suggest that UPS regulation plays a role in suppressing unnecessary defense activation in the absence of pathogens.

Besides BTH, probenazole and tiadinil (3,4-dichloro-*N*-(2-cyanophenyl)-1,2-thiazole-5-carboxamide) also induced WRKY45-dependent systemic resistance against the rice blast pathogen *M. oryzae* (Shimono et al. 2012). Rice resistance to

M. oryzae induced by BTH, tiadinil and probenazole was markedly reduced in *WRKY45*-knockdown (kd) rice plants (Shimono et al. 2012). Probenazole (trade name: Oryzemate) is another host plant defense inducer. It effectively controls rice blast (*Magnaporthe grisea*) and bacterial blight (*Xanthomonas oryzae* pv. *oryzae*) by inducing various defense genes (Oostendorp et al. 2001). Probenazole induces several defense genes. A probenazole-responsive gene, *RPR1*, has been cloned (Sakamoto et al. 1999). RPR1 contains a nucleotide binding site and leucine-rich repeats, thus sharing structural similarity with known disease resistance (R) genes. *RPR1* was induced during the systemic induced resistance (Sakamoto et al. 1999; Yasuda et al. 2006). Tiadinil induces resistance against fungal pathogens of rice (Yasuda et al. 2004; Tsubata et al. 2006). It has been developed as a commercial product by Nihon Nohyaku Co., and is practically used to control rice blast disease (Yasuda et al. 2006).

Collectively these studies suggest that the plant defense activators, which activate OsWRKY45 transcription factor – dependent priming process, may be potential tools for management of rice diseases.

6.16 Manipulation of Priming of WRKY Transcription Factors Using BABA for Crop Disease Management

DL-3-aminobutyric acid (β-aminobutyric acid, BABA) has been found to induce priming of WRKY transcription factors and trigger systemic resistance (Jakab et al. 2001). Foliar spray with BABA led to a significant reduction of lesion development in *Brassica carinata* caused by *Alternaria brassicae* (Chavan and Kamble 2013). In BABA-pretreated *Brassica* plants challenged with *A. brassicae*, priming of expression of *BcWRKY70*, *BcWRKY11* and *BcWRKY53* genes was observed. BABA application potentiated expression of defense genes *PR-1*, *PDF1.2* and *PAL* in response to the pathogen. BABA-primed expression of *BcWRKY70*, *BcWRKY11* and *BcWRKY53* genes was strongly correlated with enhanced expression of the defense genes in *Brassica* (Chavan and Kamble 2013).

BABA primes host plants to activate SA-dependent signaling system (Siegrist et al. 2000; Zimmerli et al. 2000, 2001), JA-dependent signaling system (Hamiduzzaman et al. 2005), ABA-dependent signaling system (Ton and Mauch-Mani 2004), or signaling systems independent of SA, JA, and ET signaling systems (Zimmerli et al. 2000). The BABA-induced priming phenomenon varies depending on the nature of challenging pathogen. It potentiates different signaling systems depending on the type of host-pathogen interactions (Ton and Mauch-Mani 2004). BABA treatment induces lignin accumulation in host cells (Cohen et al. 1999). It activates natural defense mechanisms of the plant, such as callose deposition, the hypersensitive response, and the formation of trailing necroses (Zimmerli et al. 2000).

BABA was found to be effective in controlling various downy mildew diseases. A single foliar spray of BABA applied to 7-day-old cauliflower seedlings protected the plants against the downy mildew pathogen (*Peronospora parasitica*) for at least 15 days (Silué et al. 2002). It controlled grapevine downy mildew caused by *Plasmopara viticola* (Cohen et al. 1999; Daire et al. 2002), tobacco downy mildew caused by *Peronospora tabacina* (Cohen 1994), lettuce downy mildew caused by *Bremia lactucae* (Pajot et al. 2001), sunflower downy mildew (*Plasmopara helianthi*=*P. halstedii*) (Lee et al. 2000), melon and cucumber downy mildew caused by *Pseudoperonospora cubensis* (Ovadia et al. 2000).

BABA was effective in controlling diseases caused by *Phytophthora* spp. It controlled tomato late blight caused by *Phytophthora infestans* (Jeun et al. 2000) and pepper blight caused by *Phytophthora capsici* (Lee et al. 2000). It also protected melons against the wilt pathogen *Fusarium oxysporum* f. sp. *melonis* and reduced disease intensity when applied as soil drench or foliar spray (Ovadia et al. 2000). BABA effectively controlled the rust (*Puccinia helianthi*) disease incidence in sunflower under the field conditions. There was about 60 % reduction in the rust disease incidence in the field trial (Amzalek and Cohen 2007). BABA induced resistance against bacterial head rot (*Pseudomonas marginalis* and *Pseudomonas fluorescens*) of broccoli (Pajot and Silué 2005). BABA induced resistance against *Tobacco mosaic virus* infection in tobacco by inducing accumulation of SA (Siegrist et al. 2000). These studies suggest that BABA may be a potential tool for management of various crop diseases by priming defense responses, probably involving priming of transcription factors.

6.17 Manipulation of *WRKY* Gene Expression Using Ergosterol for Disease Management

Ergosterol is a typical fungal sterol, which is absent in plants. It has been detected in cell walls of most fungi (Kasparovsky et al. 2003, 2004; Laquitaine et al. 2006; Lochman and Mikes 2006). Ergosterol is perceived by plant cells. It triggers production of ROS even at nano-molar concentrations in tobacco and tomato cells. It also activates mitogen-activated protein kinases on alfalfa cells (Lochman and Mikes 2006). The ergosterol induced oxidative burst and it enhanced NADPH oxidase and superoxide dismutase activities (Rossard et al. 2010). Ergosterol treatment triggered a 23-fold increase of *VvWRKY* gene expression in grape plantlets. This treatment elicited defense responses, including ROS production and phytoalexin synthesis and enhanced protection against the gray mold pathogen *Botrytis cinerea* (Laquitaine et al. 2006). These results suggest that it is possible to activate the transcription factor gene using ergosterol and induce resistance by activating the defense responses.

6.18 Manipulation of MYB Transcription Factors for Disease Management

6.18.1 Molecular Manipulation of MYB72 Transcription Factor Using Rhizobacteria to Trigger Priming and ISR for Disease Management

Several MYB (MYELOBLAST) transcription factors have been reported in plants. They are characterized by the presence of a highly conserved MYB domain at their N-termini (Du et al. 2009). MYB transcription factors contain one or more MYB domains (Stracke et al. 2001). MYB proteins are classified into subfamilies depending on the number of conserved repeats of the MYB domain they contain (Mengiste et al. 2003). A common feature of MYB proteins is the presence of a functional DNA binding domain that typically consists of one to three repeats of the MYB domain. The three MYB repeats are referred to as R1, R2, and R3. Each repeat is about 50-53 amino acids long and encodes three α-helices, with the second and third helices forming a helix-turn-helix structure which intercalates in the major groove of DNA when bound to it. MYB repeats typically contain regularly spaced tryptophan residues, which build a central tryptophan cluster in the three-dimensional helix-turn-helix fold (Du et al. 2009). MYB transcription factors bind to the *cis*-regulatory element such as MYB boxes (Laquitaine et al. 2006).

A R2R3-MYB-like transcription factor, MYB72, has been shown to be involved in triggering defense responses against broad-spectrum of pathogens in *Arabidopsis thaliana* (Van der Ent et al. 2008). The nonpathogenic rhizobacterial strain *Pseudomonas fluorescens* WCS417r induces systemic resistance (ISR) in *A. thaliana*. The transcription factor *MYB72* gene was specifically activated in the *A. thaliana* roots upon colonization by *P. fluorescens* WCS417r. T-DNA knockout mutants *myb72* were incapable of mounting ISR against the pathogens, indicating that *MYB72* is involved in activating the defense responses. However, overexpression of *MYB72* did not result in enhanced resistance against the pathogens, suggesting that *MYB72* is not sufficient for the expression of ISR. MYB72 interacts with the Ethylene Insensitive-Like3 (EIL3) transcription factor, suggesting that MYB72 acts in the ET signaling pathway. MYB72 may act upstream of ET. Exogenous application of the ethylene precursor 1-aminocyclopropane 1-carboxylate induced wild-type levels of resistance in *myb72* mutant and *P. fluorescens* activated *MYB72* in ISR-deficient, ethylene-insensitive *ein2-1* plants. These results suggest that an elicitor from *P. fluorescens* WCS417r triggers enhanced expression of *MYB72* and the activated *MYB72* gene may interact with *EIL3* transcription factor, resulting in activation of ET signaling system (Fig. 6.25; Van der Ent et al. 2008). EIL proteins act as the key transcription factors of ET-regulated gene expression and act as positive regulators of ET signaling (Guo and Ecker 2004). They bind to promoters of ET-responsive genes, such as *ERF1* and initiate a transcriptional cascade leading to the expression of ET-targeted genes (Solano et al. 1998). Collectively these results

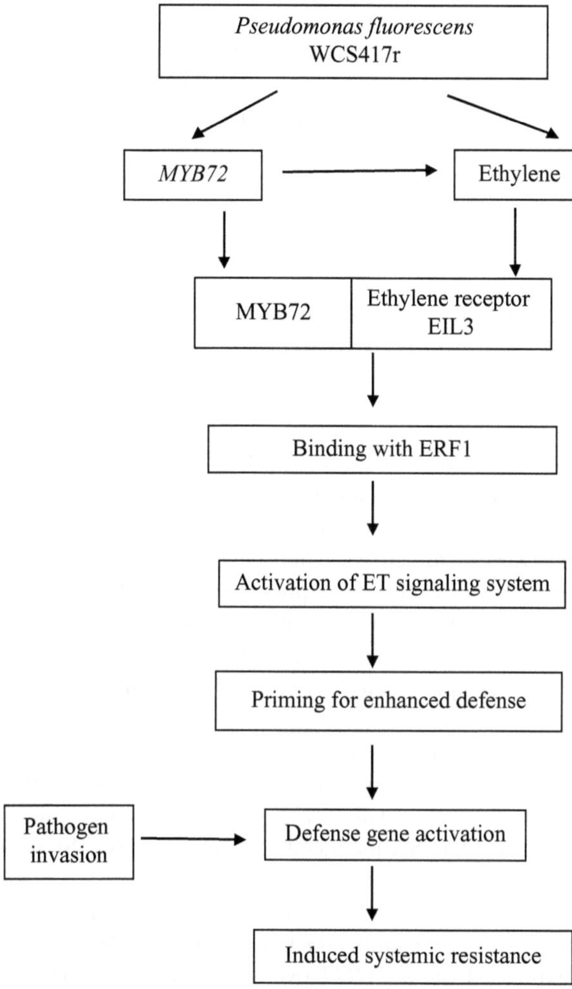

Fig. 6.25 Manipulation of MYB72 to induce systemic resistance against wide range of pathogens using rhizobacteria (Adapted from Van der Ent et al. 2008)

suggest that both *MYB72* and *EIL3* are part of signaling network involved in the plant's defense response.

MYB72-mediated ISR expression was effective in controlling the bacterial pathogens *Pseudomonas syringae* pv. *tomato* and *Xanthomonas campestris* pv. *armoraceae*, the oomycete pathogen *Hyaloperonospora parasitica*, and fungal pathogens *Alternaria brassicicola* and *Fusarium oxysporum* f. sp. *raphani* in *Arabidopsis* (Pieterse et al. 1996; Van Wees et al. 1997; Ton et al. 2002b; Van der Ent et al. 2008). The root colonization by *P. fluorescens* does not induce any significant changes in gene expression in leaves, indicating that the onset of *P. fluorescens*-mediated ISR in the leaves is not associated with a major reprogramming of the transcriptome. However, after challenge inoculation of the induced plants with the pathogen *P. syringae* pv. *tomato*, faster and stronger induction of defense genes was observed in the leaves. In contrast to constitutive activation of defense responses, priming does not require major metabolic changes when no pathogens are present (Van der Ent et al. 2008). Hence, no changes in yield potential may occur in *MYB72*-mediated induced disease resistance. These results suggest that manipulation of *MYB72* will be a potential technology to manage crop diseases.

6.18.2 Bioengineering MYB44 Transcription Factor for Management of Biotrophic/Hemibiotrophic Pathogens

AtMYB44 transcription factor belongs to the R2R3 MYB subgroup 22 transcription factor family in *Arabidopsis thaliana* (Jung et al. 2008). AtMYB44 has been shown to play a role in the plant defense response against *Pseudomonas syringae* pv. *tomato* DC3000 (Zou et al. 2013). Transgenic plants overexpressing *AtMYB44* exhibited greater levels of *PR1* gene expression, cell death, callose deposition and H_2O_2 accumulation in leaves infected with *P. syringae* pv. *tomato* DC3000 (Zou et al. 2013). The transgenic plants overexpressing *AtMYB44* showed enhanced resistance to the bacterial pathogen (Shim et al. 2013; Zou et al. 2013). The *atmyb44* T-DNA insertion mutant was more susceptible to the pathogen (Zou et al. 2013). Transgenic *Arabidopsis* plants overexpressing *AtMYB44* showed activation of *PR* genes and T-DNA insertional knockout mutation resulted in attenuated expression of them (Shim et al. 2013; Zou et al. 2013). AtMYB44 transcription factor has been shown to positively regulate SA-mediated defense responses (Shim and Choi 2013; Zou et al. 2013).

AtMYB44 appears to activate SA signaling system by activating another transcription factor, WRKY70 (Shim and Choi 2013). *WRKY70* was found to be constitutively expressed in *AtMYB44* overexpressing plants (Shim et al. 2013). AtMYB44 upregulates *WRKY70* expression. AtMYB44 binds to the *WRKY* promoter region, indicating that AtMYB44 acts as a transcriptional activator of WRKY70 by directly binding to a conserved sequence element in the *WRKY70* promoter (Shim et al. 2013). WRKY70 acts downstream of SA and *NPR1* in the SA signaling pathway (Knoth et al. 2007) WRKY70 expression is induced by SA. Induction of WRKY70 by SA is reduced in *atmyb44* and *npr1-1* mutants and is totally abolished in *atmyb44 npr1-1* double mutants, showing that WRKY70 is regulated through both NPR1 and AtMYB44. *AtMYB44* overexpression up-regulates SA-responsive *PR1* gene expression (Shim et al. 2013). AtMYB44 up-regulated *WRKY70* and SA-responsive *PR* genes, leading to enhanced resistance to the biotrophic pathogen *P. syringae* pv. *tomato* DC3000 (Fig. 6.26; Shim et al. 2013). These studies showed that AtMYB44 is a potential tool for management of bacterial pathogens.

However, transgenic plants overexpressing AtMYB44 showed susceptibility to necrotrophic fungal pathogen *Alternaria brassicicola* (Shim and Choi 2013). AtMYB44 negatively regulates expression of JA-mediated defense genes such as

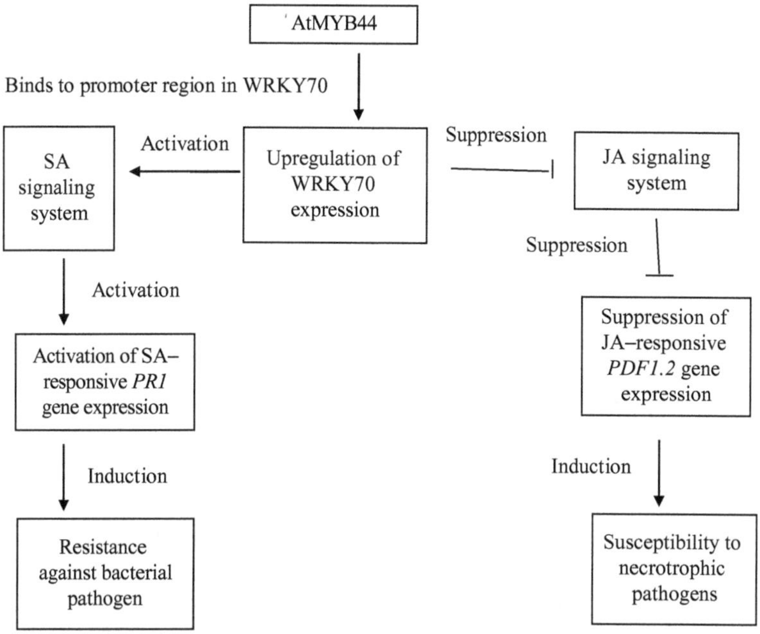

Fig. 6.26 Engineering *AtMYB44* to trigger resistance against *Pseudomonas syringae* pv. *tomato* in *Arabidopsis* but not against necrotrophic pathogens (Adapted from Shim and Choi 2013; Shim et al. 2013; Zou et al. 2013)

VSP1 and *PDF1.2* (Shim et al. 2013) and downregulates resistance to *A. brassicicola* (Shim and Choi 2013). AtMYB44 modulates antagonistic interaction between JA and SA signaling (Shim et al. 2013). AtMYB44 is induced by JA through CORONATINE INSENSITIVE1 (COI1). AtMYB44 upregulates WRKY70 expression and WRKY70 inhibits expression of JA-responsive genes (Li et al. 2004; Shim and Choi 2013; Shim et al. 2013; Fig. 6.26). JA signaling plays important role in conferring resistance against necrotrophic pathogens (Zheng et al. 2006; Méndez-Bravo et al. 2011; An and Mou 2013; Vidhyasekaran 2015) and suppression of JA signaling system may induce susceptibility. These studies suggest that AtMYB44 may be only useful for management of biotrophic/hemibiotrophic pathogens, against which SA signaling system plays an important role in conferring resistance (Thomma et al. 2001; Xu et al. 2013; Amzalek and Cohen 2007; El Rahman et al. 2012).

6.18.3 *Bioengineering OsJAMyb for Rice Blast Disease Management*

OsJAMyb, a R2R3-type MYB transcription factor, was used to develop disease-resistant rice plants. The MYB transcription factor gene *OsJAMyb* has been cloned from rice plants. It encodes a protein with 283 amino-acid residues. The expression of

OsJAMyb was more rapidly and strongly induced in the seedlings of a blast-resistant variety compared with the blast-susceptible variety after inoculation with the blast pathogen *Magnaporthe oryzae* (Cao et al. 2015). Transgenic rice plants overexpressing *OsJAMyb* were developed and these transgenic plants showed increased resistance to blast infection. Less lesion number was observed in the transgenic plants compared to wild-type plants when inoculated with *M. oryzae* (Cao et al. 2015). These results suggest that the MYB transcription factor is another potential tool to develop disease-resistant plants.

6.19 Molecular Manipulation of MYC2 Transcription Factor Using Rhizobacteria to Trigger Priming and ISR for Disease Management

MYC2 (*MYELOCYTOMATOSIS2*), which is also known as *JIN1* (*JASMONATE* INSENSITIVE1) is a nuclear localized basic helix-loop-helix-leucine zipper – type transcription factor (Boter et al. 2004; Lorenzo et al. 2004). It is a transcription factor belonging to the 133-member bHLH (basic helix-loop-helix) family characterized by its bHLH domain responsible for DNA binding and for the formation of homo-and/or heterodimers with other bHLH proteins (Wager and Browse 2012). It plays a predominant role in JA signaling pathway, activating transcription of the JA-responsive *PDF1.2* encoding the PR-12 defensin protein (Lorenzo and Solano 2005; Takahashi et al. 2007). MYC2-regulated genes also include *CHIB/PR3* and *HEL/PR4* and they are negatively regulated by MYC2 (Anderson et al. 2004; Lorenzo et al. 2004). MYC2 negatively regulates Trp and Trp-derived secondary metabolism such as indole glucosinolate biosynthesis (Dombrecht et al. 2007). MYC2 positively regulates the JA-responsive genes *VSP* and *LOX* (Lorenzo et al. 2004; Pozo et al. 2008). It is suggested that MYC2 probably acts through the transcriptional orchestration of other transcription factors, which in turn regulate downstream JA response genes involved in diverse JA-dependent plant processes (Dombrecht et al. 2007).

MYC2 plays an important role in inducing defense responses against pathogens and also against insect attack (Anderson et al. 2004; Lorenzo et al. 2004; Dombrecht et al. 2007). The *myc2* mutant Arabidopsis plants show increased resistance to fungal pathogens such as *Plectosphaerella cucumerina*, *Botrytis cinerea*, and *Fusarium oxysporum* (Anderson et al. 2004; Lorenzo et al. 2004) and the bacterial pathogen *Pseudomonas syringae* (Nickstadt et al. 2004; Laurie-Berry et al. 2006). MYC2 orthologs act as 'master switches' that regulate JA-mediated biosynthesis of secondary metabolites (Kazan and Manners 2013). *ATMYC2* has been shown to function as a downstream factor of the MKK3-MPK6 cascade in JA signaling (Takahashi et al. 2007a). Some MAP kinase cascades are involved in jasmonic acid (JA) signaling system. The *Arabidopsis* MKK3-MPK6 cascade is involved mainly in JA signaling system. This cascade negatively regulates *ATMYC2* function (Takahashi et al. 2007). It also functions as transcriptional activator in abscisic acid signaling (Abe et al. 2003). MYC2 regulates crosstalk between the signaling pathways of JA and those of other phytohormones such as ABA, SA, GA, and auxin (Kazan and Manners 2013). MYC2 is required for induced systemic resistance (ISR) triggered by beneficial soil microbes (Kazan and Manners 2013).

Several rhizobacterial strains have been shown to induce systemic resistance against a wide range of pathogens (Pieterse et al. 2002; Yan et al. 2002; Iavicoli et al. 2003; Akram et al. 2008; De Vleesschauwer et al. 2008; Vidhyasekaran 2015). Besides basal resistance responses that act at the site of pathogen infection, plants are also capable of developing systemic resistance, called "Induced Systemic Resistance (ISR)" that is effective against future pathogen attack (Bostock 2005). The induced resistance does not necessarily require a direct activation of defense responses but can also result from a faster and stronger expression of basal defense responses upon pathogen attack. This enhanced capacity to express infection-induced basal defenses is called "priming" (Conrath et al. 2006; Conrath 2009). The priming results in a faster and stronger induction of defense mechanisms after pathogen attack (Jung et al. 2009; Conrath 2011; Martinez-Medina et al. 2013; Po-Wen et al. 2013).

The rhizobacterium *Pseudomonas fluorescens* WCS417r triggered ISR in *Arabidopsis* (Pieterse et al. 1998). It elicited a substantial change in the expression of almost 100 genes locally in roots (Van der Ent et al. 2008). However, systemically in the leaves, no alteration in gene expression was observed, suggesting that the onset of ISR is not associated with detectable changes in gene expression (Verhagen et al. 2004). Analysis of the transcriptome of ISR-expressing *Arabidopsis* after challenge inoculation with the bacterial pathogen *Pseudomonas syringae pv. tomato* DC3000 revealed 81 genes with amplified expression patterns, indicating that the plants were primed to respond rapidly and strongly to the pathogen attack (Verhagen et al. 2004).

The MYC2 transcription factor has been shown to be involved in *P. fluorescens* WCS417r-induced priming to trigger ISR (Pozo et al. 2008). Colonization of roots of *Arabidopsis* plants by *P. fluoresens* WCS417r triggered expression of *MYC2* in ISR – expressing leaves. *MYC2* transcript levels were consistently raised in WCS417r-ISR expressing plants (Fig. 6.18; Pozo et al. 2008). The *npr1-1* mutant plants were unable to mount an ISR response (Pieterse et al. 1998). In these mutant plants, the level of *MYC2* mRNA was not elevated upon colonization of the roots by WCS417r (Pozo et al. 2008). These results suggest that colonizarion of roots by ISR-inducing WCS417r bacteria induces consistently enhanced expression of *MYC2* gene (Fig. 6.27).

The CACATG motif was found to be overrepresented in the promoter region of the 81 ISR-primed *P. syringae* pv. *tomato* (challenge-inoculated pathogen)-responsive genes (Pozo et al. 2008). The G-box-related motif CACATG was the only one that was significantly more overrepresented in the ISR-primed genes (Pozo et al. 2008).

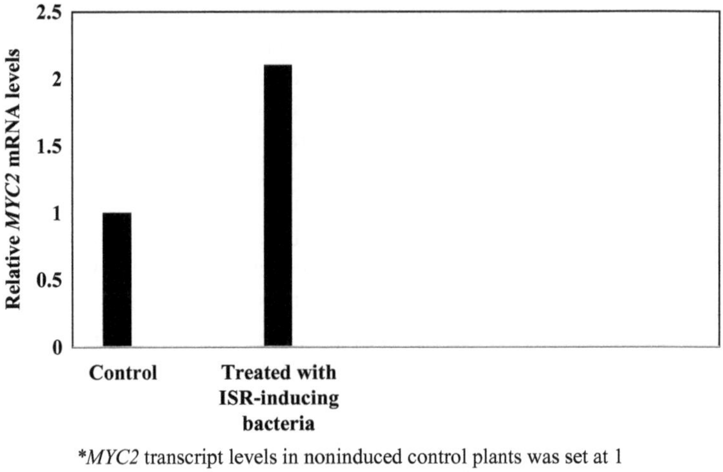

Fig. 6.27 ISR-inducing *Pseudomonas fluorescens* WCS417r strain triggers induction of MYC2 expression in *Arabidopsis* plants (Adapted from Pozo et al. 2008)

The CACATG has been reported to be a binding site for the MYC2 transcription factor (Abe et al. 1997). WCS417r-ISR resulted in a significant level of protection against both the bacterial pathogen *P. syringae* pv. *tomato* and the oomycete pathogen *Hyaloperonospora parasitica* in Arabidopsis. However, MYC2-impaired *jin1*mutants failed to develop ISR against these pathogens (Pozo et al. 2008). Priming for enhanced deposition of callose-containing papillae at sites of attempted penetration of *H. parasitica*, a typical reaction of WCS417r-ISR-expressing *Arabidopsis* plants, was significantly reduced in WCS417r-treated MYC2-impaired mutant *jin1-2* plants (Pozo et al. 2008; Van der Ent et al. 2008). Collectively these studies suggest that the rhizobacteria can activate *MYC2* expression and trigger priming and ISR.

P. fluorescens WCS417r has been developed as a biocontrol agent for management of several crop diseases. This strain applied on roots protects leaves from bacterial, fungal and oomycete pathogens in the model plant *A. thaliana*. It induced resistance against the oomycete pathogen *Hyaloperonospora parasitica* (Ton et al. 2002a), the bacterial pathogen *Pseudomonas syringae* pv. *tomato* and the fungal pathogen *Fusarium oxysporum* f. sp. *raphani* in the model plant *A. thaliana* (Pieterse et al. 1996, 1998). *P. fluorescens* WCS417r induced ISR against *Fusarium oxysporum* f. sp. *raphani* in radish (Hoffland et al. 1995). This strain triggered ISR against *Colletotrichum lindemuthianum* in bean (Meziane et al. 2005) and *Fusarium oxysporum* f. sp. *dianthi* in carnation (Van Peer et al. 1991). *P. fluorescens* WCS417r was found to induce resistance against *Fusarium oxysporum* f. sp. *lycopersici* in tomato (Duijff et al. 1998).

6.20 Molecular Manipulation of bZIP Transcription Factors for Crop Disease Management

6.20.1 Molecular Manipulation of TGA Class of bZIP Transcription Factors for Crop Disease Management

The basic leucine zipper (bZIP) proteins belong to a large family of transcription factors. The bZIP family transcription factors contain a basic region for binding DNA and a leucine zipper dimerization domain (Jakoby et al. 2002). These proteins have a conserved region rich in basic amino acid residues that binds to the target DNA and contains nuclear localization signals (NLSs) and dimerization or multimerization domains. A leucine zipper region, which consists of several heptad repeats of hydrophobic residues, is found close to the basic region. The leucine zipper region is alpha-helical and prone to dimer formation via a coiled-coil arrangement (Kuhlmann et al. 2003; Meng et al. 2005).

The TGA class of transcription factors is the important group of bZIP transcription factors involved in defense signaling. Members of the TGA family of transcription factors bind tandem repeats of a *cis*-element within the promoters called activation sequence-1 (*as-1*), which contains a TGACG motif (Lebel et al. 1998; Després et al. 2000). Several TGA transcription factors have been shown to regulate expression of defense-related genes (Kesarwani et al. 2007). This family of transcription factors recognizes the TGACG/*as-1* elements found in the promoters of a variety of plant genes, including those regulating the expression of *Arabidopsis* and tobacco *PR-1* and the *Cauliflower mosaic virus* 35S promoter (Lebel et al. 1998; Kim and Delaney 2002). The *as-1* elements are responsible for SA responsiveness of these promoters. SA treatment increases the TGACG/*as-1* binding activity. Thus, the TGA transcription factors may play an important role in SA signaling system (Zhou et al. 2000). Several SA-responsive genes are regulated by bZIP transcription factors of TGA family (Ndamukong et al. 2007). Some of the bZIP transcription factors, such as TGA2 and TGA5 in *Arabidopsis*, interact with NPR1 and recognize the *as-1 cis* element found within the promoter of several *PR* genes (Kim and Delaney 2002).

Transgenic plants overexpressing different TGA transcription factor genes have been generated to develop disease resistant plants (Kim and Delaney 2002; Fitzgerald et al. 2005). Transgenic *Arabidopsis* plants containing sense or antisense *TGA5* gene constructs were developed by Kim and Delaney (2002). None of the *TGA5* sense lines showed an apparent increase in *TGA5* transcript levels compared to wild-type plants, whereas the *TGA5*-antisense lines showed a large increase in *TGA5* transcript accumulation. Increased *TGA5* accumulation in antisense lines may be due to negative autoregulation of the *TGA5* gene (Kim and Delaney 2002). The transgenic *TGA5*- antisense lines showed reduced induction of SA-mediated

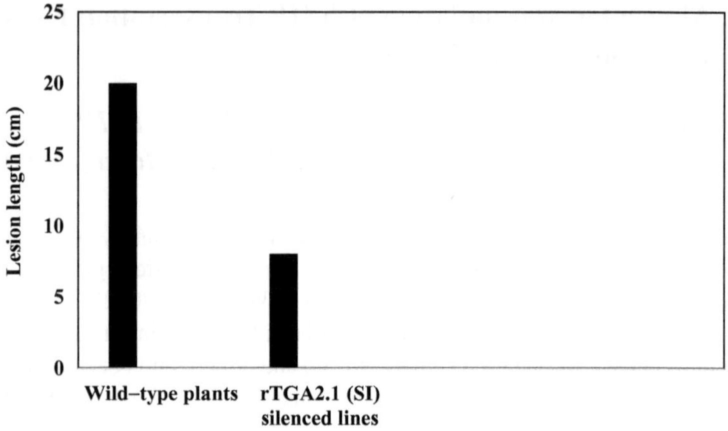

Fig. 6.28 rTGA2.1 (SI) silenced lines show enhanced resistance against the rice bacterial blight pathogen *Xanthomonas oryzae* pv. *oryzae* (Adapted from Fitzgerald et al. 2005)

expression of *PR-1* gene by the oomycete pathogen *Hyaloperonospora parasitica.* The transgenic antisense lines showed enhanced resistance to *H. parasitica* (Kim and Delaney 2002). The induced resistance by *TGA5* to the pathogen has been suggested to act independent of SA signaling system.

The rice TGA factor, rTGA2.1, has been shown to bind to defense gene promoters (Chern et al. 2001). It binds to oligonucleotides containing the *as-1* like elements from the *PR-1* gene promoter and to the promoter of the rice chitinase gene, *RCH10* (Chern et al. 2001). It appears that *rTGA2.1* negatively regulates a subset of rice defense genes (Fitzgerald et al. 2005). Transgenic rice plants that have the endogenous rTGA2.1 transcripts silenced via dsRNA-mediated silencing (SI) were generated. The loss of *rTGA2.1* activity in the SI lines resulted in reduced rice bacterial blight disease symptom development (Fig. 6.28; Fitzgerald et al. 2005).

Transgenic rice plants expressing a putative dominant-negative (DN) rTGA2.1 mutant that can no longer bind DNA were developed to exploit the negatively acting *rTGA2.1* gene for rice disease management (Fitzgerald et al. 2005). The transgenic plants showed reduced bacterial blight (*Xanthomonas oryzae* pv. *oryzae*) symptom development (Fig. 6.29; Fitzgerald et al. 2005). The resistance observed in T1 plants was stable and inheritable and the resistance was also observed in the T2 plants (Fitzgerald et al. 2005). However, the transgenic plants with reduced rTGA2.1 were reduced in both overall height and total biomass when compared with wild-type plants (Fitzgerald et al. 2005). The transgenic lines were developed using constitutive maize ubiquitin promoter and probably use of an inducible promoter would have reduced the adverse effect of these transgenic plants.

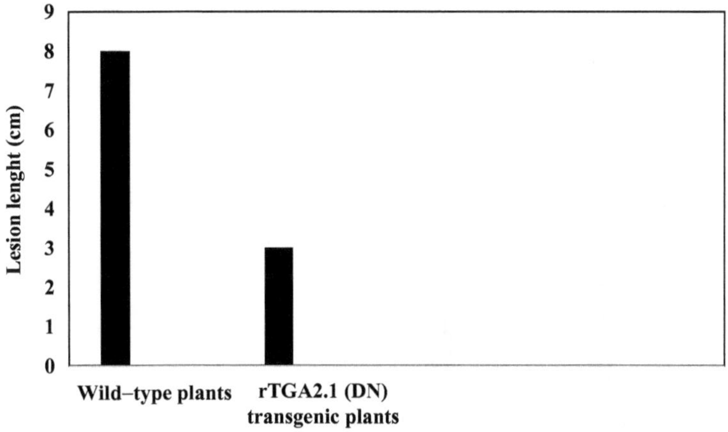

Fig. 6.29 Transgenic rice plants expressing rTGA2.1 (DN) show resistance to the rice bacterial pathogen *Xanthomonas oryzae* pv. *oryzae* (Adapted from Fitzgerald et al. 2005)

6.20.2 Molecular Manipulation of RF2a and RF2b bZIP Transcription Factors for Rice Tungro Virus Disease Management

Rice tungro disease (RTD) is an important yield constraint in all rice-growing areas (Bunawan et al. 2014). Rice tungro disease results from coinfection by *Rice tungro bacilliform virus* (RTBV) and *Rice tungro spherical virus* (RTSV) (Hibino et al. 1979). RTBV is the causative agent of RTD symptoms, whereas RTSV is required for disease transmission (Hull 1996). RTBV is a double-stranded DNA virus with a single, tissue-specific promoter that is expressed primarily in phloem tissues (Dai et al. 2006). Transcription of the RTBV DNA genome is regulated by a promoter located in the intergenic region between ORF IV and ORF I. RTBV accumulates in vascular tissues and activity of the RTBV promoter is restricted to vascular tissues (Yin et al. 1997a, b; Petruccelli et al. 2001). The promoter from RTBV is expressed only in phloem tissues in transgenic rice plants (Petruccelli et al. 2001). The phloem-specific promoter of RTBV is regulated in part by sequence-specific DNA-binding proteins that bind to BOX II, an essential *cis* element (Dai et al. 2004, 2006).

Rice transcription factors RF2a and RF2b bind to Box II, a *cis* element adjacent to the TATA box, and control gene expression from the promoter (Zhu et al. 2002; Dai et al. 2003, 2006; Ordiz et al. 2010). BOX II was found to be essential to the activity of the RTBV promoter. BoxII and its interactions with cognate rice transcription factors, including RF2a and RF2b, have been shown to be essential to the activity of the RTBV promoter and probably they may be involved in expression of the RTBV genome during virus replication in the plants (Dai et al. 2003, 2006). The C–terminal domains of both RF2a and RF2b transcription factors have been shown to activate transcription from the RTBV promoter (Liu et al. 2007b).

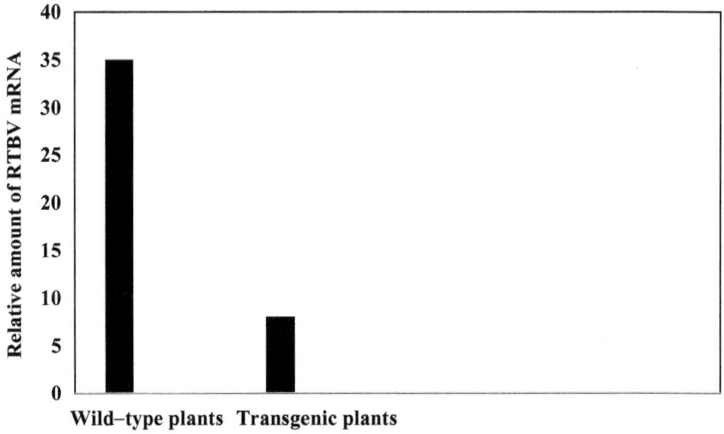

Fig. 6.30 Reduced accumulation of RTBV transcripts in transgenic rice plants with elevated levels of RF2a or RF2b after *Agrobacterium* inoculation with RTBV (Adapted from Dai et al. 2008)

RTBV appears to use the host transcription factors RF2a and RF2b during replication. The transgenic plants that produce a dominant negative mutant of the transcription factors in vascular tissues exhibited phenotypes similar to rice tungro disease symptoms. The transcription factor mutant plants were stunted and leaves were yellow-orange in color, which are characteristic symptoms of the tungro disease (Dai et al. 2008). The results suggest that stunting and leaf discoloration exhibited in rice tungro disease are caused by the lack of availability of RF2a and RF2b as a consequence of infection by RTBV.

Transgenic rice plants overexpressing RF2a or RF2b were developed. These transgenic plants were highly tolerant to virus infection and did not develop usual symptoms of the disease. The transgenic plants contained less virus than nontransgenic plants (Fig. 6.30; Dai et al. 2008). These results suggest that the transcription factors may be potential tools for developing virus-resistant plants.

6.20.3 Manipulation of Pepper bZIP Transcription Factor for Developing Disease-Resistant Plants

A pepper bZIP transcription factor gene, *CabZIP2*, has been isolated from pepper plants (Lim et al. 2015). Transcription of *CabZIP2* was found to be induced in pepper plants inoculated with the bacterial pathogen *Xanthomonas campestris* pv. *vesicatoria*. SA treatment triggered the expression of the transcription factor. Methyl jasmonate and ethylene also induced the expression of *CabZIP2* (Lim et al. 2015). *CabZIP2*- silenced pepper plants showed susceptible reaction to *X. campestris* pv. *vesicatoria*. The susceptibility was accompanied by reduced expression of defense-related genes (Lim et al. 2015). These results suggest that the *CabZIP2* gene can be exploited for engineering for disease resistance. Transgenic *Arabidopsis* plants overexpressing the pepper *CabZIP2* gene were developed and the transgenic plants showed enhanced disease resistance against the bacterial pathogen *Pseudomonas syringae* pv. *tomato* DC3000 (Lim et al. 2015). The bZIP transcription factor seems to be a potential biotechnological tool for management of bacterial diseases.

6.21 Manipulation of EREBP Transcription Factors for Crop Disease Management

6.21.1 EREBP Transcription Factor Family

Several AP2/EREBP (APETELA2 [AP2]/ETHYLENE RESPONSIVE ELEMENT BINDING PROTEIN [EREBP]) transcription factors have been found to be important biotechnological tools for developing disease-resistant plants. The AP2/EREBP superfamily is classified into the AP2 and EREBP subfamilies based on sequence similarities and the number of DNA-binding domains. AP2 subfamily members are involved in plant development, whereas EREBP subfamily members are involved in defense signaling system (Riechmann and Meyerowitz 1998). The EREBP subfamily contains two subgroups, DREB (Dehydration-responsive element-binding) and ERF (Ethylene response factor) subgroups. Among them, the ERF subgroup contains all of the AP2/EREBP genes that have been linked to disease resistance responses (Gutterson and Reuber 2004). ERFs contain a single DNA-binding domain and they bind specifically to the GCC box that is found in several promoters of the defense-related genes (Gutterson and Reuber 2004). ERFs act as both positive and negative regulators of defense responses (Koyama et al. 2003; McGrath et al. 2005; Oñate-Sánchez et al. 2007; Liu et al. 2012). The important ERF transcription factors involved in defense signaling system include OsBIERF3 and OsEREBP1 in rice (Cao et al. 2006; Jisha et al. 2015), TaERF1 and TaERF3 in wheat (Xu et al. 2007; Zhang et al. 2007), HvRAF in barley (Jung et al. 2007), Pti4 and Pti5 in tomato (Zhou et al. 1997), ERF2, ERF3, and ERF4 in tobacco (Koyama et al. 2003), ERF2 and ERF4 in canola (*Brassica napus*) (Yang et al. 2007), CaPF1 and CaERFLP 1 in pepper (Lee et al. 2004; Yi et al. 2004) and AtERF1, AtERF2, AtERF4, and AtERF14 in *Arabidopsis* (McGrath et al. 2005; Grennan 2008), Several ERF transcription factor genes have been employed to develop disease-resistant plants.

6.21.2 Pti5 Transcription Factor

The transcription factor Pti5 isolated from tomato belongs to the EREBF family of transcription factors and binds specifically to the GCC-box *cis* element present in the promoters of several defense genes. It is expressed specifically during pathogen infection and has a specific role in plant defense against pathogens (He et al. 2001). Transgenic tomato plants overexpressing *Pti5* were developed. These transgenic plants were morphologically indistinguishable from nontransgenic plants. The transgenic plants showed increased accumulation of transcripts of the defense genes *GluB* and *Catalase* when challenged with the pathogen (He et al. 2001). They showed enhanced resistance to *Pseudomonas syringae* pv. *tomato*, showing reduced disease lesions and leaf bacterial numbers as compared with nontransgenic tomato plants (He et al. 2001). The induced resistance was not comparable with the race-specific resistance observed in plants. However, the advantage with the *Pti5*-induced resistance is that the induced resistance is general resistance induced in susceptible plants lacking *R* genes.

6.21.3 Pti4 Transcription Factor

Pti4 is another tomato transcription factor belonging to ERF subgroup of EREBP family. It regulates the expression of GCC-box *cis* element present in many PR genes. It activates the SA-regulated *PR-1* and *PR-2* genes. It also activates JA- and ET-regulated genes, such as *PR3*, *PR4*, *PDF1.2*, and *Thi1.2*. The PR genes *PR3*, *PR4*, and *PDF1.2* are ET inducible, while *PDF1.2* also is inducible by JA besides ET. *Thi2.1* is induced by JA but not by ET (Gu et al. 2002). These results suggest that Pti4 acts through SA, JA, and ET signaling pathways. Transgenic *Arabidopsis* plants expressing the tomato transcription factor *Pti4* showed enhanced resistance to the powdery mildew pathogen *Erysiphe orontii* and increased tolerance to the bacterial pathogen *Pseudomonas syringae* pv. *tomato* (Gu et al. 2002). These studies suggest that the transcription factor genes can be expressed even in heterologous background. These results also suggest that these genes might be useful in engineering diverse plant species for increased broad-spectrum disease resistance.

6.21.4 GbERF2 Transcription Factor

An ERF transcription factor gene, *GbERF2*, was cloned from cotton and used to develop transgenic tobacco plants for management of tobacco diseases. *GbERF2* transcripts accumulated when the plants were exposed to exogenous ethylene treatment and *Verticillium dahliae* infection. Transgenic tobacco plants expressing the cotton transcription factor gene were developed (Zuo et al. 2007). The transgenic tobacco plants constitutively accumulated high levels of transcripts of *PR-1b*, *PR2*, and *PR4* genes. These plants showed enhanced resistance to the necrotrophic fungal pathogen *Alternaria longipes*. The *GbERF2* gene was not useful in conferring resistance to the heterotrophic bacterial pathogen *Pseudomonas syringae* pv. *tabaci* (Zuo et al. 2007). These results suggest that *GbERF2* acts through ET-signaling system and confers resistance against necrotrophic pathogens. The studies also suggest that the cotton gene can be expressed in heterologous plants to manage necrotrophic pathogens.

6.21.5 NtERF5 Transcription Factor

A transcription factor belonging to ERF family, NtERF5 has been isolated from tobacco. *NtERF5* transcription is activated in tobacco by bacterial and viral pathogens (Fischer and Dröge-Laser 2004). Transgenic tobacco plants overexpressing *NtERF5* have been generated. The transgenic plants expressed resistance to *Tobacco mosaic virus* (TMV). The virus did not spread systemically from primary infected leaves to leaves in the upper part of the transgenic plants. TMV was observed in systemically infected leaves of wild-type plants but not in the transgenic plants (Fischer and Dröge-Laser 2004). The results suggest that the amount of TMV or systemic movement of the virus, or both, is reduced in *NtERF5*-overexpressing transgenic plants (Fischer and Dröge-Laser 2004). The transgenic plants showed no constitutive *PR* gene expression. Some other NtERF5-regulated genes may be involved in suppressing viral replication and the transcription factor-activated signaling genes are yet to be identified. The studies of Fischer and Dröge-Laser (2004) have demonstrated that the transcription factor gene can be used to develop virus-resistant plants. Constitutive expression of *NtERF5* under control of the 35S promoter resulted in no visible alterations in plant growth in tobacco and hence this technology can be tried to manage virus diseases without any deleterious effect on yield characters of crops.

6.21.6 Tsi1 Transcription Factor

The tobacco transcription factor NtERF5 conferred resistance to virus disease but was ineffective in inducing resistance to bacterial diseases. Park et al. (2001) could identify an EREBP/AP2-type transcription factor gene, *Tsi1* (for *Tobacco stress-induced gene1*) which triggered resistance against bacterial diseases. Overexpression of *NtERF5* did not result in constitutive expression of *PR* genes (Fischer and Dröge-Laser 2004). By contrast, overexpression of *Tsi1* in transgenic tobacco activated the expression of several *PR* genes, such as *PR1*, *PR2*, *PR3*, osmotin, and *SAR8.2* (Park et al. 2001). The transgenic tobacco plants overexpressing *Tsi1* gene showed resistance to the bacterial pathogen *P. syringae* pv. *tabaci*. Growth of the bacteria in the transgenic plants overexpressing *Tsi1* was inhibited threefold to fivefold when compared with that in the wild-type plants. Although several lesions developed on leaves of the transgenic plants, the size and density of the lesions were much smaller than those on wild-type plants (Park et al. 2001).

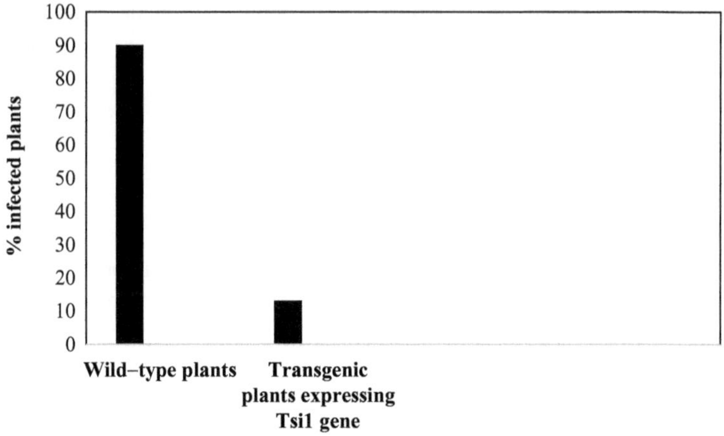

Fig. 6.31 Transgenic hot pepper plants expressing *Tsi1* gene show enhanced resistance against *Phytophthora capsici* (Adapted from Shin et al. 2002)

Transgenic hot pepper (*Capsicum annuum*) plants expressing the tobacco transcription factor gene *Tsi1* were generated (Shin et al. 2002). Overexpression of *Tsi1* in transgenic hot pepper plants triggered constitutive expression of *PR-1*, *PR-2*, *PR-4*, *PR-5*, and *PR-10* genes. The transgenic plants also showed constitutive expression of the defense genes *PinII* (*proteinase inhibitor II*), *LTP1* (*lipid transfer protein1*), and *SAR8.2* (*systemic acquired resistance 8.2*) (Shin et al. 2002). The transgenic plants showed enhanced resistance to the oomycete pathogen *Phytophthora capsici* (Fig. 6.31; Shin et al. 2002).

The transgenic plants also showed resistance to the bacterial spot disease caused by *Xanthomonas campestris* pv. *vesicatoria*. The transgenic plants had a few visible symptoms; however, control plants showed browning, withering, and yellowing of the large area due to the bacterial infection. The transgenic plants also showed resistance to *Pepper mild mottle virus* (PMMV) and *Cucumber mosaic virus* (CMV) (Shin et al. 2002). Collectively, these studies have revealed that the transcription factor Tsi1 can function in wide-range of plants and confer resistance against wide-spectrum of pathogens.

6.21.7 OsBIERF3 Transcription Factor

An EREBP transcription factor gene (*OsBIERF3*), which is induced by infection with the blast fungus *Magnaporthe oryzae* has been cloned from rice plants. Transgenic tobacco plants overexpressing the rice *OsBIERF3* gene were developed. The transgenic plants showed elevated expression of *PR-1a* gene (Cao et al. 2006). The transgenic tobacco plants showed resistance against the wild fire disease caused

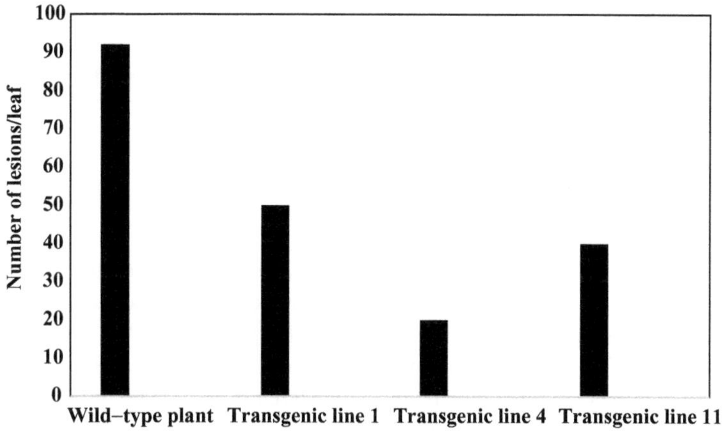

Fig. 6.32 Variability in enhanced disease resistance against *Tomato mosaic virus* in different transgenic tobacco lines overexpressing the rice *OsBIERF3* gene (Adapted from Cao et al. 2006)

by *Pseudomonas syringae* pv. *tabaci*. The lesion size was smaller and the bacterial populations in the leaf tissues were reduced in the transgenic plants as compared with the wild-type plants (Cao et al. 2006).

The tobacco plants overexpressing the rice OsBIERF3 showed resistance to *Tomato mosaic virus* (ToMV). The lesion size on the leaves of the wild type plants was around 2.5 mm in diameter, but the lesion size on the leaves of the transgenic plants was only about 1.0 mm in diameter. The lesion numbers on the leaves were also lower in the transgenic plants than that in wild-type plants. However, the disease resistance against ToMV in the various transgenic lines varied significantly. The transgenic line 4 showed almost 80 % reduction in disease intensity, while the transgenic line showed only 30 % reduction in disease intensity (Fig. 6.32; Cao et al. 2006). Some transgenic lines had single copy of *OsBIERF3* gene, while others had two or more copies of the transgene in their genome. These observations suggest that proper selection of transgenic lines is necessary to develop disease resistant cultivars.

6.21.8 OsERF922 Transcription Factor

Rice OsERF922, an AP2/ERF type transcription factor was rapidly and strongly induced by *Magnaporthe oryzae* infection in rice plants. It was found to be localized to the nucleus. It negatively regulates induction of defense responses in rice plants. The expression of OsERF922 was knocked down by means of RNAi (Liu et al. 2012). The RNAi plants showed enhanced resistance against *M. oryzae*. The elevated disease resistance of the RNAi plants was associated with increased expression of *PR*, *PAL*, and other genes encoding phytoalexin biosynthetic enzymes and without *M. oryzae* infection (Liu et al. 2012). These results suggest that the ERF transcription factor genes, which negatively regulate defense responses, can also be used for developing disease-resistant plants by using RNAi technology.

6.21.9 CaPF1 Transcription Factor

An ERF transcription factor (*CaPF1* [*Capsicum annuum* pathogen and freezing tolerance-related protein 1]) was isolated from hot pepper (*Capsicum annuum*) leaves. Transgenic tobacco plants expressing the pepper *CaPF1* were generated (Yi et al. 2004). The transgenic plants constitutively expressed *PR-2*, *PR-3*, *PR-4*, and *PR-5* genes. These transgenic plants showed resistance to the wild fire pathogen *P. syringae* pv. *tabaci* (Yi et al. 2004). The *CaPF1* gene was also able to confer freezing tolerance in *Arabidopsis* (Yi et al. 2004). These studies indicate that this gene may be useful in developing crop plants with tolerance to both biotic and abiotic stress in plants.

6.21.10 OPBP1 Transcription Factor

An AP2/EREBP-like transcription factor, OPBP1 (osmotin promoter binding protein 1) has been isolated from tobacco. The *OPBP1* gene was induced by elicitor treatment. Both ethephon and methyl jasmonate treatments triggered the expression of *OPBP1* gene in tobacco. Transgenic tobacco plants overexpressing *OPBP1* gene accumulated high levels of *PR-1a* and *PR-5d* genes (Guo et al. 2004). These studies suggest that the transcription factor *OPBP1* may operate through ET/JA signaling pathway. The transgenic tobacco plants overexpressing *OPBP1* showed enhanced resistance to the black shank disease caused by *Phytophthora parasitica* var. *nicotianae* and also the wild fire pathogen *P. syringae* pv. *tabaci* (Guo et al. 2004). The transgenic plants overexpressing *OPBP1* gene also exhibited tolerance to salt stress (Guo et al. 2004). Thus the transcription factor may contribute to resistance to wide-spectrum of pathogens and also to abiotic stress.

6.21.11 HvRAF Transcription Factor

An ERF-type transcription factor, HvRAF, has been isolated from young seedlings of barley. *HvRAF* expression was induced in barley seedlings by treatment with SA, ET, and JA (Jung et al. 2007). Transgenic *Arabidopsis* plants overexpressing the barley *HvRAF* gene were developed. Several defense-related genes including *PR-1*,

PR-5, and *PDF1.2* were activated in the transgenic plants. The transgenic plants also showed enhanced resistance to the bacterial pathogen *Ralstonia solanacearum*. The transcription factor conferred tolerance to high salinity, besides conferring resistance to the pathogen (Jung et al. 2007). These studies suggest that the *HvRAF* transcription factor gene may be a potential tool to manage crop diseases and abiotic stresses.

6.21.12 ERF1 Transcription Factor

ERF1 is an *Arabidopsis* ERF transcription factor, which acts as a regulator of ethylene responses after pathogen attack in *Arabidopsis* (Berrocal-Lobo et al. 2002). ERF1 appears to be a downstream component of ET and JA defense responses. Transgenic *Arabidopsis* plants overexpressing *ERF1* showed enhanced resistance to necrotrophic fungal pathogens *Botrytis cinerea* and *Plectosphaerella cucumeris* (Berrocal-Lobo et al. 2002). These transgenic plants also showed enhanced resistance to the wilt pathogens *Fusarium oxysporum* f. sp. *conglutinans* and *F. oxysporum* f. sp. *lycopersici* (Berrocal-Lobo and Molina 2004). *Arabidopsis* resistance to *F. oxysporum* appears to require the ET, JA, and SA signaling pathways and the *NPR1* gene (Berrocal-Lobo and Molina 2004). In contrast, transgenic *Arabidopsis* plants overexpressing *ERF1* did not increase tolerance against the bacterial pathogen *Pseudomonas syringae* pv. *tomato* (Berrocal-Lobo et al. 2002). Collectively, these results suggest that the ERF1 transcription factor can be exploited to develop crop plants with enhanced resistance to broad-spectrum of necrotrophic fungal pathogens.

6.21.13 OsEREBP1 Transcription Factor

OsEREBP1, a rice AP2/ERF transcription factor has been used for developing disease-resistant rice plants (Jisha et al. 2015). Transgenic rice plants constitutively expressing *OsEREBP1*gene driven by maize *ubiquitin* promoter were developed. Overexpression of *OsEREBP1* caused increased expression of several genes related to JA and ABA biosynthesis. The transgenic rice plants showed enhanced resistance against the bacterial blight pathogen *Xanthomonas oryzae* pv. *oryzae* (Fig. 6.33; Jisha et al. 2015). A 50–100 fold decrease in the bacterial population was observed in the transgenic lines compared to the susceptible wild-type plants. Constitutive expression of *OsEREBP1* in rice driven by maize *ubiquitin* promoter did not affect normal plant growth (Jisha et al. 2015). The results suggest that the osEREBP1 is a potential tool for engineering for disease resistance.

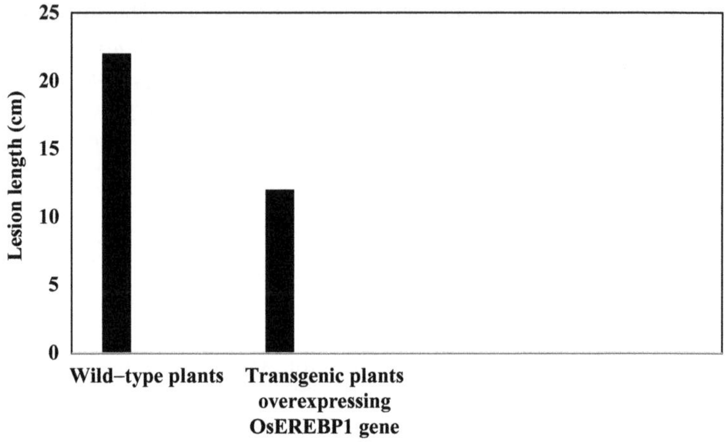

Fig. 6.33 Transgenic rice plants overexpressing *OsEREBP1* show enhanced resistance against rice bacterial blight (Adapted from Jisha et al. 2015)

6.22 Manipulation of NAC Transcription Factors for Crop Disease Management

6.22.1 NAC Transcription Factors in Plant Defense Responses

NAC transcription factors belong to one of the largest families of plant-specific transcription factors playing important role in plant defense responses (Nakashima et al. 2007; Chen et al. 2013; McLellan et al. 2013; Nuruzzaman et al. 2013; Liu et al. 2014; Shan et al. 2015; Sun et al. 2015; Wang et al. 2015). The name 'NAC' is derived from three genes that were initially discovered to contain a particular domain (the NAC domain): NAM (for No Apical Meristem), ATAF1/2 (for Arabidopsis thaliana Activating Factor1/2), and CUC2 (for Cup-Shaped Cotyledon) (Chen et al. 2013; Nuruzzaman et al. 2013; Garapati et al. 2015). Several NAC genes have been detected in rice, soybean, tobacco, grape, citrus, poplar and Arabidopsis (OOka et al. 2003; Rushton et al. 2008; Hu et al. 2010; Nuruzzaman et al. 2010; Le et al. 2011; Nuruzzaman et al. 2013; Sun et al. 2015). Several studies predicted presence of 152 NAC-transcription factors in potato (Singh et al. 2013), 147 in grapevine (Wang et al. 2013), 110 in apple (Su et al. 2013), 145 in *Gossypium raimondii* (Shang et al. 2013), 74 in rice (Nuruzzaman et al. 2010), and 180 in *Serratia italica* (Puranik et al. 2013).

Several NAC transcription factors have been shown to be involved in plant-pathogen interactions. Rice seedlings infected with *Rice stripe virus* (RSV) and *Rice tungro spherical virus* (RTSV) showed increased expression of six *OsNAC* genes (Nuruzzaman et al. 2010). ATAF2, the NAC domain transcription was induced in response to *Tobacco mosaic virus* (TMV) infection in *Arabidopsis thaliana*

(Wang et al. 2009b). Expression of *CaNAC1* was preferentially associated with incompatible interactions between chili pepper and pathogens (Oh et al. 2005). Changes in expression of the rice NAC transcription factor gene *OsNAC19* have been reported in rice plants infected by the blast pathogen *M. oryzae* (Lin et al. 2007). Transcription of OsNAC111 was up-regulated after *M. oryzae* inoculation (Yokotani et al. 2014). Rice transcription factors *ONAC122* and *ONAC131* expression was induced after infection by *M. oryzae* (Sun et al. 2013). Expression of both the genes was faster and higher in the incompatible interaction compared with the compatible interaction. Expression of both the genes was also induced by treatment with SA, methyl jasmonate, or 1-aminocyclopropane-1-carboxylic acid (a precursor of ethylene). Plants silenced for *ONAC122* or *ONAC131* expression showed enhanced susceptibility and expression of various defense-related genes was also down-regulated in these *ONAC122* or *ONAC131* silenced plants (Sun et al. 2013). These results suggest that the rice NAC transcriptional factors play important role in plant defense responses against *M. oryzae*.

NAC transcription factors have been shown to be involved in defense signaling against viral pathogens (Selth et al. 2005; Yoshii et al. 2009; Donze et al. 2014). The Arabidopsis NAC transcription factor ATAF2 was found to be induced in response to *Tobacco mosaic virus* (TMV) infection. Transgenic *Arabidopsis* plants overexpressing *ATAF2* showed resistance to TMV and the overexpression of *ATAF2* significantly reduced virus accumulation (Wang et al. 2009b). The transgenic plants showed increased expression of the defense-related genes *PR1*, *PR2*, and *PDF1.2* (Wang et al. 2009b). The NAC transcription factor, GRAB (Geminivirus RepA Binding), alters Geminivirus multiplication (Xie et al. 1999). *Turnip crinkle virus* (TCV) interacts with a transcription factor, TIP (TCV-Interacting Protein), of *Arabidopsis thaliana*, via its coat protein (CP) (Donze et al. 2014). Viral coat proteins are considered as the viral elicitors involved in PAMP-triggered immunity (Culver and Dawson 1991; Culver et al. 1994; Bendahmane et al. 1995; Gilardi et al. 2004; Matsumoto et al. 2008; Vidhyasekaran 2014). The interaction between coat protein of *Turnip crinkle virus* and TIP transcription factor has been shown to correlate with the disease resistance response in *Arabidopsis* (Donze et al. 2014). A *Turnip crinkle virus* mutant with mutated coat protein failed to interact with TIP transcription factor. The CP-mutant TCV caused more severe symptoms in the susceptible Arabidopsis ecotype. Loss of ability of coat protein to bind TIP transcription factor results in inability of the mutant virus to suppress the defense responses (Donze et al. 2014). The results suggest the TIP transcription factor plays an important role in virus disease resistance and the viral coat protein may inhibit the basal immune responses to virus invasion by binding to the NAC transcription factor TIP.

Many NAC transcription factors have been shown to regulate defense genes either positively or negatively. *TaNAC8* gene is involved in plant defense response against stripe rust pathogen in wheat (Xia et al. 2010a, b). The NAC transcription factors ANACO19 and ANACO55 regulate JA-signaled defense responses in *Arabidopsis* (Bu et al. 2008). OsNAC111 positively regulates the expression of defense-related genes in rice (Yokotani et al. 2014). The HvNAC6 transcription factor has been shown to be a positive regulator of *B. graminis* f. sp. *hordei* penetration resistance in barley and *Arabidopsis* (Jensen et al. 2007). The *Arabidopsis* ATAF1, a NAC transcription factor, is a negative regulator of defense responses against necrotrophic fungal and bacterial pathogens (Wang et al. 2009a). ATAF1 negatively regulates resistance to *Botrytis cinerea* in *Arabidopsis* (Wu et al. 2010). A NAC domain protein interacts with tomato leaf curl virus replication accessory protein and enhances viral replication (Selth et al. 2005). ATAF2, the NAC domain transcription factor was found to suppress systemic host defense against *Tobacco mosaic virus* in *Arabidopsis* (Wang et al. 2009b). ATAF1 has been shown to attenuate ABA signaling in *Arabidopsis* and triggers resistance against the fungal pathogen *Blumeria graminis* f. sp. *tritici* (Jensen et al. 2008). ABA suppresses the fungal penetration resistance in *Arabidopsis*. ATAF1 plays important role in suppression of action of ABA, which is involved in inducing susceptibility (Jensen et al. 2008). The pathogen inoculation resulted in decreased endogenous ABA levels in an ATAF1-dependent manner and the ABA biosynthetic mutant *aao3* showed increased penetration resistance to *B. graminis* f. sp. *hordei* (Jensen et al. 2008). The results suggest that the NAC transcription factor ATAF1 mediates efficient penetration resistance in *Arabidopsis* upon *B. graminis* f. sp. *hordei* attack by attenuating ABA signaling, which is otherwise involved in inducing susceptibility.

A tomato NAC gene *SlSRN1* (*Solanum lycopersicum Stress-related NAC1*) positively regulates defense responses (Liu et al. 2014). Expression of *SlSRN1* was significantly induced by infection with the fungal pathogen *Botrytis cinerea* or the bacterial pathogen *Pseudomonas syringae* pv. *tomato* DC3000, Expression of *SlSRN1* was also induced by SA, JA and 1-amino cyclopropane-1-carboxylic acid (Liu et al. 2014).

Some NAC transcription factors have dual roles as positive or negative regulators of response to different pathogens (Seo et al. 2010). ATAF1 and its barley homolog HvNAC6 enhanced resistance to *Blumeria graminis* f.sp. *hordei* (Chen et al. 2013) but attenuated resistance to *Botrytis cinerea*, *Alternaria brassicicola* and *Pseudomonas syringae* (Wang et al. 2009a; Wu et al. 2010).

6.22.2 Engineering NAC Transcription Factors for Disease Management

OsNAC111 transcription factor was engineered to develop disease-resistant plants (Yokotani et al. 2014). Transgenic rice plants overexpressing *OsNAC111* showed increased resistance to the rice blast pathogen *Magnaporthe oryzae*. The transgenic plants showed increased expression of two chitinase genes and one β-1,3-glucanase gene and the expression of these genes showed reduced expression in transgenic rice plants in which *OsNAC111* function was suppressed by a chimeric repressor (*OsNAC111-SRDx*). OsNAC111 activated transcription factor promoters of the chitinase and β-1,3-glucanase genes in rice cells (Yokotani et al. 2014). These results suggest that the transcription factor OsNAC111 is another biotechnological tool to develop disease-resistant rice plants.

Another rice NAC transcription factor, OsNAC6 was also engineered to develop disease resistant plants. Transgenic rice plants over-expressing *OsNAC6* exhibited increased tolerance to the rice blast disease (Nakashima et al. 2007). However, these transgenic plants over-expressing *OsNAC6* constitutively exhibited growth retardation and low grain yield (Nakashima et al. 2007). Probably, use of inducible promoter may help to reduce the adverse effect of constitutive expression of NAC genes in the transgenic plants.

6.22.3 Manipulation of NAC Transcription Factor Genes for Crop Disease Management

Some of the NAC transcription factors negatively regulate plant disease resistance and such transcription factors have been exploited for inducing resistance by gene silencing. TaNAC1 acts as a negative regulator of stripe rust (*Puccinia striiformis* f. sp. *tritici*) resistance in wheat (Wang et al. 2015). *TaNAC1* was strongly expressed in wheat roots in response to infection by the stripe rust pathogen. SA, methyl jasmonate, and ethylene treatments also triggered expression of *TaNAC1* (Wang et al. 2015). TaNAC1 overexpression in *Arabidopsis thaliana* plants showed enhanced susceptibility and attenuated systemic-acquired resistance (SAR) to *Pseudomonas syringae* pv. *tomato* DC3000. *TaNAC1* overexpression also suppressed the expression levels of resistance-related genes *PR1* and *PR2* (Wang et al. 2015). These results suggest that the *TaNAC1* is a negative regulator of defense responses.

To exploit the *TaNAC1* gene for management of wheat stripe rust, the gene TaNAC1 was knocked down by using *Barley stripe mosaic virus*-induced gene silencing method. The *TaNAC1* gene-silenced wheat plants showed enhanced resistance against the stripe rust pathogen (Wang et al. 2015). The results suggest silencing of the negative regulator transcription factors may be a useful strategy in developing disease-resistant plants.

6.22.4 NAC Transcription Factor Enhances ABA Biosynthesis and Promotes Disease Resistance

Barley HvNAC6 acts as a positive regulator of basal resistance in barley against the powdery mildew pathogen *Blumeria graminis* f. sp. *hordei*. The induction of disease resistance depends on HvNAC6 - modulated abscisic acid (ABA) signaling system (Chen et al. 2013), *HvNAC6* expression was silenced using RNA interference (RNAi). The transgenic *HvNAC6* RNAi plants displayed reduced *HvNAC6* transcript levels. These *HvNAC6* RNAi barley plants were more susceptible to *B. graminis* f. sp. *hordei* than wild-type plants. Application of exogenous ABA increased basal resistance against the powdery mildew pathogen in wild-type plants.

However, application of ABA did not increase the basal resistance against the pathogen in *HVNAC6* RNAi plants (Chen et al. 2013). These results suggest that ABA-induced disease resistance depends on the NAC transcription factor HvNAC6.

ABA is derived from oxidative cleavage of the 11,12 double bond of 9-*cis* epoxy carotenoids (neoxanthin and/or violaxanthin). The 9-*cis*-epoxycarotenoid dioxygenase (NCED) enzyme cleaves 9-*cis* xanthophylls to xanthoxin (Tan et al. 2003; Wasilewska et al. 2008). Six genes encoding NCED have been identified. The T-DNA insertion adjacent to *9-cis-epoxycarotenoid dioxygenase5* (*NCED5*) caused a massive increase in transcript level and enhanced ABA levels (Fan et al. 2009). The NCED is the first committed step for ABA synthesis (Schwartz et al. 2003; Tan et al. 2003; Chen et al. 2006; Wasilewska et al. 2008; Fan et al. 2009). The expression of *HvNCED1* and *HvNCED2* was compromised in the *HvNAC6* RNAi barley lines (Chen et al. 2013). The results suggest that HvNAC6 is involved in triggering accumulation of ABA, which promotes basal resistance against the powdery mildew pathogen. Exogenous application of ABA appears to be a potential strategy to manipulate NAC transcription factor for crop disease management.

6.23 Engineering *NtWIF* Transcription Factor Gene for Crop Disease Management

NtWIF is a transcription factor activated upon by wound-induced protein kinase (WIPK) in tobacco plants. NtWIF is directly phosphorylated by the WIPK (Chung and Sano 2008). The transcription factor induces accumulation of SA and triggers induction of *PR* genes (Waller et al. 2006). Transgenic tobacco plants overexpressing *NtWIF* exhibited constitutive accumulation of transcripts for PR genes *PR-1a* and *PR-2*. SA levels were 50-fold higher in the transgenic plants than those in wild-type plants (Waller et al. 2006). These results suggest that *NtWIF* influences SA biosynthesis and acts downstream of WIPK and SA in tobacco. *NtWIF* did not alter the levels of JA, suggesting that it acts independently of JA signaling system. The transgenic tobacco plants expressing *NtWIF* showed resistance to *Tobacco mosaic virus* (TMV) (Waller et al. 2006).

6.24 Engineering AT-Hook Motif-Containing Transcription Factor Gene (*CaATL1*) for Crop Disease Management

A transcription factor detected in chilli pepper, CaATL1 (*Capsicum annuum AT-hook-Like gene 1*), has been found to be a potential tool to manage diseases (Kim et al. 2007). The transcription factor contained an AT-hook motif. The AT-hook is a small DNA-binding protein motif (Aravind and Landsman 1998). The AT-hook

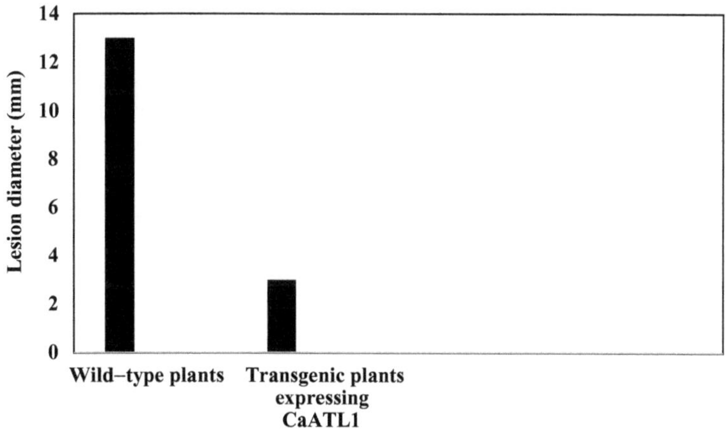

Fig. 3.34 Induction of disease resistance against *Phytophthora capsici* in transgenic tomato plants expressing *CaATL1* transcription factor from chilli pepper (*Capsicum annuum*) (Adapted from Kim et al. 2007)

motif-containing proteins are known to bind to their target elements, commonly AT-rich sequences (Reisdorf-Cren et al. 2002). The AT-hook motifs may serve as accessory DNA-binding domains for several transcription factors (Aravind and Landsman 1998). The CaATL1 protein contains one AT-hook motif (*RRPRGRPPGS*) which is a conserved motif of the AT-rich element binding factors (Kim et al. 2007). Strong induction of *CaATL1* transcripts was observed in pepper plants inoculated with bacterial and viral pathogens, particularly in resistant interactions. The defense-related signal molecules SA and Et also induced the accumulation of *CaATL1* mRNA in pepper leaves (Kim et al. 2007). These results suggest that *CaATL1* can be exploited to induce resistance against diseases.

Transgenic tomato plants constitutively expressing *CaATL1* under the control of the *Cauliflower mosaic virus* 35S promoter were generated. The T2 *CaATL1* expressing tomato plants did not show any visible phenotypic differences compared with the wild-type plants. These transgenic plants showed resistance to the bacterial speck pathogen *Pseudomonas syringae* pv. *tomato*. The leaves of the transgenic plants showed even up to ten-fold lower bacterial growth than control plants (Kim et al. 2007).

The transgenic plants also showed resistance to the root rot pathogen *Phytophthora capsici* (Fig. 3.34; Kim et al. 2007). *P. capsici*-inoculated leaves of the control plants displayed water-soaked wilt within 2 days of inoculation. In contrast, the *CaATL1* transgenic plants exhibited significantly delayed disease development (Kim et al. 2007). These studies provide strong evidence that the *CaATL1* can be used to develop commercially important crops with broad-spectrum disease resistance.

References

Abbruscato P, Nepusz T, Mizzi L, Del Corvo M, Morandini P, Fumasoni I, Michel C, Paccanaro A, Guiderdoni E, Schaffrath U, Morel J-B, Piffanelli P, Faivre-Rampant O (2012) *OsWRKY22*, a monocot *WRKY* gene, plays a role in the resistance response to blast. Mol Plant Pathol 13:828–841

Abe H, Yamaguchi-Shinozaki Y, Urao T, Iwasaki T, Hosokawa D, Shinozaki K (1997) Role of Arabidopsis MYC and MYB homologs in drought – and abscisic acid-regulated gene expression. Plant Cell 9:1859–1868

Abe H, Urao T, Seki M, Shinozaki K, Yamaguchi-Shinozaki Y (2003) Arabidopsis AtMYC2 (bHLH) and AtMYB2 (MYB) function as transcriptional activators in abscisic acid signaling. Plant Cell 15:63–78

Agarwal P, Reddy MP, Chikara J (2011) WRKY: its structure, evolutionary relarionship, DNA-binding selectivity, role in stress tolerance and development of plants. Mol Biol Rep 38:3883–3896

Akagi A, Fukushima S, Okada K, Jiang C-J, Yoshida R, Nakayama A et al (2014) WRKY45-dependent priming of diterpenoid phytoalexin biosynthesis in rice and the role of cytokinin in triggering the reaction. Plant Mol Biol 86:171–183

Akram A, Ongena M, Duby F, Dommes J, Thonart P (2008) Systemic resistance and lipoxygenase-related defence response induced in tomato by *Pseudomonas putida* strain BTP1. BMC Plant Biol 8:113

Allen GJ, Chu SP, Schumacher K, Shimazaki CT, Vafeados D, Kemper A, Hawke SD, Tallman G, Tsien RY, Harper JF, Chory J, Schroeder JI (2000) Alteration of stimulus-specific guard cell calcium oscillations and stomatal closing in *Arabidopsis det3* mutant. Science 289:2338–2342

Alves MS, Dadalto SP, Goncalves AB, de Souza GB, Barros VA, Fietto LG (2014) Transcription factor functional protein-protein interactions in plant defense responses. Proteomics 2:85–106

Amzalek E, Cohen Y (2007) Comparative efficacy of systemic acquired resistance-inducing compounds against rust infection in sunflower plants. Phytopathology 97:179–186

An C, Mou Z (2013) The function of the Mediator complex in plant immunity. Plant Signaling Behav 8(3), e23182

Anderson JP, Badruzsaufari E, Schenk PM, Manners JM, Desmond OJ, Ehlert C, Maclean DJ, Ebert PR, Kazan K (2004) Antagonistic interaction between abscisic acid and jasmonate-ethylene signaling pathways modulates defense gene expression and disease resistance in *Arabidopsis*. Plant Cell 16:3460–3479

Andreasson E, Jenkins T, Brodersen P, Thorgrimsen S, Petersen NH, Zhu S, Qiu JL, Micheelsen P, Rocher A, Petersen M, Newman MA, Bjørn Nielsen H, Hirt H, Somssich I, Mattsson O, Mundy J (2005) The MAP kinase substrate MKS1 is a regulator of plant defense responses. EMBO J 24:2579–2589

Aravind L, Landsman D (1998) AT-hook motifs identified in a wide variety of DNA-binding proteins. Nucleic Acids Res 26:4413–4421

Asada K (2006) Production and scavenging of reactive oxygen species in chloroplasts and their functions. Plant Physiol 141:391–396

Asai T, Tena G, Plotnikova J, Willmann MR, Chiu W-L, G6mez-G6mez L, Boller T, Ausubel FM, Sheen J (2002) MAP kinase signalling cascade in *Arabidopsis* innate immunity. Nature 415:977–983

Bae H, Kim MS, Sicher RC, Bae H-J, Bailey BA (2006) Necrosis- and ethylene-inducing peptide from *Fusarium oxysporum* induces a complex cascade of transcripts associated with signal transduction and cell death in *Arabidopsis*. Plant Physiol 141:1056–1067

Bahrini I, Ogawa T, Kobayashi F, Kawahigashi H, Handa H (2011a) Overexpression of the pathogen-inducible wheat *TaWRKY45* gene confers disease resistance to multiple fungi in transgenic wheat plants. Breeding Sci 61:319–326

Bahrini I, Sugisawa M, Kikuchi R, Ogawa T, Kawahigashi H, Ban T, Handa H (2011b) Characterization of a wheat transcription factor, *TaWRKY45*, and its effect on *Fusarium* head blight resistance in transgenic wheat plants. Breeding Sci 61:121–129

Balbi V, Devoto A (2008) Jasmonate signalling network in *Arabidopsis thaliana*: Crucial regulatory nodes and new physiological scenarios. New Phytologist 177:301–318

Barilli E, Rubiales D, Amalfitano C, Evidente A (2015) BTH and BABA induce resistance in pea against rust (*Uromyces pisi*) involving differential phytoalexin accumulation. Planta. doi 10.1007/s00425-015-2339-8 (in Press)

Beckers GJM, Conrath U (2007) Priming for stress resistance: from the lab to the field. Curr Opin Plant Biol 10:425–431

Beckers GJM, Jaskiewicz M, Liu Y, Underwood WR, He Y, Zhang S, Conrath U (2009) Mitogen-activated protein kinases 3 and 6 are required for full priming of stress responses in *Arabidopsis thaliana*. Plant Cell 21:944–953

Bendahmane A, Köhn BA, Dedi C, Baulcombe DC (1995) The coat protein of potato virus X is a strain-specific elicitor of Rx1-mediated virus resistance in potato. Plant J 8:933–941

Bender J (2004) Chromatin-based silencing mechanisms. Curr Opin Plant Biol 7:521–526

Bengtsson M, Wulff E, Lyngs Jørgensen HJ, Pham A, Lübeck M, Hockenhull J (2009) Comparative studies on the effects of a yucca extract and acibenzolar-S-methyl (ASM) on inhibition of *Venturia inaequalis* in apple leaves. Eur J Plant Pathol 124:187–198

Berger SL (2002) Histone modifications in transcriptional regulation. Curr Opin Genet Dev 2:142–148

Berrocal-Lobo M, Molina A (2004) Ethylene response factor 1 mediates Arabidopsis resistance to the soilborne fungus *Fusarium oxysporum*. Mol Plant-Microbe Interact 17:763–770

Berrocal-Lobo M, Molina A, Solano R (2002) Constitutive expression of ETHYLENE-RESPONSE-FACTOR1 in *Arabidopsis* confers resistance to several necrotrophic fungi. Plant J 29:23–32

Beyer K, Binder A, Boller T, Collinge M (2001) Identification of potato genes induced during colonization by *Phytophthora infestans*. Mol Plant Pathol 2:125–134

Bhattarai KK, Atamian HS, Kaloshian I, Eulgem T (2010) WRKY72 transcription factors contribute to basal immunity in tomato and Arabidopsis as well as gene-for-gene resistance mediated by the tomato *R* gene *Mi-1*. Plant J 63:229–240

Birkenbihl RP, Diezel C, Somssich IE (2012) Arabidopsis WRKY33 is a key transcriptional regulator of hormonal and metabolic responses toward *Botrytis cinerea* infection. Plant Physiol 159:266–285

Blume B, Nürnberger T, Nass N, Scheel D (2000) Receptor-mediated increase in cytoplasmic free calcium required for activation of pathogen defense in parsley. Plant Cell 12:1425–1440

Bostock RM (2005) Signal crosstalk and induced resistance: straddling the line between cost and benefit. Annu Rev Phytopathol 43:545–580

Boter M, Ruiz-Rivero O, Abdeen A, Prat S (2004) Conserved MYC transcription factors play a key role in jasmonate signalling both in tomato and Arabidopsis. Genes Dev 18:1577–1591

Bouché N, Scharlat A, Snedden WA, Bouchez D, Fromm H (2002) A novel family of calmodulin-binding transcription activators in multicellular organisms. J Biol Chem 277:21851–21861

Britton LM, Gonzales-Cope M, Zee BM, Garcia BA (2011) Breaking the histone code with quantitative mass spectrometry. Expert Rev Proteomics 8:631–643

Brown RL, Kazan K, McGrath KC, Maclean DJ, Manners JM (2003) A role for the GCC-box in jasmonate-mediated activation of the *PDF1.2* gene of *Arabidopsis*. Plant Physiol 132:1020–1032

Bu Q, Jian H, Li CB, Zhai Q, Zhang J, Wu X, Sun J, Xie Q, Li C (2008) Role of the *Arabidopsis thaliana* NAC transcription factors ANACo19 and ANACo55 in regulating jasmonic acid signaled defense responses. Cell Res 18:756–767

Bunawan H, Dusik L, Bunawan SN, Amin NM (2014) Rice tungro disease: From identification to disease control. World Appl Sci J 31:1221–1226

Camañes G, Pastor V, Cerezo M, Garcia-Andrade J, Vicedo B, Garcia-Agustin P, Flors V (2012) A deletion in *NRT2.1* attenuates *Pseudomonas syringae*-induced hormonal perturbation, resulting in primed plant defenses. Plant Physiol 158:1054–1066

Cao X, Aufsatz W, Zilberman D, Mette MF, Jacobsen SE (2003) Role of the DRM and CMT3 methyltransferases in RNA-directed DNA methylation. Curr Biol 13:2212–2217

Cao Y, Wu Y, Zheng Z, Song F (2006) Overexpression of the rice EREBP-like gene *OsBIERF3* enhances disease resistance and salt tolerance in transgenic tobacco. Physiol Mol Plant Pathol 202–211

Cao W-L, Chu R-Z, Zhang Y, Luo J, Su Y-Y, Xie L-J, Zhang H-S, Wang J-F, Bao Y-M (2015) *OsJAMyb*, a R2R3-type MYB transcription factor, enhanced blast resistance in transgenic rice. Physiol Mol Plant Pathol 92:154–160

Century K, Reuber TL, Ratcliffe OJ (2008) Regulating the regulators: The future prospects for transcription-factor-based agricultural biotechnology products. Plant Physiol 147:20–29

Cevik V, Kidd BN, Zhang P, Hill C, Kiddle S, Denby KJ, Holub EB, Cahill DM, Manners JM, Schenk PM, Beynon J, Kazan K (2012) MEDIATOR25 acts as an integrative hub for the regulation of jasmonate-responsive gene expression in Arabidopsis. Plant Physiol 160:541–555

Chavan V, Kamble A (2013) β-Aminobutyric acid primed expression of WRKY and defence genes in *Brassica carianta* against *Alternaria* blight. J Phytopathol 161:859–865

Chen C, Chen Z (2002) Potentiation of developmentally regulated defense response by AtWRKY18, a pathogen-induced *Arabidopsis* transcription factor. Plant Physiol 129:706–716

Chen W, Provart NJ, Glazebrook J, Katagiri F, Chang HS, Eulgem T, Mauch F, Luan S, Zou G, Whitham SA, Budworth PR, Tao Y, Xie Z, Chen X, Lam S, Kreps JA, Harper JF, Si-Ammour A, Mauch-Mani B, Heinlein M, Kobayashi K, Hohn T, Dangl JL, Wang X, Zhu T (2002) Expression profile matrix of *Arabidopsis* transcription factor genes suggests their putative functions in response to environmental stresses. Plant Cell 14:559–567

Chen Y, Fangfang JI, Xie H, Liang J (2006) Overexpression of the regulator of G-protein signalling protein enhances ABA-mediated inhibition of root elongation and drought tolerance in *Arabidopsis*. J Expt Bot 57:2101–2110

Chen X, Hu Y, Zhou DX (2011) Epigenetic gene regulation by plant jumonji group of histone demethylase. Biochim Biophys Acta 1809:421–426

Chen YJ, Perera V, Christiansen MW, Holme IB, Gregersen PL, Grant MR, Collinge DB, Lyngkjær MF (2013) The barley HvNAC6 transcription factor affects ABA accumulation and promotes basal resistance against powdery mildew. Plant Mol Biol 83:577–590

Cheng Z, Sun L, Qi T, Zhang B, Peng W, Liu Y, Xie D (2011) The bHLH transcription factor MYC3 interacts with the jasmonate ZIM-domain proteins to mediate jasmonate response in Arabidopsis. Mol Plant 4:279–288

Cheng H, Liu H, Deng Y, Xiao J, Li X, Wang S (2015) The WRKY45-2 WRKY13 WRKY42 transcriptional regulatory cascade is required for rice resistance to fungal pathogen. Plant Physiol 167:1087–1099

Cheong YH, Moon BC, Kim JK, Kim CY, Kim MC, Kim IH, Park CY, Kim JC, Park BO, Koo SC, Yoon HW, Chung WS, Lim CO, Lee SY, Cho MJ (2003) BWMK1, a rice mitogen-activated protein kinase, locates in the nucleus and mediates pathogenesis-related gene expression by activation of a transcription factor. Plant Physiol 132:1961–1972

Chern M-S, Fitzgerald H, Yadav RC, Canlas PE, Dong X, Ronald PC (2001) Evidence for a disease-resistance pathway in rice similar to the *NPR1*-mediated signaling pathway in *Arabidopsis*. Plant J 27:101–113

Chern M, Canlas PE, Ronald PC (2008) Strong suppression of systemic acquired resistance in *Arabidopsis* by NRR is dependent on its ability to interact with NPR1 and its putative repressive domain. Molecular Plant 1:552–559

Choi MS, Kim MC, Yoo JH, Moon BC, Koo SC, Park BO, Lee JH, Koo YD, Han HJ, Lee SY, Chung WS, Lim CO, Cho MJ (2005) Isolation of a calmodulin-binding transcription factor from rice (*Oryza sativa* L.). J Biol Chem 280:40820–40831

Chujo T, Takai R, Akimoto-Tomiyama C, Ando S, Minami E, Nagamura Y, Kaku H, Shibuya N, Yasuda M, Nakashita H, Umemura K, Okada K, Nojiri H, Yamane H (2007) Involvement of the elicitor-induced gene *OsWRKY53* in the expression of defense-related genes in rice. Biochim Biophys Acta 1769:497–505

Chujo T, Kato T, Yamada K, Takai R, Akimoto-Tomiyama C, Minami H, Nagamura Y, Shibuya N, Yasuda M, Nakashita H, Umemura K, Okada A, Okada K, Nojiri H, Yamane H (2008)

Characterization of an elicitor-induced rice WRKY gene, *OsWRKY71*. Biosci Biotechnol Biochem 72:240–245
Chujo T, Sugioka N, Masuda Y, Shibuya N, Takemura T, Okada K, Nojiri H, Yamane H (2009) Promoter analysis of the elicitor-induced WRKY gene *OsWRKY53*, which is involved in defense responses in rice. BioSci Biotechnol Biochem 73:1901–1904
Chujo T, Miyamoto K, Shimogawa T, Shimizu T, Otake Y, Yokotani N, Nishizawa Y, Shibuya N, Nojiri H, Yamane H, Minami E, Okada K (2013) OsWRKY28, a PAMP-responsive transrepressor, negatively regulates innate immune responses in rice against blast fungus. Plant Mol Biol 82:23–37
Chujo T, Miyamoto K, Ogawa S, Masuda Y, Shimizu T, Kishi-Kaboshi M, Takahashi A, Nishizawa Y, Minami E, Norjiri H, Yamane H, Okada K (2014) Overexpression of phosphomimic mutated OsWRKY53 leads to enhanced blast resistance in rice. PLOS ONE 9(6), e98737. doi:10.1371/journal.pone.0098737
Chung K-M, Sano H (2008) Finding a missing link in MAP kinase cascade. Plant Signaling & Behavior 3:606–608
Clay NK, Adio AM, Denoux C, Jander G, Ausubel FM (2009) Glucosinolate metabolites required for an *Arabidopsis* innate immune response. Science 323:95–101
Coego A, Ramirez V, Jose Gil M, Flors V, Mauch-Mani B, Vera P (2005) An *Arabidopsis* homeodomain transcription factor, OVEREXPRESSOR OF CATIONIC PEROXIDASE 3, mediates resistance to infection by necrotrophic pathogens. Plant Cell 17:2123–2137
Cohen Y (1994) 3-Aminobutyric acid induces systemic resistance against *Peronospora tabacina*. Physiol Mol Plant Pathol 44:273–288
Cohen Y, Reuveni M, Baider A (1999) Local and systemic activity of BABA (DL-3-aminobutyric acid) against *Plasmopara viticola* in grapevines. Eur J Plant Pathol 105:351–361
Conrath U (2009) Priming of induced plant defense responses. Adv Bot Res 51:361–395
Conrath U (2011) Molecular aspects of defence priming. Trends Plant Sci 16:524–531
Conrath U, Pieterse CMJ, Mauch-Mani B (2002) Priming in plant-pathogen interactions. Trends Plant Sci 7:210–216
Conrath U, Beckers GJM, Flors V, Garcia-Agustin P, Jakab G, Mauch F, Newman M-A, Pieterse CMJ, Poinssot B, Pozo MJ, Pugin A, Schaffrath U, Ton J, Wendehenne D, Zimmerli L, Mauch-Mani B (2006) Priming: Getting ready for battle. Mol Plant-Microbe Interact 19:1062–1071
Culver JN, Dawson WO (1991) Tobacco mosaic virus elicitor coat protein genes produce a hypersensitive phenotype in transgenic *Nicotiana sylvestris* plants. Mol Plant-Microbe Interact 4:458–463
Culver JN, Stubbs G, Dawson WO (1994) Structure-function relationship between tobacco mosaic virus coat protein and hypersensitivity in *Nicotiana sylvestris*. J Mol Biol 242:130–138
Cumming RC, Andon NL, Haynes PA, Park M, Fischer WH, Schubert D (2004) Protein disulfide bond formation in the cytoplasm during oxidative stress. J Biol Chem 279:21749–21758
Dai S, Petruccelli S, Ordiz MI, Zhang Z, Chen S, Beachy RN (2003) Functional analysis of RF2a, a rice transcription factor. J Biol Chem 278:36396–36402
Dai S, Zhang Z, Chen S, Beachy RN (2004) RF2b, a rice bZIP transcription activator, interacts with RF2a and is involved in symptom development of rice tungro disease. Proc Natl Acad Sci USA 101:687–692
Dai S, Zhang Z, Bick J, Beachy RN (2006) Essential role of the Box II *cis* element and cognate host factors in regulating the promoter of rice tungro bacilliform virus. J Gen Virol 87:715–722
Dai S, Wei X, Alfonso AA, Pei L, Duque UG, Zhang Z, Babb GM, Beachy RN (2008) Transgenic rice plants that overexpress transcription factors RF2a and RF2b are tolerant to rice tungro virus replication and disease. Proc Natl Acad Sci USA 105:21012–21016
Daire X, Poinssot B, Bentéjac M, Silué D, Pugin A (2002) Stimulation of the natural defence mechanisms of the grapevine against pathogens: encouraging results for downy mildew. Phytoma 548:24–26
Dang F, Wang Y, She J, Lei Y, Liu Z, Eulgem T, Lai Y, Lin J, Yu L, Lei D, Guan D, Li X, Yuan Q, He S (2014) Overexpression of *CaWRKY27*, a subgroup IIe transcription factor of *Capsicum*

annuum, positively regulates tobacco resistance to *Ralstonia solanacearum* infection. Physiologia Plantarum 150:397–411

De Vleesschauwer D, Djavaheri M, Bakker PAHM, Höfte M (2008) *Pseudomonas fluorescens* WCS374r-induced systemic resistance in rice against *Magnaporthe oryzae* is based on pseudobactin-mediated priming for a salicylic acid-repressable multifaceted defense response. Plant Physiol 148:1996–2012

Delker C, Zolman BK, Miersch O, Wasternack C (2007) Jasmonate biosynthesis in *Arabidopsis thaliana* requires peroxisomal β-oxidation enzymes – Additional proof by properties of *pex6* and *aim1*. Phytochemistry 68:1542–1650

Dellagi A, Helibronn J, Avrova AO, Montesano M, Palva ET, Stewart HE, Toth IK, Cooke DE, Lyon GD, Birch PR (2000) A potato gene encoding a WRKY-like transcription factor is induced in interactions with *Erwinia carotovora* subsp. *atroseptica* and *Phytophthora infestans* and is coregulated with a class I endochitinase expression. Mol Plant-Microbe Interact 13:1092–1101

Deluc L, Barrieu F, Marchive C, Lauvergeat V, Decendit A, Richard T, Carde JP, Merillon JM, Hamdi S (2006) Characterization of a grapevine R2R3-MYB transcription factor that regulates the phenylpropanoid pathway. Plant Physiol 140:499–511

Denoux C, Galletti R, Mammarella N, Gopalan S, Werck D, De Lorenzo G, Ferrari S, Ausubel FM, Dewdney J (2008) Activation of defense response pathways by OGs and Flg22 elicitors in Arabidopsis seedlings. Mol Plant 1:423–445

Desikan R, Hancock JT, Ichimura K, Shinozaki K, Neill SJ (2001) Harpin induces activation of the *Arabidopsis* mitogen-activated protein kinases AtMPK4 and AtMPK6. Plant Physiol 126:1579–1587

Desikan R, Hancock JT, Bright J, Harrison J, Weir I, Hooley R, Neill SJ (2005) A role for ETR1 in hydrogen peroxide signaling in stomatal guard cells. Plant Physiol 137:831–834

Deslandes L, Olivier J, Theulieres F, Hirsch J, Dong XF, Bittner-Eddy P, Beynon J, Marco Y (2002) Resistance to *Ralstonia solanacearum* in *Arabidopsis thaliana* is conferred by the recessive *RRS1-R* gene, a member of a novel family of resistance genes. Proc Natl Acad Sci USA 99:2404–2409

Després C, DeLong C, Glaze S, Liu E, Fobert PR (2000) The Arabidopsis NPR1/NIM1 protein enhances the DNA binding activity of a subgroup of the TGA family of bZIP transcription factors. Plant Cell 12:279–290

Després C, Chubak C, Rochon A, Clark R, Bethune T, Desveaux D, Fobert PR (2003) The Arabidopsis *NPR1* disease resistance protein is a novel cofactor that confers redox regulation of DNA binding activity to the basic domain/leucine zipper transcription factor TGA1. Plant Cell 15:2181–2191

Djamei A, Pitzschke A, Nakagami H, Rajh I, Hirt H (2007) Trojan horse strategy in *Agrobacterium* transformation: Abusing MAPK defense signaling. Science 318:453–456

Dombrecht B, Xue GP, Sprague SJ, Kirkegaard JA, Ross JJ, Reid JB, Fitt GP, Sewelam N, Schenk PM, Manners JM, Kazan K (2007) MYC2 differentially modulates diverse jasmonate-dependent functions in *Arabidopsis*. Plant Cell 19:2225–2245

Dong J, Chen C, Chen Z (2003) Expression profiles of the Arabidopsis WRKY gene superfamily during plant defense response. Plant Mol Biol 51:21–37

Donze T, Qu F, Twigg P, Morris TJ (2014) *Turnip crinkle virus* coat protein inhibits the basal immune response to virus invasion in *Arabidopsis* by binding to the NAC transcription factor TIP. Virology 449:207–214

Dröge-Laser W, Kaiser A, Lindsay WP, Halker BA, Loake GJ, Doerner P, Dixon RA, Lamb C (1997) Rapid stimulation of a soybean protein-serine kinase which phosphorylates a novel bZIP DNA-binding protein, G/HBF-1 during the induction of early transcription-dependent defenses. EMBO J 16:726–738

Du H, Zhang L, Liu L, Tang X-F, Yang W-J, Wu Y-M, Huang Y-B, Tang Y-X (2009) Biochemical and molecular characterization of plant MYB transcription factor family. Biochemistry (Moscow) 74:1–11

Duijff BJ, Poulain D, Olivain C, Alabouvette C, Lemanceau P (1998) Implication of systemic induced resistance in the suppression of *Fusarium* wilt of tomato by *Pseudomonas fluorescens* WCS417r and by nonpathogenic *Fusarium oxysporum* Fo47. Eur J Plant Pathol 104:903–910

El Rahman TA, El Oirdi M, Gonzalez-Lamothe R, Bouarab K (2012) Necrotrophic pathogens use the salicylic acid pathway to promote disease development in tomato. Mol Plant Microbe Interact 25:1584–1593

Eulgem T, Somssich IE (2007) Networks of WRKY transcription factors in defense signaling. Curr Opin Plant Biol 10:366–371

Eulgem T, Rushton PJ, Schmelzer E, Hahlbrock K, Somssich IE (1999) Early nuclear events in plant defence signaling: rapid gene activation by WRKY transcription factors. EMBO J 18:4689–4699

Eulgem T, Rushton PJ, Robatzek S, Somssich IE (2000) The WRKY superfamily of plant transcription factors. Trends Plant Sci 5:199–206

Faize M, Faize L, Koike N, Ishizaka M, Ishii H (2004) Acibenzolar-S-methyl-induced resistance to Japanese pear scab is associated with potentiation of multiple defense responses. Phytopathology 94:604–612

Fan J, Hill L, Crooks C, Doerner P, Lamb C (2009) Abscisic acid has a key role in modulating diverse plant-pathogen interactions. Plant Physiol 150:1750–1761

Fan H, Wang F, Gao H, Wang L, Xu J, Zhao Z (2011) Pathogen-induced *MdWRKY1* in 'Qinguan' apple enhances disease resistance. J Plant Biol 54:150–158

Fedoroff N (2006) Redox regulatory mechanisms in cellular stress responses. Ann Bot (Lond) 98:289–300

Ferrari S, Plotnokova JM, De Lorenzo G, Ausubel FM (2003) *Arabidopsis* local resistance to *Botrytis cinerea* involves salicylic acid and camalexin and requires *EDS4* and *PAD2*, but not *SID2*, *EDS5* or *PAD4*. Plant J 35:193–205

Finkler A, Kaplan B, Fromm H (2007) Ca^{2+}-responsive cis-elements in plants. Plant Signaling & Behavior 2:17–19

Fischer U, Dröge-Laser W (2004) Overexpression of *Nt*ERF5, a new member of the tobacco ethylene response transcription factor family enhances resistance to *Tobacco mosaic virus*. Mol Plant-Microbe Interact 17:1162–1171

Fitzgerald HA, Canlas PE, Chern M-S, Ronald PC (2005) Alteration of TGA factor activity in rice results in enhanced tolerance to *Xanthomonas oryzae* pv. *oryzae*. Plant J 43:335–347

Flors V, Ton J, van Doorn R, Garcia-Agustin P, Mauch-Mani B (2008) Interplay between JA, SA, and ABA signaling during basal and induced resistance against *Pseudomonas syringae* and *Alternaria brassicicola*. Plant J 54:81–92

Fobert PR, Després C (2005) Redox control of systemic acquired resistance. Curr Opin Plant Biol 8:378–382

Forman HJ, Fukuto JM, Torres M (2004) Redox signaling: thiol chemistry defines which reactive oxygen species can act as second messengers. Am J Physiol Cell Physiol 287:C246–C256

Frye CA, Tang D, Innes RW (2001) Negative regulation of defense responses in plants by a conserved MAPKK kinase. Proc Natl Acad Sci USA 98:373–378

Fujimoto SY, Ohta M, Usui A, Shinshi H, Ohme-Takagi M (2000) Arabidopsis ethylene-responsive element binding factors act as transcriptional activators or repressors of GCC box-mediated gene expression. Plant Cell 12:393–404

Fukuoka S, Saka N, Mizukami Y, Koga H, Yamanouchi U, Yoshioka Y, Hayashi N, Ebana K, Mizobuchi R, Yano M (2015) Gene pyramiding enhances durable blast disease resistance in rice. Scientific Reports 5 Article number 7773 doi: 10.1038/srep07773

Galon Y, Aloni R, Nachmias D, Snir O, Feldmesser E, Scrase-Field S, Boyce JM, Bouché N, Knight MR, Fromm H (2010) Calmodulin-binding transcription activator 1 mediates auxin signaling and responds to stresses in Arabidopsis. Planta 232:165–178

Garapati P, Xue GP, Munné-Bosch S, Balazadeh S (2015) Transcription factor ATAF1 in *Arabidopsis* promotes senescence by direct regulation of key chloroplast maintenance and senescence transcriptional cascades. Plant Physiol 168:1122–1139

Garcion C, Métraux J-P (2006) Salicylic acid. In: P Hedden, SG Thomas, eds, Plant Hormone Signaling. Annual Reviews Vol. 24:229-257 Blackwell Press, Oxford

Garcion C, Lohmann A, Lamodiere E, Catinot J, Buchala A, Doermann P, Métraux J-P (2008) Characterization and biological function of the *ISOCHORISMATE SYNTHASE2* gene of the Arabidopsis. Plant Physiol 147:1279–1287

Genger RK, Jurkowski GI, McDowell JM, Lu H, Jung HW, Greenberg JT, Bent AF (2008) Signaling pathways that regulate the enhanced disease resistance of *Arabidopsis* "Defense, No death" mutants. Mol Plant-Microbe Interact 21:1285–1296

Gfeller A, Liechti R, Farmer EE (2010) Arabidopsis jasmonate signaling pathway. Sci Signal 3(109):cm4

Gilardi P, Garcia-Luque I, Serra MT (2004) The coat protein of tobamovirus acts as elicitor of both *L2* and *L4* gene-mediated resistance in *Capsicum*. J Gen Virol 85:2077–2085

Goto S, Sasakura-Shimoda F, Suetsugu M, Selvaraj MG, Hayashi N, Yamazaki M, Ishitani M, Shimono M, Sugano S, Matsushita A, Tanabata T, Takatsuji H (2015) Development of disease-resistant rice by optimized expression of WRKY45. Plant Biotechnol J 13:753–765

Grant JJ, Loake GJ (2000) Role of reactive oxygen intermediates and cognate redox signaling in disease resistance. Plant Physiol 8:629–643

Greenberg MVC, Ausin I, Chan SWL, Cokus SJ, Cuperus JT, Feng S, Law JA, Chu C, Pellegrini M, Carrington JC, Jacobsen SE (2011) Identification of genes required for *de novo* DNA methylation in *Arabidopsis*. Epigenetics 6:344–354

Grennan AK (2008) Ethylene response factors in jasmonate signaling and defense response. Plant Physiol 146:1457–1458

Gu YQ, Wildermuth MC, Chakravarthy S, Loh YT, Yang C, He X, Han Y, Martin GB (2002) Tomato transcription factors pti4, pti5, and pti6 activate defense responses when expressed in *Arabidopsis*. Plant Cell 14:817–831

Guo H, Ecker JR (2004) The ethylene signaling pathway; new insights. Curr Opin Plant Biol 7:40–49

Guo Z, Chen XJ, Wu XL, Ling JQ, Xu P (2004) Overexpression of the AP2/EREBP transcription factor OPBP1 enhances disease resistance and salt tolerance in tobacco. Plant Mol Biol 55:607–618

Gutterson N, Reuber TL (2004) Regulation of disease resistance pathways by AP2/ERF transcription factors. Curr Opin Plant Biol 7:465–471

Hamiduzzaman MM, Jakeb G, Barnavon L, Neuhaus JM, Mauch-Mani B (2005) β-Aminobutyric acid-induced resistance against downy mildew in grapevine acts through the potentiation of callose formation and jasmonic acid signaling. Mol Plant-Microbe Interact 18:819–829

Han M, Ryu H-S, Kim C-Y, Park D-S, Ahn Y-K, Jeon J-S (2013) *OsWRKY30* is a transcription activator that enhances rice resistance to the *Xanthomonas oryzae* pathovar oryzae. J Plant Biol 56:258–265

Hancock J, Desikan R, Harrison J, Bright J, Hooley R, Neill S (2006) Doing the unexpected: proteins involved in hydrogen peroxide perception. J Expt Bot 57:1711–1718

Harmon AC (2003) Calcium-regulated protein kinases of plants. Gravitational and Space Biology Bulletin 16(2):83–90

Harmon AC, Gribskov M, Harper JF (2000) CDPKs – a kinase for every Ca^{2+} signal? Trends Plant Sci 5:154–159

Hashimoto K, Eckert C, Anschūtz U, Scholz M, Held K, Waadt R, Reyer A, Hippler M, Becker D, Kudla J (2012) Phosphorylation of calcineurin B-like (CBL) calcium sensor proteins by their CBL-interacting protein kinases (CIPKs) is required for full activity of CBL-CIPK complexes toward their target proteins. J Biol Chem 287:7956–7968

He P, Warren RF, Zhao TH, Shan LB, Zhu LH, Tang XY, Zhou JM (2001) Overexpression of Pti5 in tomato potentiates pathogen-induced defense gene expression and enhances disease resistance to *Pseudomonas syringae* pv. *tomato*. Mol Plant Microbe Interact 14:1453–1457

Henderson IR, Deleris A, Wong W, Zhong X, Chin HG, Horwitz GA, Kelly KA, Pradhan S, Jacobsen SE (2010) The *de novo* cytosine methyltransferase DRM2 requires intact UBA domains and a catalytically mutated paralog DRM3 during RNA-directed DNA methylation in *Arabidopsis thaliana*. PLoS Genet 6, e1001182

Herman MAB, Davidson JK, Smart CD (2008) Induction of plant defense gene expression by plant activators and *Pseudomonas syringae* pv. *tomato* in greenhouse-grown tomatoes. Phytopathology 98:1226–1232

Hibino H, Saleh N, Roechan M (1979) Transmission of two kinds of rice tungro-associated viruses by insect vectors. Phytopathology 69:1266–1268

Higashi K, Ishiga Y, Inagaki Y, Toyoda K, Shiraishi T, Ichinose Y (2008) Modulation of defense signal transduction by flagellin-induced WRKY41 transcription factor in *Arabidopsis thaliana*. Mol Genet Genomics 279:303–312

Hoffland E, Pieterse CMJ, Bik L, Van Pelt JA (1995) Induced systemic resistance in radish is not associated with accumulation of pathogenesis-related proteins. Physiol Mol Plant Pathol 46:309–320

Hong JC, Timur Momol M, Pingsheng JP, Olson SM, Colee J, Jones JB (2011) Management of bacterial wilt in tomatoes with thymol and acibenzolar-S-methyl. Crop Prot 30:1340–1346

Hu R, Qi G, Kong Y, Kong D, Gao Q, Zhou G (2010) Comprehensive analysis of NAC domain transcription factor gene family in *Populus trichocarpa*. BMC Plant Biol 10:145

Hu Y, Dong Q, Yu D (2012) *Arabidopsis* WRKY46 coordinates with WRKY70 and WRKY53 in basal resistance against pathogen *Pseudomonas syringae*. Plant Sci Vol 185–188:288–297

Huang J, Wang H, Xie X, Zhang D, Liu Y, Guo G (2010) Roles of DNA methyltransferases in *Arabidopsis* development. African J Biotechnol 9:8506–8514

Hukkanen AT, Kokko HI, Buchala AJ, McDougall GJ, Stewart D, Kárenlampi SO, Karjalainen RO (2007) Benzothiadiazole induces the accumulation of phenolics and improves resistance to powdery mildew in strawberries. J Agric Food Chem 55:1862–1870

Hull R (1996) Molecular biology of rice tungro viruses. Annu Rev Phytopathol 34:275–297

Hwang S-H, Yie SW, Hwang D-J (2011) Heterologous expression of *OsWRKY6* gene in *Arabidopsis* activates the expression of defense related genes and enhances resistance to pathogens. Plant Sci 181:316–323

Iavicoli A, Boutet E, Buchala A, Metraux JP (2003) Induced systemic resistance in *Arabidopsis thaliana* in response to root inoculation with *Pseudomonas fluorescens* CHAO. Mol Plant-Microbe Interact 16:851–858

Inoue H, Hayashi N, Matsushita A, Xinqiong L, Nakayama A, Sugano S, Jiang CJ, Takatsuji H (2013) Blast resistance of CC-NB-LRR protein Pb1 is mediated by WRKY45 through protein-protein interaction. Proc Natl Acad Sci USA 110:9577–9582

Ishihama N, Yamada R, Yoshioka M, Katou S, Yoshioka H (2011) Phosphorylation of the *Nicotiana benthamiana* WRKY8 transcription factor by MAPK functions in the defense response. Plant Cell 23:1153–1170

Iwai T, Seo S, Mitsuhara I, Ohashi Y (2007) Probenazole-induced accumulation of salicylic acid confers resistance to *Magnaporthe grisea* in adult rice plants. Plant Cell Physiol 48:915–924

Jakab G, Cottier V, Toquin V, Rigoli G, Zimmerli L, Metraux JP, Mauch-Mani B (2001) β-aminobutyric acid-induced resistance in plants. Eur J Plant Pathol 107:29–37

Jakoby M, Weisshaar B, Droge-Laser W, Vicente-Carbajosa J, Tiedemann J, Kroj T, Parcy F (2002) bZIP transcription factors in *Arabidopsis*. Trends Plant Sci 7:106–111

Jaskiewicz M, Conrath U, Peterhänsel C (2011) Chromatin modification acts as a memory for systemic acquired resistance in the plant stress response. EMBO Rep 12:50–55

Jensen MK, Rung JH, Gregersen PL, Gjetting T, Fuglsang AT, Hansen M, Joehnk N, Lyngkjaer MF, Collinge DB (2007) The HvNAC6 transcription factor: a positive regulator of penetration resistance in barley and Arabidopsis. Plant Mol Biol 65:137–150

Jensen MK, Hagedorn PH, de Torres-Zabala M, Grant MR, Rung JH, Collinge DB, Lyngkjaer MF (2008) Transcriptional regulation by an NAC (NAM-ATAF1,2-CUC2) transcription factor attenuates ABA signaling for efficient basal defence towards *Blumeria graminis* f. sp. *hordei* in Arabidopsis. Plant J 56:867–880

Jeun YC, Siegrist J, Buchenauer H (2000) Biochemical and cytological studies on mechanisms of systemic induced resistance in tomato plants against *Phytophthora infestans*. J Phytopathol 148:129–140

Jiang Y, Duan Y, Yin J, Ye S, Zhu J, Zhang F, Lu W, Luo K (2014) Genome-wide identification and characterization of the *Populus* WRKY transcription factor family and analysis of their expression in response to biotic and abiotic stresses. J Exp Bot 65:6629–6644

Jingyuan Z, Xuexiao Z, Zhenchuan M, Bingyan X (2011) A novel pepper (*Capsicum annuum* L.) WRKY gene, *CaWRKY30*, is involved in pathogen stress responses. J Plant Biol 54:329–337

Jisha V, Dampanaboina L, Vadassery J, Mithöfer A, Kappara S, Ramanan R (2015) Overexpression of an AP2/ERF type transcription factor *OsEREBP1* confers biotic and abiotic stress tolerance in rice. PLOS ONE 10(6), e0127831. doi:10.1371/journal.pone.0127831

Johansson A, Staal J, Dixelius C (2006) Early responses in the *Arabidopsis- Verticillium longisporum* pathosystem are dependent on *NDR1*, JA- and ET-associated signals via cytosolic NPR1 and *RFO1*. Mol Plant Microbe Interact 19:958–969

Johnson C, Mhatre A, Arias J (2008) NPR1 preferentially binds to the DNA-inactive form of *Arabidopsis* TGA2. Biochim Biophys Acta 1779:583–589

Joo JH, Wang S, Chen JG, Jones AM, Fedoroff NV (2005) Different signaling and cell death roles of heterotrimeric G protein α and β subunits in the *Arabidopsis* oxidative stress response to ozone. Plant Cell 17:957–970

Journot-Catalino N, Somssich IE, Roby D, Kroj T (2006) The transcription factors WRKY11 and WRKY17 act as negative regulators of basal resistance in *Arabidopsis thaliana*. Plant Cell 18:3289–3302

Jung J, Won SY, Suh SC, Kim H, Wing R, Jeong Y, Hwang I, Kim M (2007) The barley ERF-type transcription factor HvRAF confers enhanced pathogen resistance and salt tolerance in *Arabidopsis*. Planta 225:575–588

Jung C, Seo JS, Han SW, Koo YJ, Kim CH, Song SI, Nahm BH, Choi YD, Cheong JJ (2008) Overexpression of AtMYB44 enhances stomatal closure to confer abiotic stress tolerance in transgenic Arabidopsis. Plant Physiol 146:623–625

Jung HW, Tschaplinski TJ, Wang L, Glazebrook J, Greenberg JT (2009) Priming in systemic plant immunity. Science 324:89–91

Kalde M, Barth M, Somssich IE, Lippok B (2003) Members of the *Arabidopsis* WRKY group III transcription factors are part of different plant defense signaling pathways. Mol Plant-Microbe Interact 16:295–305

Kaneda T, Taga Y, Takai R, Iwano M, Matsui H, Takayama S, Isogai A, Che F-S (2009) The transcription factor OsNAC4 is a key positive regulator of plant hypersensitive cell death. EMBO J 28:926–936

Kang CH, Jung WY, Kang YH, Kim JY, Kim DG, Jeong JC, Baek DW, Jin JB, Lee JY, Kim MO, Chung WS, Mengiste T, Koiwa H, Kwak SS, Bahk JD, Lee SY, Nam JS, Yun DJ, Cho MJ (2006) AtBAG6, a novel calmodulin-binding protein, induces programmed cell death in yeast and plants. Cell Death and Differentiation 13:84–95

Kasparovsky T, Milat M-L, Humbert C, Blein J-P, Havel L, Mikes V (2003) Elicitation of tobacco cells with ergosterol activates a signal pathway including mobilization of internal calcium. Plant Physiol Biochem 41:495–501

Kasparovsky T, Blein JP, Mikes V (2004) Ergosterol elicits burst in tobacco cells via phospholipase A2 and protein kinase C signal pathway. Plant Physiol Biochem 42:429–435

Kavum MA, Jung HJ, Park JI, Ahmed NU, Saha G, Yang TJ, Nou IS (2015) Identification and expression analysis of WRKY family genes under biotic and abiotic stresses in *Brassica rapa*. Mol Genet Genomics 290:79–95

Kazan K, Manners JM (2008) Jasmonate signaling: toward an integrated view. Plant Physiol 146:1459–1468

Kazan K, Manners JM (2013) MYC2: the master in action. Mol Plant. doi:10.1093/mp/sss128

Keates SE, Kostman TA, Anderson JD, Bailey BA (2003) Altered gene expression in three plant species in response to treatment with Nep1, a fungal protein that causes necrosis. Plant Physiol 132:1610–1622

Kesarwani M, Joo J, Dong X (2007) Genetic interactions of TGA transcription factors in the regulation of pathogenesis-related genes and disease resistance in *Arabidopsis*. Plant Physiol 144:336–346

Kim HS, Delaney TP (2002) Overexpression of *TGA5*, which encodes a bZIP transcription factor that interacts with NIM1/NPR1, confers SAR-independent resistance in *Arabidopsis thaliana* to *Peronospora parasitica*. Plant J 32:151–163

Kim CY, Zhang S (2004) Activation of a mitogen-activated protein kinase cascade induces WRKY family of transcription factors and defense genes in tobacco. Plant J 38:142–151

Kim KC, Fan B, Chen Z (2006) Pathogen-induced Arabidopsis WRKY7 is a transcriptional repressor and enhances susceptibility to *Pseudomonas syringae*. Plant Physiol 142:1180–1192

Kim S-Y, Kim Y-C, Seong ES, Lee Y-H, Park M, Choi D (2007) The chili pepper CaATL1: an AT-hook motif-containing transcription factor implicated in defence responses against pathogens. Mol Plant Pathol 8:761–771

Kim K-C, Lai Z, Fan B, Chen Z (2008) *Arabidopsis* WRKY38 and WRKY62 transcription factors interact with histone deacetylase 19 in basal defense. Plant Cell 20:2357–2371

Kitajima S, Koyama T, Ohme-Takagi M, Shinshi H, Sato F (2000) Characterization of gene expression of NsERFs, transcription factors of basic PR genes from *Nicotiana sylvestris*. Plant Cell Physiol 41:817–824

Knoth C, Ringler J, Dangl JL, Eulgem T (2007) *Arabidopsis* WRKY70 is required for full *RPP4*-mediated disease resistance and basal defense against *Hyaloperonospora parasitica*. Mol Plant Microbe Interact 20:120–128

Kohler A, Schwindling S, Conrath U (2002) Benzothiadiazole-induced priming for potentiated responses to pathogen infection, wounding, and infiltration of water into leaves requires the *NPR1/NIM1* gene in *Arabidopsis*. Plant Physiol 128:1046–1056

Kojo K, Yaeno T, Kusumi K, Matsumura H, Fujisawa S, Terauchi R, Iba K (2006) Regulatory mechanisms of ROI generation are affected by rice *spl* mutations. Plant Cell Physiol 47:1035–1044

Kombrink E (2012) Chemical and genetic exploration of jasmonate biosynthesis and signaling paths. Planta 236:1351–1366

Korolev N, David DR, Elad Y (2008) The role of phytohormones in basal resistance in *Trichoderma*-induced systemic resistance to *Botrytis cinerea* in *Arabidopsis thaliana*. Biocontrol 53:667–683

Koyama T, Okada T, Kitajima S, Ohme-Takagi M, Shinshi H, Sato F (2003) Isolation of tobacco ubiquitin-conjugating enzyme cDNA in a yeast two-hybrid system with tobacco ERF3 as bait and its characterization of specific interaction. J Expt Bot 54:1175–1181

Kravchuk Z, Vicedo B, Flors V, Camañes G, González-Bosch C, Garcia-Agustin P (2011) Priming for JA-dependent defenses using hexanoic acid is an effective mechanism to protect Arabidopsis against *B. cinerea*. J Plant Physiol 186:359–366

Kudla J, Batistić O, Hashimoto K (2010) Calcium signals: The lead currency of plant information processing. Plant Cell 22:541–563

Kuhlmann M, Horvay K, Strathmann A, Heinekamp T, Fischer U, Böttner S, Dröge-Laser W (2003) The α-helical D1 domain of the bZIP transcription factor BZI-1 interacts with the ankyrin-repeat protein ANK1, and is essential for BZI-1 function, both in auxin signalling and pathogen response. J Biol chem 278:8786–8794

Kwaaitaal M, Huisman R, Maintz J, Reinstädler A, Panstruga R (2011) Ionotropic glutamate receptor (iGluR)-like channels mediate MAMP-induced calcium influx in *Arabidopsis thaliana*. Biochem J 440:355–365

Lai Z, Vinod KM, Zheng Z, Fan B, Chen Z (2008) Roles of *Arabidopsis* WRKY3 and WRKY4 transcription factors in plant responses to pathogens. BMC Plant Biol 8:68

Laquitaine L, Gomès E, Francois J, Marchive C, Pascal S, Hamdi S, Atanassova R, Delrot S, Coutos-Thévenot P (2006) Molecular basis of ergosterol-induced protection of grape against *Botrytis cinerea*: Induction of type 1 LTP promoter activity, WRKY, and stilbene synthase gene expression. Mol Plant-Microbe Interact 19:1103–1112

Laurie-Berry N, Joardar V, Street JH, Kunkel BN (2006) The *Arabidopsis thaliana JASMONATE INSENSITIVE 1* gene is required for suppression of salicylic acid-dependent defenses during infection by *Pseudomonas syringae*. Mol Plant-Microbe Interact 19:789–800

Lawton KA, Friedrich I, Hunt M, Weymann K, Delaney T, Kessmann H, Staub T, Ryals J (1996) Benzothiadiazole induces disease resistance in *Arabidopsis* by activation of the systemic acquired resistance signal transduction pathway. Plant J 10:71–82

Le DT, Nishiyama R, Watanabe Y, Mochida K, Yamaguchi-Shinozaki K, Shinozaki K, Phan Tran L-S (2011) Genome-wide survey and expression analysis of the plant-specific NAC transcription factor family in soybean during development and dehydration stress. DNA Res 18:263–276

Le Hénanff G, Profizi C, Courteaux B, Rabenoelina F, Gérard C, Clément C, Baillieul F, Cordelier S, Dhondt-Cordelier S (2013) Grapevine NAC1 transcription factor as a convergent node in developmental processes, abiotic stresses, and necrotrophic/biotrophic pathogen tolerance. J Exp Biol 64:4877–4893

Lebel E, Heifetz P, Thorne L, Uknes S, Ryals J, Ward E (1998) Functional analysis of regulatory sequences controlling *PR-1* gene expression in Arabidopsis. Plant J 16:223–233

Lecourieux D, Ranjeva R, Pugin A (2006) Calcium in plant defence-signalling pathways. New Phytol 171:249–269

Lee YK, Hong JK, Hippe-Sanwald S, Hwang BK (2000) Histological and ultrastructural comparisons of compatible, incompatible and DL-β-amino-*n* –butyric acid-induced resistance responses of pepper stems to *Phytophthora capsici*. Physiol Mol Plant Pathol 57:269–280

Lee MW, Qi M, Yang Y (2001) A novel jasmonic acid-inducible rice *myb* gene associates with fungal infection and host cell death. Mol Plant-Microbe Interact 14:527–535

Lee JH, Hong JP, Oh SK, Lee S, Choi D, Kim WT (2004) The ethylene-responsive factor like protein 1 (CaERFLP1) of hot pepper (*Capsicum annuum* L.) interacts in vitro with both GCC and DRE/CRT sequence with different binding affinities: possible biological roles of CaERFLP1 in response to pathogen infection and high salinity conditions in transgenic tobacco plants. Plant Mol Biol 42:335–344

Lee H, Ko YJ, Cha J-Y, Park R, Ahn I, Hwang D-J (2013) The C-terminal region of *Os*WRKY30 is sufficient to confer enhanced resistance to pathogen and activate the expression of defense-related genes. Plant Biotech Rep 7:221–230

Lehtonen NT, Akita M, Frank W, Reski R, Valkonen JPT (2012) Involvement of a class III peroxidase and the mitochondrial protein TSPO in oxidative burst upon treatment of moss plants with a fungal elicitor. Mol Plant Microbe Interact 25:363–371

León J, Lawton MA, Raskin I (1995) Hydrogen peroxide stimulates salicylic acid biosynthesis in tobacco. Plant Physiol 108:1673–1678

Li J, Brader G, Palva ET (2004) The WRKY70 transcription factor: a node of convergence for jasmonate-mediated and salicylate-mediated signals in plant defense. Plant Cell 16:319–331

Li J, Brader G, Kariola T, Tapio Palva E (2006) WRKY70 modulates the selection of signaling pathways in plant defense. Plant J 46:477–491

Li H, Xu Y, Xiao Y, Zhu Z, Xie X, Zhao H, Wang Y (2010) Expression and functional analysis of two genes encoding transcription factors, *VpWRKY1* and *VpWRKY2*, isolated from Chinese wild *Vitis pseudoreticulata*. Planta 232:1325–1337

Li C, He X, Luo X, Xu L, Liu L, Min L, Jin L, Zhu L, Zhang X (2014) Cotton WRKY1 mediates the plant defense-to-development transition during infection of cotton by *Verticillium dahliae* by activating JASMONATE ZIM-DOMAIN1 expression. Plant Physiol 166:2179–2194

Li C, Li D, Shao F, Lu S (2015a) Molecular cloning and expression analysis of WRKY transcription factor genes in *Salvia miltiorrhiza*. BMC Genomics 16:200. doi:10.1186/s12864-015-1411-x

Li J, Wang J, Wang N, Guo X, Gao Z (2015b) *GhWRKY44*, a WRKY transcription factor of cotton, mediates defense responses to pathogen infection in transgenic *Nicotiana benthamiana*. Plant Cell Tissue Organ Culture 121:127–140

Libault M, Wan J, Czechowski T, Xu D, Udvardi M, Stacey G (2007) Identification of 118 *Arabidopsis* transcription factor and 30 ubiquitin-ligase genes responding to chitin, a plant-defense elicitor. Mol Plant Microbe interact 20:900–911

Lim CW, Lim S, Baek W, Han S-W, Lee SC (2015) Expression and functional roles of the pepper pathogen-induced bZIP transcription factor, CabZIP2, in enhanced disease resistance to bacterial pathogen infection. Mol Plant Microbe Interact 28:825–833

Lin R, Zhaom W, Mengm X, Wang M, Peng Y (2007) Rice gene *OsNAC19* encodes a novel NAC-domain transcription factor and responds to infection by *Magnaporthe grisea*. Plant Sci 172:120–130

Ling J, Jiang W, Zhang Y, Yu H, Mao Z, Gu X, Huang S, Xie B (2011) Genome-wide analysis of WRKY gene family in Cucumis sativus. BMC Genomics 12:471. doi:10.1186/1471-2164-12-471

Lippok B, Birkenbihl RP, Rivory G, Brummer J, Schmelzer E, Logemann E, Somssich IE (2007) Expression of *AtWRKY33* encoding a pathogen- or PAMP-responsive WRKY transcription factor is regulated by a composite DNA motif containing W box elements. Mol Plant-Microbe Interact 20:420–429

Liu Y, Nakayama N, Schiff M, Dinesh-Kumar SP (2004) Involvement of MEK1, MAPKK, NTF6 MAPK, WRKY/MYB transcription factors, COI1 and CTR1 in *N*-mediated resistance to tobacco mosaic virus. Plant J 38:800–809

Liu XQ, Bai XQ, Qian Q, Wang XJ, Chen MS, Chu CC (2005) OsWRKY03, a rice transcriptional activator that functions in defense signaling pathway upstream of OsNPR1. Cell Res 15:593–603

Liu X, Bai X, Wang X, Chu C (2007a) OsWRKY71, a rice transcription factor, is involved in rice defense response. J Plant Physiol 164:969–979

Liu Y, Dai S, Beachy RN (2007b) Role of the C-terminal domains of rice (*Oryza sativa*) bZIP proteins RF2a and RF2b in regulating transcription. Biochem J 405:243–249

Liu Y, Kong X, Pan J, Li D (2010) VIP1: linking *Agrobacterium*-mediated transformation to plant immunity? Plant Cell Rep 29:805–812

Liu R, Chen L, Jia Z, Lü B, Shi H, Shao W, Dong H (2011) Transcription factors AtMYB44 regulates induced expression of the *ETHYLENE INSENSITIVE2* gene in *Arabidopsis* responding to a harpin protein. Mol Plant-Microbe interact 24:377–389

Liu D, Chen X, Liu J, Ye J, Guo Z (2012) The rice transcription factor *OsERF922* negatively regulates resistance to *Magnaporthe oryzae* and salt tolerance. J Exp Bot 53:3899–3911

Liu B, Quang Z, Zhang Y, Li X, Hong Y, Huang L, Liu S, Zhang H, Li D, Song F (2014) Tomato NAC transcription factor SlSRN1 positively regulates defense response against biotic stress but negatively regulates abiotic stress response. PLos One 9(7), e102067. doi:10.1371/journal.pone.0102067

Liu S, Kracher B, Ziegler J, Birkenbihl RP, Somssich IE (2015) Negative regulation of ABA signaling by WRKY33 is critical for Arabidopsis immunity towards *Botrytis cinerea* 2100. Elife 15:4. doi:10.7554/elife.07295

Lochman J, Mikes V (2006) Ergosterol treatment leads to the expression of a specific set of defence-related genes in tobacco. Plant Mol Biol 62:43–51

Lopez AMQ, Lucas JA (2002) Effects of plant defence activators on anthracnose disease of cashew. Eur J Plant Pathol 108:409–420

Lorenzo O, Solano R (2005) Molecular players regulating the jasmonate signaling network. Curr Opin Plant Biol 8:532–540

Lorenzo O, Piqueras R, Sánchez-Serrano JJ, Solano R (2003) ETHYLENE RESPONSE FACTOR 1 integrates signals from ethylene and jasmonate pathways in plant defense. Plant Cell 15:165–178

Lorenzo O, Chico JM, Sánchez-Serrano JJ, Solano R (2004) JASMONATE-INSENSITIVE1 encodes a MYC transcription factor essential to discriminate between different jasmonate-regulated defense responses in Arabidopsis. Plant Cell 16:1938–1950

Luan S, Kudla J, Rodriguez-Concepcion M, Yalovsky S, Gruissem W (2002) Calmodulins and calcineurin B-like proteins: Calcium sensors for specific signal response coupling in plants. Plant Cell 14 suppl:S389–S400

Luna E, Bruce TJA, Roberts MR, Flors V, Ton J (2012) Next generation systemic acquired resistance. Plant Physiol 158:844–853

Luo H, Song F, Goodman RM, Zheng Z (2005a) Up-regulation of OsBIHD1, a rice gene encoding BELL homeodomain transcriptional factor, in disease resistance responses. Plant Biol 7:459–468

Luo H, Song F, Zheng Z (2005b) Overexpression in transgenic tobacco reveals different roles for the rice homeodomain gene *OsBIHD1* in biotic and abiotic stress responses. J Exp Bot 56:2673–2682

Mabrouk Y, Badr E, Rakha F, Ghazy A-H (2008) Differential expression of potato *WRKY1* gene upon *Erwinia carotovora* infection and bioinformatics analysis for a novel tomato WRKY-like gene. Austr J Basic Appl Sci 2:30–36

Maeo K, Hayashi S, Kojima-Suzuki H, Morikami A, Nakamura D (2001) Role of conserved residues of the WRKY domain in the DNA-binding of tobacco WRKY family proteins. BioSci Biotechnol Biochem 65:2428–2436

Maffi D, Iriti M, Pigni M, Vannini C, Faoro F (2011) *Uromyces appendiculatus* infection in BTH-treated bean plants: ultrastructural details of a lost fight. Mycopathologia 171:209–221

Maleck K, Levine A, Eulgem T, Morgan A, Schmid J, Lawton KA, Dangl JL, Dietrich RA (2000) The transcriptome of *Arabidopsis thaliana* during systemic acquired resistance. Nat Genet 26:403–410

Mandal B, Mandal S, Csinos AS, Martinez N, Culbreath AK, Pappu HR (2008) Biological and molecular analyses of the acibenzolar-S-methyl-induced systemic acquired resistance in flue-cured tobacco against *Tomato spotted wilt virus*. Phytopathology 98:196–204

Mao P, Duan M, Wei C, Li Y (2007) WRKY62 transcription factor acts downstream of cytosolic NPR1 and negatively regulates jasmonate responsive gene expression. Plant Cell Physiol 48:833–842

Mao G, Meng X, Liu Y, Zheng Z, Chen Z, Zhang S (2011) Phosphorylation of a WRKY transcription factor by two pathogen-responsive MAPKs drives phytoalexin biosynthesis in *Arabidopsis*. Plant Cell 23:1639–1653

Marchive C, Mzid R, Deluc L, Barrieu F, Pirrello J, Gauthier A, Corio-Costet M-F, Regard F, Cailleteau B, Hamdi S, Lauvergeat V (2007) Isolation and characterization of a *Vitis vinifera* transcription factor, VvWRKY1, and its effect on responses to fungal pathogens in transgenic tobacco plants. J Exp Bot 58:1999–2010

Marchive C, Léon C, Kappel C, Coutos-Thévenot P, Corio-Costet M-F, Delrot S, Lauvergeat V (2013) Over-expression of VvWRKY1 in grapevines induces expression of jasmonic acid pathway-related genes and confers higher tolerance to the downy mildew. PLoS ONE 8(1), e54185. doi:10.1371/journal.pone.0054185

Martinez-Medina A, Fernández I, Sánchez-Guzmán MJ, Jung SC, Pascual JA, Pozo MJ (2013) Deciphering the hormonal signalling network behind the systemic resistance induced by *Trichoderma harzianum* in tomato. Front Plant Sci 4:206

Mateo A, Muhlenbock P, Rustérucci C, Chang CC, Miszalski Z, Karpinska B, Parker JE, Mullineaux PM, Karpinski S (2004) *LESION SIMULATING DISEASE 1* is required for acclimation to conditions that promote excess excitation energy. Plant Physiol 136:2818–2830

Matsumoto K, Sawada H, Matsumoto K, Hamada H, Yoshimoto E, Ito T, Takeuchi S, Tsuda S, Suzuki K, Kobayashi K, Kiba A, Okuno T, Hikichi Y (2008) The coat protein gene of tobamovirus P_0 pathotype is a determinant for activation of temperature-insensitive L^{1a}-gene-mediated resistance in *Capsicum* plants. Arch Virol 155:645–650

Matsushita A, Inoue H, Goto S, Nakayama A, Sugano S, Hayashi N, Takatsuji H (2012) The nuclear ubiquitin proteasome degradation affects WRKY45 function in the rice defense program. Plant J 73:302–313

McGrath KC, Dombrecht B, Manners JM, Schenk PM, Edgar CI, Maclean DJ, Scheible W-R, Udvardi MK, Kazan K (2005) Repressor- and activator-type ethylene response factors functioning in jasmonate signaling and disease resistance identified via a genome-wide screen of Arabidopsis transcription factor gene expression. Plant Physiol 139:949–959

McLellan H, Boevink PC, Armstrong MR, Pritchard L, Gomez S, Morales J, Whisson SC, Beynon JL, Birch PR (2013) An RxLR effector from *Phytophthora infestans* prevents relocalisation of two plant NAC transcription factors from the endoplasmic reticulum to the nucleus. PLoS Pathog 9(10), e1003670. doi:10.1371/journal.ppat.1003670

Mei C, Qi M, Sheng G, Yang Y (2006) Inducible overexpression of a rice allene oxide synthase gene increases the endogenous jasmonic acid level, PR gene expression, and host resistance to fungal infection. Mol Plant Microbe Interact 19:1127–1137

Memelink J, Verpoorte R, Kijne JW (2001) ORCAnization of jasmonate-responsive gene expression in alkaloid metabolism. Trends Plant Sci 6:212–219

Méndez-Bravo A, Calderón-Vázquez C, Ibarra-Laclette E, Raya-González J, Ramirez-Chávez E, Molina-Torres J, Guevara-Garcia AA, López-Bucio J, Herrera-Estrella L (2011) Alkamides activate jasmonic acid biosynthesis and signaling pathways and confer resistance to *Botrytis cinerea* in *Arabidopsis thaliana*. PLoS ONE 6(11), e27251

Meng X-B, Zhao W-S, Lin R-M, Wang M, Peng Y-L (2005) Identification of a novel rice bZIP-type transcription factor gene, *OsbZIP1*, involved in response to infection of *Magnaporthe grisea*. Plant Mol Biol Rep 23:301a–301m

Mengiste T, Chen X, Salmeron J, Dietrich R (2003) The *Botrytis Susceptible 1* gene encodes an R2R3MYB transcription factor protein that is required for biotic and abiotic stress responses in Arabidopsis. Plant Cell 15:2551–2565

Menke FLH, Kang H-G, Chen Z, Mee Park J, Kumar D, Klessig DF (2005) Tobacco transcription factor WRKY1 is phosphorylated by the MAP kinase SIPK and mediates HR-like cell death in tobacco. Mol Plant-Microbe Interact 10:1027–1034

Meziane H, Van Der Sluis I, Van Loon LC, Höfte M, Bakker PAHM (2005) Determinants of *Pseudomonas putida* WCS358 involved in inducing systemic resistance in plants. Mol Plant Pathol 6:177–185

Miao Y, Zentgraf U (2007) The antagonist function of *Arabidopsis* WRKY53 and ESR/ESP in leaf senescence is modulated by the jasmonic and salicylic acid equilibrium. Plant Cell 19:819–830

Midoh N, Iwata M (1996) Cloning and characterization of a probenazole- inducible gene for an intracellular pathogenesis-related protein in rice. Plant Cell Physiol 37:9–18

Momol MT, Olson SM, Funderburk JE, Stavisky J, Marois JJ (2004) Integrated management of tomato spotted wilt on field-grown tomatoes. Plant Dis 88:882–890

Moreau M, Tian M, Klessig DF (2012) Salicylic acid binds NPR3 and NPR4 to regulate NPR1-dependent defense responses. Cell Res. doi:10.1038/cr 2012.100

Mou Z, Fan W, Dong X (2003) Inducers of plant systemic acquired resistance regulate NPR1 function through redox changes. Cell 113:935–944

Mukherjee M, Larrimore KE, Ahmed NJ, Bedick TS, Barghouthi NT, Traw MB, Barth C (2010) Ascorbic acid deficiency in *Arabidopsis* induces constitutive priming that is dependent on hydrogen peroxide, salicylic acid, and the *NPR1* gene. Mol Plant-Microbe Interact 23:340–351

Murata Y, Pei ZM, Mori IC, Schroeder J (2001) Abscisic acid activation of plasma membrane Ca^{2+} channels in guard cells requires cytosolic NAD(P)H and is differentially disrupted upstream and downstream of reactive oxygen species production in *abi1-1* and *abi2-1* protein phosphatase 2C mutants. Plant Cell 13:2513–2523

Murmu J, Wilton M, Allard G, Pandeya R, Desveaux D, Singh J, Subramaniam R (2013) *Arabidopsis* GOLDEN2-LIKE (GLK) transcription factors activate jasmonic acid (JA)-dependent disease susceptibility to the biotrophic pathogen *Hyaloperonospora arabidopsidis*, as well as JA-dependent plant immunity. Mol Plant Pathol 15:174–184

Murray SL, Ingle RA, Petersen LN, Denby KJ (2007) Basal resistance against *Pseudomonas syringae* in *Arabidopsis* involves WRKY53 and a protein with homology to a nematode resistance protein. Mol Plant Microbe Interact 20:1431–1438

Mzid R, Marchive C, Blancard D, Deluc L, Barrieu F, Corio=Costet MF, Drira N, Hamdi S, Lauvergeat V (2007) Overexpression of *VvWRKY2* in tobacco enhances broad resistance to necrotrophic fungal pathogens. Physiol Plant 131:434–447

Nakagami H, Soukupová H, Schikora A, Zársk V, Hirt H (2006) A mitogen-activated protein kinase kinase kinase mediates reactive oxygen species homeostasis in *Arabidopsis*. J Biol Chem 281:38697–38704

Nakashima K, Tran LS, VanNguyen D, Fujita M, Maruyama K, Todaka D, Ito Y, Hayashi N, Shinozaki K, Yamaguchi-Shinozaki K (2007) Functional analysis of a NAC-type transcription

factor OsNAC6 involved in abiotic and biotic stress-responsive gene expression in rice. Plant J 51:617–630
Nakashita H, Yoshioka K, Yasuda M, Nitta T, Arai Y, Yoshida S, Yamaguchi I (2002) Probenazole induces systemic acquired resistance in tobacco through salicylic acid accumulation. Physiol Plant Pathol 61:197–203
Nakayama A, Fukushima S, Goto S, Matsushita A, Shimono M, Sugano S, Jiang C-J, Akagi A, Yamazaki M, Inoue H, Takatsuji H (2013) Genome-wide identification of WRKY45-regulated genes that mediate benzothiadiazole-induced defense responses in rice. BMC Plant Biol 13:150
Ndamukong I, Abdallat AA, Thurow C, Fode B, Zander M, Weigel R, Gatz C (2007) SA-inducible Arabidopsis glutaredoxin interacts with TGA factors and suppresses JA-responsive *PDF1.2* transcription. Plant J 50:128–139
Nickstadt A, Thomma BPHJ, Feussner I, Kangasjarvi J, Zeier J, Loeffer C, Scheel D, Berger S (2004) The jasmonate-insensitive mutant *jin1* shows increased resistance to biotrophic as well as necrotrophic pathogens. Mol Plant Pathol 5:425–434
Nishizawa Y, Saruta M, Nakazono K, Nishio Z, Soma M, Yoshida T, Nakajima E, Hibi T (2003) Characterization of transgenic rice plants over-expressing the stress-inducible β-glucanase gene *Gns1*. Plant Mol Biol 51:143–152
Nobuta K, Okrent RA, Stoutemyer M, Rodibaugh N, Kempema L, Wildermuth MC, Innes RW (2007) The GH3 acyl adenylase family member PBS3 regulates salicylic acid-dependent defense responses in Arabidopsis. Plant Physiol 144:1144–1156
Nurmberg PL, Knox KA, Yun B, Morris PC, Shafiei R, Hudson A, Loake GJ (2007) The developmental selector *AS1* is an evolutionarily conserved regulator of the plant immune response. Proc Natl Acad Sci USA 104:18795–18800
Nuruzzaman M, Manimekalai R, Sharoni AM, Satoh K, Kondoh H, Ooka H, Kikuchi S (2010) Genome-wide analysis of NAC transcription factor family in rice. Gene 465:30–44
Nuruzzaman M, Sharoni AM, Kikuchi S (2013) Roles of NAC transcription factors in the regulation of biotic and abiotic stress responses in plants. Front Microbiol 4: article 248, pp 16. doi:10.3389/fmicb.2013.00248
Ogawa D, Nakajima N, Seo S, Mitsuhara I, Kamada H, Ohashi Y (2006) The phenylalanine pathway is the main route of salicylic acid biosynthesis in Tobacco mosaic virus-infected tobacco leaves. Plant Biotechnol 23:395–398
Oh SK, Lee S, Yu SH, Choi D (2005) Expression of a novel NAC domain-containing transcription factor (*CaNAC1*) is preferentially associated with incompatible interactions between chili pepper and pathogens. Planta 222:876–887
Oh SK, Baek KH, Park JM, Yi SY, Yu SH, Kamoun S, Choi D (2008) *Capsicum annuum* WRKY protein CaWRKY1 is a negative regulator of pathogen defense. New Phytol 177:977–989
Ohme-Takagi M, Shinshi H (1995) Ethylene-inducible DNA binding proteins that interact with an ethylene-responsive element. Plant Cell 7:173–182
Ohta M, Matsui K, Hiratsu K, Shinshi H, Ohme-Takagi M (2001) Repression domains of class II ERF transcriptional repressors share an essential motif for active repression. Plant Cell 13:1959–1968
Oñate-Sánchez L, Anderson JP, Young J, Singh KB (2007) AtERF14, a member of the ERF family of transcription factors, plays a nonredundant role in plant defense. Plant Physiol 143:400–409
Onions J, Hermann S, Grundstrom T (2000) A novel type of calmodulin interaction in the inhibition of basic helix-loop-helix transcription factors. Biochemistry 39:4366–4374
Ooka H, Satoh K, Doi K, Nagata T, Otomo Y, Murakami K, Matsubara K, Osalo N, Kawai J, Carninci P, Hayashizaki Y, Suzuki K, Kojima K, Takahara Y, Suzuki K, Kojima K, Takahara Y, Yamamoto K, Kikuchi S (2003) Comprehensive analysis of NAC family genes in *Oryza sativa* and *Arabidopsis thaliana*. DNA Res 10:239–247
Oostendorp M, Kunz W, Dietrich B, Staub T (2001) Induced disease resistance in plants by chemicals. Eur J Plant Pathol 107:19–28

Ordiz MI, Yang J, Barbazuk WB, Beachy RN (2010) Functional analysis of the activation domain of RF2a, a rice transcription factor. Plant Biotechnol J 8:835–844

Orober M, Siegrist J, Buchenaur H (2002) Mechanisms of phosphate-induced disease resistance in cucumber. Eur J Plant Pathol 108:345–353

Quang J, Gill G (2009) SUMO engages multiple corepressors to regulate chromatin structure and transcription. Epigenetics 4:440–444

Ovadia A, Biton R, Cohen Y (2000) Induced resistance to downy mildew and Fusarium wilt in cucurbits. Acta Hortic 510:55–59

Pajot E, Silué D (2005) Evidence that DL-3-aminobutyric acid and acibenzolar-*S*-methyl induce resistance against bacterial head rot disease of broccoli. Pest Management Sci 61:1110–1114

Pajot E, Le Corre D, Silué D (2001) Phytogard® and DL-β aminobutyric acid (BABA) induce resistance to downy mildew (*Bremia lactucae*) in lettuce (*Lactuca sativa* L.). Eur J Plant Pathol 107:861–869

Pandey SP, Somssich IE (2009) The role of WRKY transcription factors in plant immunity. Plant Physiol 150:1648–1655

Park CJ, Ronald PC (2012) Cleavage and nuclear localization of the rice XA21 immune receptor. Nat Commun 3:920

Park JM, Park CJ, Lee SB, Ham BK, Shin R, Paek KH (2001) Overexpression of the tobacco *Tsi1* gene encoding an EREBP/AP2-type transcription factor enhances resistance against pathogen attack and osmotic stress in tobacco. Plant Cell 13:1035–1046

Park CY, Lee JH, Yoo JH, Moon BC, Choi MS, Kang YH, Lee SM, Kim HS, Kang KY, Chung WS, Lim CO, Cho MJ (2005) WRKY group IId transcription factors interact with calmodulin. FEBS Lett 579:1545–1550

Park HC, Kim ML, Lee SM, Bahk JD, Yun D-J, Lim CO, Hong JC, Lee SY, Cho MJ, Chung WS (2007) Pathogen-induced binding of the soybean zinc finger homoeodomain proteins GmZF-HD1 and GmZF-HD2 to two repeats of ATTA homeodomain binding site to calmodulin isoform 4 (*GmCaM4*) promoter. Nuc Acids Res 35:3612–3623

Peng Y, Bartley LE, Chen X, Dardick C, Chern M, Ruan R, Canlas PE, Ronald PC (2008) OsWRKY62 is a negative regulator of basal and *Xa21*-mediated defense against *Xanthomonas oryzae* pv. *oryzae* in rice. Mol Plant 1:446–458

Peng Y, Bartley LE, Canlas P, Ronald PC (2010) OsWRKY IIa transcription factors modulate rice innate immunity. Rice 3:36–42

Peng X, Hu Y, Tang X, Zhou P, Deng X, Wang H, Guo Z (2012) Constitutive expression of rice *WRKY30* gene increases the endogenous jasmonic acid accumulation, PR gene expression and resistance to fungal pathogens in rice. Planta 236:1485–1498

Petruccelli S, Dai S, Carcamo R, Yin Y, Chen S, Beachy RN (2001) Transcription factor RF2a alters expression of the rice tungro bacilliform virus promoter in transgenic tobacco plants. Proc Natl Acad Sci USA 98:7635–7640

Pieterse CMJ, van Wees SCM, Hoffland E, Van Pelt JA, Van Loon LC (1996) Systemic resistance in *Arabidopsis* induced by biocontrol bacteria is independent of salicylic acid accumulation and pathogenesis-related gene expression. Plant Cell 8:1225–1237

Pieterse CMJ, van Wees SCM, van Pelt JA, Knoester M, Laan R, Gerrits H, Weisbeek PJ, van Loon JJA (1998) A novel signaling pathway controlling induced systemic resistance in Arabidopsis. Plant Cell 10:1571–1580

Pieterse CMJ, van Wees SCM, Ton J, Van Pelt JA, Van Loon LC (2002) Signaling in rhizobacteria-induced systemic resistance in *Arabidopsis thaliana*. Plant Biol 4:535–544

Popescu SC, Popescu GV, Bachan S, Zhang Z, Seay M, Gerstein M, Snyder M, Dinesh-Kumar SP (2007) Differential binding of calmodulin-related proteins to their targets revealed through high-density Arabidopsis protein microarrays. Proc Natl Acad Sci USA 104:4730–4735

Po-Wen C, Singh P, Zimmerli L (2013) Priming of the Arabidopsis pattern-triggered immunity response upon infection by necrotrophic *Pectobacterium carotovorum* bacteria. Mol Plant Pathol 14:58–70

Pozo MJ, Van Der Ent S, Van Loon LC, Pieterse CMJ (2008) Transcription factor MYC2 is involved in priming for enhanced defense during rhizobacteria-induced systemic resistance in *Arabidopsis thaliana*. New Phytol 180:511–523

Pré M, Atallah M, Champion A, De Vos M, Pieterse CMJ, Memelink J (2008) The AP2/ERF domain transcription factor ORA59 integrates jasmonic acid and ethylene signals in plant defense. Plant Physiol 147:1347–1357

Puranik S, Sahu PP, Mandal SN, Venkata Suresh B, Parida SK, Prasad M (2013) Comprehensive genome-wide survey, genomic constitution and expression profiling of the NAC transcription factor family in foxtail millet (*Setaria italica* L.). PLoS ONE 8, e64594

Qiu D, Xiao J, Ding X, Xiong M, Cai M, Cao Y, Li X, Xu C, Wang S (2007) OsWRKY13 mediates rice disease resistance by regulating defense-related genes in salicylate- and jasmonate-dependent signaling. Mol Plant Microbe Interact 20:492–499

Qiu D, Xiao J, Xie W, Liu H, Li X, Xiong L, Wang S (2008a) Rice gene network inferred from expression profiling of plants overexpressing OsWRKY13, a positive regulator of disease resistance. Mol Plant 1:538–551

Qiu JL, Fiil BK, Petersen K, Nielsen HB, Botanga CJ, Thorgrimsen S, Palma K, Suarez-Rodriguez MC, Sandbech-Clausen S, Lichota J, Brodersen P, Grasser KD, Mattson O, Glazebrook J, Mundy J, Petersen M (2008b) *Arabidopsis* MAP kinase 4 regulates gene expression through transcription factor release in the nucleus. EMBO J 27:2214–2221

Qiu JL, Zhou L, Yun B, Nielsen H, Fiil BK, Petersen K, MacKinlay J, Loake GJ, Mundy J, Morris PC (2008c) *Arabidopsis* mitogen-activated protein kinase kinases MKK1 and MKK2 have overlapping functions in defense signaling mediated by MEKK1, MPK4, and MKS1. Plant Physiol 148:212–222

Qu LJ, Zhu YX (2006) Transcription factor families in Arabidopsis: Major progress and outstanding issues for future research. Curr Opin Plant Biol 9:544–549

Raffaele S, Rivas S, Roby D (2006) An essential role for salicylic acid in AtMYB30-mediated control of the hypersensitive cell death program in Arabidopsis. FEBS Lett 580:3498–3504

Raffaele S, Vailleau F, Leger A, Joubes J, Miersch O, Huard C, Blee E, Mongrand S, Domergue F, Roby D (2008) A MYB transcription factor regulates very-long-chain fatty acid biosynthesis for activation of the hypersensitive cell death response in *Arabidopsis*. Plant Cell 20:752–767

Rainaldi M, Yamniuk AP, Murase T, Vogel HJ (2007) Calcium-dependent and -independent binding domain of tobacco MAPK phosphatase-1. J Biol Chem 282:6031–6042

Ranf S, Eschen-Lippold L, Pecher P, Lee J, Scheel D (2011) Interplay between calcium signaling and early signaling elements during defense responses to microbe-or damage-associated molecular patterns. Plant J 68:100–113

Reddy VS, Reddy AS (2004) Proteomics of calcium-signaling components in plants. Phytochemistry 65:1745–1776

Reddy VS, Ali GS, Reddy AS (2003) Characterization of a pathogen-induced calmodulin-binding protein: mapping of four Ca^{2+}-dependent calmodulin-binding domains. Plant Mol Biol 52:143–159

Reddy ASN, Ali GS, Celesnik H, Day IS (2011a) Coping with stresses: roles of calcium- and calcium/calmodulin-regulated gene expression. Plant Cell 23:2010–2032

Reddy ASN, Ben-Hur A, Day IS (2011b) Experimental and computational approaches for the study of calmodulin interactions. Phytochemistry 72:1007–1019

Reimann M, Haga K, Shimizu T, Okada K, Ando S, Mochizuki S, Nishizawa Y, Yamanouchi U, Nick P, Yano M, Minami E, Takano M, Yamane H, Iino M (2013) Identification of rice Allene Oxide Cyclase mutants and the function of jasmonate for defence against *Magnaporthe oryzae*. Plant J 74:226–238

Reisdorf-Cren M, Carrayol E, Terce-Laforgue T, Hirel B (2002) A novel HMG A-like protein binds differentially to the AT-rich regions binds differentially to the AT-rich regions located in the far distal and proximal parts of a soybean glutamine synthetase gene (GS15) promoter. Plant Cell Physiol 43:1006–1016

Ren X-J, Huang W-D, Li W-Z, Yu D-Q (2010) Tobacco transcription factor WRKY4 is a modulator of leaf development and disease resistance. Biol Plantarum 54:684–690

Riechmann JL, Meyerowitz EM (1998) The AP2/EREBP family of transcription factors. Biol Chem 379:633–646

Robatzek S, Somssich IE (2001) A new member of the Arabidopsis WRKY transcription factor family, *AtWRKY6*, is associated with both senescence- and defence-associated processes. Plant J 28:123–133

Robatzek S, Somssich IE (2002) Targets of AtWRKY6 regulation during plant senescence and pathogen defense. Genes Dev 16:1139–1149

Rochon A, Boyle P, Wignes T, Fobert PR, Després C (2006) The coactivator function of *Arabidopsis* NPR1 requires the core of its BTB/POZ domain and the oxidation of C-terminal cysteines. Plant Cell 18:3670–3685

Rohilla R, Singh US, Singh RL (2002) Mode of action of acibenzolar-S-methyl against sheath blight of rice, caused by *Rhizoctonia solani* Kühn. Pest Manag Sci 58:63–69

Ross CA, Liu YL, Shen QJ (2007) The WRKY gene family in rice (*Oryza sativa*). J Integ Plant Biol 49:827–842

Rossard S, Roblin G, Atanassova R (2010) Ergosterol triggers characteristic elicitation steps in *Beta vulgaris* leaf tissues. J Exp Bot 61:1807–1816

Rudd JJ, Franklin-Tong VE (2001) Unravelling response-specificity in Ca^{2+} signaling in plant cells. New Phytol 151:7–33

Rushton PJ, Torres JT, Parniske M, Wernert P, Hahlbrock K, Somssich IE (1996) Interaction of elicitor-induced DNA-binding proteins with elicitor response elements in the promoters of parsley *PR1* genes. EMBO J 15:5690–5700

Rushton PJ, Bokowiec MT, Han S, Zhang H, Brannock JF, Chen X, Laudeman TW, Timko MP (2008) Tobacco transcription factors: novel insights into transcriptional regulation in the Solanaceae. Plant Physiol 147:280–295

Rushton PJ, Somssich IE, Ringler P, Shen QJ (2010) WRKY transcription factors. Trends Plant Sci 15:247–258

Ruston DL, Tripathi P, Rabara RC, Lin J, Ringler P, Boken AK, Langum TJ, Smidt L, Boomsma DD, Emme NJ, Chen X, Finer JJ, Shen QJ, Ruston PJ (2012) WRKY transcription factors; key components in abscisic acid signalling. Plant Biotechnol J 10:2–11

Sagi M, Fluhr R (2006) Production of reactive oxygen species by plant NADPH oxidases. Plant Physiol 336–340

Sakamoto K, Tada Y, Yokozeki Y, Akagi H, Hyashi N, Fujimura T, Ichikava N (1999) Chemical induction of disease resistance in rice is correlated with the expression of a gene encoding a nucleotide binding site and leucine-rich repeats. Plant Mol Biol 40:847–855

Sanders D, Pelloux J, Brownlee C, Harper JF (2002) Calcium at the cross-roads of signaling. Plant Cell 14:S401–S417

Sawada H, Shim IS, Usui K (2006) Induction of benzoic acid 2-hydroxylase and salicylic acid biosynthesis-modulation by salt stress in rice seedlings. Plant Sci 171:263–270

Schluttenhofer C, Yuan L (2015) Regulation of specialized metabolism by WRKY transcription factors. Plant Physiol 167:295–306

Schommer C, Palatnik JF, Aggarwal P, Chételat A, Cubas P, Farmer EE, Nath U, Weigel D (2008) Control of jasmonate biosynthesis and senescence by miR319 targets. PLoS Biol 6(9), e230

Schwartz SH, Tan BC, McCarty DR, Welch W, Zeevaart JA (2003) Substrate specificity and kinetics for VP14, a carotenoid cleavage dioxygenase in the ABA biosynthetic pathway. Biochim Biophys Acta 1619:9–14

Schweizer P, Schiagenhauf E, Schaffrath U, Dudler R (1999) Different patterns of host genes are induced in rice by *Pseudomonas syringae*, a biological inducer of resistance, and the chemical inducer benzothiadiazole (BTH). Eur J plant Pathol 105:659–665

Selth LA, Dogra SC, Rasheed MS, Healy H, Randles JW, Rezaian MA (2005) A NAC domain protein interacts with tomato leaf curl virus replication accessory protein and enhances viral replication. Plant Cell 17:311–325

Seo PJ, Park C-M (2010) MYB96-mediated abscisic acid signals induce pathogen resistance response by promoting salicylic acid biosynthesis in *Arabidopsis*. New Phytol 186:471–483

Seo PJ, Kim MJ, Park JY, Kim SY, Jeon J, Lee YH, Kim J, Park CM (2010) Cold activation of a plasma membrane-tethered NAC transcription factor induces a pathogen response in *Arabidopsis*. Plant J 61:661–671

Shan W, Chen J-Y, Kuang J-F, Lu W-J (2015) Banana fruit NAC transcription factor MaNAC5 cooperates with MaWRKYs to enhance the expression of pathogenesis-related genes against *Colletotrichum musae*. Mol Plant Pathol doi:. doi:10.1111/mpp.12281

Shang H, Li W, Zou C, Yuan Y (2013) Analyses of the NAC transcription factor gene family in *Gossypium raimondii* Ulbr.: chromosomal location, structure, phylogeny, and expression patterns. J Integr Plant Biol 55:663–676

Shi W, Hao L, Li J, Liu D, Guo X, Li H (2014a) The *Gossypium hirsutum* WRKY gene *GhWRKY39-1* promotes pathogen defense responses and mediates salt stress tolerance in transgenic *Nicotiana benthamiana*. Plant Cell Reports 33:483–498

Shi W, Liu D, Hao L, Wu C, Guo X, Li H (2014b) GhWRKY39, a member of the WRKY transcription factor family in cotton, has a positive role in disease resistance and salt stress tolerance. Plant Cell Tissue Organ Cult 118:17–32

Shim JS, Choi YD (2013) Direct regulation of WRKY70 by AtMYB44 in plant defense responses. Plant Signal Behav 8, e20783. doi:10.4161/psb.24509

Shim JS, Jung C, Lee S, Min K, Lee YW, Choi Y, Lee JS, Song JT, Kim JK, Choi YD (2013) AtMYB44 regulates WRKY70 expression and modulates antagonistic interaction between salicylic acid and jasmonic acid signaling. Plant J 73:483–495

Shimono M, Sugano S, Nakayama A, Jiang C-J, Ono K, Toki S, Takatsuji H (2007) Rice WRKY45 plays a crucial role in benzothiadiazole-inducible blast resistance. Plant Cell 19:2064–2076

Shimono M, Koga H, Akagi A, Hayashi N, Goto S, Sawada M, Kurihara T, Matsushita A, Sugano S, Jiang CJ, Kaku H, Inoue H, Takatsuji H (2012) Rice WRKY45 plays important roles in fungal and bacterial disease resistance. Mol Plant Pathol 13:83–94

Shin R, Park JM, An JM, Paek KH (2002) Ectopic expression of *Tsi1* in transgenic hot pepper plants enhances host resistance to viral, bacterial, oomycete pathogens. Mol Plant-Microbe Interact 15:983–989

Siegrist J, Orober M, Buchenauer H (2000) β-Aminobutyric acid-mediated enhancement of resistance in tobacco to tobacco mosaic virus depends on the accumulation of salicylic acid. Physiol Mol Plant Pathol 56:95–106

Sillero JC, Rojas-Molina MM, Avila CM, Rubiales D (2012) Induction of systemic acquired resistance against rust, ascochyta blight and broomrape in faba bean by exogenous application of salicylic acid and benzothiadiazole. Crop Prot 34:55–69

Silué D, Pajot E, Cohen Y (2002) Induction of resistance to downy mildew (*Peronospora parasitica*) in cauliflower by DL-B-amino-*n*-butanoic acid (BABA). Plant Pathol 51:97–102

Singh K, Foley RC, Onate-Sanchez L (2002) Transcription factors in plant defense and stress responses. Curr Opin Plant Biol 5:430–436

Singh AK, Sharma V, Pal AK, Acharya V (2013) Ahuja PS. Genome-wide organization and expression profiling of the NAC transcription factor family in potato (Solanum tuberosum L) DNA Res 20:403–423

Slaughter A, Daniel X, Flors V, Luna E, Hohn B, Mauch-Mani B (2012) Descendants of primed *Arabidopsis* plants exhibit resistance to biotic stress. Plant Physiol 158:835–843

Snedden WA, Fromm H (2001) Calmodulin as a versatile calcium signal transducer in plants. New Phytol 151:35–66

Solano R, Stepanova A, Chao Q, Ecker JR (1998) Nuclear events in ethylene signaling: a transcriptional cascade mediated by ETHYLENE –INSENSITIVE3 and ETHYLENE-RESPONSE FACTOR1. Gen Dev 12:3703–3714

Son GH, Wan J, Kim HJ, Nguyen XC, ChungW-S HJC, Stacey G (2012) Ethylene-responsive element-binding factor 5, ERF5, is involved in chitin-induced innate immunity response. Mol Plant Microbe Interact 25:48–60

Stracke R, Werber M, Weisshaar B (2001) The *R2R3-MYB* gene family in *Arabidopsis thaliana.* Curr Opin Plant Biol 4:447–456

Su H, Zhang S, Yuan X, Chen C, Wang XF, Hao YJ (2013) Genome-wide analysis and identification of stress-responsive genes of the NAM-ATAF1,2-CUC2 transcription factor family in apple. Plant Physiol Biochem 71:11–21

Sun F, Liu P, Xu J, Dong H (2010) Mutation in RAP26L, a transactivator of the ERF transcription factor family, enhances *Arabidopsis* resistance to *Pseudomonas syringae*. Physiol Mol Plant Pathol 74:295–302

Sun J-Q, Jiang H-L, Li C-Y (2011) Systemin/jasmonate-mediated systemic defense signaling in tomato. Mol Plant 4:607–615

Sun L, Zhang H, Li D, Huang L, Hong Y, Ding XS, Nelson RS, Zhou X, Song F (2013) Functions of rice NAC transcription factors, ONAC122 and ONAC131, in defense responses against *Magnaporthe grisea.* Plant Mol Biol 81:41–58

Sun L, Huang L, Hong Y, Zhang H, Song F, Li D (2015) Comprehensive analysis suggests overlapping expression of rice ONAC transcription factors in abiotic and biotic stress responses. Int J Mol Sci 16:4306–4326

Taga Y, Takai R, Kaneda T, Matsui H, Isogai A, Che FS (2009) Role of OsHSP90 and IREN, Ca^{2+} dependent nuclease, in plant hypersensitive cell death induced by transcription factor OsNAC4. Plant Signal Behav 4:740–742

Takabatake R, Karita E, Seo S, Mitsuhara I, Kuchitsu K, Ohashi Y (2007) Pathogen-induced calmodulin isoforms in basal resistance against bacterial and fungal pathogens in tobacco. Plant Cell Physiol 48:414–423

Takahashi F, Yoshida R, Ichimura K, Mizoguchi T, Seo S, Yonezawa M, Maruyama K, Yamaguchi-Shinozaki K, Shinozaki K (2007) The mitogen-activated protein kinase cascade MKK3-MPK6 is an important part of the jasmonate signal transduction pathway in Arabidopsis. Plant Cell 19:805–818

Takatsuji H (2014) Development of disease-resistant rice using regulatory components of induced disease resistance. Front Plant Sci 5:630. doi:10.3389/fpls.2014.00630

Tally A, Oostendorp M, Lawton K, Staub T, Bassi B (1999) Commercial development of elicitors of induced resistance to pathogens. In: Agrawal AA, Tuzun S, Bent E (eds) Inducible plant defenses against pathogens and herbivores. APS Press, St Paul, pp 357–369

Tan B-C, Joseph LM, Deng W-T, Liu L, Li Q-B, Cline K, McCarty DR (2003) Molecular characterization of the *Arabidopsis* 9-*cis* epoxycarotenoid dioxygenase gene family. Plant J 35:44–56

Tang D, Christiansen KM, Innes RW (2005) Regulation of plant disease resistance, stress responses, cell death, and ethylene signaling in Arabidopsis by the EDR1 protein kinase. Plant Physiol 138:1018–1026

Tao Z, Liu H, Qiu D, Zhou Y, Li X, Xu C, Wang S (2009) A pair of allelic *WRKY* genes play opposite roles in rice-bacteria interactions. Plant Physiol 151:936–948

Thomma B, Penninckx I, Broekaert WF, Cammue BPA (2001) The complexity of disease signaling in *Arabidopsis*. Curr Opin Immunol 13:63–68

Ton J, Mauch-Mani B (2004) β-amino-butyric acid-induced resistance is based on ABA-dependent priming for callose. Plant J 38:119–130

Ton J, De Vos M, Robben C, Buchala A, Metraux JP, Van Loon LC, Pieterse CM (2002a) Characterization of Arabidopsis enhanced disease susceptibility mutants that are affected in systemically induced resistance. Plant J 29:11–21

Ton J, Van Pelt JA, Van Loon LC, Pieterse CM (2002b) Differential effectiveness of salicylate-dependent and jasmonate/ethylene-dependent induced resistance in *Arabidopsis*. Mol Plant Microbe Interact 15:27–34

Torres MA, Jones JDG, Dangl JL (2006) Reactive oxygen species signaling in response to pathogens. Plant Physiol 141:373–378

Trejo-Saavedra DL, Garcia-Neria MA, Rivera-Bustamante RF (2013) Benzothiadiazole (BTH) induces resistance to Pepper golden mosaic virus (PepGMV) in pepper (Capsicum annuum L.). Biol Res 46:333–340

Tsubata K, Kuroda K, Yamamoto Y, Yasokawa N (2006) Development of a novel plant activator for rice diseases, tiadinil. J Pestic Sci 31:161–162

Tsuda K, Somssich I (2015) Transcriptional networks in plant immunity. New Phytol 206:932–947

Ulker B, Somssich IE (2004) WRKY transcription factors: from DNA binding towards biological function. Curr Opin Plant Biol 7:491–498

Umemura K, Satou J, Iwata M, Uozumi N, Koga J, Kawano T, Koshiba T, Anzai H, Mitomi M (2009) Contribution of salicylic acid glucosyltransferase, OsSGT1, to chemically induced disease resistance in rice plants. Plant J 57:463–472

Van der Ent S, Verhagen BWM, Van Doorn R, Bakker D, Verlaan MG, Pel MJC, Joosten RG, Proveniers MCG, Van Loon LC, Ton J, Pieterse CMJ (2008) *MYB72* is required in early signaling steps of rhizobacteria-induced systemic resistance in *Arabidopsis*. Plant Physiol 146:1293–1304

van Hulten M, Pelser M, van Loon LC, Pieterse CMJ, Ton J (2006) Costs and benefits of priming for defense in Arabidopsis. Proc Natl Acad Sci USA 103:5602–5607

Van Peer R, Niemann GJ, Schippers B (1991) Induced resistance and phytoalexin accumulation in biological control of fusarium wilt of carnation by *Pseudomonas* sp. strain WCS417r. Phytopathology 81:728–734

van Verk MC, Pappaaioannou D, Neeleman L, Bol JF, Linthorst HJM (2008) A novel WRKY transcription factor is required for induction of *PR-1a* gene expression by salicylic acid and bacterial elicitors. Plant Physiol 140:1983–1995

van Verk MC, Bol JF, Linthorst HJM (2011) WRKY transcription factors involved in activation of SA biosynthesis genes. BMC Plant Biol 11:89. doi:10.1186/1471-2229-11-89

Van Wees SCM, Pieterse CMJ, Trijssenaar A, Van't Westende YAM, Hartog F, Van Loon LC (1997) Differential induction of systemic resistance in Arabidopsis by biocontrol bacteria. Mol Plant Microbe Interact 10:716–724

Vandenabeele S, Van Der Kelen K, Dat J, Gadjev I, Boonefaes T, Morsa S, Rottiers P, Slooten L, Van Montagu M, Zabeau M, Inzé D, Van Breusegem F (2003) A comprehensive analysis of hydrogen peroxide-induced gene expression in tobacco. Proc Natl Acad Sci USA 100:16113–16118

Verhagen BWM, Glazebrook J, Zhu T, Chang H-S, Van Loon LC (2004) The transcriptome of Rhizobacteria-induced systemic resistance in Arabidopsis. Mol Plant Microbe Interact 17:895–908

Vidhyasekaran P (2007) Fungal pathogenesis in plants and crops: molecular biology and host defense mechanisms, IInd edn. CRC Press/Taylor Francis Group, Boca Raton, p 510

Vidhyasekaran P (2014) PAMP signals in plant innate immunity: signal perception and transduction. Springer, Dordrecht, pp 442

Vidhyasekaran P (2015) Plant hormone signaling systems in plant innate immunity. Springer, Dordrecht, pp 458

Vranová E, Inzé D, Van Brueusegem F (2002) Signal transduction during oxidative stress. J Exp Bot 53:1227–1236

Wager A, Browse J (2012) Social network: JAZ protein interactions expand our knowledge of jasmonate signaling. Fron Plant Sci 3:Article 41

Waller F, Müller A, Chung K-M, Yap Y-K, Nakamura K, Weiler E, Sano H (2006) Expression of a WIPK-activated transcription factor results in increase of endogenous salicylic acid and pathogen resistance in tobacco plants. Plant Cell Physiol 47:1169–1174

Wan J, Zhang S, Stacey G (2004) Activation of a mitogen-activated protein kinase pathway in *Arabidopsis* by chitin. Mol Plant Pathol 5:125–135

Wang D, Amonsiripanitch N, Dong X (2006) A genomic approach to identify regulatory nodes in the transcriptional network of systemic acquired resistance in plants. PloS Pathog 2, e123

Wang H, Hao J, Chen X, Hao Z, Wang X, Lou Y, Peng Y, Guo Z (2007) Overexpression of rice WRKY89 enhances ultraviolet B tolerance and disease resistance in rice plants. Plant Mol Biol 65:799–815

Wang X, Basnayake BM, Zhang H, Li G, Li W, Virk N, Mengiste T, Song F (2009a) The Arabidopsis ATAF1, a NAC transcription factor, is a negative regulator of defense responses against necrotrophic fungal and bacterial pathogens. Mol Plant Microbe Interact 22:1227–1238

Wang X, Goregaoker SP, Culver JN (2009b) Interaction of *Tobacco mosaic virus* replicase protein with a NAC domain transcription factor is associated with the suppression of systemic host defenses. J Virol 83:9720–9730

Wang Q, Wang M, Zhang X, Hao B, Kaushik SK, Pan Y (2011) WRKY gene family evolution in *Arabidopsis thaliana*. Genetics 139:973–983

Wang W-H, Yi X-Q, Han A-D, Liu T-W, Chen J, Wu F-H, Dong X-J, He J-X, Pei Z-M, Pei Z-M, Zheng H-L (2012) Calcium-sensing receptor regulates stomatal closure through hydrogen peoxide and nitric oxide in response to extracellular calcium in *Arabidopsis*. J Expt Bot 63:177–190

Wang N, Zheng Y, Xin H, Fang L, Li S (2013) Comprehensive analysis of NAC domain transcription factor gene family in *Vitis vinifera*. Plant Cell Rep 32:61–75

Wang L, Zhu W, Fang L, Sun X, Su L, Liang Z, Wang N, Londo JP, Li S, Xin H (2014) Genome-wide identification of *WRKY* family genes and their response to cold stress in *Vitis vinifera*. BMC Plant Biol 14:103. doi:10.1186/1471-2229-14-103

Wang F, Lin R, Feng J, Chen W, Qiu D, Xu S (2015) TaNAC1 acts as a negative regulator of stripe rust resistance in wheat, enhances susceptibility to *Pseudomonas syringae*, and promotes lateral root development in transgenic *Arabidopsis thaliana*. Front Plant Sci 6:108. doi:10.3389/fpls.2015 00108

Wasilewska A, Vlad F, Sirichandra C, Redko Y, Jammes F, Valon C, Frei dit Frey F, Leung J (2008) An update on abscisic acid signaling in plants and more. Molecular Plant 1:198–217

Wasternack C (2007) Jasmonates: an update on biosynthesis, signal transduction and action in plant stress response, growth and development. Ann Bot (Lond) 100:681–697

Wasternack C, Hause B (2013) Jasmonates: biosynthesis, perception, signal transduction and action in plant stress response, growth and development. An update to the 2007 review in Annals of Botany. Ann Bot 111:1021–1058

Wei T, Ou B, Li J, Zhao Y, Guo D, Zhu Y, Chen Z, Gu H, Li C, Qin G, Qu LJ (2013) Transcriptional profiling of rice early response to *Magnaporthe oryzae* identified OsWRKYs as important regulators in rice blast resistance. PLoS One 8, e59720

Wen N, Chu Z, Wang S (2003) Three types of defense-responsive genes are involved in resistance to bacterial blight and fungal blast diseases in rice. Mol Genet Genomics 269:331–339

Wiermer M, Feys BJ, Parker JE (2005) Plant immunity: the EDS1 regulatory node. Curr Opin Plant Biol 8:383–389

Wildermuth MC, Dewdney J, Wu G, Ausubel FM (2001) Isochorismate synthesis is required to synthesize salicylic acid for plant defense. Nature 414:562–565

Wu Y, Deng Z, Lai J, Zhang Y, Yang C, Yin B, Zhao Q, Zhang L, Li Y, Yang C, Xie O (2010) Dual function of *Arabidopsis* ATAF1 in abiotic and biotic stress responses. Cell Res 19:1279–1290

Xia N, Zhang G, Liu XY, Deng L, Cai GL, Zhang Y, Wang XJ, Zhao J, Huang LL, Kang ZS (2010a) Characterization of a novel wheat NAC transcription factor gene involved in defense response against stripe rust pathogen infection and abiotic stresses. Mol Biol Rep 37:3703–3712

Xia N, Zhang G, Sun YF, Zhu L, Xu LS, Chen XM, Liu B, Yu YT, Wang X-J, Huang L-L, Kang Z-S (2010b) TaNAC8, a novel NAC transcription factor gene in wheat, responds to stripe rust pathogen infection and abiotic stresses. Physiol Mol Plant Pathol 74:394–402

Xiao J, Cheng H, Li X, Xiao J, Xu C, Wang S (2013) Rice WRKY13 regulates cross talk between abiotic and biotic stress signaling pathways by selectively binding to different *cis*-elements. Plant Physiol 163:1868–1882

Xie Q, Sanz-Burgos AP, Guo H, Garcia JA, Gutiérrez C (1999) GRAB proteins, novel members of the NAC domain family, isolated by their interaction with a geminivirus protein. Plant Mol Biol 39:647–656

Xing DH, Lai ZB, Zheng ZY, Vinod KM, Fan BF, Chen ZX (2008) Stress- and pathogen-induced *Arabidopsis* WRKY48 is a transcriptional activator that represses plant basal defense. Mol Plant 1:459–470

Xu X, Chen C, Fan B, Chen Z (2006) Physical and functional interactions between pathogen-induced *Arabidopsis* WRKY18, WRKY40, and WRKY60 transcription factors. Plant Cell 18:1310–1326

Xu Z-S, Xia L-Q, Chen M, Cheng X-G, Zhang R-Y, Li L-C, Zhao Y-X, Lu Y, Ni Z-Y, Liu L, Qiu Z-G, Ma Y-Z (2007) Isolation and molecular characterization of the *Triticum aestivum* L ethylene-responsive factor1 (*TaERF1*) that increases multiple stress tolerance. Plant Mol Biol 65:719–732

Xu J, Audenaert K, Hofte M, De Vleesschauwer D (2013) Abscisic acid promotes susceptibility to the rice blight pathogen *Xanthomonas oryzae* pv. *oryzae* by suppressing salicylic acid-mediated defenses. PLoS One 8(6), e67413

Yamada S, Kano A, Tamaoki D, Miyamoto A, Shishido H, Miyoshi S, Taniguchi S, Akimitsu K, Gomi K (2012) Involvement of OsJAZ8 in jasmonate-induced resistance to bacterial blight in rice. Plant Cell Physiol 53:2060–2072

Yamaguchi Y, Huffaker A, Bryan AC, Tax FE, Ryan CA (2010) PEPR2 is a second receptor for the Pep1 and Pep2 peptides and contributes to defense responses in *Arabidopsis*. Plant Cell 22:508–522

Yamanouchi U, Yano M, Lin H, Ashikari M, Yamada K (2002) A rice spotted leaf gene, *Spl7*, encodes a heat stress transcription factor protein. Proc Natl Acad Sci USA 99:7530–7535

Yan Z, Reddy MS, Ryu CM, McInroy JA, Wilson M, Kloepper JW (2002) Induced systemic protection against tomato late blight elicited by plant growth-promoting rhizobacteria. Phytopathology 92:1329–1333

Yang B, Srivastava S, Deyholos MK, Kav NNV (2007) Transcriptional profiling of canola (*Brassica napus* L) responses to the fungal pathogen *Sclerotinia sclerotiorum*. Plant Sci 173:156–171

Yang P, Wang Z, Fan B, Chen C, Chen Z (1999) A pathogen-and salicylic acid-induced WRKY DNA-binding activity recognizes the elicitor response element of tobacco class I chitinase gene promoter genes controlling expression of defense responses in Arabidopsis. Plant J 18:141–149

Yang T, Poovaiah BW (2000) An early ethylene up-regulated gene encoding a calmodulin-binding protein involved in plant senescence and death. J Biol Chem 275:38467–38473

Yang T, Poovaiah BW (2002) A calmodulin-binding/CGCG box DNA-binding protein family involved in multiple signaling pathways in plants. J Biol Chem 277:45049–45058

Yang Y, Shah J, Klessig DF (1997) Signal perception and transduction in plant defense responses. Genes & Dev 11:1621–1639

Yasuda M, Kusajima M, Nakajima M, Akutsu K, Kudo T, Yoshida S, Nakashita H (2006) Thiadiazole carboxylic acid moiety of tiadinil, SV-03, induces systemic acquired resistance in tobacco without salicylic acid accumulation. J Plant Sci 31:329–334

Yasuda M, Nakashita H, Yoshida S (2004) Tiadinil, a novel class of activator of systemic acquired resistance, induces defense gene expression and disease resistance in tobacco. J Pest Sci 29:46–49

Yi SY, Kim JH, Joung YH, Lee S, Kim WT, Yu SH, Choi D (2004) The pepper transcription factor *CaPF1* confers pathogen and freezing tolerance in Arabidopsis. Plant Physiol 136:2862–2874

Yin YL, Chen L, Beachy R (1997a) Promoter elements required for phloem specific gene expression from the RTBV promoter in rice. Plant J 12:1179–1188

Yin Y, Zhu Q, Dai S, Lamb C, Beachy RN (1997b) RF2a, a bZIP transcriptional activator of the phloem-specific rice tungro bacilliform virus promoter, functions in vascular development. EMBO J 16:5247–5259

Yoda P, Ogawa M, Yamaguchi Y, Koizumi N, Kusano T, Sano H (2002) Identification of early-responsive genes associated with the hypersensitive response to tobacco mosaic virus and characterization of a WRKY-type transcription factor in tobacco plants. Mol Genet Genomics 267:154–161

Yokotani N, Tsuchida-Mayama T, Ichikawa H, Mitsuda N, Ohme-Takagi M, Kaku H, Minami E, Nishizawa Y (2014) OsNAC111, a blast disease-responsive transcription factor in rice, positively regulates the expression of defense-related genes. Mol Plant-Microbe Interact 27:1027–1034

Yoo JH, Park CY, Kim JC, Heo WD, Cheong MS, Park HC, Kim MC, Moon BC, Choi MS, Kang YA, Lee JH, Kim HS, Lee SM, Yoon H-W, Lim CO, Yun DJ, Lee SY, Chung WS, Cho MJ (2005) Direct interaction of a divergent CaM isoform and the transcription factor, MYB2, enhances salt tolerance in *Arabidopsis*. J Biol Chem 280:3697–3706

Yoshii M, Shimizu T, Yamazaki M, Higashi T, Miyao A, Hirochika H, Omura T (2009) Disruption of a novel gene for a NAC-domain protein in rice confers resistance to rice dwarf virus. Plant J 57:615–625

Yoshii M, Yamazaki M, Rakwal R, Kishi-Kaboshi M, Miyao A, Hirochika H (2010) The NAC transcription factor RIM1 of rice is a new regulator of jasmonate signaling. Plant J 61:804–815

Yu D, Chen C, Chen Z (2001) Evidence for an important role of WRKY DNA binding proteins in the regulation of *NPR1* gene expression. Plant Cell 13:1527–1539

Yu F, Huaxia Y, Lu W, Wu C, Cao X, Guo X (2012) *GhWRKY15*, a member of the WRKY transcription factor family identified from cotton (*Gossypium hirsutum* L.), is involved in disease resistance and plant development. BMC Plant Biol 12:144

Zander M, La Camera S, Lamotte O, Métraux JP, Gatz C (2010) *Arabidopsis thaliana* class-II TGA transcription factors are essential activators of jasmonic acid/ethylene-induced defense responses. Plant J 61:200–210

Zarei A, Körbes AP, Younessi P, Montiel G, Champion A, Memelink J (2011) Two GCC boxes and AP2/ERF-domain transcription factor ORA59 in jasmonate/ethylene-mediated activation of the PDF1.2 promoter in Arabidopsis. Plant Mol Biol 75:321–331

Zhang X (2008) The epigenetic landscape of plants. Science 320:489–492

Zhang M, Yuan T (1998) Molecular mechanisms of calmodulin's functional versatility. Biochem Cell Biol 76:313–323

Zhang Y, Fan M, Kinkema M, Li X, Dong X (1999) Interfaction of NPR1 with basic leucine zipper protein transcription factors that bind sequences required for salicylic acid induction of the *PR-1* gene. Proc Natl Acad Sci USA 96:6523–6528

Zhang Y, Tessaro MJ, Lassner M, Li X (2003) Knockout analysis of *Arabidopsis* transcription factors *TGA2*, *TGA5*, and *TGA6* reveals their redundant and essential roles in systemic acquired resistance. Plant Cell 15:2647–2653

Zhang H, Zhang D, Chen J, Yang Y, Huang Z, Huang D, Wang X-C, Huang R (2004) Tomato stress-responsive factor TSRF1 interacts with ethylene responsive element GCC box and regulates pathogen resistance to *Ralstonia solanacearum*. Plant Mol Biol 55:825–834

Zhang Z, Yao W, Dong N, Liang H, Liu H, Huang R (2007) A novel ERF transcription activator in wheat and its induction kinetics after pathogen and hormone treatments. J Exp Bot 58:2993–3003

Zhang J, Peng Y, Guo Z (2008) Constitutive expression of pathogen-inducible OsWRKY31 enhances disease resistance and root growth and auxin response in transgenic rice plants. Cell Res 18:508–521

Zhang L, Xi D, Luo L, Meng F, Li Y, Wu CA, Guo X (2011) Cotton GhMPK2 is involved in multiple signaling pathways and mediates defense responses to pathogen infection and oxidative stress. FEBS J 278:1267–1378

Zheng Z, Qamar SA, Chen Z, Mengiste T (2006) Arabidopsis WRKY33 transcription factor is required for resistance to necrotrophic fungal pathogens. Plant J 48:596–605

Zheng Z, Mosher SL, Fan B, Klessig DF, Chen Z (2007) Functional analysis of Arabidopsis WRKY25 transcription factor in plant defense against *Pseudomonas syringae*. BMC Plant Biol 7:2

Zhou J, Tang T, Martin GB (1997) The Pto kinase conferring resistance to tomato bacterial speck disease interacts with proteins that bind a *cis*-element of pathogenesis-related genes. EMBO J 16:3207–3218

Zhou J-M, Trifa Y, Silva H, Pontier D, Lam E, Shah J, Klessig DF (2000) NPR1 differentially interacts with members of the TGA/OBF family of transcription factors that bind an element of the *PR-1* gene required for induction by salicylic acid. Mol Plant-Microbe Interact 13:191–202

Zhou C, Zhang L, Duan J, Miki B, Wu K (2005) Histone deacetylase19 is involved in jasmonic acid and ethylene signaling of pathogen response in Arabidopsis. Plant Cell 17:1196–1204

Zhou J, Zhang H, Yang Y, Zhang Z, Zhang H, Hu X, Chen J, Wang X-C, Huang R (2008) Abscisic acid regulates TSRF1-mediated resistance to *Ralstonia solanacearum* by modifying the expression of GCC box-containing genes in tobacco, J Expt Bot erm353VI

Zhu Q, Ordiz MI, Dabi T, Beachy RN, Lamb C (2002) Rice TATA binding protein interacts functionally with transcription factor IIB and the RF2a bZIP transcriptional activator in an enhanced plant *in vitro* transcription system. Plant Cell 14:795–803

Zhu Z, Shi J, Cao J, He M, Wang Y (2012) VpWRKY3, a biotic and abiotic stress-related transcription factor from the Chinese wild *Vitis pseudoreticulata*. Plant cell Rep 31:2109–2120

Ziadi S, Barbedette S, Godard JF, Monot C, Le Corre D, Silue D (2008) Production of pathogenesis-related proteins in the cauliflower (*Brassica oleracea* var *botrytis*)-downy mildew (*Peronospora parasitica*) pathosystem treated with acibenzolar-S-methyl. Plant Pathol 50:579–586

Zimmerli L, Jakab G, Metraux JP, Mauch-Mani B (2000) Potentiation of pathogen-specific defense mechanisms in *Arabidopsis* by β-aminobutyric acid. Proc Natl Acad Sci USA 97:12920–12925

Zimmerli L, Metraux J-P, Mauch-Mani B (2001) β-Aminobutyric acid-induced protection of *Arabidopsis* against the necrotrophic fungus *Botrytis cinerea*. Plant Physiol 126:517–523

Zou B, Jia Z, Tian S, Wang X, Gou Z, Lü B, Dong H (2013) AtMYB44 positively modulates disease resistance to *Pseudomonas syringae* through the salicylic acid signalling pathway in Arabidopsis. Funct Plant Biol 40:304–331

Zuo K-J, Qin J, Zhao J-Y, Ling H, Zhang L-D, Cao Y-F, Tang K-X (2007) Over-expression of *GbERF2* transcription factor in tobacco enhances brown spots disease resistance by activating expression of downstream genes. Gene 391:80–90

MIX
Papier aus verantwortungsvollen Quellen
Paper from responsible sources
FSC® C105338

If you have any concerns about our products,
you can contact us on
ProductSafety@springernature.com

In case Publisher is established outside the EU,
the EU authorized representative is:
Springer Nature Customer Service Center GmbH
Europaplatz 3, 69115 Heidelberg, Germany

Printed by Libri Plureos GmbH
in Hamburg, Germany